Lecture Notes in Computer Sci

Commenced Publication in 1973
Founding and Former Series Editors:
Gerhard Goos, Juris Hartmanis, and Jan van Leeuwen

Phong Q. Nguyen (Ed.)

Progress in Cryptology – VIETCRYPT 2006

First International Conference on Cryptology in Vietnam
Hanoi, Vietnam, September 25-28, 2006
Revised Selected Papers

 Springer

Volume Editor

Phong Q. Nguyen
Ecole Normale Supérieure
Département d'Informatique
45, rue d'Ulm, 75230 Paris Cedex 05, France
E-mail: Phong.Nguyen@ens.fr

Library of Congress Control Number: 2006938421

CR Subject Classification (1998): E.3, G.2.1, D.4.6, K.6.5, K.4, F.2.1-2, C.2

LNCS Sublibrary: SL 4 – Security and Cryptology

ISSN   0302-9743
ISBN-10   3-540-68799-8 Springer Berlin Heidelberg New York
ISBN-13   978-3-540-68799-3 Springer Berlin Heidelberg New York

Springer is a part of Springer Science+Business Media

springer.com

© Springer-Verlag Berlin Heidelberg 2006
Printed in Germany

Typesetting: Camera-ready by author, data conversion by Scientific Publishing Services, Chennai, India
Printed on acid-free paper   SPIN: 11958239   06/3142   5 4 3 2 1 0

Preface

These are the proceedings of VIETCRYPT 2006, the first international conference on cryptology hosted in Vietnam. The conference was organized by FPT Software, in cooperation with Vietnam's Institute of Mathematics. It was held in the beautiful city of Hanoi, September 25–28, 2006. This conference would certainly not have been possible without Phan Dinh Dieu, the General Chair. I also wish to thank Nguyen Quoc Khanh, Nguyen Duy Lan and Phan Duong Hieu for their invaluable help in organizing the conference.

The Program Committee, consisting of 36 members from 17 countries, considered 78 papers (from 19 countries) and selected 24 for presentation. These proceedings include the revised versions of the 24 papers accepted by the Program Committee. These papers were selected from all the anonymous submissions to the conference on the basis of originality, quality and relevance to cryptography. Revisions were not checked and the authors bear full responsibility for the contents of their papers. The conference program also included two invited talks: it was a great honor to have Tatsuaki Okamoto and Jacques Stern as invited speakers. Their talks were entitled, respectively, "On Pairing-Based Cryptosystems" and "Cryptography in Financial Transactions: Current Practice and Future Directions."

The selection of papers was a difficult and challenging task. Each submission was reviewed by at least three referees. I wish to thank the Program Committee members, who did an excellent job, and devoted much effort and valuable time to read and select the papers. In addition, I gratefully acknowledge the help of a large number of colleagues who reviewed submissions in their areas of expertise. They are all listed here and I apologize for any inadvertent omission. I also wish to thank Springer for publishing the proceedings in the *Lecture Notes in Computer Science* series.

All paper submissions to VIETCRYPT 2006 were handled electronically, using the amazing iChair software developed at the École Polytechnique Fédérale de Lausanne (EPFL) by Thomas Baignères and Matthieu Finiasz. I also wish to thank Jacques Beigbeder for installing iChair at the ENS.

Finally, I would like to thank all the authors who submitted papers.

October 2006   Nguyễn Phong Quang

VIETCRYPT 2006

International Conference on Cryptology in Vietnam

September 25 – 28, 2006, Hanoi, Vietnam

Organized by

FPT Software

in cooperation with

Vietnam's Institute of Mathematics

Organization Committee

Nguyen Quoc Khanh .... FPT Software, Vietnam
Nguyen Lam Phuong .... FPT Software, Vietnam - Chair
Phan Van Hoa .... FPT Software, Vietnam

General Chair

Phan Dinh Dieu, Vietnam National University, Vietnam

Program Chair

Nguyen Phong Quang, École Normale Supérieure and CNRS, France

Program Committee

Masayuki Abe .... NTT Information Sharing Platform Laboratories, Japan
Feng Bao .... Institute for Infocomm Research, Singapore
Alex Biryukov .... University of Luxembourg, Luxembourg
Daniel Bleichenbacher .... Switzerland
Xavier Boyen .... Voltage, USA
Jung Hee Cheon .... Seoul National University, South Korea
Ed Dawson .... Queensland University of Technology, Australia
Marc Fischlin .... Technische Universität Darmstadt, Germany
Craig Gentry .... Stanford University, USA
Ha Huy Khoai .... Institute of Mathematics, Vietnam
Shai Halevi .... IBM Research, USA
Antoine Joux .... DGA and University of Versailles, France
Pascal Junod .... Nagravision, Switzerland
Jonathan Katz .... University of Maryland, USA
Kwangjo Kim ...... Information and Communications University, South Korea

Lars Knudsen .... Technical University of Denmark, Denmark
Neal Koblitz .... University of Washington, USA
Kaoru Kurosawa .... Ibaraki University, Japan
Arjen K. Lenstra .... EPFL, Switzerland
Ilya Mironov .... Microsoft Research, USA
Chanathip Namprempre .... Thammasat University, Thailand
Mats Naslund .... Ericsson, Sweden
Nguyen Duy Lan .... CSIRO ICT Centre, Australia
Nguyen Quoc Khanh .... FPT Corporation, Vietnam
Kazuo Ohta .... University of Electro-Communications, Japan
Pascal Paillier .... Gemalto, France
Kenny Paterson .... Royal Holloway University of London, UK
Phan Duong Hieu .... University College London, UK
Bart Preneel .... Katholieke Universiteit Leuven, Belgium
C. Pandu Rangan .... IIT Madras, India
Matt Robshaw .... France Telecom R&D, France
Phil Rogaway .... UC Davis, USA and Chiang Mai University, Thailand
Nigel Smart .... University of Bristol, UK
Mike Szydlo .... RSA, USA
Tsuyoshi Takagi .... Future University, Japan
Xiaoyun Wang .... Tsinghua University, China

External Reviewers

Michel Abdalla	Yuuichi Kokubun	Vincent Rijmen
Joosang Baek	Yuichi Komano	Peter Ryan
Colin Boyd	Nam-Seok Kwak	Bagus Santoso
Reinier Broker	Reynald Lercier	Dong-Gyu Seon
Michael Cheng	Joseph Liu	Ji Sun Shin
Sherman Chow	Daegun Ma	Masaaki Shirase
Yvonne Cliff	Alexander May	Isamu Teranishi
Matthez Dailey	Satoshi Miyagawa	Soren Steffen Thomsen
Jintai Ding	Kunihiko Miyazaki	Dongvu Tonien
Ratna Dutta	Sourav Mukhopadhyay	Pim Tuyls
Jean-Charles Faugère	Dang Nguyen Duc	Frederik Vercauteren
Rosario Gennaro	Karl Norrman	Charlotte Vikkelsoe
Rob Granger	DaeHun Nyang	Duc Liem Vo
Yoshikazu Hanatani	Miyako Ohkubo	David Woodruff
Matt Henricksen	Seiji Okuaki	Yongdong Wu
Mattias Johansson	Dag Arne Osvik	Kazuki Yoneyama
Yutaka Kawai	Dan Page	Eun Sun Yoo
Phongsak	Kun Peng	HyoJin Yoon
Keeratiwintakorn	Duong Quang Viet	

Table of Contents

Public-Key Encryption

Probabilistic Multivariate Cryptography[*]

Aline Gouget[1] and Jacques Patarin[2]

[1] Gemalto, 34 rue Guynemer, F-92447 Issy-les-Moulineaux, France
[2] University of Versailles, 45 avenue des Etats-Unis, F-78035 Versailles, France

Abstract. In public key schemes based on multivariate cryptography, the public key is a finite set of m (generally quadratic) polynomial equations and the private key is a trapdoor allowing the owner of the private key to invert the public key. In existing schemes, a signature or an answer to an authentication is valid if *all* the m equations of the public key are satisfied. In this paper, we study the idea of *probabilistic multivariate cryptography*, i.e., a signature or an authentication value is valid when at least α equations of the m equations of the public key are satisfied, where α is a fixed parameter of the scheme. We show that many new public key signature and authentication schemes can be built using this concept. We apply this concept on some known multivariate schemes and we show how it can improve the security of the schemes.

1 Introduction

The security of most of the public key schemes relies on the difficulty of solving one of the two problems that are currently considered to be hard, i.e., the problem of factoring large integers and the problem of computing discrete logarithms. However, the techniques for solving these two famous problems improve continually. Then, it becomes very important to find alternative problems and to proceed further to the study of known candidates that are considered to be minors until now. Furthermore, some new attractive properties may be achieved by using alternative difficult problems.

One possibility for secure public key schemes is based on the problem of solving *multivariate nonlinear equations* over small finite fields. In multivariate cryptography, the public key is a set $\mathcal{A}$ of m polynomial equations in n variables over a small finite field K. Public key schemes for encryption, signature or authentication can be built with such public keys. Most of the time, the equations are chosen quadratic since solving quadratic systems is already $\mathcal{NP}$-complete and also hard on average.

1.1 Related Work

Since the introduction of the first multivariate schemes [7,15,9] in 1985, many schemes have been proposed. Most of these schemes have been broken but several

---

[*] This work has been partially financially supported by the European Commission through the IST Program under Contract IST-2002-507932 ECRYPT.

P.Q. Nguyen (Ed.): VIETCRYPT 2006, LNCS 4341, pp. 1–18, 2006.

schemes are still unbroken. Recently, C. Wolf and B. Preneel proposed a taxonomy [25] of public key schemes based on the problem of multivariate quadratic equations. They grouped the known schemes into a taxonomy of only four schemes: Matsumoto Imai (C^*) [15], Hidden Field Equations (HFE) [18], Stepwise Triangular Systems (STS) [24] and Unbalanced Oil and Vinegar (UOV) [10].

Some of these schemes [15,24] are broken. However, from these four basic schemes, it is possible to design more schemes by applying a *perturbation* in order to improve the security of the basic scheme. For instance, the scheme C^{*--} which is a variant of the C* scheme using the perturbation *minus* (i.e., a part of the public key is kept secret) is still unbroken.

The security of unbroken schemes is most of the time an open problem since it consists in checking that all known attacks do not apply. However, multivariate schemes have attractive properties that cannot be reached using classical public key schemes based on factorization or discrete logarithm. For instance, it becomes possible to get very short signatures or very fast computations. Furthermore, the study of multivariate schemes is interesting from a theoretical point of view since it leads to the study of some new specific problems.

A notion close to the idea of *probabilistic multivariate cryptography* presented in this paper is given in [1] but the context is different since it is the IP problem-based traitor tracing.

1.2 Outline

In Section 2, we first present the general problem of multivariate polynomials and the public key of multivariate schemes. Then, we compare how this public key is used in classical (non-probabilistic) schemes and in probabilistic schemes. In Section 3, we explain how a probabilistic scheme can be built from a classical trapdoor. This construction will sometimes also hide the trapdoor in a much better way than in a classical construction. In Sections 4 and 5, we present some explicit probabilistic multivariate schemes: in Section 4, we present an adaptation of the multivariate scheme C^* in a probabilistic way (several variants of the proposed scheme are discussed in Appendix D), and in Section 5, we present an adaptation of the multivariate scheme UOV. In Section 6, we give some security arguments for the proposed schemes. Finally, we conclude in Section 7.

2   Public Key of Multivariate Schemes

In this section, we first recall the general difficult problem underlying multivariate cryptography. Next, we briefly describe public key schemes in the context of *classical* multivariate cryptography (i.e. the multivariate cryptography of the state of the art). Then, we describe the public key protocols in the context of *probabilistic* multivariate cryptography.

2.1 Problem of Polynomial Equations in Finite Fields

Let K be a finite field. Let $\mathcal{A} = (a_1, \ldots, a_m)$ be a system of $m \in \mathbb{N}$ polynomials in $n \in \mathbb{N}$ variables with degree $d \in \mathbb{N}$. Given $y = (y_1, \ldots, y_m) \in K^m$, the problem is to find a solution $x = (x_1, \ldots, x_n) \in K^n$ of the equation system $y_i = a_i(x_1, \ldots, x_n), 1 \le i \le m$.

Most of the time, the polynomial equations of a multivariate cryptographic scheme are quadratic (i.e. $d = 2$) since the problem of solving such system is $\mathcal{NP}$-complete and hard on average. In this case, the problem is called *Multivariate Quadratic Equations* problem and for every i, $1 \le i \le m$, the polynomial a_i has the form:

$$a_i = \sum_{1 \le j \le n} \sum_{1 \le k \le n} \gamma_{i,j,k} x_j x_k + \sum_{1 \le j \le n} \delta_{i,j} x_j + \xi_i \ ,$$

where the coefficients $\gamma_{i,j,k}$, $\delta_{i,j}$ and ξ_i are elements of K.

2.2   Classical Multivariate Schemes

A classical multivariate scheme relies on the knowledge of a trapdoor $T_{\mathcal{A}}$ in connection with a system $\mathcal{A}$ of m polynomial equations in n variables over a finite field K.

The public key is the system $\mathcal{A}$ and the private key is the trapdoor $T_{\mathcal{A}}$ that allows to compute, for any given value $y = (y_1, \ldots, y_m)$, a value $x = (x_1, \ldots, x_n)$ such that, $y_i = a_i(x_1, \ldots, x_n)$ for every i, $1 \le i \le m$ (or equivalently such that $y = \mathcal{A}(x)$).

On the one hand, the computation of x such that $y = \mathcal{A}(x)$ must be easy using the trapdoor $T_{\mathcal{A}}$, and on the other hand, the computation of x without the knowledge of the trapdoor $T_{\mathcal{A}}$ must be computationally difficult (i.e. the number of operations must be greater than 2^{80}).

Multivariate Signature. Given a message M, one can compute the hash value y of the message M, i.e. $y = H(M)$, where H is a collision resistant hash function. Then, given a hash value y of a message M, a *signature* of the message M is a value x such that $y = \mathcal{A}(x)$. Only the owner of the private key can compute such a value x, and any verifier can check that $y = \mathcal{A}(x)$ for the hash value y of a given message M, its signature x and the public key $\mathcal{A}$.

Multivariate Authentication. An *authentication* between a prover and a verifier works as follows. The verifier sends a challenge y to the prover. Then, by using the trapdoor $T_{\mathcal{A}}$, the prover computes the value x such that $y = \mathcal{A}(x)$, and he sends x to the verifier. At last, the verifier computes $\mathcal{A}(x)$ and the authentication protocol is valid if and only if the equality $y = \mathcal{A}(x)$ holds.

Multivariate Public Key Encryption. For an *encryption scheme*, anybody can encrypt a message x by using the public key $\mathcal{A}$, that is, anybody can computes the ciphertext $y = \mathcal{A}(x)$. Furthermore, only the owner of the private key $T_{\mathcal{A}}$ can decrypt the value $y = \mathcal{A}(x)$ and recovers the value x.

Then, in classical multivariate schemes, *all* the m equations of the system $y = \mathcal{A}(x)$ must be satisfied in order to validate a protocol.

2.3   Probabilistic Multivariate Schemes

In this paper, we focus on authentication protocols and signature schemes (it may also be possible to build probabilistic encryption scheme but this is a difficult problem that we will not study here).

In a *probabilistic multivariate scheme*, the public key is a system $\mathcal{A}$ of m polynomial equations in n variables. A signature (resp. a response to a challenge) will be valid if **at least** α equations of the system $\mathcal{A}$ are satisfied where α is a fixed parameter of the scheme (or more generally, if at least α_1 of the m_1 first equations of $\mathcal{A}$ are satisfied, and at least α_2 equations of the m_2 next equations of $\mathcal{A}$ are satisfied etc., and at least α_ℓ of the m_ℓ last equations of $\mathcal{A}$ are satisfied, where $\alpha_1, \ldots, \alpha_\ell, m_1, \ldots, m_\ell$ and ℓ are well chosen integers with $m_1 + \cdots + m_\ell = m$).

General Description When $m_1 = m$. Let K be a finite field. The public key $\mathcal{A}$ is a system of m polynomial equations of the form $y_i = a_i(x_1, \ldots, x_n)$ where $1 \leq i \leq m$, and $x_1, \ldots, x_n, y_1, \ldots, y_m$ are variables defined over K and $a_1, \ldots, a_m$ are polynomials of degree d with coefficients in K.

The construction of a probabilistic multivariate scheme relies on the existence of a trapdoor $T_{\mathcal{A}}$ such that, given a value y, it is possible with a probability close to 1, to find a value x such that at least α equations of the m equations of $\mathcal{A}$ are satisfied. The parameter α is fixed (e.g. if $K = GF(2)$ then we have $\alpha > \frac{m}{2}$). In exchange, the probability to find a value x (such that α equations of $\mathcal{A}$ are satisfied) without the knowledge of $T_{\mathcal{A}}$ must be very close to 0.

Assuming that such a trapdoor exists, one can construct a probabilistic multivariate scheme for signature or authentication. A value y is either generated by the prover and called a *challenge* in an authentication protocol or the hash value of the message M to be signed (i.e. $y = H(M)$ where H is a hash function assumed to be not only collision resistant but also near-collision resistant, i.e., we assume that it is difficult to find y and y' such that $H(y) \oplus H(y')$ has low Hamming weight[1]) in a signature scheme. Then, a value x such that at least α equations of the m equations of $\mathcal{A}$ are satisfied is a valid authentication value or a valid signature.

In this paper, we only consider the construction of probabilistic multivariate schemes based on known trapdoors. However, it would be very interesting (but certainly very difficult) to find new basic trapdoors and it may be easier to find a basic trapdoor for probabilistic multivariate schemes than for classical multivariate schemes.

3   Probabilistic Schemes Using a Classical Trapdoor

3.1   General Construction

Let $\mathcal{B}$ denote the public key of a classical multivariate scheme and $T_{\mathcal{B}}$ denote the trapdoor associated to $\mathcal{B}$. For simplicity, we set the finite field K to be $GF(2)$.

---

[1] Assuming this additional condition on H is one possibility to avoid existential forgery; alternative techniques will be presented in the full version of this paper.

Construction of the Public Key $\mathcal{A}$. Recall that $\mathcal{B} = (b_1, \ldots, b_m)$ is a system of $m \in \mathbb{N}$ polynomial equations in $n \in \mathbb{N}$ variables with degree $d \in \mathbb{N}$. Let $\mathcal{C} = (c_1, \ldots, c_m)$ be a system of $m \in \mathbb{N}$ polynomial equations in $n \in \mathbb{N}$ variables such that $c_i(x_1, \ldots, x_n) = 0$ with probability κ, where $\kappa > \frac{1}{2}$ (e.g. in Section 4, the quadratic polynomials c_i are chosen such that $\kappa = \frac{3}{4}$). The public key $\mathcal{A}$ is defined to be the set of m equations of the form:

$$y_i = b_i(x_1, \ldots, x_n) + c_i(x_1 \ldots, x_n) = a_i(x_1, \ldots, x_n)$$

where $1 \leq i \leq m$.

Remark 1. The system $\mathcal{C}$ can be used to *mask* the algebraic structure of any classical system $\mathcal{B}$. For instance, in Section 4, we use the C^* scheme and in Section 5, we use the Oil and Vinegar scheme. It is also possible to use for example a FLASH scheme, i.e. the C^{*--} scheme [20] or the HFE scheme [18].

Authentication Scheme

1. The verifier randomly chooses a challenge $y = (y_1, \ldots, y_m)$ in $\in K^m$ and sends it to the prover.
2. The prover follows three steps:
   (a) For every $i \in [1; m]$, the value y_i is replaced by $y_i \oplus 1$ with probability β (where $\beta \neq 0^2$ is a fixed parameter)). Then, the prover gets the value $y' = (y'_1, \ldots, y'_m)$. In average, βm values of y are modified to get y'.
   (b) Using the trapdoor $T_{\mathcal{B}}$, the prover computes the value $x = (x_1, \ldots, x_n)$ such that for every $i \in [1; m]$, we have $y'_i = b_i(x_1, \ldots, x_n)$.
   (c) The prover checks that for at least α integers i of $[1; m]$, the equation

   $$y_i = b_i(x_1, \ldots, x_n) + c_i(x_1, \ldots, x_n)$$

   is satisfied. If not, then the prover restart at the beginning of step 2, else the prover sends $x = (x_1, \ldots, x_n)$ to the verifier.
3. Finally, the verifier checks that at least α among the m equations of the form:

$$y_i \stackrel{?}{=} a_i(x_1, \ldots, x_n)$$

where $1 \leq i \leq m$ are satisfied.

The general execution of a probabilistic scheme is summarized in Figure 1.

Remark 2. In practice, the indices i such that $y_i \neq y'_i$ are chosen with a pseudo-random algorithm that depends only of $(y_1, \ldots, y_m)$ such that for every i, $1 \leq i \leq m$, we have $y_i \neq y'_i$ with probability β and of the current run. Then, if the challenge $y = (y_1, \ldots, y_m)$ is given twice, then the prover will always answer with the same $x = (x_1, \ldots, x_n)$. Here the aim is to prevent the attacker from replaying the same challenge several times in order to get information of the system $\mathcal{C}$.

---

^{2} The reason why β must be different from 0 will be explained in Section 3.2.

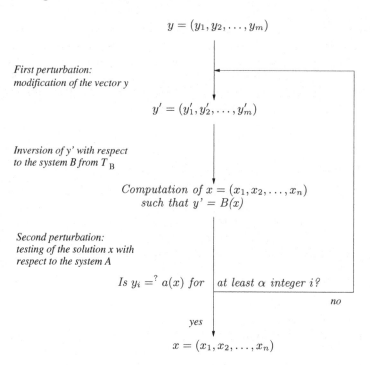

$$y = (y_1, y_2, \ldots, y_m)$$

First perturbation:
modification of the vector y

$$y' = (y'_1, y'_2, \ldots, y'_m)$$

Inversion of y' with respect
to the system B from T_B

Computation of $x = (x_1, x_2, \ldots, x_n)$
such that $y' = B(x)$

Second perturbation:
testing of the solution x with
respect to the system A

Is $y_i \overset{?}{=} a(x)$ for at least α integer i?

no

yes

$$x = (x_1, x_2, \ldots, x_n)$$

Fig. 1. Example of a probabilistic scheme

Signature Scheme. One possibility to construct a probabilistic multivariate scheme based on a known trapdoor is to assume the knowledge of a near-collision hash function H and to replace the challenge y sent by the verifier into the authentication protocol by the hash value $y = H(M)$ of the message M to be signed. This condition on H is to avoid the following attack. Assume that $(M, y = H(M), x)$ is a valid tuple such that there are $\alpha + a$ equations satisfied with $a > 0$. Then, one can construct a new pair (y', x) by changing up to a component in y. Thus, if H is not near-collision resistant, then an attacker will be able to construct a valid tuple $(M', y' = H(M'), x)$.

Alternative solutions will be presented in the full version of this paper.

3.2 The Parameter β Must Be Different from 0

Recall that β is the probability that a bit y_i of the received challenge y is modified by the prover (before inverting the system). The role of the perturbation system $\mathcal{C}$ is to mask the algebraic structure of the system $\mathcal{B}$ (the aim is to prevent the attacker from accessing the system $\mathcal{B}$). However, in order to prevent the attacker to reconstruct the system $\mathcal{C}$, and then, to retrieve the system $\mathcal{B}$, the parameter β must be chosen in a better way.

Suppose that $\beta = 0$. Then, for every pair (x, y) the attacker would know that all the equations of $\mathcal{B}$ are satisfied by (x, y) with probability 1. If $\beta = 0$, then from $\mathcal{O}(n^2)$ pairs (x, y), an attacker will be able to reconstruct the system $\mathcal{B}$ with probability 1 by Gaussian reductions (on the quadratic coefficients of the equations of $\mathcal{B}$). In this case the difficulty of breaking the system is equivalent to the difficulty of breaking the original trapdoor associated to the system $\mathcal{B}$. Thus $\mathcal{C}$ has no interest anymore since it can be removed. Thus, we have $\beta \neq 0$.

When β is different from zero, the attacker has to deal with several cases:

– if a relation $y_i = a(x)$ is valid, then:
  1. y_i equals y'_i and $c_i(x) = 0$ happens with probability $(1 - \beta)(1 - \kappa)$ (on average);
  2. y_i is different from y'_i and $c_i(x) = 1$ happens with probability $\beta\kappa$ (on average);
– if a relation y_i is different from $a(x)$, then:
  1. y_i equals y'_i and $c_i(x) = 1$ happens with probability $(1-\beta)\kappa$ (on average);
  2. y_i is different from y'_i and $c_i(x) = 0$ happens with probability $\beta(1 - \kappa)$ (on average).

Then, the value of the parameter β must be chosen in accordance with the value of κ (recall that the value κ is fixed by choosing the polynomials c_i, $1 \leq i \leq m$).

3.3 Relation Between the Parameters α, β and κ

Recall that α is the number of equations of the public key that must be satisfied to validate an authentication or a signature. The parameters β and κ concern the two perturbations involved in a multivariate probabilistic scheme based on a known trapdoor: the value β is the probability that a bit of the received challenge y is modified by the prover before inverting the system, and the value κ is the probability that a polynomial equation of the perturbation system $\mathcal{C}$ equals 1.

The value α depends on the probability that the equation $y_i \overset{?}{=} a_i(x_1, \ldots, x_n)$, $1 \leq i \leq m$, is not satisfied, that is, α depends on the two values β and κ. There are (on average) κm integers $i \in [1; m]$ such that $y'_i = b_i(x_1, \ldots, x_n) + c_i(x_1, \ldots, x_n)$ and the prover has changed βm values of y. Thus, the parameter α must be chosen such that:

$$\alpha \simeq (\kappa - \beta)m \ .$$

Since the probability κ is fixed by choosing the polynomials c_i, $1 \leq i \leq m$, the values of α and β must be chosen in accordance with the value of κ. Notice that we must choose α such that $\alpha > \frac{m}{2}$ in order to prevent that a random value is valid with a probability $\frac{1}{2}$ and β must be different from 0.

3.4 Size of the Public Key

Assume that the equations of the public key look as random equations of degree d for an adversary who do not have the secret key. We have $\frac{m}{2} < \alpha \leq m$. Let λ

be the value defined by $\alpha = \lambda m$. Then, in order to ensure a security in 2^{80}, the number m of equations of a public key must be chosen such that:

$$m\left(1 + \lambda\frac{\ln\lambda}{\ln 2} + (1-\lambda)\frac{\ln(1-\lambda)}{\ln 2}\right) \simeq 80 \ .$$

Details of this approximation are given in Appendix A.

Example 1. For $\lambda = \frac{3}{4}$, we get $m \simeq 423$, and for $\lambda = \frac{9}{10}$, we get $m \simeq 150$ equations.

Remark 3. As a consequence, the public key is larger in a probabilistic scheme than in a non-probabilistic where at least about 80 equations are required.

4   The Probabilistic Multivariate Scheme $C^* + LL'$

The Matsumoto-Imai scheme (also called C^*) was presented in [15] and cryptanalysed in [17,3]; the description of the scheme C^* is briefly recalled in Appendix C. We present a probabilistic variation of the C^* scheme, called $C^* + LL'$ where no attack is known; another way to repair the C^* scheme is for example the FLASH scheme of [20].

In this section, we keep the notation of Section 3, the public key $\mathcal{A} = \mathcal{B} + \mathcal{C}$ will be constructed such that $\mathcal{B}$ is a public key of a C^* scheme and $\mathcal{C}$ is a set of product of linear forms ($\mathcal{B}$ and $\mathcal{C}$ are kept secret).

4.1   Construction of the Public Key $\mathcal{A}$

Let $K = GF(2)$. Let $\mathcal{B}$ be the public key of a C^* scheme, that is, $\mathcal{B}$ is a set of n quadratic equations in n variables over $GF(2)$ of the form

$$y_i = b_i(x_1, \ldots, x_n)$$

where $1 \leq i \leq n$ and $x_1, \ldots, x_n, y_1 \ldots, y_n$ are elements of K. The trapdoor associated to $\mathcal{B}$ is denoted by $T_\mathcal{B}$. Notice that both $\mathcal{B}$ and $T_\mathcal{B}$ are kept secret.

Let $L_1, \ldots, L_n, L'_1, \ldots, L'_n$ be $2n$ secret linear forms in the variables $x_1, \ldots, x_n$. For every i, $1 \leq i \leq n$, let $c_i = L_i \cdot L'_i$. Then, the public key $\mathcal{A}$ of the scheme $C^* + LL'$ is the set of the n quadratic equations in n variables of the form:

$$y_i = b_i(x_1, \ldots, x_n) + L_i(x_1, \ldots, x_n) \cdot L'_i(x_1, \ldots, x_n) = a_1(x_1, \ldots, x_n)$$

where $1 \leq i \leq n$.

Remark 4. The classification of quadratic forms over $GF(q)$ (for q odd or even) is well-known; it is given for example in [13] pp. 278-289 and recalled in Appendix B. We are interested here in the case q even since q is generally a power of two. Then, we have only one or two canonic forms when n is fixed and non degenerated, so we have at least $2n$ possible canonic forms when q is fixed.

4.2 The Scheme $C^* + LL'$

As usual, $y = (y_1, \ldots, y_n)$ is the challenge of an authentication scheme or the hash value of the message to be signed in a signature scheme. The value $x = (x_1, \ldots, x_n)$ will be a successful authentication value or a valid signature if at least α equations of $\mathcal{A}$ are satisfied. Recall that $y' = (y'_1, \ldots, y'_n)$ is the modified challenge computed at the first step of the computation of the value x (see Section 3).

We do not describe precisely the authentication protocol of the $C^* + LL'$ since it is straightforward from Section 3 and the description above of the $C^* + LL'$ public key. We only discuss the parameters of the scheme.

For every $i, 1 \le i \le m$, we have $L_i(x_1, \ldots, x_n) \cdot L'_i(x_1, \ldots, x_n) = 0$ with probability $\kappa = \frac{3}{4}$. Then, we have $y'_i = b_i(x_1, \ldots, x_n) + L_i(x_1, \ldots, x_n) \cdot L'_i(x_1, \ldots, x_n)$ with a probability $\frac{3}{4}$. Next, we have $y_i = y'_i$ with a probability $(1 - \beta)$. Thus, we deduce that we have $y_i = f_i(x_1, \ldots, x_n) + L_i(x_1, \ldots, x_n) \cdot L'_i(x_1, \ldots, x_n)$ with a probability greater than or equal to $\frac{3}{4} - \beta$.

Then, the expectation value of the number N of equations of $\mathcal{A}$ that are satisfied is greater than or equal to $\left(\frac{3}{4} - \beta\right) n \simeq \alpha$. For a given $(y_1, \ldots, y_n)$, if N is lower than α, then we can try again at step 1 by computing another $(y'_1, \ldots, y'_n)$ with again about βn values changed from $(y_1, \ldots, y_n)$ chosen with a deterministic pseudo-random algorithm that depends only of $(y_1, \ldots, y_n)$ and of the current run. After a few tries, we get a solution $(x_1, \ldots, x_n)$ with at least α equations of $\mathcal{B}$ that are satisfied, i.e., a valid signature or a valid answer to a challenge.

Remark 5. For a security greater than or equal to 2^{80}, we need $n \ge 423$ when β is small. For instance, with $\beta = \frac{1}{10}$ and $n \simeq 500$, no attack of this scheme exists to the best of our knowledge.

Many variants of the scheme $C^* + LL'$ are described in Appendix D.

5 The Probabilistic Multivariate Scheme $UOV + LL'$

The scheme *Oil and Vinegar* was introduced in [19] and it was broken in [12]. Next, a generalisation of the original scheme, called *Unbalanced Oil and Vinegar (UOV)*, was introduced in [10]; the scheme UOV is not broken for well-chosen parameters. In this section, we will be able to use more possible parameters since some attacks valid for UOV will not work any more for UOV+LL'. The scheme *UOV* is briefly recalled in Appendix C.

The scheme UOV+LL' proceeds exactly as the scheme $C^* + LL'$ except that the C^* equations are changed with UOV equations. Since this UOV+LL' scheme looks particularly interesting, we describe the construction of the public key and the scheme and we give some remarks on its efficiency.

5.1 Construction of the Public Key $\mathcal{A}$

Let $K = GF(2)$ and $\mathcal{B}$ be the public key of a UOV scheme, i.e., $\mathcal{B}$ is a set of m quadratic equations in n variables $(x_1, \ldots, x_n)$ over $GF(2)$. Each equation of $\mathcal{B}$

is of the form $y_i = f_i(x_1, \ldots, x_n)$ where $1 \leq i \leq m$, $x_1, \ldots, x_n, y_i \in K$, and f_i is a quadratic function.

There are $n - p$ *oil variables* denoted by $o_1, \ldots, o_{n-p} \in K$ and p *vinegar variables* denoted by $v_1, \ldots, v_p \in K$ and there is a secret affine and invertible transformation s such that $(x_1, \ldots, x_n) = s(o_1, \ldots, o_{n-p}, v_1, \ldots, v_p)$ and such that each y_i of $\mathcal{B}$ written in the $o_1, \ldots, o_{n-p}, v_1, \ldots, v_p$ variables (instead of $x_1, \ldots, x_n$ variables) is of the form:

$$y_i = \sum_{j=1}^{n-p}\sum_{k=1}^{p} \gamma_{i,j,k} o_j v_k + \sum_{j=1}^{p}\sum_{k=1}^{p} \mu_{i,j,k} v_j v_k + \sum_{j=1}^{n-p} \delta_{i,j} o_j + \sum_{j=1}^{p} \nu_{i,j} v_j + \xi_i$$

where $1 \leq i \leq m$ and $\gamma_{i,j,k}$, $\mu_{i,j,k}$, $\delta_{i,j}$, $\nu_{i,j}$ and ξ_i are elements of K. Notice that we do not have any term in $a_i a_j$: we can have oil $\times$ vinegar, vinegar $\times$ vinegar but never oil $\times$ oil.

Let $L_1, \ldots, L_m, L'_1, \ldots, L'_m$ be $2m$ secret linear forms in $x_1, \ldots, x_n$ (or equivalently in the variables $a_1, \ldots, a_h, b_1, \ldots, b_v$). Let $\mathcal{A}$ be the set of the m quadratic equations of the form $y_i = f_i(x_1, \ldots, x_n) + L_i(x_1, \ldots, x_n) \cdot L'_i(x_1, \ldots, x_n)$. The set $\mathcal{A}$ will be the public key of the scheme UOV+LL' (while f_i, $\mathcal{B}$, L_i, L'_i and s are kept secret).

5.2   The Scheme $UOV + LL'$

Recall that y is the challenge in an authentication scheme, or the hash value of the message to be signed in a signature scheme. The value x is a valid signature or a successful authentication if at least α equations of $\mathcal{A}$ are satisfied, with $\alpha \simeq \left(\frac{3}{4} - \beta\right) m$, where β is a fixed parameter (for example, we can choose $\beta \simeq \frac{1}{10}$).

Computation of the Value x. In order to compute $x = (x_1, \ldots, x_n)$ with the secrets, the prover proceeds as follows.

1. For every $i \in [1; m]$, the value y_i is replaced by $y_i \oplus 1$ with probability β and then the value $y' = (y'_1, \ldots, y'_m)$ is obtained.
2. The prover randomly chooses the *vinegar variables* $v_1, \ldots, v_p$.
3. The prover computes the values $a_1, \ldots, a_m$ such that:

$$\forall i, 1 \leq i \leq m, y'_i = f_i(x_1, \ldots, x_n) = f_i(s(o_1, \ldots, o_{n-p}, v_1, \ldots, v_p))$$

Here we have a linear system of m equations in the variables $o_1, \ldots, o_{n-p}$. If we have no solution we try again with other random vinegar values $v_1, \ldots, v_p$.

For all i, $1 \leq i \leq m$, we have $y'_i = f_i(x_1, \ldots, x_n) + L_i(x_1, \ldots, x_n) \cdot L'_i(x_1, \ldots, x_n)$ with a probability $\frac{3}{4}$. Moreover, with a probability $(1 - \beta)$, we have $y_i = y'_i$. Thus, with a probability greater than or equal to $\frac{3}{4} - \beta$ we have $y_i = f_i(x_1, \ldots, x_n) + L_i(x_1, \ldots, x_n) \cdot L'_i(x_1, \ldots, x_n)$. Then, the expectation value of the number N of equations of $\mathcal{A}$ that are satisfied is greater than or equal to $\left(\frac{3}{4} - \beta\right) \simeq \alpha$. If we have $N < \alpha$, then we can try again with new random vinegar variables.

Remark 6. The random variables $v_i, \ldots, v_p$ and the indices i such that $y_i \neq y'_i$ are chosen with a pseudo-random algorithm that depends only of y and of the current run. Thus, if twice the same challenge $(y_1, \ldots, y_m)$ is given, the prover will always answer with the same $(x_1, \ldots, x_n)$.

Remark 7. If we compare UOV and $UOV + LL'$, we can notice that in $UOV + LL'$ we do not need any more to have $v \geq 2m$ in order to avoid the Shamir-Kipnis attack of [12]. Moreover in the equations of UOV, we have oil × oil, oil × vinegar and vinegar × vinegar, so the scheme might be more secure for smaller values of the parameters.

Notice that the variations given in Appendix D for $C^* + LL'$ are also possible variants for $UOV + LL'$.

6 Security Arguments

In this section, we discuss the three main techniques generally used to attack multivariate schemes and we explain why our schemes should resist these attacks.

6.1 Gröbner Bases

Gröbner bases are used as a general attack method for any multivariate cryptographic schemes. There are several algorithms for computing Gröbner basis including Buchberger's, F4[5] and F5[6].

When using the perturbation LL' in a probabilistic multivariate scheme, we involve $2n$ linear forms. That comes to add n additional momomials to the basic set of monomials deduced from the public key of a basic multivariate scheme. Then, the perturbation LL' increases the complexity of the computation of the Gröbner basis.

Moreover recall that for the proposed schemes (e.g. $C^* + LL'$ or UOV+LL'), nobody knows how to invert the system (the knowledge of the secret key does not allow to inverse the system). Thus, we do not expect to be able to inverse the system even with Gröbner basis.

6.2 Rank Attack on One Quadratic Equation

The idea of exploiting the rank to attack a multivariate scheme was first used by T. Patarin [17] to separate branches in the Matsumoto-Imai scheme. Next, C. Wolf et al. [24] used a similar idea to attack the STS scheme.

For example, in the C^* scheme, the rank is near the maximum, i.e. near n, and the effect of the perturbation LL' when adding to a basic multivariate scheme is that the rank is eiher increased by one, decreased by one or unchanged. Therefore, the rank of $C^* + LL'$ will be very near the maximum as for random quadratic equations with high probability. Thus, we do not expect this attack to work here.

6.3   Differential Cryptanalysis (i.e. Rank of the Polar Form Attack)

Differential cryptanalysis for multivariate schemes was recently introduced by P.-A. Fouque, L. Granboulan and J. Stern in [8] to attack the scheme PMI (Perturbated Matsumoto-Imai) which is a variant of the scheme C^* using the internal perturbation of Ding [4]. The key point of the attack is that the dimension of the kernel can be used to identify elements that cancel the perturbation. More precisely, the attack consists first on the reconstruction of the linear space $\mathcal{K}$ where there is no noise.

In the case of the probabilistic multivariate scheme $C^* + LL'$, there is no set equivalent to the set $\mathcal{K}$. Indeed, in the PMI scheme, the perturbation is a set of r quadratic equations where r is a small value and the set $\mathcal{K}$ is of dimension $n - r$. In the scheme $C^* + LL'$, the perturbation is a set of n quadratic equations construct by using $2n$ random linear forms. The dimension of the perturbation of the $C^* + LL'$ scheme is n with high probability and then there is no set $\mathcal{K}$ to recover. Thus the attack described in [8] does not directly apply on the scheme $C^* + LL'$.

Furthermore, it may be not possible to distinguish a public key of the scheme $C^* + LL'$ from a random set of quadratic equations by using the technique proposed in [8] since the first part of the attack requires $\mathcal{O}(q^r)$ computations and in the PMI scheme, the value r must be small since the secret key computation part costs $\mathcal{O}(q^r)$ whereas in the $C^* + LL'$ scheme, we have $r = n$ and $q^r \geq 2^{80}$.

7   Conclusion

Probabilistic Multivariate Cryptography is a new concept in public key cryptography with many possible schemes. It opens new opportunities and new questions that we think are interesting, both from a practical and from a theoretical point of view. In this paper we have presented some new public key schemes (C^* + LL' and UOV + LL' for example) based on this idea of probabilistic Multivariate Cryptography with some explicit examples for the parameters. These schemes were built from the transformation of non-probabilistic multivariate schemes to probabilistic multivariate schemes in order to get more security or more efficiency. An interesting problem is to find a trapdoor for probabilistic multivariate schemes which allows directly to find an approximation of the solution associated to the challenge or the message to be signed. Another interesting problem is to find probabilistic multivariate schemes for encryption (not only for signatures or authentications).

References

1. J. Bringer, H. Chabanne, and E. Dottax. Perturbing and Protecting a Traceable Block Cipher. Cryptology ePrint Archive, Report 2006/064., 2006.
2. N. Courtois. The Security of Hidden Field Equations (HFE). In *Progress in Cryptology - CT-RSA 2001*, volume LNCS 2020, pages 266–281.

3. P. Delsarte, Y. Desmedt, A.M. Odlyzko, and P. Piret. Fast Cryptanalysis of the Matsumoto-Imai Public Key Scheme. In *Advances in Cryptology - Eurocrypt'84*, volume LNCS 209, pages 142–149.

4. J. Ding. A New Variant of the Matsumoto-Imai Cryptosystem Through Perturbation. In *Public Key Cryptography PKC 2004*, volume LNCS 2947, pages 305–318.

5. J.-C. Faugère. A new efficient algorithm for computing Grobner basis (F4). In *Journal of Pure and Applied Algebra*, pages 61–88, 1999.

6. J.-C. Faugère. A new efficient algorithm for computing Grobner basis without reduction to zero (F5). In *Proceedings of ISSAC, ACM Press*, pages 75–83, 2002.

7. H. Fell and W. Diffie. Analysis of a public key approach based on polynomial substitution. In *Advances in Cryptology - Crypto'85*, volume 218, pages 340–349.

8. P-A. Fouque, L. Granboulan, and J. Stern. Differential Cryptanalysis for Multivariate Schemes. In *Advances in Cryptology Eurocrypt'05*, volume LNCS 3494, pages 341–353.

9. H. Imai and T. Matsumoto. Algebraic Methods for Constructing Asymetric Cryptosystems. In *Algebraic Algorithms and Error-Correctings Codes – AAECC*, pages 108–119, 1985.

10. A. Kipnis, J. Patarin, and L. Goubin. Unbalanced Oil and Vinegar Signature Schemes. In *Advances in Cryptology - Eurocrypt'99*, volume LNCS 1592, pages 206–222.

11. A. Kipnis and A. Shamir. Cryptanalysis of the HFE Public Key Cryptosystem by Relinearization. In *Advances in Cryptology - Crypto'99*, volume LNCS 1666, pages 19–30.

12. A. Kipnis and A. Shamir. Cryptanalysis of the Oil & Vinegar Signature Scheme. In *Advances in Cryptology - Crypto'98*, volume LNCS 1462, pages 257–266.

13. R. Lidl and H. Niederreiter. *Finite fields*, volume 20 of *Encyclopedia of Mathematics and its applications*. Cambridge University Press, 1997.

14. F. J. MacWilliams and N. J. A. Sloane. *The theory of error-correcting codes*. Elsevier, North-Holl., 1977.

15. T. Matsumoto and H. Imai. Public quadratic polynomial-tuples for efficient signature-verification and message-encryption. In *Advances in Cryptology - Eurocrypt'88*, volume LNCS 330, pages 419–453.

16. J. Patarin. Asymmetric Cryptography with a Hidden Monomial. In *Advances in Cryptology - Crypto'96*, volume LNCS 1109, pages 45–60.

17. J. Patarin. Cryptanalysis of the Matsumoto and Imai Public Key Scheme of Eurocrypt'88. In *Advances in Cryptology - Crypto'95*, volume LNCS 963, pages 248–261.

18. J. Patarin. Hidden Fields Equations (HFE) and Isomorphisms of Polynomials (IP): Two New Families of Asymmetric Algorithms. In *Advances in Cryptology - Eurocrypt'96*, volume LNCS 1070, pages 33–48.

19. J. Patarin. The Oil and Vinegar Signature Scheme. Presented at the Dagstuhl Workshop on Cryptography, 1997.

20. J. Patarin, N. Courtois, and L. Goubin. FLASH, a Fast Multivariate Signature Algorithm. In *Progress in Cryptology - CT-RSA 2001*, volume LNCS 2020, pages 298–307.

21. J. Patarin, N. Courtois, and L. Goubin. QUARTZ, 128-Bit Long Digital Signatures. In *Progress in Cryptology - CT-RSA 2001*, volume LNCS 2020, pages 282–297.

22. J. Patarin, L. Goubin, and N. Courtois. C^*_{-+} and HM: Variations around two Schemes of T. Matsumoto and H. Imai. In *Advances in Cryptology - Asiacrypt'98*, volume 1514, pages 35–49.

23. A. Shamir. Efficient Signature Schemes Based on Birational Permutations. In *Advances in Cryptology - Crypto'93*, volume LNCS 773, pages 1–12.

24. C. Wolf, A. Braeken, and B. Preneel. Efficient cryptanalysis of RSE(2)PKC and RSSE(2)PKC. In *In Conference on Security in Communication Networks – SCN 2004*, volume LNCS 3352, pages 145–151.
25. C. Wolf and B. Preneel. Taxonomy of Public Key Schemes based on the problem of Multivariate Quadratic equations. Cryptology ePrint Archive, Report 2005/077.

A   Size of the Public Key

Let $K = GF(2)$. We want to evaluate the minimum number of equations of a public key in order to ensure a security in 2^{80}. Notice that, we also assume that the equations look as random equations of degree d for an adversary who do not have the secret key.

Given a hash value of a message or a challenge $y \in K^m$, an adversary can choose a random value $x \in K^n$ for the signature or the authentication value. For each try, the attacker has a probability $\frac{1}{2^m} \sum_{i=\alpha}^{m} \binom{m}{i}$ to have α or more satisfied equations. Then, m must be chosen such that:

$$\frac{1}{2^m} \sum_{i=\alpha}^{m} \binom{m}{i} \leq \frac{1}{2^{80}} .$$

We have $\frac{m}{2} < \alpha \leq m$. Let λ be the value defined by $\alpha = \lambda m$. If λ is sufficiently different from $\frac{1}{2}$, then the dominant term in $\sum_{i=\alpha}^{m} \binom{m}{i}$ is $\binom{m}{\alpha}$. More precisely, we can overvalue $\sum_{i=\alpha}^{m} \binom{m}{i}$ by a geometric sum with the first term $\binom{m}{\alpha}$. Thus, we want to evaluate:

$$\frac{1}{2^m} \binom{m}{\alpha} = \frac{1}{2^m} \cdot \frac{m!}{\alpha!(m-\alpha)!} = \frac{1}{2^m} \frac{m!}{(\lambda m)!\,(m(1-\lambda))!} .$$

From stirling formula $n! \sim n^n \exp^{-n} \sqrt{2\pi n}$, we get:

$$\frac{1}{2^m} \binom{m}{\alpha} \approx \frac{1}{2^m} \frac{m^m \exp^{-m} \sqrt{2\pi m}}{(\lambda m)^{\lambda m} \exp^{-\lambda m} \sqrt{2\pi \lambda m} \cdot (m(1-\lambda))^{m(1-\lambda)} \exp^{-m(1-\lambda)} \sqrt{2\pi m(1-\lambda)}} .$$

After simplifications, we get:

$$\frac{1}{2^m} \binom{m}{\alpha} \approx \frac{1}{2^{m\left(1+\lambda \frac{\ln \lambda}{\ln 2}+(1-\lambda)\frac{\ln(1-\lambda)}{\ln 2}\right)} \sqrt{2\pi m \lambda (1-\lambda)}} .$$

In first approximation, this will be about $\frac{1}{2^{80}}$ when $m\left(1 + \lambda \frac{\ln \lambda}{\ln 2} + (1-\lambda) \frac{\ln(1-\lambda)}{\ln 2}\right) \simeq 80$.

B   Classification of Quadratic Forms over $GF(q)$

The classification of quadratic forms over $GF(q)$ (for q odd or even) is well-known; it is given for example in [13] pp. 278–289. We are interested here in the case q even since q is generally a power of two. Then, we recall here the two main theorems for the case q even.

Theorem 1 ([13] p.287). *Let $GF(q)$ be a finite field with q even. Let $f \in GF(q)[x_1, \ldots, x_n]$ be a non degenerate quadratic form. If n is odd, then f is equivalent to:*

$$x_1 x_2 + x_3 x_4 + \ldots, x_{n-2} x_{n-1} + x_n^2 .$$

If n is even, then f is equivalent to one of the two forms:

1. $x_1 x_2 + x_3 x_4 + \ldots, x_{n-1} x_n$
2. $x_1 x_2 + x_3 x_4 + \ldots, x_{n-1} x_n + x_{n-1}^2 + a x_n^2$

where $a \in GF(q)$ satisfies $Tr_{GF(q)}(a) = 1$.

Theorem 2 ([13] p.288). *Let $GF(q)$ be a finite field with q even. Let $b \in GF(q)$.*

For odd n, the number of solutions of the equation

$$x_1 x_2 + x_3 x_4 + \ldots + x_{n-2} x_{n-1} + x_n^2 = b$$

in $GF(q)^n$ is q^{n-1}.

For even n, the number of solutions of the equation

$$x_1 x_2 + x_3 x_4 + \ldots + x_{n-1} x_n = b$$

in $GF(q)^n$ is $q^{n-1} + \nu(b) q^{\frac{n-2}{2}}$, with $\nu(b) = -1$ if $b \neq 0$ and $\nu(0) = q - 1$.

For even n and $a \in GF(q)$ with $Tr_{GF(q)}(a) = 1$, the number of solutions of the equation

$$x_1 x_2 + x_3 x_4 + \ldots + x_{n-1} x_n + x_{n-1}^2 + a x_n^2 = b$$

in $GF(q)^n$ is $q^{n-1} - \nu(b) q^{\frac{n-2}{2}}$, with $\nu(b) = -1$ if $b \neq 0$ and $\nu(0) = q - 1$.

Then, we have only one or two canonic forms when n is fixed and non-degenerate, so we have at most $2n$ possible canonic forms when q is fixed. This number is generally too small to give any useful information in our schemes, for example when the transformation LL' is applied.

C   Basic Trapdoors

C.1   Matsumoto-Imai Scheme (C^*)

Let $K = \mathbb{F}_q$ be a finite field and $\mathbb{E}$ be an extension field of dimension n over K. Let Φ be an isomorphism from $\mathbb{E}$ to K^n. Let f be the function defined over $\mathbb{E}$ by

$$f : x \longmapsto x^{1+q^{\theta}},$$

where $\theta \in \mathbb{N}$. If the finite field K has characteristic 2 and $\gcd(q^n - 1, q^\theta + 1) = 1$, then f is a bijection. Furthermore, the restriction on θ allows an efficient inversion of the function f. Indeed, $f^{-1}(y) = y^{h'}$, where h' is the inverse of $1 + q^\theta$ modulo $q^n - 1$.

The public key is the function $A := x \mapsto T \circ \Phi \circ f \circ \Phi \circ S(x)$. The hardness of the Matsumoto-Imai scheme is based on the IP-problem, that is, the difficulty of finding transformations S and T for given polynomials equations P and P'.

C.2   The Scheme UOV

Let $K = \mathbb{F}_q$ be a small finite field. Let m, n and p be three positive integers. The hash value y of the message to be signed is an element of K^m, and the signature x is an element of K^n.

The public key is a set $\mathcal{A}$ of m polynomials in n variables of the form:

$$y_i = f_i(x_1, \ldots, x_n), \quad 1 \leq i \leq m \ .$$

There exists a bijective affine function $s : K^n \to K^n$ such that:

$$(x_1, \ldots, x_n) = s(o_1, \ldots, o_{n-p}, v_1, \ldots, v_p)$$

and such that for every i, $1 \leq i \leq m$:

$$y_i = \sum_{j=1}^{n-p} \sum_{k=1}^{p} \gamma_{i,j,k} o_j v_k + \sum_{j=1}^{p} \sum_{k=1}^{p} \mu_{i,j,k} v_j v_k + \sum_{j=1}^{n-p} \delta_{i,j} o_j + \sum_{j=1}^{p} \nu_{i,j} b_j + \xi_i$$

Note that the vinegar variables v_i's are combined quadratically while the oil variables o_i's are only combined with vinegar variables in a quadratic way. Therefore assigning random values to the vinegar variables results in a system of linear equations in the oil variables which can be solved, for instance, by using gaussian elimination.

D   Variants of the Scheme $C^* + LL'$

First Variant: $C^* + LL' + L''L'''$. The first variant consists in replacing the linear product LL' by the linear product $LL' + L''L'''$ (as a consequence, the value of the parameter κ is modified). We keep the same notations, that is, $\mathcal{B}$ is a public key of a C^* scheme and $\mathcal{A}$ is the set of n equations of the form:

$$y_i = b_i(x_1, \ldots, x_n) + c_i(x_1, \ldots, x_n) = a_i(x_1, \ldots, x_n) \ ,$$

where c_i, $1 \leq i \leq n$, is a product a linear forms which is defines as follows.

Let L_i, L'_i, L''_i and L'''_i, $1 \leq i \leq n$, be $4n$ secret linear forms in the n variables $x_1, \ldots, x_n$. The set $\mathcal{C}$ is defined by the set of n equations of the form:

$$y_i = b_i(x_1, \ldots, x_n) + L_i(x_1, \ldots, x_n) L'_i(x_1, \ldots, x_n) + L''_i(x_1, \ldots, x_n) L'''_i(x_1, \ldots, x_n)$$

where $1 \leq i \leq n$.

The value of the parameter κ is the probability that the equation $L_i \cdot L'_i + L''_i \cdot L'''_i \overset{?}{=} 0$ is satisfied, that is, $\kappa = \frac{10}{16}$. according to the figure 2.

Since we have $\frac{1}{2} < \kappa = \frac{10}{16} < \frac{3}{4}$, the scheme $C^* + LL' + L''L'''$ will generally be less efficient than the scheme $C^* + LL'$. However, it may be difficult to distinguish the public key of $C^* + LL' + L''L'''$ from random quadratic equations than the public key of $C^* + LL'$, and thus, for C^* public key $\mathcal{B}$, the scheme $C^* + LL' + L''L'''$ may be more secure than the scheme $C^* + LL'$.

More generally, we know [13] the exact numbers of solutions $x_1, \ldots, x_n$ of any quadratic form $q(x_1, \ldots, x_n) = 0$. For instance, the number of $(x_1, \ldots, x_n) \in \mathbb{F}_2^n$

L_i	L'_i	L''_i	L'''_i	$L_i \cdot L'_i + L''_i \cdot L'''_i$	L_i	L'_i	L''_i	L'''_i	$L_i \cdot L'_i + L''_i \cdot L'''_i$
1	1	1	1	0	0	1	1	1	1
1	1	1	0	1	0	1	1	0	0
1	1	0	1	1	0	1	0	1	0
1	1	0	0	1	0	1	0	0	0
1	0	1	1	1	0	0	1	1	1
1	0	1	0	0	0	0	1	0	0
1	0	0	1	0	0	0	0	1	0
1	0	0	0	0	0	0	0	0	0

Fig. 2. Truth table of $L_i L'_i + L''_i L'''_i$

such that $x_1 x_2 + x_3 x_4 + \cdots + x_{n-1} x_n = 0$ with n even is $2^{n-1} + 2^{\frac{n-2}{2}}$, i.e. $2^{n-1} \left(1 + \frac{1}{2^{\frac{n}{2}}} \right)$ instead of 2^{n-1} for an average quadratic form of n variables.

Second Variant: Decomposing $\mathcal{A}$ in Sets of Equations with Various Probability. Instead of having about 423 equations $C^* + LL'$ in $\mathcal{A}$, we can have, for example, 40 equations that come from a C^{*--} scheme (all these equations will have to be satisfied) and 160 equations that come from a $C^* + LL'$ scheme (at least 120 equations will have to be satisfied). Many other choices of parameters are possible.

Third Variant: Public Key of Degree 3 Instead of 2. When using a public key formed with quadratic polynomials, it is not possible to prevent the attacker that observe an equation $y_i \neq a(x)$ from distinguishing between the first case $[y_i = y'_i$ and $L_i(x) \cdot L'_i(x) = 1]$ and the second case $[y_i \neq y'_i$ and $L_i(x) \cdot L'_i(x) = 0]$. Indeed, we have $y_i = y'_i$ with probability $(1 - \beta)$ and we have $L_i(x) \cdot L'_i(x) = 0$ with probability κ. Then, to prevent the attacker from distinguishing between case 1 and case 2, we have to choose the values of β and κ such that $(1 - \beta)(1 - \kappa) = \beta \kappa$. Furthermore, we have $\alpha \approx (\kappa - \beta)n \geq \frac{m}{2}$. That comes to choose the values of κ and β such that $\kappa + \beta = 1$ and $\kappa - \beta > \frac{1}{2}$.

These conditions imply that $\kappa > \frac{3}{4}$. When the public key has degree 2 then, the higher value of κ is $\frac{3}{4}$ (c.f. the weight distribution of quadratic forms). If $\kappa = \frac{3}{4}$, then there is no solution β fulfilling both $\kappa + \beta = 1$ and $\kappa - \beta > \frac{1}{2}$.

This property can be achieved by using public key of degree 3. In a C^* scheme, a monomial $b = a^{1+q^\theta}$ is hidden by affine transformations. In [16], the possibility of replacing $b = a^{1+q^\theta}$ by $b = a^{1+q^\theta+q^\varphi}$ is studied; the public key has degree 3 instead of 2. The attack of the scheme C^* given in [17] does not apply directly on the scheme "C^* of degree 3". However the scheme is insecure as it is shown in [16]. We use the scheme "C^* of degree 3" as a basic scheme to construct a probabilistic multivariate scheme.

Let $\mathcal{B}$ be the public key of a scheme "C^* of degree 3", that is $\mathcal{B}$ is a set of n equations in n variables of degree 3 over K of the form $y_i = b_i(x_1, \ldots, x_n)$ where $1 \leq i \leq n$ and $x_1, \ldots, x_n, y_1, \ldots, y_n$ are elements of K. The trapdoor associated to $\mathcal{B}$ is denoted by $T_\mathcal{B}$.

Let $L_1, \ldots, L_n, L_1', \ldots, L_n', L_1'', \ldots, L_n''$ be $3n$ secret linear forms in the variables $x_1, \ldots, x_n$. Then, the public key $\mathcal{A}$ is the set of the n equations of degree 3 in n variables $y_i = b_i(x_1, \ldots, x_n) + L_i(x_1, \ldots, x_n)L_i'(x_1, \ldots, x_n)L_i'(x_1, \ldots, x_n)$, $1 \leq i \leq n$.

Parameter κ. For all i, $1 \leq i \leq n$, we have $L_i(x_1, \ldots, x_n) = 0$ with a probability $\frac{1}{2}$ and we also have $L_i'(x_1, \ldots, x_n) = 0$ and $L_i''(x_1, \ldots, x_n) = 0$ with a probability $\frac{1}{2}$. Thus, we have $L_1(x_1, \ldots, x_n)L_1'(x_1, \ldots, x_n)L_1'(x_1, \ldots, x_n) = 0$ with probability $\kappa = \frac{7}{8}$.

Parameters α and β. Recall that α, β and κ must fulfill the relation $\alpha \simeq (\kappa - \alpha)n = \frac{3}{4}n \geq \frac{n}{2}$. By choosing $\beta = 1 - \kappa = \frac{1}{8}$, an attacker would not be able to distinguish between the two possible cases when a relation of the public key is not satisfied.

Short 2-Move Undeniable Signatures

Jean Monnerat[*] and Serge Vaudenay

EPFL, Switzerland
http://lasecwww.epfl.ch/

Abstract. Attempting to reach a minimal number of moves in cryptographic protocols is a quite classical issue. Besides the theoretical interests, minimizing the number of moves can clearly facilitate practical implementations in environments with communication constraints. In this paper, we offer a solution to this problem in the context of undeniable signatures with interactive verification protocols by proposing a way to achieve these protocols in 2 moves. To this goal, we review a scheme we proposed at Asiacrypt 2004 whose property is the full scalability of the signature length against security. We slightly modify (to make it nontransferable) a 2-move version of this scheme which was mentioned in the original article without a proof of security. In the random oracle model, we prove the security of the modified version against an active adversary and precisely assess the security in terms of the signature length. To the best of our knowledge, this scheme is the first 2-move undeniable signature scheme with a security proof.

Keywords: Undeniable signatures, 2-move protocols.

1 Introduction

The concept of undeniable signature was introduced by Chaum and van Antwerpen [6] in 1989. The difference between this kind of signature and a classical one is that the verification of a signature cannot be achieved without the cooperation of the signer (originally, for privacy motivations). Namely, by interacting with a verifier in a so-called confirmation (resp. denial) protocol the signer is able to prove the validity (resp. invalidity) of a given message-signature pair. This property opposes to the universal verifiability of classical digital signatures and allows the signer to have a control on the spread of his signatures. Further applications of undeniable signatures such as licensing software or auctions were proposed in the literature. Since then, lots of contributions and new schemes have been published, among them are [3,5,8,9,13,14,17,18,19].

At Eurocrypt 2005, Kurosawa et al. [13] proposed a variant of the scheme of Chaum [5] with 3-move confirmation and denial protocols in the random oracle model. Although this scheme does not achieve non-transferability, it is the first one presenting 3-move protocols with a security proof. Until this scheme proposal, all provably secure interactive undeniable signature schemes were composed of zero-knowledge confirmation and denial protocols which required at

---

[*] Supported by a grant of the Swiss National Science Foundation, 200020-109133.

P.Q. Nguyen (Ed.): VIETCRYPT 2006, LNCS 4341, pp. 19–36, 2006.
© Springer-Verlag Berlin Heidelberg 2006

least 4 moves. Non-interactive variants of undeniable signatures can be obtained as shown in [12,15] using a so-called designated verifier technique by using classical techniques for non-transferability. In this setting, the signature is only intended to one designated recipient. To ensure that this one cannot convince another party of the validity of the signature, it is required that the recipient could have been able (with his secret key) to produce the signature. When this can be done perfectly, we say that the scheme satisfies perfect non-transferability. In this case, such (designated verifier) signatures cannot satisfy the non-repudiation property.

The main contribution of this article is to show how to achieve a scheme with interactive protocols having a minimal number of rounds. To this end, we revisit a 2-move variant of the MOVA undeniable signature we mentioned in [17] (without any security proof). In order to achieve perfect non-transferability, we modify the protocols of the MOVA scheme by adding a trapdoor one-way permutation with a secret key associated to the verifier. This differs from the commonly used techniques of trapdoor commitments which does not seem appropriate for a 2-move protocol. In the random oracle model, we provide some formal security proofs on the different required properties related to the confirmation and denial protocols such as the soundness, zero-knowledge and non-transferability. We redo the invisibility and unforgeability analysis in settings where the attacker has access to signing, confirmation and denial oracles. This provides precise security bounds and explain how to select MOVA parameters.

In the next section, we recall the definition of an undeniable signature. Section 3 is devoted to the security model of an undeniable signature. Then, we present the 4-move and modified 2-move versions of the MOVA scheme [17] in Section 4. We prove security properties of the modified 2-move version in the subsequent section. Finally, Section 6 concludes this paper.

2   Undeniable Signature

We consider two players who are the signer ($\mathbf{S}$) and the verifier ($\mathbf{V}$). Let $k \in \mathbf{N}$ be a security parameter, $\mathcal{M}$ be the message space and Σ be the signature space. An undeniable signature scheme is composed of the four following algorithms.

Setup. The setup is composed of two probabilistic polynomial time algorithms $\mathsf{Setup}^{\mathbf{S}}$ and $\mathsf{Setup}^{\mathbf{V}}$ producing the signer's key pair $(\mathcal{K}_{\mathrm{p}}^{\mathbf{S}}, \mathcal{K}_{\mathrm{s}}^{\mathbf{S}}) \leftarrow \mathsf{Setup}^{\mathbf{S}}(1^k)$ and the verifier's key pair $(\mathcal{K}_{\mathrm{p}}^{\mathbf{V}}, \mathcal{K}_{\mathrm{s}}^{\mathbf{V}}) \leftarrow \mathsf{Setup}^{\mathbf{V}}(1^k)$.

Sign. Let $m \in \mathcal{M}$ be a message to sign. On the input of the signer's secret key $\mathcal{K}_{\mathrm{s}}^{\mathbf{S}}$, the (probabilistic) polynomial time algorithm Sign generates a signature $\sigma \leftarrow \mathsf{Sign}(m, \mathcal{K}_{\mathrm{s}}^{\mathbf{S}})$ of m (which lies in Σ). We say that (m, σ) is *valid* if there exists a random tape such that $\mathsf{Sign}(m, \mathcal{K}_{\mathrm{s}}^{\mathbf{S}})$ outputs σ. Otherwise, we say that (m, σ) is *invalid*.

Confirm. Let $(m, \sigma) \in \mathcal{M} \times \Sigma$ be a supposedly valid message-signature pair. Confirm is an interactive protocol between $\mathbf{S}$ and $\mathbf{V}$ i.e., a pair of interactive probabilistic polynomial time algorithms $\mathsf{Confirm}_{\mathbf{S}}$ and $\mathsf{Confirm}_{\mathbf{V}}$ such that

m, σ, $\mathcal{K}_p^S$, $\mathcal{K}_p^V$ is input of both, $\mathcal{K}_s^S$ is the auxiliary input of $\mathsf{Confirm_S}$, $\mathcal{K}_s^V$ is the auxiliary input of $\mathsf{Confirm_V}$. At the end of the protocol, $\mathsf{Confirm_V}$ outputs a boolean value which tells whether σ is accepted as valid signature of m.

Deny. Let $(m, \sigma') \in \mathcal{M} \times \Sigma$ be an alleged invalid message-signature pair. Deny is an interactive protocol between $\mathbf{S}$ and $\mathbf{V}$ i.e., a pair of interactive probabilistic polynomial time algorithms $\mathsf{Deny_S}$ and $\mathsf{Deny_V}$ such that m, σ', $\mathcal{K}_p^S$, $\mathcal{K}_p^V$, is input of both, $\mathcal{K}_s^S$ is the auxiliary input of $\mathsf{Deny_S}$, $\mathcal{K}_s^V$ is the auxiliary input of $\mathsf{Deny_V}$. At the end of the protocol, $\mathsf{Deny_V}$ outputs a boolean value which tells whether σ' is accepted as invalid signature.

An execution of the confirmation (resp. denial) protocol will be denoted by $\mathsf{Confirm_{S,V}}(\star)$ (resp. $\mathsf{Deny_{S,V}}(\star)$), where $\star$ is the common input of the players.

3 Security Model

This section is devoted to the different security notions which are required for an undeniable signature to be secure. We consider four basic security notions related to the confirmation and denial protocols which are the *completeness*, the *soundness, zero-knowledge*, and the *non-transferability*. The last one ensures that a malicious verifier is not able to convince any third party of the validity of the statement (e.g., a given message signature is valid) proven in the protocol. The non-transferability notion may be important in some applications where the validity of the proof itself is valuable (like for licensing software).

Security notions about the undeniable signature are considered as well. We require *non-repudiation* by resisting adaptive existential forgery attacks. Furthermore, since the motivation of undeniable signature was to avoid the universal verifiability (like for classical signatures), it is important that a scheme satisfies the *invisibility* property. We will consider an active attacker who has access to some oracles and who will have to distinguish a valid message-signature pair from a randomly picked one.

We recall the definition of the statistical distance between two distributions.

Definition 1. *The statistical distance Δ between two random variables X_1 and X_2 with range $\mathcal{X}$ is $\Delta(X_1, X_1) := \frac{1}{2} \sum_{x \in \mathcal{X}} |\Pr[X_1 = x] - \Pr[X_2 = x]|$.*

Completeness. *Given random key pairs $(\mathcal{K}_p^S, \mathcal{K}_s^S) \leftarrow \mathsf{Setup}^S(1^k)$, $(\mathcal{K}_p^V, \mathcal{K}_s^V) \leftarrow \mathsf{Setup}^V(1^k)$, for any valid (resp. invalid) message-signature pair $(m, \sigma) \in \mathcal{M} \times \Sigma$, the confirmation (resp. denial) protocol $\mathsf{Confirm_{S,V}}(m, \sigma, \mathcal{K}_p^S, \mathcal{K}_p^V)$ (resp. $\mathsf{Deny_{S,V}}(m, \sigma, \mathcal{K}_p^S, \mathcal{K}_p^V)$) outputs 1 with probability 1 when $\mathbf{S}$ and $\mathbf{V}$ correctly follow all steps of the protocol.*

Soundness. *Given random key pairs $(\mathcal{K}_p^S, \mathcal{K}_s^S) \leftarrow \mathsf{Setup}^S(1^k)$, $(\mathcal{K}_p^V, \mathcal{K}_s^V) \leftarrow \mathsf{Setup}^V(1^k)$, for any invalid (resp. valid) message-signature pair $(m, \sigma) \in \mathcal{M} \times \Sigma$ and any cheating signer $\mathbf{S}^*$ (modelled as a probabilistic polynomial time interactive algorithm with access to $\mathcal{K}_s^S$), the probability that the protocol $\mathsf{Confirm_{S^*,V}}(m, \sigma, \mathcal{K}_p^S, \mathcal{K}_p^V)$ (resp. $\mathsf{Deny_{S^*,V}}(m, \sigma, \mathcal{K}_p^S, \mathcal{K}_p^V)$) succeeds is negligible with respect to k.*
The success probability of $\mathbf{S}^*$ is denoted by $\mathsf{Succ_{S^*}^{sd\text{-}con}}$ (resp. $\mathsf{Succ_{S^*}^{sd\text{-}den}}$).

Straight-Line Zero-Knowledge. *Let us consider some random key pairs generated as follows*

$$(\mathcal{K}_{\mathrm{p}}^{\mathbf{S}}, \mathcal{K}_{\mathrm{s}}^{\mathbf{S}}) \leftarrow \mathsf{Setup}^{\mathbf{S}}(1^k), \quad (\mathcal{K}_{\mathrm{p}}^{\mathbf{V}}, \mathcal{K}_{\mathrm{s}}^{\mathbf{V}}) \leftarrow \mathsf{Setup}^{\mathbf{V}}(1^k).$$

The confirmation (resp. denial) protocol is zero-knowledge if there exists a probabilistic polynomial time oracle machine $\mathcal{B}$ called simulator such that for any probabilistic polynomial verifier $\mathbf{V}^*$ (with or without $\mathcal{K}_{\mathrm{s}}^{\mathbf{V}}$) and any valid (resp. invalid) pair $(m, \sigma) \in \mathcal{M} \times \Sigma$, $\mathcal{B}^{\mathbf{V}^*}$ outputs a transcript which is indistinguishable from the transcript of the protocol $\mathsf{Confirm}_{\mathbf{S}, \mathbf{V}^*}(m, \sigma, \mathcal{K}_{\mathrm{p}}^{\mathbf{S}}, \mathcal{K}_{\mathrm{p}}^{\mathbf{V}})$ (resp. $\mathsf{Deny}_{\mathbf{S}, \mathbf{V}^*}(m, \sigma, \mathcal{K}_{\mathrm{p}}^{\mathbf{S}}, \mathcal{K}_{\mathrm{p}}^{\mathbf{V}})$), where $\mathbf{S}$ is the honest signer. We assume that $\mathcal{B}$ and $\mathbf{V}^*$ share the same information (e.g., $\mathcal{K}_{\mathrm{s}}^{\mathbf{V}}$ if any). Namely, when $\mathbf{V}^*$ has access to some random oracles, $\mathcal{B}$ can see the queries (and answers) as well. Moreover, we say that the protocol is straight-line zero-knowledge if $\mathcal{B}$ does not need to rewind $\mathbf{V}^*$.

Non-Transferability. *Let us consider some random key pairs generated as follows*

$$(\mathcal{K}_{\mathrm{p}}^{\mathbf{S}}, \mathcal{K}_{\mathrm{s}}^{\mathbf{S}}) \leftarrow \mathsf{Setup}^{\mathbf{S}}(1^k), \quad (\mathcal{K}_{\mathrm{p}}^{\mathbf{V}}, \mathcal{K}_{\mathrm{s}}^{\mathbf{V}}) \leftarrow \mathsf{Setup}^{\mathbf{V}}(1^k).$$

The confirmation (resp. denial) protocol is said non-transferable if there exists a probabilistic polynomial time interactive machine $\mathcal{B}$ with input $\mathcal{K}_{\mathrm{s}}^{\mathbf{V}}$ such that for any computationally unbounded verifier $\tilde{\mathbf{V}}$, any pair $(m, \sigma) \in \mathcal{M} \times \Sigma$, the transcript of $\mathsf{Confirm}_{\mathcal{B}, \tilde{\mathbf{V}}}(m, \sigma, \mathcal{K}_{\mathrm{p}}^{\mathbf{S}}, \mathcal{K}_{\mathrm{p}}^{\mathbf{V}})$ (resp. $\mathsf{Deny}_{\mathcal{B}, \tilde{\mathbf{V}}}(m, \sigma, \mathcal{K}_{\mathrm{p}}^{\mathbf{S}}, \mathcal{K}_{\mathrm{p}}^{\mathbf{V}})$) is indistinguishable from that of the protocol $\mathsf{Confirm}_{\mathbf{S}, \tilde{\mathbf{V}}}(m, \sigma, \mathcal{K}_{\mathrm{p}}^{\mathbf{S}}, \mathcal{K}_{\mathrm{p}}^{\mathbf{V}})$ (resp. $\mathsf{Deny}_{\mathbf{S}, \tilde{\mathbf{V}}}(m, \sigma, \mathcal{K}_{\mathrm{p}}^{\mathbf{S}}, \mathcal{K}_{\mathrm{p}}^{\mathbf{V}})$). When $\tilde{\mathbf{V}}$ has access to some random oracles, $\mathcal{B}$ does not see any queries (nor answers) made to them. However, $\mathcal{B}$ is assumed to be given a bit telling whether (m, σ) is valid or not.

We consider here the two following notions of indistinguishability.

Perfect Zero-Knowledge (resp. Non-Transferability). Both transcript distributions are identical.

Statistical Zero-Knowledge (resp. Non-Transferability). The statistical distance between the two transcript distributions is negligible.

We note that the definition of non-transferability allows to avoid some attacks in which the verifier $\mathbf{V}^*$ identified with $\mathcal{K}_{\mathrm{p}}^{\mathbf{V}}$ forwards messages to the honest signer which were generated by a hidden verifier $\tilde{\mathbf{V}}$. Namely, our definition assures that $\mathbf{V}^*$ with knowledge of $\mathcal{K}_{\mathrm{s}}^{\mathbf{V}}$ could simulate the answer of $\mathbf{S}$ (without any help from $\mathbf{S}$) so that $\tilde{\mathbf{V}}$ does not have evidence of the proof validity.

Our definition of non-transferability is similar to that proposed by Camenisch and Michels [4] with the main difference that our version assumes that $\tilde{\mathbf{V}}$ is computationally unbounded. We can thus assume that $\tilde{\mathbf{V}}$ makes no queries to the signing and confirmation/denial oracles. Therefore, the non-transferability of the protocols presented below will also hold with respect to the Camenisch-Michels definition.

We note that the above definition of zero-knowledge is black-box which means that we require the existence of one "universal" simulator having an oracle access to the verifier which is able to produce an indistinguishable transcript for any verifier. More details about the black-box zero-knowledge notion are given in [10].

In the standard model, Barak et al. [1] proved that zero-knowledge proofs of an NP-complete language (possibly non-black-box) requires at least 3 moves. To overcome this limitation, the notion of zero-knowledge was extended in the random oracle model (for more details, see [2]) in which the queries to the random oracles are controlled by the simulator, i.e., it can simulate the output of the oracles provided that the output distribution is correct. Recently, Pass [21] proposed the notion of *deniable zero-knowledge* in the random oracle. The difference with classical zero-knowledge in the random oracle is that the simulator is no longer allowed to simulate the output of the random oracles, but is only able to observe the queries made to the random oracles as well as the corresponding answers. This actually means that the simulator's transcript really corresponds to the view of the verifier. In this model, Pass [21] showed that 2 moves are necessary to achieve zero-knowledge for NP and proposed a general 2-move protocol for NP which is not very convenient for practical purposes. In our results, proofs of zero-knowledge in the random oracle will be deniable as well.

Existential Unforgeability. We consider the standard security notion of existential forgery under an adaptive chosen-message attack as defined by Goldwasser et al. [11] for classical digital signatures. This notion is similar to Kurosawa-Heng [13] and is adapted as follows.

An undeniable signature scheme is secure against an existential forgery under adaptive chosen-message attack if there exists no probabilistic polynomial time algorithm $\mathcal{F}$ which wins the following game with a non-negligible probability.

Game: *$\mathcal{F}$ receives a public key $\mathcal{K}_p^S$ from $(\mathcal{K}_p^S, \mathcal{K}_s^S) \leftarrow \mathsf{Setup}^S(1^k)$ and a verifier's key pair $(\mathcal{K}_p^V, \mathcal{K}_s^V) \leftarrow \mathsf{Setup}^V(1^k)$. Then, $\mathcal{F}$ can query some chosen messages to a signing oracle, some chosen pairs $(m, \sigma) \in \mathcal{M} \times \Sigma$ to a confirmation (and denial) protocol oracle and interact with it in a confirmation (denial) protocol where the oracle plays the role of the signer. All these queries must be polynomially bounded in k and can be sent adaptively. $\mathcal{F}$ wins the game if it outputs a valid pair $(m^*, \sigma^*) \in \mathcal{M} \times \Sigma$ such that m^* was not queried to the signing oracle.*

The success probability of $\mathcal{F}$ in this game is denoted by $\mathsf{Succ}_{\mathcal{F}}^{\mathsf{ef\text{-}cma}}$.

Invisibility. We use a similar definition as Kurosawa-Heng [13]. Consider first a probabilistic polynomial time algorithm $\mathcal{D}$ called *invisibility distinguisher* and the two following games with respect to a bit b.

Game$^{\mathsf{inv\text{-}cma\text{-}b}}$: *$\mathcal{D}$ receives $\mathcal{K}_p^S$ from $(\mathcal{K}_p^S, \mathcal{K}_s^S) \leftarrow \mathsf{Setup}^S(1^k)$ and a verifier's key pair $(\mathcal{K}_p^V, \mathcal{K}_s^V) \leftarrow \mathsf{Setup}^V(1^k)$, it can query some chosen messages to a signing oracle and some chosen message-signature pairs $(m, \sigma) \in \mathcal{M} \times \Sigma$ to some oracles running the confirmation and denial protocols. After a given* *time, $\mathcal{D}$ chooses one message $m^* \in \mathcal{M}$ which was not queried to the signing oracle and submits it to the challenger. If $b = 0$, he sets $\sigma^* = \mathsf{Sign}(m^*, \mathcal{K}_s^{\mathsf{S}})$. Otherwise, σ^* is picked uniformly at random in Σ. $\mathcal{D}$ receives σ^*. After that, the distinguisher can query the signing, confirmation, and denial oracles again provided that m^* is not a query of the signing oracle and (m^*, σ^*) is not a query of the confirmation or denial protocols. Finally, $\mathcal{D}$ outputs a guess bit b'.*

We define the advantage of the distinguisher as follows

$$\mathsf{Adv}_{\mathcal{D}}^{\mathsf{inv\text{-}cma}} := \left| \Pr\left[b' = 1 \text{ in } \mathbf{Game}^{\mathsf{inv\text{-}cma\text{-}1}} \right] - \Pr\left[b' = 1 \text{ in } \mathbf{Game}^{\mathsf{inv\text{-}cma\text{-}0}} \right] \right|,$$

where probabilities are over the random tapes of the involved algorithms. An undeniable signature scheme is said to be invisible under a chosen-message attack *if there exists no probabilistic polynomial time algorithm $\mathcal{D}$ such that the advantage $\mathsf{Adv}_{\mathcal{D}}^{\mathsf{inv\text{-}cma}}$ is non-negligible.*

Note that this definition is similar to that of Galbraith et al. [8] except that the distinguisher is not allowed to query m^* to the signing oracle in our definition. The invisibility notion of Galbraith et al. cannot be satisfied when the signature is deterministic (which is the case for MOVA). This will be discussed in Remark 6.

4   MOVA Scheme

In this section, we present the scheme proposed in [17] as well as the underlying principles. This scheme generalizes the MOVA scheme [18] proposed earlier in 2004 in a very natural way and therefore will be called MOVA as well.

4.1   Preliminaries

We first recall some definitions, useful lemmas, and mathematical problems from [17] related to the interpolation of group homomorphisms.

Let G and H be two Abelian groups. Given $S := \{(x_1, y_1), \ldots, (x_s, y_s)\} \subseteq G \times H$, we say that the set of points S *interpolates in a group homomorphism* if there exists a group homomorphism $f : G \longrightarrow H$ such that $f(x_i) = y_i$ for $i = 1, \ldots, s$. We say that a set of points $B \subseteq G \times H$ *interpolates in a group homomorphism with another set of points* $A \subseteq G \times H$ if $A \cup B$ interpolates in a group homomorphism.

Lemma 2 ([17]). *Let G, H be two finite Abelian groups. We denote by d the order of H and by p the smallest prime factor of d.*

1. *Let $x_1, \ldots, x_s \in G$ which span a subgroup denoted by G'. The following properties are equivalent. In this case, we say that $x_1, \ldots, x_s$ H-generate G.*
   (a) For all $y_1, \ldots, y_s \in H$, there exists at most one group homomorphism $f : G \longrightarrow H$ such that $f(x_i) = y_i$ for $i = 1, \ldots, s$.
   (b) $G' + dG = G$.
2. *Let $x_1, \ldots, x_s \in G$ which H-generate G. The mapping $g : G \times \mathbf{Z}_d^s \to G$ which is defined by $g(r, a_1, \ldots, a_s) := dr + a_1 x_1 + \cdots + a_s x_s$ is balanced.*

3. *Given a set of s points $S = \{(x_1, y_1), \ldots, (x_s, y_s)\}$, such that $x_1, \ldots, x_s$ H-generate G. We assume that there exists a function $f : G \longrightarrow H$ such that*

$$\rho := \Pr_{(r, a_1, \ldots, a_s) \in_U G \times \mathbf{Z}_d^s} [f(dr + a_1 x_1 + \cdots + a_s x_s) = a_1 y_1 + \cdots + a_s y_s] > \frac{1}{p}.$$

The set of points S interpolates in a group homomorphism.

Although, our treatment uses arbitrary G, H, d, p, the implementation analysis of [16] suggests that parameters $G = \mathbf{Z}_n^*$ (for n product of two primes), $d = p = 2$ lead to the most efficient protocols for the signer. The homomorphisms are the Legendre symbols in G.

n-S-GHI Problem (Group Homomorphism Interpolation Prob. [17])
Parameters: Two Abelian groups G and H, a set of s points $S \subseteq G \times H$, and $n \in \mathbf{N}$.
Instance Generation: n elements $x_1, \ldots, x_n \in_U G$ are picked uniformly at random.
Problem: Find $y_1, \ldots, y_n \in H$ such that $\{(x_1, y_1), \ldots, (x_n, y_n)\}$ interpolates with S in a group homomorphism.
The success probability of an n-S-GHIP solver $\mathcal{A}$ will be denoted by $\mathsf{Succ}_{\mathcal{A}}^{n\text{-}S\text{-GHIP}}$.

n-S-GHID Problem (n-S-GHI Decisional Problem)
Parameters: Two Abelian groups G and H, a set of s points $S \subseteq G \times H$ and $n \in \mathbf{N}$.
Instance Generation: The instance T is generated according to one of the two following ways and is denoted T_0 or T_1 respectively. T_0 is a set of points $\{(x_1, y_1), \ldots, (x_n, y_n)\} \in (G \times H)^n$ picked uniformly at random such that it interpolates with S in a group homomorphism. T_1 is picked uniformly at random in $(G \times H)^n$.
Problem: Decide whether the instance T is of type T_0 or T_1.
The advantage of an n-S-GHID distinguisher $\mathcal{D}$ is given by

$$\mathsf{Adv}_{\mathcal{D}}^{n\text{-}S\text{-GHID}} := |\Pr[b = 0 \mid T \text{ is of type } T_0] - \Pr[b = 0 \mid T \text{ is of type } T_1]|,$$

where b denotes the output bit of $\mathcal{D}$.

The S-GHI (resp. S-GHID) problem defined in [17] corresponds to the 1-S-GHI (resp. 1-S-GHID) problem. We consider the n-S-GHI and n-S-GHID problems for sets S which interpolate in a unique group homomorphism. Hence, S defines a homomorphism. The n-S-GHI problem consists in computing it on n elements. The n-S-GHID problem consists in deciding whether a set of points T is in its graph.

4.2 Interactive Proofs

The original version of the MOVA scheme makes use of two 4-move interactive proofs, namely one for the confirmation protocol and one for the denial protocol.

In the first proof, a prover proves that a set of points interpolates in a group homomorphism known by himself. In the second one, the prover knows a group homomorphism which interpolates in a set of points S and proves that a second set of points T does not interpolate in this group homomorphism. These two proofs, taken from [17], are given below. Again, G, H denote two Abelian groups and $d := |H|$ is the order of H with smallest prime factor p. The group homomorphism which is known by the prover is denoted by f. The security parameter of the following proofs is an integer denoted by ℓ.

GHIproof$_\ell(S)$
Parameters: G, H, d
Input: ℓ, $S = \{(g_1, e_1), \ldots, (g_s, e_s)\} \subseteq G \times H$
  1: The verifier picks $r_i \in_U G$ and $a_{i,j} \in_U \mathbf{Z}_d$ uniformly at random for $i = 1, \ldots, \ell$ and $j = 1, \ldots, s$. He computes $u_i = dr_i + a_{i,1}g_1 + \cdots + a_{i,s}g_s$ and $w_i = a_{i,1}e_1 + \cdots + a_{i,s}e_s$ for $i = 1, \ldots, \ell$. He sends $u_1, \ldots, u_\ell$ to the prover.
  2: The prover computes $v_i = f(u_i)$ for $i = 1, \ldots, \ell$. He sends to the verifier a commitment to $v_1, \ldots, v_\ell$.
  3: The verifier sends all r_i's and $a_{i,j}$'s to the prover.
  4: The prover checks that the u_i's computations are correct. He then opens his commitment.
  5: The verifier checks that $v_i = w_i$ for $i = 1, \ldots, \ell$.

coGHIproof$_\ell(S, T)$
Parameters: G, H, d, p
Input: ℓ, $S = \{(g_1, e_1), \ldots, (g_s, e_s)\}$, $T = \{(x_1, z_1), \ldots, (x_t, z_t)\}$
  1: The verifier picks $r_{i,k} \in_U G$, $a_{i,j,k} \in_U \mathbf{Z}_d$, and $\lambda_i \in_U \mathbf{Z}_p$ uniformly at random for $i = 1, \ldots, \ell$, $j = 1, \ldots, s$, $k = 1, \ldots, t$. He computes $u_{i,k} := dr_{i,k} + \sum_{j=1}^{s} a_{i,j,k}g_j + \lambda_i x_k$ and $w_{i,k} := \sum_{j=1}^{s} a_{i,j,k}e_j + \lambda_i z_k$. Set $u := (u_{1,1}, \ldots, u_{\ell,t})$ and $w := (w_{1,1}, \ldots, w_{\ell,t})$. He sends u and w to the prover.
  2: The prover computes $v_{i,k} := f(u_{i,k})$ and $y_k := f(x_k)$ for $i = 1, \ldots, \ell$, $k = 1, \ldots, t$. Since $w_{i,k} - v_{i,k} = \lambda_i(z_k - y_k)$, he should be able[1] to find every λ_i if the verifier is honest since $z_k \neq y_k$ for at least one k. Otherwise, he sets λ_i to a random value. He then sends a commitment to $\lambda = (\lambda_1, \ldots, \lambda_\ell)$ to the verifier.
  3: The verifier sends all $r_{i,k}$'s and $a_{i,j,k}$'s to the prover.
  4: The prover checks that u and w were correctly computed. He then opens the commitment to λ.
  5: The verifier checks that the prover could find the right λ.

In the original article [17], a 2-move variant for these two protocols was suggested without a proof. The variant is achieved by removing the two messages

---

[1] Note that this requires to select H in which one can extract discrete logarithms lying in the restricted set $\{0, 1, \ldots, p - 1\}$. In practice, this may not be a problem since we prefer $p = 2$ as shown in [16].

sent in the middle of the protocol for achieving the zero-knowledge property through the commitment scheme. In order to maintain zero-knowledge, the verifier sends a kind of commitment on a seed which generates the challenges to the prover. This commitment can only be opened by the prover after this one solved the challenges. We notably modify the original 2-move protocols by adding a trapdoor one-way permutation with associated secret key $\mathcal{K}_s^V$. Namely, we consider the permutation $\mathsf{TPOW}_{\mathcal{K}_p^V}(\cdot)$ and its inverse $\mathsf{TPOW}_{\mathcal{K}_s^V}(\cdot)^{-1}$. We denote $\mathsf{Succ}_{\mathcal{A}}^{\mathsf{inv\text{-}tp}}$ the probability that an adversary $\mathcal{A}$ can compute $\mathsf{TPOW}_{\mathcal{K}_s^V}^{-1}(y)$ given a random y, without knowing $\mathcal{K}_s^V$. For the sake of simplicity, we use the same notation for both protocols. The 2-move variant of GHIproof is given here.

2-GHIproof$_\ell(S)$
Parameters: G, H, d
Input: ℓ, $S = \{(g_1, e_1), \ldots, (g_s, e_s)\} \subseteq G \times H$

1: The verifier picks seedC $\in_U \{0,1\}^{k_c}$ uniformly at random, and by applying a pseudorandom generator GenC on this seed, generates values $r_i \in G$ and $a_{i,j} \in \mathbf{Z}_d$ for $i = 1, \ldots, \ell$ and $j = 1, \ldots, s$. He computes $u_i = dr_i + a_{i,1}g_1 + \cdots + a_{i,s}g_s$, $w_i = a_{i,1}e_1 + \cdots + a_{i,s}e_s$ for $i = 1, \ldots, \ell$, and $\vartheta_c = \mathsf{TPOW}_{\mathcal{K}_p^V}(\text{seedC})$. Using a cryptographic hash function $H_c : \{0,1\}^* \to \{0,1\}^{k_c}$, the verifier computes $h_c := H_c(w_1, \ldots, w_\ell) \oplus \text{seedC}$. He sends $u_1, \ldots, u_\ell$, h_c and ϑ_c to the prover.
2: The prover computes the values $v_i = f(u_i)$ for $i = 1, \ldots, \ell$ and seedC$' = H_c(v_1, \ldots, v_\ell) \oplus h_c$. He checks that $\vartheta_c = \mathsf{TPOW}_{\mathcal{K}_p^V}(\text{seedC}')$ and that GenC(seedC$'$) generates values $a_{i,j}$'s and r_i's such that $u_i := dr_i + a_{i,1}g_1 + \cdots + a_{i,s}g_s$ for $i = 1, \ldots, \ell$. He sends seedC$'$ to the verifier.
3: The verifier checks that seedC$' = $ seedC.

The interactive proof **coGHIproof** can be transformed in a 2-move protocol in a similar way. Namely, the verifier picks seedD $\in \{0,1\}^{k_d}$, and uses a pseudorandom generator GenD to generate the $r_{i,k}$'s, $a_{i,j,k}$'s, and λ_i's, and $\vartheta_d = \mathsf{TPOW}_{\mathcal{K}_p^V}(\text{seedD})$. He then sends the corresponding u, w, $h_d := H_d(\lambda_1, \ldots, \lambda_\ell) \oplus$ seedD, and ϑ_d, where $H_d : \{0,1\}^* \to \{0,1\}^{k_d}$ is a cryptographic hash function. In step 2 of the protocol, the prover retrieves seedD$'$, and checks whether $\vartheta_d = \mathsf{TPOW}_{\mathcal{K}_p^V}(\text{seedD}')$ and GenD(seedD$'$) generates the right u, w. Then, he sends seedD$'$.

Note that the complexity of both protocols are comparable to their 4-move variants.

4.3 MOVA Description

Below, we briefly present the MOVA scheme. For a more detailed description, we refer to [17].

Setup. The signer chooses two Abelian groups Xgroup and Ygroup and a secret group homomorphism Hom : Xgroup $\to$ Ygroup. He picks seedK $\in$

$\{0,1\}^{k_s}$ and using a pseudorandom generator GenK generates Lkey values $\text{Xkey}_1, \ldots, \text{Xkey}_{\text{Lkey}} \in \text{Xgroup}$. Then, he computes $\text{Ykey}_i := \text{Hom}(\text{Xkey}_i)$ for $i = 1, \ldots, \text{Lkey}$.

Public Key. $\mathcal{K}_p^S := (\text{Xgroup}, \text{Ygroup}, d, \text{seedK}, (\text{Ykey}_1, \ldots, \text{Ykey}_{\text{Lkey}}), \text{para})$, where the set para $= (\text{Lkey}, \text{Lsig}, \text{Icon}, \text{Iden}, k_c, k_d, k_s)$ is composed of integer parameters.

Secret Key. $\mathcal{K}_s^S := \text{Hom}$.

The main goal of the setup is to ensure that the points $(\text{Xkey}_i, \text{Ykey}_i)$'s uniquely characterize Hom to avoid that several secret keys correspond to the same public key. This is necessary to guarantee the non-repudiation of the signature scheme. For this, one can either put many enough points or produce an interactive or non-interactive zero-knowledge proof of unique interpolation. These additional setup variants are described in [17]. In fact, the different setup variants ensure that $\text{Xkey}_1, \ldots, \text{Xkey}_{\text{Lkey}}$ Ygroup-generate Xgroup. In this case, we say that the public key is *valid*.

Signature Generation. Let $m \in \{0,1\}^*$ be a message. Applying a pseudorandom generator GenS on the message m, the signer generates Lsig values $\text{Xsig}_1, \ldots, \text{Xsig}_{\text{Lsig}} \in \text{Xgroup}$. He then computes $\text{Ysig}_i := \text{Hom}(\text{Xsig}_i)$ for $i = 1, \ldots, \text{Lsig}$. The signature σ is $(\text{Ysig}_1, \ldots, \text{Ysig}_{\text{Lsig}})$.

Confirmation Protocol. Given a message-signature pair (m, σ) as input and an integer Icon a security parameter, the signer (prover) and the verifier retrieve the values Xkey_i's, Xsig_j's from the message and the public key. The signer checks the validity of the signature. If this one is valid, the signer and the verifier run $\mathbf{GHIproof}_{\text{Icon}}(S)$ on the set

$$S = \{(\text{Xkey}_i, \text{Ykey}_i) | i = 1, \ldots, \text{Lkey}\} \cup \{(\text{Xsig}_j, \text{Ysig}_j) | j = 1, \ldots, \text{Lsig}\}.$$

Otherwise, the signer aborts.

Denial Protocol. Given an alleged invalid message-signature pair (m, σ) as input and an integer Iden a security parameter, we denote the signature $\sigma = (\text{Zsig}_1, \ldots, \text{Zsig}_{\text{Lsig}})$. The signer and the verifier retrieve the Xkey_i's and Xsig_j's. The signer checks the invalidity of (m, σ). If this one is really invalid, they run the protocol $\mathbf{coGHIproof}_{\text{Iden}}(S, T)$ on the sets

$$S = \{(\text{Xkey}_i, \text{Ykey}_i) | i = 1, \ldots, \text{Lkey}\} \quad T = \{(\text{Xsig}_j, \text{Zsig}_j) | j = 1, \ldots, \text{Lsig}\}.$$

The 2-move version of MOVA is exactly as above except that $\mathbf{GHIproof}$ and $\mathbf{coGHIproof}$ are replaced by $\mathbf{2\text{-}GHIproof}$ and $\mathbf{2\text{-}coGHIproof}$ respectively.

5   Security of the 2-Move MOVA Scheme

Here, we prove that the 2-move modified version of the MOVA scheme satisfies the security properties mentioned in Section 3. The proofs of resistance against forgery attacks and invisibility were inspired from [13].

Theorem 3. *Let* $S = \{(\text{Xkey}_1, \text{Ykey}_1), \ldots, (\text{Xkey}_{\text{Lkey}}, \text{Ykey}_{\text{Lkey}})\}$ *and* e *denote the natural logarithm base. Assuming that* GenC, GenS, GenD, H_d, *and* H_c *are random oracles, that signer's public key is valid, and that* TPOW *is a trapdoor one-way permutation, the* MOVA *scheme with 2-move confirmation and denial protocols satisfies the following security properties.*

1. *The confirmation (resp. denial) protocol is complete.*
2. *Let* p *be the smallest prime factor of* d*. The confirmation (resp. denial) protocol is sound: for any invalid (valid) message-signature pair, any cheating signer* $\mathbf{S}^*$ *limited to* q_{H_c} *(resp.* q_{H_d}*) queries to* H_c *(resp.* H_d*), is such that the probability* $\text{Succ}_{\mathbf{S}^*}^{\text{sd-con}} < \text{Succ}^{\text{inv-tp}} + q_{H_c} p^{-\text{Icon}}$ *(resp.* $\text{Succ}_{\mathbf{S}^*}^{\text{sd-den}} < \text{Succ}^{\text{inv-tp}} + q_{H_d} p^{-\text{Iden}}$*), where* $\text{Succ}^{\text{inv-tp}}$ *is the maximum of* $\text{Succ}_{\mathcal{A}}^{\text{inv-tp}}$ *among all adversaries* $\mathcal{A}$ *which have similar complexity as* $\mathbf{S}^*$*.*
3. *The confirmation (resp. denial) protocol is perfect non-transferable.*
4. *The confirmation (resp. denial) protocol is statistical black-box straight-line zero-knowledge.*
5. *Assume that for any solver* $\mathcal{B}$ *with a given complexity, we have*

$$\text{Succ}_{\mathcal{B}}^{\text{Lsig-}S\text{-GHIP}} \leq \varepsilon.$$

Then, any forger $\mathcal{F}$ *with similar complexity using* q_S *signing queries and* q_V *queries to the confirmation/denial oracle wins the existential forgery game under an adaptive chosen-message attack with a probability*

$$\text{Succ}_{\mathcal{F}}^{\text{ef-cma}} \leq e(1 + q_S)(1 + q_V)\varepsilon.$$

6. *Assume that for any algorithm* $\mathcal{B}$ *with a given complexity, we have*

$$\text{Adv}_{\mathcal{B}}^{\text{Lsig-}S\text{-GHID}} \leq \varepsilon \quad \text{and} \quad \text{Succ}_{\mathcal{B}}^{\text{Lsig-}S\text{-GHIP}} \leq \varepsilon'.$$

Then, any distinguisher $\mathcal{D}$ *with similar complexity using* q_S *signing queries and* q_V *queries to the confirmation/denial oracle wins the invisibility game under a chosen-message attack with advantage*

$$\text{Adv}_{\mathcal{D}}^{\text{inv-cma}} \leq e(1 + q_S)(\varepsilon + 2e(1 + q_V)\varepsilon').$$

Remark 4. The soundness and zero-knowledge of the confirmation and denial protocols as well as the invisibility and the resistance to existential forgery attacks hold in the random oracle model.

Remark 5. Similarly to [14], the efficiency of the security reduction for the existential forgery can be improved (factor $(1 + q_V)^{-1}$ is removed) by replacing GHI problem by its *gap* variant [20]. This problem consists in solving the GHI problem using an access to an oracle which solves the GHID problem. This one helps to simulate the confirmation and denial oracles.

Proof. Below we prove Theorem 3. Completeness is omitted since it is obvious.

Soundness of Confirmation. Let $\mathbf{S}^*$ be a cheating prover who wants to confirm the validity of an invalid signature $\sigma = (\text{Zsig}_1, \dots, \text{Zsig}_{\text{Lsig}})$. Note that $\mathbf{S}^*$ is fed with the signer secret key $\mathcal{K}_{\text{s}}^{\text{S}}$. Without loss of generality, we can assume that $\mathbf{S}^*$ always responds correctly to the verifier whenever he queries seedC to GenC. Indeed, he can check that seedC is the preimage of ϑ_c by TPOW and answer seedC to the challenge if correct. (With an honest verifier, there is no need to check whether the challenge is valid.) Hence, the verifier always accepts when the prover queries seedC to GenC. Similarly, we can assume that $\mathbf{S}^*$ always responds correctly to the verifier whenever he queries the right w to H_c because he can deduce seedC from h_c afterwards. Note that when $\mathbf{S}^*$ interacts with an honest verifier, the verifier only accepts if $\mathbf{S}^*$ outputs seedC.

We transform $\mathbf{S}^*$ into an algorithm to invert the trapdoor permutation as follows.

1. We receive a random challenge ϑ_c, whose preimage by TPOW is denoted seedC.
2. We generate the key material for the MOVA signature and generate some random values r_i's and $a_{i,j}$'s. We deduce some u_i's and w_i's and pick a random h_c. Then (u, h_c, ϑ_c) is a challenge for the prover. We simulate GenC as follows: for any query except seedC (we can check whether a value is seedC by checking that its image by TPOW is ϑ_c) we simulate a random oracle as usual i.e., we maintain a list of elements queried to GenC with corresponding answers and simulate according to this list. If the query is new, we simply pick the answer at random and add the pair in the list. For the query seedC we stop the overall simulation and yield seedC: the inversion of ϑ_c succeeded. We simulate H_c as follows: for any query except $w = (w_1, \dots, w_{\text{Icon}})$ we simulate a random oracle (like for GenC). For the query w we stop: the inversion of ϑ_c failed.
3. We run $\mathbf{S}^*$ according to our simulation rules. If $\mathbf{S}^*$ outputs some value, we check whether it is seedC. If it is, we output it, otherwise we fail.

The algorithm succeeds to invert the trapdoor permutation at the condition that either (event A) $\mathbf{S}^*$ succeeds without even querying seedC to GenC nor w to H_c, or (event B) that $\mathbf{S}^*$ queries seedC to GenC without querying w to H_c beforehand. Let C be the event that $\mathbf{S}^*$ queries w to H_c before querying seedC to GenC. Since the simulation is perfect, $\Pr[A \cup B] + \Pr[C]$ is the probability that $\mathbf{S}^*$ passes the protocol with an honest verifier. We have $\Pr[A \cup B] \leq \text{Succ}^{\text{inv-tp}}$. Below we show an upper bound for $\Pr[C]$. To this, we consider a simulator $\mathcal{B}$ which plays with $\mathbf{S}^*$ to win the following game:

Game: A challenger picks elements r_i's and $a_{i,j}$'s uniformly at random and compute $u_i = dr_i + \sum_{j=1}^{\text{Lkey}} a_{i,j} \text{Xkey}_j + \sum_{j=\text{Lkey}+1}^{\text{Lkey}+\text{Lsig}} a_{i,j} \text{Xsig}_{j-\text{Lkey}}$ and $w_i := \sum_{j=1}^{\text{Lkey}} a_{i,j} \text{Ykey}_j + \sum_{j=\text{Lkey}+1}^{\text{Lkey}+\text{Lsig}} a_{i,j} \text{Zsig}_{j-\text{Lkey}}$. The simulator $\mathcal{B}$ receives the u_i's and wins the game if he finds all the values w_i's.

$\mathcal{B}$ simply forwards the received challenges u_i's and picks h_c and ϑ_c uniformly at random in $\{0,1\}^{k_c}$. $\mathcal{B}$ simulates the oracle H_c as above, except that he guesses when the w_i's are queried. For this, he just picks an integer $\ell \in \{1, \ldots, q_{H_c}\}$ uniformly and stops the simulation at the ℓth query made to H_c. The simulator then answers the values w_i's. Note that $\mathbf{S}^*$ cannot query seedC to GenC when event C occurs. The simulation is perfect in the C case provided that ℓ is correctly guessed. Thus, we have $\Pr[D] \geq 1/q_{H_c} \cdot \Pr[C]$, where D denotes the event of winning the above game. By the assertion 3 of Lemma 2, $\Pr[D] \leq p^{-\text{Icon}}$. Thus, $\Pr[C] \leq q_{H_c} p^{-\text{Icon}}$. So, the confirmation cannot succeed with probability larger than $\text{Succ}^{\text{inv-tp}} + q_{H_c} p^{-\text{Icon}}$.

Soundness of Denial. This proof works in a very similar way as for the confirmation. The only difference is that we replace GenC by GenD, H_c by H_d, Icon by Iden, k_c by k_d, seedC by seedD.

Non-transferability of Confirmation. We describe a simulator $\mathcal{B}$ interacting with $\tilde{\mathbf{V}}$. First, $\mathcal{B}$ launches $\tilde{\mathbf{V}}$ and receives the first message (which should be $u = (u_1, \ldots, u_{\text{Icon}})$, h_c, and ϑ_c). If (m, σ) is valid, the simulator computes $\text{seedC}' = \text{TPOW}_{\mathcal{K}_s^V}^{-1}(\vartheta_c)$ and using GenC generates coefficients a'_{ij} and r'_i and corresponding u'_i and w'_i for $i = 1, \ldots, \text{Icon}$ and $j = 1, \ldots, \text{Lkey} + \text{Lsig}$. Then, $\mathcal{B}$ checks whether $u'_i = u_i$ for $i = 1, \ldots, \text{Icon}$, $\text{seedC}' = H_c(w'_1, \ldots, w'_{\text{Icon}}) \oplus h_c$. If it is the case, $\mathcal{B}$ outputs the transcript $(h_c, w, \vartheta_c, \text{seedC}')$. Otherwise, it outputs $(h_c, w, \vartheta_c, \text{abort})$. If (m, σ) is invalid, the simulator outputs abort. Note that an honest signer would check exactly the same equalities (in a different way) and would answer exactly in the same way. Hence, the non-transferability is perfect.

Non-transferability of Denial. This proof is similar.

Straight-Line Zero-Knowledge of Confirmation. If $\mathbf{V}^*$ is given $\mathcal{K}_s^V$, the simulation can be done perfectly as for the non-transferability. Now, we consider that $\mathbf{V}^*$ (and the simulator $\mathcal{B}$) is not given $\mathcal{K}_s^V$. $\mathcal{B}$ runs the verifier $\mathbf{V}^*$ and looks at the queries made by $\mathbf{V}^*$ to the oracle GenC. $\mathcal{B}$ puts these q_{GenC} queries seedC_k for $1 \leq k \leq q_{\text{GenC}}$ as well as the corresponding answers of GenC in memory. The simulator then receives the first message M of $\mathbf{V}^*$. If this one has not a correct format, the simulator outputs the transcript (M, abort). Otherwise, the simulator checks whether one answer among those queries seedC_k's made to GenC generates the challenges u_i's correctly and the image of this query by TPOW is equal to ϑ_c. If it is not the case, $\mathcal{B}$ outputs the transcript $(u_1, \ldots, u_{\text{Icon}}, h_c, \vartheta_c, \text{abort})$. Otherwise, the simulator is able to compute the right w_i's from this answer (the right r_i's and $a_{i,j}$'s) using the homomorphic property of Hom, namely $w_i = \text{Hom}(u_i) = \sum_{j=1}^{\text{Lkey}} a_{i,j} \text{Ykey}_j + \sum_{j=\text{Lkey}+1}^{\text{Lkey}+\text{Lsig}} a_{i,j} \text{Ysig}_{j-\text{Lkey}}$ for $i = 1, \ldots, \text{Icon}$. From the w_i's, $\mathcal{B}$ computes $\text{seedC}^* := h_c \oplus H_c(w_1, \ldots, w_{\text{Icon}})$ and checks whether seedC^* generates the right r_i's and $a_{i,j}$'s. In the positive case, $\mathcal{B}$ outputs the transcript $(u_1, \ldots, u_{\text{Icon}}, h_c, \vartheta_c, \text{seedC}^*)$. In the negative case, it outputs the following transcript $(u_1, \ldots, u_{\text{Icon}}, h_c, \vartheta_c, \text{abort})$.

It remains to show that the two transcript distributions are statistically indistinguishable. When the first message has not a correct format, the two transcripts are clearly identical. Let consider the case where the verifier did not query any seedC$_k$ which produces the challenges u_i's and whose image by TPOW leads to ϑ_c. In this case, the honest prover will not abort the protocol only if he retrieves a seedC $= H(w_1, \ldots, w_{\text{Icon}}) \oplus h_c$ which generates the challenges u_i's and ϑ_c. This occurs only if the verifier $\mathbf{V}^*$ was able to guess that the output values of the query seedC to the oracle GenC generate the right r_i's and a_{ij}'s. Since GenC is a random oracle, no polynomial time verifier $\mathbf{V}^*$ can succeed to do that with a non-negligible probability. We still have to consider the case where the verifier queried a seedC$_k$ which produces the challenges u_i's and ϑ_c. We see that the two transcripts are always identical, since the simulator clearly knows the answer of the honest prover by learning the right w_i's. Therefore, we can conclude that the two transcript distributions are statistically indistinguishable.

Straight-Line Zero-Knowledge of Denial. This proof is similar.

Unforgeability. Let $\mathcal{F}$ be a forger who succeeds to existentially forge a signature under an adaptive chosen-message attack with a non-negligible probability ε. We will construct an algorithm $\mathcal{B}$ which solves the Lsig-S-GHI problem with $S := \{(\text{Xkey}_1, \text{Ykey}_1), \ldots, (\text{Xkey}_{\text{Lkey}}, \text{Ykey}_{\text{Lkey}})\}$ using the forger $\mathcal{F}$ and $\mathcal{K}_s^{\mathbf{V}}$. At the beginning, $\mathcal{B}$ receives the challenges $x_1, \ldots, x_{\text{Lsig}} \in \text{Xgroup}$ of the Lsig-S-GHI problem. Then, $\mathcal{B}$ runs the forger and simulates the queries to the random oracle GenS, q_S queries to the signing oracle Sign and q_V queries to the denial/confirmation oracle Ver. We can assume that all messages sent to Sign resp. Ver were previously queried to GenS (since the oracle Sign resp. Ver has to make such queries anyway). $\mathcal{B}$ simulates the oracles GenS and Sign as follows:

GenS. For each message m queried to GenS, $\mathcal{B}$ maintains a list of each message and corresponding answer $(m, \text{Xsig}_1, \ldots, \text{Xsig}_{\text{Lsig}})$. If the message was already queried, $\mathcal{B}$ outputs the corresponding answer in the list. Otherwise, he picks $a_{i,j} \in_U \mathbf{Z}_d$ and $r_i \in_U \text{Xgroup}$ uniformly at random for $1 \leq i \leq \text{Lsig}$, $1 \leq j \leq \text{Lkey}$. With probability q, he answers $\text{Xsig}_i := dr_i + \sum_{j=1}^{\text{Lkey}} a_{i,j} \text{Xkey}_j$ for $i = 1, \ldots, \text{Lsig}$. We call it type-1 answer. With probability $1 - q$, the answer is $\text{Xsig}_i := dr_i + x_i + \sum_{j=1}^{\text{Lkey}} a_{i,j} \text{Xkey}_j$ for $i = 1, \ldots, \text{Lsig}$. We call it type-2 answer. For each message, $\mathcal{B}$ keeps the coefficients $a_{i,j}$'s and r_i's and answer type in memory. Note that the simulation is perfect by the assertion 2 of Lemma 2, since the public key is valid.

Sign. For a message m, if the answer to the GenS query of m was of type-1, then $\mathcal{B}$ answers $\text{Ysig}_i := \sum_{j=1}^{\text{Lkey}} a_{i,j} \text{Ykey}_j$ for $i = 1, \ldots, \text{Lsig}$. Otherwise, it aborts the simulation.

Let (m_i, σ_i) denote the ith query to Ver for $1 \leq i \leq q_V$ and $(m_{qv+1}, \sigma_{qv+1})$ denote the $\mathcal{F}$ output. In order to simulate the answers of the queries made to Ver, $\mathcal{B}$ guesses the smallest i such that (m_i, σ_i) is a valid forged pair (i.e., m was not queried to Sign). To this, $\mathcal{B}$ simply picks ℓ uniformly at random in $\{1, \ldots q_V + 1\}$. $\mathcal{B}$ deals with the ith query as follows:

$i < \ell$. To any query (m_i, σ_i), $\mathcal{B}$ checks whether m_i was submitted to Sign. If it is the case, $\mathcal{B}$ is able to decide whether (m_i, σ_i) is valid and simulates the appropriate protocol. Otherwise, $\mathcal{B}$ guesses that (m_i, σ_i) is invalid and simulate the appropriate protocol. The simulation is done as the simulator in the proof of non-transferability of the confirmation (resp. denial) protocol.

$i = \ell$. Let $(m_\ell, \sigma_\ell) := (m_\ell, \text{Ysig}_1, \ldots, \text{Ysig}_{\text{Lsig}})$. If the corresponding Xsig_i's were of type-1, $\mathcal{B}$ aborts. Otherwise, when ℓ was correctly guessed $\text{Ysig}_i = y_i + \sum_{j=1}^{\text{Lkey}} a_{i,j} \text{Ykey}_j$ and $\mathcal{B}$ is able to deduce the y_i's of the Lsig-S-GHI problem.

It remains to compute the probability that $\mathcal{B}$ retrieves the y_i's and did not abort. This event occurs if $\mathcal{B}$ is able to simulate all Sign queries, guess the right ℓ and use the message m_ℓ to deduce the y_i's. Therefore, $\Pr[\mathcal{B} \text{ succeeds}|\mathcal{F} \text{ succeeds}] = q^{q_S}(1-q)/(q_V+1)$. As for the full-domain hash technique [7] and as in [13], the optimal $q_{\text{opt}} = q_S/(q_S + 1)$. Thus, the success probability is greater or equal to $(1/e(1 + q_S)(1 + q_V))\varepsilon$.

Invisibility. Let $\mathcal{D}$ be a distinguisher which breaks the invisibility of the MOVA scheme with an advantage ε. We construct an algorithm $\mathcal{B}$ which solves the Lsig-S-GHID problem by using $\mathcal{D}$ and $\mathcal{K}_s^{\mathbf{V}}$. At the beginning, $\mathcal{B}$ is challenged with a tuple $\{(x_1, y_1), \ldots, (x_{\text{Lsig}}, y_{\text{Lsig}})\} \in (\text{Xgroup} \times \text{Ygroup})^{\text{Lsig}}$ for which it has to decide whether $\text{Hom}(x_i) = y_i$ for all $1 \leq i \leq \text{Lsig}$ or if this tuple was picked at random. Like for the proof of the existential forgery, the simulator $\mathcal{B}$ runs $\mathcal{D}$ and simulates the queries to the random oracle GenS, q_S queries to the signing oracle Sign and the queries to the denial/confirmation oracle Ver. We can assume that each message queried to Sign or Ver was previously queried to the random oracle GenS. We assume that no query m to Ver was submitted to Sign beforehand. (Otherwise, we can just simulate them with $\mathcal{K}_s^{\mathbf{V}}$.) Let Forge be the event in which $\mathcal{D}$ sends a valid message-signature pair to Ver. We first remove all instances for which the event Forge occurs. So, we can now assume that $\mathcal{D}$ never submits any valid pair (m, σ) to Ver such that m was not previously submitted to Sign. $\mathcal{B}$ simulates the oracles just like in the proof of unforgeability with $\ell = q_V + 1$ (we excluded valid forged pairs).

After a given time, the distinguisher $\mathcal{D}$ sends a message m^* to the challenger of the invisibility game which is simulated by $\mathcal{B}$. We can assume that m^* was queried to GenS (otherwise $\mathcal{B}$ simulates a new query). If the answer of m^* to GenS was of type-1, $\mathcal{B}$ aborts the simulation. Otherwise, it sends the challenge signature $(\text{Ysig}_1^*, \ldots, \text{Ysig}_{\text{Lsig}}^*)$ where $\text{Ysig}_i^* := y_i + \sum_{j=1}^{\text{Lkey}} a_{i,j} \text{Ykey}_j$ for $1 \leq i \leq \text{Lsig}$. Then, $\mathcal{D}$ continues to query the oracles which are simulated by $\mathcal{B}$ as above.

Finally, $\mathcal{D}$ outputs a guess bit b'. The simulator $\mathcal{B}$ outputs the same bit b' as guess bit to the Lsig-S-GHID challenger or a random bit when $\mathcal{B}$ aborted.

Using the homomorphic property of Hom, we deduce that the set $\{(x_i, y_i)\}_{i=1}^{\text{Lsig}}$ interpolates in a group homomorphism with the set of points S if and only if $(m^*, \text{Ysig}_1^*, \ldots, \text{Ysig}_{\text{Lsig}}^*)$ is a valid message-signature pair. Hence, when the simulator does not abort and the event Forge does not occur, $\mathcal{B}$ perfectly simulates the invisibility games. It remains to compute the advantage of $\mathcal{B}$.

For a bit b, we denote A_b the probability event that $\mathcal{B}$ does not abort when the challenge to $\mathcal{B}$ was of the form T_b (thus, $\mathcal{B}$ simulates the game **Game**$^{\text{inv-cma-}b}$ to $\mathcal{D}$). Note that the probability $\Pr[A_1] = \Pr[A_0]$ can be bounded in an optimal way as in the proof of existential forgery attacks, namely, by choosing q adequately we get $\Pr[A_1] \geq (1/e(1 + q_S))$. We now define the events B_b and D_b which occur when $\mathcal{B}$ and $\mathcal{D}$ respectively outputs the bit 0 when the challenge was of the form T_b. Note that if A_b happens, both events B_b and D_b occurs simultaneously. Let us denote ε_0 resp. ε_1, the probability for $\mathcal{D}$ to output 0 in the game **Game**$^{\text{inv-cma-0}}$ resp. **Game**$^{\text{inv-cma-1}}$. We now estimate $\Pr[B_0|A_0]$ and $\Pr[B_1|A_1]$ with respect to ε_0 and ε_1. To this end, we notice that the event $B_0|A_0$ resp. $B_1|A_1$ occurs simultaneously with the event where $\mathcal{D}$ outputs 0 in the game **Game**$^{\text{inv-cma-0}}$ resp. **Game**$^{\text{inv-cma-1}}$, provided that the event Forge does not occur. Hence, applying the difference lemma of Shoup [22] leads to

$$|\Pr[B_b|A_b] - \varepsilon_b| \leq \Pr[\mathsf{Forge}]$$

for $b = 0, 1$. From this, we can deduce that $\Pr[B_0|A_0] \geq \varepsilon_0 - \Pr[\mathsf{Forge}]$ and $\Pr[B_1|A_1] \leq \varepsilon_1 + \Pr[\mathsf{Forge}]$. Without loss of generality, we can assume that $\Pr[B_0] \geq \Pr[B_1]$. The advantage of $\mathcal{B}$ is then equal to

$$\Pr[B_0] - \Pr[B_1] = \Pr[\neg A_0] \cdot (\Pr[B_0|\neg A_0] - \Pr[B_1|\neg A_1])$$
$$+ \Pr[A_0] \cdot (\Pr[B_0|A_0] - \Pr[B_1|A_1]).$$

Since $\Pr[B_0|\neg A_0] = \Pr[B_1|\neg A_1] = 1/2$ and $\varepsilon_0 - \varepsilon_1 = \mathsf{Adv}_{\mathcal{D}}^{\text{inv-cma}}$, we finally have

$$\mathsf{Adv}_{\mathcal{B}}^{\text{Lsig-}S\text{-GHID}} \geq \frac{1}{(1 + q_S)e} \left(\mathsf{Adv}_{\mathcal{D}}^{\text{inv-cma}} - 2\Pr[\mathsf{Forge}] \right).$$

We can conclude by noting that Forge occurs with a probability bounded by $e(1 + q_S)(1 + q_V)\varepsilon'$ by assertion 5. □

Remark 6. MOVA scheme can be made probabilistic so that the invisibility notion defined in [8] is satisfied. To this, it suffices to append some randomness r to the message to sign and to add r in the signature. The drawback is that the signature enlarges.

Consequences for the Signature Parameters. One of the main advantage of MOVA scheme as stated in [17] is the fully scalable signature size. It was argued that one could potentially consider signatures of size of 20 bits, but the corresponding security level was not precisely quantified. Namely, the efficiency of the security reduction in [17] is not detailed and the security model did not consider queries to the confirmation/denial oracle. Our security reduction provides a more precise result. Assuming that any solver with same computational resource as a given forger cannot solve Lsig-S-GHI problem with a probability significatively greater than $|\mathsf{Ygroup}|^{-\mathsf{Lsig}}$, the assertion 5 of Theorem 3 shows that we have $\mathsf{Succ}_{\mathcal{F}}^{\text{ef-cma}} \leq |\mathsf{Ygroup}|^{-\mathsf{Lsig}} e(q_S + 1)(q_V + 1)$. Note that the assumption can be reached by scaling Xgroup adequately, namely without any modification of the signature size. This is the case when Hom is the Legendre symbol $(\cdot/p)$ defined on an RSA modulus $n = pq$. A signature size of Lsig ≥ 52 bits achieves a success probability for the existential forgeability of at most 2^{-20} with $q_S = 2^{10}$ and $q_V = 2^{20}$. Similarly, assuming that $\mathsf{Adv}_{\mathcal{B}}^{\text{Lsig-}S\text{-GHID}} \approx 0$ for any $\mathcal{B}$ with similar complexity as the invisibility distinguisher $\mathcal{D}$, assertion 6 of Theorem 3 shows that $\mathsf{Adv}_{\mathcal{D}}^{\mathsf{inv-cma}} \approx 2e^2 q_S q_V 2^{-\mathsf{Lsig}}$, which leads to $\mathsf{Adv}_{\mathcal{D}}^{\mathsf{inv-cma}} \approx 2^{-18}$. Results for the soundness can be obtained with $\mathsf{Succ}^{\mathsf{inv-tp}} \approx 0$. For the 2-move verification protocols, we can achieve a soundness probability of 2^{-20} with $\mathsf{Icon} = \mathsf{Iden} = 60$, $q_{H_c} = q_{H_d} = 2^{40}$.

6   Conclusion

We revisited a 2-move variant of the MOVA undeniable signature scheme which was proposed without any proof. By using a trapdoor one-way permutation adequately, we were able to make the verification protocols non-transferable. All the other required security properties are thoroughly analyzed in the random oracle model, thereby allowing to quantify the security of the different properties in terms of the signature parameters. So, as far as we know, this is the first time a provably secure undeniable signature scheme with 2-move confirmation and denial protocols is obtained. This result shows that minimal number of moves in an undeniable signature with interactive protocols can be reached in practice.

References

1. B. Barak, Y. Lindell, and S. P. Vadhan. Lower Bounds for Non-Black-Box Zero Knowledge. In *44th Annual IEEE Symposium on Foundations of Computer Science, FOCS '03*, pages 384–393. IEEE Computer Society, 2003.
2. M. Bellare and P. Rogaway. Random Oracles are Practical: A Paradigm for Designing Efficient Protocols. In *1st ACM Conference on Computer and Communications Security*, pages 62–73. ACM Press, 1993.
3. J. Boyar, D. Chaum, I. Damgård, and T. P. Pedersen. Convertible Undeniable Signatures. In *Advances in Cryptology – CRYPTO '90*, volume 537 of *Lecture Notes in Computer Science*, pages 189–205. Springer-Verlag, 1991.
4. J. Camenisch and M. Michels. Confirmer Signature Schemes Secure against Adaptive Adversaries. In *Advances in Cryptology – EUROCRYPT '00*, volume 1807 of *Lecture Notes in Computer Science*, pages 243–258. Springer-Verlag, 2000.
5. D. Chaum. Zero-Knowledge Undeniable Signatures. In *Advances in Cryptology – EUROCRYPT '90*, volume 473 of *Lecture Notes in Computer Science*, pages 458–464. Springer-Verlag, 1990.
6. D. Chaum and H. van Antwerpen. Undeniable Signatures. In *Advances in Cryptology – CRYPTO '89*, volume 435 of *Lecture Notes in Computer Science*, pages 212–217. Springer-Verlag, 1990.
7. J.-S. Coron. On the Exact Security of Full Domain Hash. In *Advances in Cryptology – CRYPTO '00*, volume 1880 of *Lecture Notes in Computer Science*, pages 229–235. Springer-Verlag, 2000.
8. S. D. Galbraith and W. Mao. Invisibility and Anonymity of Undeniable and Confirmer Signatures. In *Topics in Cryptology – CT–RSA '03*, volume 2612 of *Lecture Notes in Computer Science*, pages 80–97. Springer-Verlag, 2003.
9. R. Gennaro, H. Krawczyk, and T. Rabin. RSA-Based Undeniable Signatures. In *Advances in Cryptology – CRYPTO '97*, volume 1294 of *Lecture Notes in Computer Science*, pages 132–149. Springer-Verlag, 1997.

10. O. Goldreich. *Foundations of Cryptography, Volume I Basic Tools.* Cambridge University Press, 2001.
11. S. Goldwasser, S. Micali, and R. L. Rivest. A Digital Signature Scheme Secure Against Adaptive Chosen-Message Attacks. *SIAM Journal on Computing*, 17(2):281–308, 1988.
12. M. Jakobsson, K. Sako, and R. Impagliazzo. Designated Verifier Proofs and Their Applications. In *Advances in Cryptology – EUROCRYPT '96*, volume 1070 of *Lecture Notes in Computer Science*, pages 143–154. Springer-Verlag, 1996.
13. K. Kurosawa and S.-H. Heng. 3-Move Undeniable Signature Scheme. In *Advances in Cryptology – EUROCRYPT '05*, volume 3494 of *Lecture Notes in Computer Science*, pages 181–197. Springer-Verlag, 2005.
14. F. Laguillaumie and D. Vergnaud. Short Undeniable Signatures Without Random Oracles: the Missing Link. In *Progress in Cryptology – INDOCRYPT '05*, volume 3797 of *Lecture Notes in Computer Science*, pages 283–296. Springer-Verlag, 2005.
15. H. Lipmaa, G. Wang, and F. Bao. Designated Verifier Signature Schemes: Attacks, New Security Notions and a New Construction. In *Automata, Languages and Programming: 32nd International Colloquium, ICALP '05*, volume 3580 of *Lecture Notes in Computer Science*, pages 459–471. Springer-Verlag, 2005.
16. J. Monnerat, Y. A. Oswald, and S. Vaudenay. Optimization of the MOVA Undeniable Signature Scheme. In *Progress in Cryptology – MYCRYPT '05*, volume 3715 of *Lecture Notes in Computer Science*, pages 196–209. Springer-Verlag, 2005.
17. J. Monnerat and S. Vaudenay. Generic Homomorphic Undeniable Signatures. In *Advances in Cryptology – ASIACRYPT '04*, volume 3329 of *Lecture Notes in Computer Science*, pages 354–371. Springer-Verlag, 2004.
18. J. Monnerat and S. Vaudenay. Undeniable Signatures Based on Characters: How to Sign with One Bit. In *Public Key Cryptography – PKC '04*, volume 2947 of *Lecture Notes in Computer Science*, pages 69–85. Springer-Verlag, 2004.
19. W. Ogata, K. Kurosawa, and S.-H. Heng. The Security of the FDH Variant of Chaum's Undeniable Signature Scheme. In *Public Key Cryptography – PKC '05*, volume 3386 of *Lecture Notes in Computer Science*, pages 328–345. Springer-Verlag, 2005. Extended version available on: Cryptology ePrint Archive, Report 2004/290, http://eprint.iacr.org/.
20. T. Okamoto and D. Pointcheval. The Gap-Problems: A New Class of Problems for the Security of Cryptographic Schemes. In *Public Key Cryptography – PKC '01*, volume 1992 of *Lecture Notes in Computer Science*, pages 104–118. Springer-Verlag, 2001.
21. R. Pass. On Deniability in the Common Reference String and Random Oracle Model. In *Advances in Cryptology – CRYPTO '03*, volume 2729 of *Lecture Notes in Computer Science*, pages 316–337. Springer-Verlag, 2003.
22. V. Shoup. Sequences of Games: A Tool for Taming Complexity in Security Proofs. Cryptology ePrint Archive, Report 2004/332, 2004. http://eprint.iacr.org/.

Searching for Compact Algorithms: CGEN

M.J.B. Robshaw

France Telecom Research and Development
38–40, rue du Général Leclerc
92794 Issy les Moulineaux, Cedex 9, France
matt.robshaw@orange-ft.com

Abstract. In this paper we describe an AES-like pseudo-random number generator called CGEN. Initial estimates suggest that the computational resources required for its implementation are sufficiently modest for it be suitable for use in RFID tags.

1 Introduction

There has been much recent interest in extending cryptographic techniques to resource-constrained devices. Much of this work has centered around the design of exotic primitives and protocols suitable for demanding environments such as *radio frequency identification* (RFID) tags. There is, however, something unsettling about building a security infrastructure on entirely new design techniques that might be vulnerable to innovative attack. However, at the same time, many standardised algorithms are not suited to deployment on the cheapest tags. This gap between extreme proposals and more conventional primitives has made the design of lightweight, but trusted, algorithms an active topic of research.

In this paper we propose an algorithm that is designed to (partially) fill this gap and we introduce a *pseudo-random number generator* (PRNG) that we call CGEN for "compact generator". Some reasons for concentrating on this type of primitive are provided in Section 1.1, but our goal is a PRNG that would occupy 1000-1500 gate equivalents (GE) in silicon, make modest power demands, and yet fit this middle ground of *dedicated-but-strong* cryptography.

The properties we seek for a PRNG essentially coincide with those we expect for a good stream or a block cipher in an appropriate mode. Unfortunately, however, there are currently no widely-trusted stream ciphers that are suitable for compact hardware implementation, though there are efforts to try and rectify this with the eSTREAM initiative [11]. Two proposals in particular—the more traditional GRAIN [16] and the innovative TRIVIUM [5]—appear to offer considerable promise. By contrast, we do have a well-trusted block cipher—the AES [21]— and, up to a number of encryptions dictated by birthday attacks, the output from this block cipher in counter mode would be adequate for our purposes. However, even when considering recent impressive implementation results [12], the hardware requirements of the AES remain excessive.

In this paper we build on recent trends in block cipher design but, for the application we have in mind, we observe that a block cipher offers more than we

P.Q. Nguyen (Ed.): VIETCRYPT 2006, LNCS 4341, pp. 37–49, 2006.

need. In fact the restricted scope for attacks against the proposed PRNG, particularly when used in its intended environment, suggest that much of the usual block cipher machinery can be removed. Our goal, therefore, is an algorithm whose dedicated purpose[1] is to act as a PRNG and our proposal is best viewed as a complement to algorithms like GRAIN and TRIVIUM, but one that uses typical block cipher techniques in its design.

1.1   Why a (New) Compact Pseudo-random Number Generator?

The recent surge of interest in computationally-restricted devices has generated many proposals for a wide range of protocols. Often, a hidden requirement is that there be an efficient way to generate random bits, but it is not always clear how this might be best achieved and a tag-specific PRNG may be more practical. Another reason for looking at a PRNG can be found when considering public-key schemes such as the GPS identification scheme [14,22]. A common optimisation in many commitment-challenge-response schemes is the use of *coupons*. These consist of pre-computed commitments that are stored securely on the tag and used one-by-one. Depending on storage limitations and the specifics of the scheme, a further optimisation can sometimes be attained by regenerating (parts of) the coupon with an efficient and compact PRNG as specified in ISO-9798 [15]. Thus, a compact PRNG might be used to generate a modest number of coupons for use within a larger scheme. More generally, for a variety of constrained applications, the ability to efficiently generate pseudorandom bits from a fixed, per-device, secret seed could well be of great interest.

Block ciphers, or constructions built around block ciphers [17], can provide a convenient solution and a benchmark is provided by the AES for which a compact implementation requires around 3600 GE [12]. Despite the name, the *Tiny Encryption Algorithm* TEA, and extensions [25,26], require more resources than we would like, and our attention shifts to two dedicated block cipher proposals MCRYPTON [18], which has many design features in common with CGEN and will be discussed in Sections 2.2 and 3.2, and SEA, the *Scalable Encryption Algorithm* [24] which achieves a small implementation footprint at some cost to performance. Other approaches to designing a PRNG might rely on the use of dedicated hash functions such as MD5 [23] or SHA-1 [20]. But independently of issues about the suitability of these primitives for new applications, the resource cost of implementing such algorithms is high. We therefore believe that a PRNG that stays close to established design principles, but provides the necessary functionality at a reduced cost, may be attractive.

In general terms there might be two approaches to the design of a new PRNG. We might try and develop a "provably secure" scheme that would relate any advantage in predicting the output from the PRNG to an advantage in solving some underlying hard problem. While such an approach has typically been viewed as difficult in symmetric cryptography, particularly when reasonable performance

---

[1] Since CGEN is essentially built around a block cipher one could use it in this way after some adjustments.

is required, there has been much progress in this direction with the designs of QUAD [2] and VSH [8]. This is a trend we might expect to develop in the coming years. Here though, we design a scheme whose security depends on bit-level design and analysis. Just like the AES, however, its construction allows us to make some (positive) claims about the security offered by the scheme.

1.2 The Intended Environment and Implications

The essential nature of an RFID application is of cheap tags being deployed widely, used relatively infrequently, and then disposed. This characterisation doesn't apply to all RFID-based applications, but it is a reasonable description of those that use cheaper tags.

We will see in what follows that CGEN is essentially a block cipher in counter mode. However, since the secret key is a fixed, per-device seed, CGEN lacks a key schedule. There is also no need to provide decryption, and CGEN operates on a restricted space of input plaintexts. All of these factors contribute to a reduced hardware footprint, but what remains is still cryptographically strong.

The block size of the underlying cipher is 64 bits thus, as the number of outputs approaches 2^{32}, the birthday paradox implies that CGEN would be distinguishable from a perfect random number generator. However, CGEN is only intended to be used at most 2^{16} times and this has implications on the capabilities of the attacker and the attacks that can be mounted.

An attacker knows the inputs to at most 2^{16} different invocations of CGEN since these are generated by a counter. This immediately limits the number of plaintexts and plaintext pairs available for cryptanalysis. There will be no opportunity for choosing the input, and the secret key is fixed so there are no opportunities for some of the more sophisticated block cipher attacks such as those that rely on related keys. And finally, depending on the application, it is possible that the output from CGEN might never leave the device. Despite these limitations on the capabilities of the attacker, we do require CGEN to offer a good level of security and this is discussed at some length in Section 3.

We note that hardware-level attacks to recover the secret seed from the tag, or denial of service attacks that consume all available iterations of CGEN and thereby exhaust the tag can always be attempted. However, these threats are more relevant to the application than the algorithm. Thus they lie outside the scope of this paper which will be solely concerned with cryptanalytic threats.

2 The Proposal: CGEN

One approach to the design of a PRNG might be to use "very simple operations very often". The approach within CGEN is to use slightly heavier components more sparingly. In what follows, it is obvious that we stay very close to the design principles embodied in the AES. These are techniques that work, and ones with which we are familiar. Essentially CGEN is a block cipher running in counter mode. However, since the intended environment uses a fixed, per-device,

a_{00}	a_{01}	a_{02}	a_{03}
a_{10}	a_{11}	a_{12}	a_{13}
a_{20}	a_{21}	a_{22}	a_{23}
a_{30}	a_{31}	a_{32}	a_{33}

1. For counter value $c_i = [c_{i0}||c_{i1}||c_{i2}||c_{i3}]$, for $0 \leq j \leq 3$, set $a_{0j} = a_{0j} \oplus c_{ij}$.
2. Do one round of `mixtable`.
3. Combine state with seed: for $0 \leq i, j \leq 3$ set $a_{ij} = a_{ij} \oplus s_{16+4i+j}$.
4. Do three rounds of `mixtable`.
5. Combine state with seed: for $0 \leq i, j \leq 3$ set $a_{ij} = a_{ij} \oplus s_{4i+j}$.
6. Do two rounds of `mixtable`.
7. Combine state with seed: for $0 \leq i, j \leq 3$ set $a_{ij} = a_{ij} \oplus s_{16+4i+j}$.
8. Do two rounds of `mixtable`.
9. Combine state with seed: for $0 \leq i, j \leq 3$ set $a_{ij} = a_{ij} \oplus s_{4i+j}$.
10. Do two rounds of `mixtable`.
11. Combine state with seed: for $0 \leq i, j \leq 3$ set $a_{ij} = a_{ij} \oplus s_{16+4i+j}$.
12. Output v_i where $v_i = [a_{00}|| \cdots ||a_{03}||a_{10}|| \cdots ||a_{13}|| \cdots\cdots ||a_{33}]$.
13. If $c_i = 2^{16} - 1$ the PRNG is no longer used. Otherwise increment c_i by one.

Fig. 1. An overview of CGEN where the counter is initialised to zero. The 64-bit state is considered as a (4×4)-array which, for $0 \leq i, j \leq 3$, is initialised with $a_{ij} = s_{4i+j}$.

secret seed there is a much reduced (effectively non-existent) key schedule. One significant deviation from the AES is the relative size of elements in the state array and the s-box; in the AES these are the same but in CGEN they are not. This provides some advantages over other compact AES-like proposals.

Since the state is often the most space-consuming aspect, we adopt a state of size 64 bits for CGEN. Each tag using CGEN possesses its own secret 128-bit seed which need never be shared with any other entity. Compromise of the seed leads to a compromise of CGEN and the seed s is represented as $s = s_0 \ldots s_{31}$ where each s_i is four bits long. At each iteration of CGEN (see Figure 1) a 64-bit value v_i is generated from some counter value $c_i = i$ and we have that $v_i = \text{CGEN}(c_i, s)$ for $0 \leq i < 2^{16}$. The counter limit can be set to some other value if appropriate.

2.1   The `mixtable` Operation

Given the success of the elegant construction behind the AES, it is appealing to consider similar approaches. While we will see considerable similarity to the AES, our construction uses an s-box that spans two table entries instead of one. This has numerous security advantages over the obvious alternative of using a reduced 4-bit s-box that covers a single array element. Such reduced s-boxes can be found in natural small-scale variants of the AES [7] and in the design of the nibble-based block cipher MCRYPTON [18]. The primary observation is

	-0	-1	-2	-3	-4	-5	-6	-7	-8	-9	-a	-b	-c	-d	-e	-f
0-	ba	54	2f	74	53	d3	d2	4d	50	ac	8d	bf	70	52	9a	4c
1-	ea	d5	97	d1	33	51	5b	a6	de	48	a8	99	db	32	b7	fc
2-	e3	9e	91	9b	e2	bb	41	6e	a5	cb	6b	95	a1	f3	b1	02
3-	cc	c4	1d	14	c3	63	da	5d	5f	dc	7d	cd	7f	5a	6c	5c
4-	f7	26	ff	ed	e8	9d	6f	8e	19	a0	f0	89	0f	07	af	fb
5-	08	15	0d	04	01	64	df	76	79	dd	3d	16	3f	37	6d	38
6-	b9	73	e9	35	55	71	7b	8c	72	88	f6	2a	3e	5e	27	46
7-	0c	65	68	61	03	c1	57	d6	d9	58	d8	66	d7	3a	c8	3c
8-	fa	96	a7	98	ec	b8	c7	ae	69	4b	ab	a9	67	0a	47	f2
9-	b5	22	e5	ee	be	2b	81	12	83	1b	0e	23	f5	45	21	ce
a-	49	2c	f9	e6	b6	28	17	82	1a	8b	fe	8a	09	c9	87	4e
b-	e1	2e	e4	e0	eb	90	a4	1e	85	60	00	25	f4	f1	94	0b
c-	e7	75	ef	34	31	d4	d0	86	7e	ad	fd	29	30	3b	9f	f8
d-	c6	13	06	05	c5	11	77	7c	7a	78	36	1c	39	59	18	56
e-	b3	b0	24	20	b2	92	a3	c0	44	62	10	b4	84	43	93	c2
f-	4a	bd	8f	2d	bc	9c	6a	40	cf	a2	80	4f	1f	ca	aa	42

Fig. 2. For a straw-man proposal of CGEN we use the tweaked ANUBIS s-box [1]

that small s-boxes (operating on nibbles) cannot offer the same level of local resistance to differential and linear cryptanalysis as larger ones. A secondary observation is that when using a larger s-box that spans multiple entries, the s-box is not limited to providing *confusion*, but also contributes to *diffusion* within the state since it provides some mixing across the cells of an array. This can provide a faster avalanche of change, as measured (say) by the expected weight of a differential trail starting with one active array element which, in the case of our construction, is 6.75 active s-boxes over two rounds instead of five as for the AES. We will explore this issue more in Section 3.2. We will also see, in Section 3.3, that the conflict between the size of the array elements and the s-boxes may help protect against structural attacks that are currently among the most effective against AES-like structures.

Like the AES, the `mixtable` operation consists of s-box look-ups and column mixes. In the substitution layer each s-box instantiates an 8-bit to 8-bit permutation, with S-boxes in different rows of the arrays being offset against each other.

$$a_{00}||a_{01} = S[a_{00}||a_{01}], \quad a_{02}||a_{03} = S[a_{02}||a_{03}],$$
$$a_{13}||a_{10} = S[a_{10}||a_{13}], \quad a_{11}||a_{12} = S[a_{11}||a_{12}],$$
$$a_{20}||a_{21} = S[a_{20}||a_{21}], \quad a_{22}||a_{23} = S[a_{22}||a_{23}],$$
$$a_{33}||a_{30} = S[a_{33}||a_{30}], \quad a_{31}||a_{32} = S[a_{31}||(a_{32} \oplus 0x1)].$$

The AES s-box is one option though an s-box that is less demanding in hardware might be preferred (see Section 4). Given this latter consideration, we suggest to use the tweaked s-box from the ANUBIS block cipher [1] for a straw-man version of CGEN. While the construction of this s-box is nibble-based, something that might

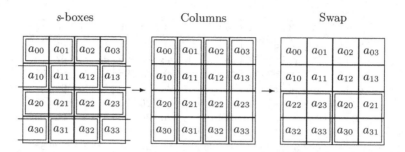

Fig. 3. Pictorial representation of the `mixtable` operation used within CGEN

be viewed as being too compatible with the CGEN structure, its implementation requirements help us to achieve our hardware goal. The *s*-box is given in Figure 2.

The column mix can be written as four parallel invocations of a (4×4) matrix M over $GF(2^4)$. To remain close to other proposals in the literature [7], we use the field representation given by $p(x) = x^4 + x + 1$ and the following array of nibbles in hexadecimal notation;

$$M = \begin{pmatrix} 2\,3\,1\,1 \\ 1\,2\,3\,1 \\ 1\,1\,2\,3 \\ 3\,1\,1\,2 \end{pmatrix}.$$

We then swap $a_{20}||a_{21}$ with $a_{22}||a_{23}$ while we swap $a_{30}||a_{31}$ with $a_{32}||a_{33}$. Note that decryption is not required in CGEN and so the form of the inverse of the matrix M is not of interest. The `mixtable` operation is illustrated in Figure 3.

2.2   Design Issues

The use of *s*-boxes that span two array elements will be discussed further in Section 3. It will be noted that at each round `0x1` is exclusive-ored to array position a_{32}. This is motivated by considering the following array pattern that would propagate across any number of rounds of `mixtable` without the exclusive-or of `0x1`, though the specific values to A and B would change between rounds.

A	B	A	B
B	A	B	A
A	B	A	B
B	A	B	A

Over the entirety of CGEN such symmetries are not a concern since they would be destroyed by the inclusion of the seed, particularly the second insertion of seed material whose form conflicts with the combination of seed and counter at the start. Thus the value of breaking such potential symmetry is not clear, particularly given the subsequent impact on implementation. However, it may be prudent to avoid giving even a few rounds of predictable behavior for free, particularly when something as simple as flipping one bit within the array seems to suffice. While exclusive-oring a round counter would help to reduce round symmetries even further, it would consume more resources. Thus the ideal resolution of this issue will likely depend on implementation requirements.

The length of the input counter c_i can be increased at little additional computational cost. While this would give a greater operational lifetime to CGEN, there are advantages in not using CGEN too often (see Section 1.2). Figure 1 illustrates a straw-man proposal for CGEN but the number of rounds of computation can be adjusted up, or down, as dictated by future analysis.

3   Security Issues

Existing analysis identifies promising elements in the design of CGEN with one advantage lying in the interplay between the 8-bit s-box and the 4-bit state mixing. In Section 1.2 we outlined some of the ways the cryptanalyst would be constrained in practice; the output might not always be available, when it is available the amount of output is small, and at the same time the inputs to CGEN have a restricted form. However, despite such practical limits on the attacker, we consider a range of attacks inspired by conventional block cipher cryptanalysis.

3.1   Brute-Force Attacks

We conjecture that there is no analytic attack on CGEN, respecting the limit of 2^{16} inputs and the inability to change key, that is more effective than the range of brute-force attacks that take no advantage of the internal structure. Such attacks include the fact that when used a limited number of times the output from CGEN can always be guessed with probability that remains close to 2^{-64} and an attacker can recover the seed in 2^{-128} operations. However, as for other cryptographic algorithms, there are time-memory trade-offs. For example [4] with a pre-computation of 2^{96} operations and 2^{56} memory, one seed from a set of 2^{32} CGEN implementations can be recovered in 2^{80} operations, provided the output for a specific counter value is available for every case. Other trade-offs apply and the environment will often limit the application of such attacks.

3.2   Differential and Linear Attacks

Differential [3] and linear cryptanalysis [19] are powerful techniques that are applicable, at least in principle, to a wide-range of primitives including block

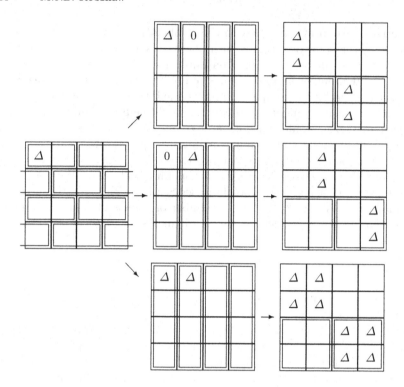

Fig. 4. An illustration of one-round difference propagation in CGEN starting from a single active array position

ciphers, stream ciphers, message authentication codes, hash functions, and of particular interest to us here, PRNGs. These attacks are often referred to as *statistical attacks* and the cryptanalyst exploits multiple interactions with the algorithm, often at the cost of large quantities of data, with the goal of identifying statistical patterns that allow key information to be recovered.

As we have observed, the intended environment of use for CGEN limits the number of texts available to the adversary. Regarding differential attacks, for example, the limited number of possible counter values constrains the amount of data available. This is likely to hinder statistical attacks, particularly since an adversary may be required to intercept many, if not all, the outputs from CGEN. Nevertheless, it is instructive to consider the cryptanalytic properties of the proposed construction. To illustrate, Figure 4 demonstrates three different possibilities for a differential trail over one round that starts with one active array position.

For a good 8-bit s-box[2], the first case occurs with a probability close to $\frac{1}{16}$, the second with a probability close to $\frac{1}{16}$, and the third with a probability close

---

[2] For the particular case of the AES s-box, the experimentally-derived probabilities are $\frac{228}{3840}$, $\frac{222}{3840}$, and $\frac{3390}{3840}$ respectively which are close to our estimates.

Table 1. Lower bounds to the number of active *s*-boxes and upper bounds to the probability of differential characteristics in CGEN when using the AES or ANUBIS *s*-box

	# Active s-boxes	Probabilty with AES s-box	Probabilty with ANUBIS s-box
2-round bounds:	5	2^{-30}	2^{-25}
3-round bounds:	8	2^{-48}	2^{-40}
Basic 4-round bounds:	10	2^{-60}	2^{-50}
Improved 4-round bounds:	15	2^{-90}	2^{-75}

to $\frac{7}{8}$. Using this, the expected number of active *s*-boxes over two rounds of CGEN, when starting with a single active array position, is

$$\approx \frac{1+4}{16} + \frac{1+4}{16} + \frac{7(1+6)}{8} \approx 6.75.$$

Thus, while five active *s*-boxes are guaranteed to be active over two successive rounds, it may be as many as seven. Note that we are counting the number of active *s*-boxes, not the number of active array positions.

The lower bound of five active *s*-boxes immediately yields a bound of 10 active *s*-boxes over four rounds. This is sufficient for our purposes, but we now show how to improve on this. The reasoning comes in two parts.

1. If any column of the array has two or less active nibbles at round i, then the action of the matrices and the subsequent swap ensure that there must be two columns with active nibbles in round $i + 1$. The active nibbles that arise in this way cannot both be assimilated within the same *s*-box look-up. Thus, over rounds $i + 1$ and $i + 2$ these two columns must contribute at least five active *s*-boxes each. Looking back from round i, we see that round $i - 1$ and round i themselves must also contribute at least five active *s*-boxes. Thus, in total, over rounds $i - 1$ to $i + 2$ there must be at least 15 active *s*-boxes.
2. Now consider a column of the array at round i with three or more active nibbles. Due to the swap at the end of round $i - 1$, there must have been two columns with active nibbles in round $i - 1$. These active nibbles cannot both be assimilated within the same *s*-box. Thus, over rounds $i - 2$ and $i - 1$ these two columns must contribute at least five active *s*-boxes each. Looking forward, rounds i and $i + 1$ must also contribute at least five active *s*-boxes. Thus, over rounds $i - 2$ to $i + 1$, there must be at least 15 active *s*-boxes.

The bounds on the evolution of a differential trail for the proposed PRNG are given in Table 1. We also include the maximum probability of any specific xor-differential characteristic over the stated number of rounds when using *s*-boxes from the AES and ANUBIS [1]. A more hardware-efficient *s*-box, such as that proposed in ANUBIS, might not offer the same level of resistance to differential cryptanalysis as the AES *s*-box but it should still offer sufficient resistance to differential cryptanalysis.

We note that the bound on the *number* of active s-boxes over four rounds is not as good as the 25 attained for a scaled-down version of the AES. Nor does it match the 16 active s-boxes guaranteed by four rounds of MCRYPTON [18]. However this is not as important as the resulting differential probability. Since we are using an 8-bit s-box, we can attain better local resistance to differential cryptanalysis than we would by using a 4-bit s-box. Thus, for any particular differential characteristic, the probability of 2^{-90} over four rounds for the construction shown here using the 8-bit AES s-box should be compared to $(2^{-2})^{25} = 2^{-50}$, the probability that would be achieved over four rounds for an AES-style s-box over $GF(2^4)$ or to $(2^{-2})^{16} = 2^{-64}$, the probability achieved over four rounds of MCRYPTON. Even moving to an eight-bit s-box that is less resistant to differential and linear cryptanalysis than the AES s-box, such as the tweaked s-box used in ANUBIS, gives better bounds (2^{-75}).

Note that the existence of differential characteristics of the form $(\Delta||0) \rightarrow (0||\Delta)$ (or *vice versa*) across the s-box suggest there is little point using a specific operation to offset array elements by a single column position. This is required in the AES and is achieved using the `ShiftRows` operation. Instead we rely on the action of the s-box to mix across adjacent columns.

In summary, if we use an s-box with good local resistance to differential cryptanalysis, then weaknesses due to the poor evolution of an xor-differential characteristic are not anticipated. Further research will likely consider the effectiveness of differentials rather than characteristics. Duality with linear cryptanalysis means that an exploitable bitwise correlation in CGEN is unlikely.

3.3   Structural Attacks

For the AES a variety of structural attacks such as the *square* [10] and *bottleneck* [13] attacks have been identified. Apart from brute-force time-memory trade-offs, these provide the most effective attacks against reduced-round versions of the AES. Given the regular structure of CGEN, similar attacks might be anticipated. However, when using a good 8-bit s-box such structural attacks appear to be hindered by the structure of CGEN.

Informally, an s-box spanning two table elements means that nibble-based properties are destroyed while byte-based properties are destroyed by the column operations. For instance, after the first 16 counter values every value occurs once, and only once, in position a_{03}. All other table entries would have been constant. This is the starting position for the classical form of the square attack. However a single invocation of the s-boxes will destroy this structure and the strong nibble-based structure of the input will not be evident in the output.

Alternatively, we might imagine a structure created by the first 256 counter values for which every possible 256-bit value appears in two adjacent nibbles $[a_{02}||a_{03}]$. Such a property would be preserved over an s-box invocation that instantiates a permutation. However, the column operation on nibbles offers an intrinsic conflict with the s-box operation on element-pairs. So, while analysis shows that related structure can be preserved a few rounds into CGEN, it is quickly overwhelmed.

3.4   Algebraic Attacks

Algebraic attacks on the AES have been conjectured [9] but, with the techniques at hand today, are generally viewed to be unlikely [6]. With a smaller state space, there might be some concerns that algebraic attacks could become a threat, particularly if we were to use the AES s-box or to use smaller 4-bit s-boxes. However, we resist the temptation to use smaller s-boxes and Section 4 proposes that low-cost implementations of CGEN avoid the AES s-box. This helps to avoid inversion over $GF(2^8)$ in the AES s-box that lies at the root of conjectured algebraic attacks.

4   Performance Estimates

While an accurate assessment of the hardware requirements for CGEN is required, we make some crude estimates. To begin to understand the anticipated hardware performance profile, we turn to estimates for the full AES given in [12]. This AES implementation is dominated by the s-box and RAM requirements which are claimed to be 395 and 2337 *gate equivalents* (GE) respectively.

It is not clear how close to optimal the s-box requirements for the AES implementation in [12] are, but other s-boxes are likely to be more attractive. For instance, the tweaked s-box for ANUBIS is claimed to require 100–120 GE for a hardware implementation [1]. While even more compact s-boxes may be available, it is worth observing that the four-bit s-box used in MCRYPTON still requires 107 GE [18]. So while the systematic development of ultra-compact s-boxes would be an interesting line of research, s-boxes of a similar size to that proposed for ANUBIS might be a reasonable expectation.

An estimate for the RAM requirements of CGEN can be made by observing that we have half the state of the AES and we dispense with a key-schedule (the key in the tag is fixed). Following [12] we assume that key material is stored in EEPROM and will not figure in the RAM estimates. This suggests, therefore, that we can estimate the RAM requirements for CGEN to be around one quarter of those required for the AES, namely 600 GE. The column mixing in the AES implementation is achieved using a byte-oriented architecture that requires 252 GE [12]. It would be interesting to consider the trade-offs in retaining a byte-oriented approach in CGEN, which may give a faster processing time, or to use a nibble-based architecture which may consume roughly half the gates.

Thus, provided we choose a suitable s-box, a crude estimate for the hardware requirements of the s-box, RAM, column operations, and key mixing of CGEN might be around $110 + 600 + 126 + 90 = 926$ GE. Allowing 10% to 20% for the inherent over-simplifications in such estimates and for overheads in the implementation, we might crudely estimate a hardware requirement of around 1100–1200 GE. With all operational characteristics remaining unchanged, the reduction in the number of gates will lead to an improved power consumption profile. Likewise, the number of operations required for CGEN when compared to the AES, suggest that CGEN will have reduced on-tag clocking requirements than implementations of the AES [12]. So while our focus has been on the hardware footprint required by CGEN, when compared to the AES improvements to the peak and average power consumption as well as the processing time for CGEN can be expected.

5   Conclusions

In this paper we have introduced CGEN; a lightweight PRNG. The main goal of the paper is to promote ongoing research into lightweight symmetric cryptography and to consider alternative design approaches to those offered by stream ciphers. While CGEN is built closely around the principles underlying the AES, it is not intended to offer the security of the AES. Instead it is intended to offer *sufficient* security for a restricted range of applications. This allows us to make a trade-off and even though we use block cipher design techniques, CGEN appears to be far less resource-demanding than trusted block cipher alternatives.

A full independent security and performance analysis is strongly encouraged. Since this paper provides no more than a basic analysis, it is possible that closer evaluation will reveal security or implementation oversights. It would then be interesting to consider what changes might be made to CGEN to help achieve our goal. Nevertheless, it seems that the goal of a simultaneously strong and compact PRNG is attainable and such a proposal would be of interest to application developers working within constrained environments such as RFID tags.

References

1. P.S.L.M. Barreto and V. Rijmen. The Anubis Block Cipher (tweaked version). Available via paginas.terra.com.br/informatica/paulobarreto/.
2. C. Berbain, H. Gilbert, and J. Patarin. QUAD: A Practical Stream Cipher with Provable Security. In S. Vaudenay, editor, *Proceedings of Eurocrypt 2006*, LNCS, volume 4004, pages 109–128, Springer-Verlag, 2006.
3. E. Biham and A. Shamir. *Differential Cryptanalysis of the Data Encryption Standard*. Springer Verlag, 1993.
4. A. Biryukov, S. Mukhopadhyay, and P. Sarkar. Improved Time-memory Trade-offs with Multiple Data. In B. Preneel and S. Tavares, editors, *Proceedings of SAC 2005*, LNCS, volume 3897, pages 110-127, Springer Verlag.
5. C. de Cannière and B. Preneel. Trivium Specifications. Available via www.ecrypt.eu.org.
6. C. Cid and G. Leurent. An Analysis of the XSL Algorithm. In B. Roy, editor, Proceedings of Asiacrypt 2005, LNCS, volume 3788, pages 333–352, Springer-Verlag, 2005.
7. C. Cid, S. Murphy, and M.J.B. Robshaw. Small Scale Variants of the AES. In H. Gilbert and H. Handschuh, editors, *Proceedings of FSE 2005*, LNCS, volume 3557, pages 145–162, Springer-Verlag, 2005.
8. S. Contini, A.K. Lenstra, and R. Steinfeld. VSH, an Efficient and Provable Collision-Resistant Hash Function. In S. Vaudenay, editor, *Proceedings of Eurocrypt 2006*, LNCS, volume 4004, pages 165–182, Springer-Verlag, 2006.

9. N. Courtois and J. Pieprzyk. Cryptanalysis of Block Ciphers with Overdefined Systems of Equations. In Y. Zheng, editor, *Proceedings of Asiacrypt 2002*, LNCS, volume 2501, pages 267–287, Springer-Verlag, 2002.

10. J. Daemen, L. Knudsen, and V. Rijmen. The Block Cipher Square. In E. Biham, editor, *Proceedings of FSE '97*, LNCS, volume 1267, pages 149–165, Springer-Verlag, 1997.

11. ECRYPT Network of Excellence. The Stream Cipher Project: eSTREAM. Available via `www.ecrypt.eu.org/stream`.

12. M. Feldhofer, S. Dominikus, and J. Wolkerstorfer. Strong Authentication for RFID Systems Using the AES algorithm. In M. Joye and J.-J. Quisquater, editors, *Proceedings of CHES 2004*, LNCS, volume 3156, pages 357–370, Springer Verlag, 2004.

13. H. Gilbert and M. Minier. A Collision Attack on Seven Rounds of Rijndael. In NIST, editors, *Proceedings of the 3rd Advanced Encryption Standard Conference*, pages 230–241, April, 2000. Available via `csrc.nist.gov`.

14. M. Girault. An Identity-based Identification Scheme Based on Discrete Logarithms Modulo a Composite Number. In I. Damgård, editor, *Proceedings of Eurocrypt '90*, LNCS, volume 473, pages 481–486, Springer-Verlag, 1990.

15. ISO/CEI 9798-5:2004. Information Technology - Security techniques - Entity authentication - Part 5: Mechanisms using zero-knowledge techniques. ISO/IEC 2004.

16. M. Hell, T. Johansson and W. Meier. Grain - A Stream Cipher for Constrained Environments. Available via `www.ecrypt.eu.org`.

17. J. Kelsey, B. Schneier, and N. Ferguson. Yarrow-160: Notes on the Design and Analysis of the Yarrow Cryptographic Pseudorandom Number Generator. In H. Heyes and C. Adams, editors, *Proceedings of SAC 1999*, LNCS, volume 1758, pages 13–33, Springer-Verlag, 1999.

18. C. Lim and T. Korkishko. mCrypton - A Lightweight Block Cipher for Security of Low-cost RFID Tags and Sensors. In J. Song, T. Kwon, and M. Yung, editors, *Workshop on Information Security Applications - WISA'05*, LNCS, volume 3786, pages 243-258, Springer-Verlag, 2005.

19. M. Matsui. First Experimental Cryptanalysis of the Data Encryption Standard. In Y. Desmedt, editor, *Proceedings of Crypto '94*, LNCS, volume 839, pages 1–11, Springer-Verlag, 1994.

20. National Institute of Standards and Technology. FIPS 180-2: Secure Hash Standard, August 2002. Available via `csrc.nist.gov`.

21. National Institute of Standards and Technology. FIPS 197: Advanced Encryption Standard, November 2001. Available via `csrc.nist.gov`.

22. G. Poupard and J. Stern. Secuity Analysis of a Practical "on the fly" Authentication and Signature Generation. In K. Nyberg, editor, *Proceedings of Eurocrypt '98*, LNCS, volume 1403, pages 422–436, Springer-Verlag, 1998.

23. R.L. Rivest. RFC 1321: The MD5 Message-Digest Algorithm, April 1992. Available via `www.ietf.org/rfc/rfc1321.txt`.

24. F.-X. Standaert, G. Piret, N. Gershenfeld, and J.-J. Quisquater. SEA: A Scalable Encryption Algorithm for Small Embedded Applications. In J. Domingo-Ferrer, J. Posegga, and D. Schreckling, editors, *Smart Card Research and Applications, Proceedings of CARDIS 2006*, LNCS, volume 3928, pages 222–236, Springer-Verlag.

25. D. Wheeler and R. Needham. TEA, a Tiny Encryption Algorithm. In B. Preneel, editor, *Proceedings of FSE 1994*, LNCS, volume 1008, pages 363–366, Springer-Verlag, 1994.

26. D. Wheeler and R. Needham. TEA extensions. October, 1997. (Also Correction to XTEA. October, 1998.) Available via `www.ftp.cl.cam.ac.uk/ftp/users/djw3/`.

On Pairing-Based Cryptosystems

Tatsuaki Okamoto

NTT Laboratories, Nippon Telegraph and Telephone Corporation
1-1 Hikarino-oka, Yokosuka, 239-0847 Japan
okamoto.tatsuaki@lab.ntt.co.jp

Abstract. The pairing technique that uses the (Weil and Tate) pairings over elliptic (or hyperelliptic) curves represents a great breakthrough in cryptography. This paper surveys this new trend in cryptography, and emphasizes the design of efficient cryptographic primitives that are provably secure in the standard model (i.e., without the random oracle model).

1 Introduction

Elliptic curves have been applied to practical cryptographic designs for two decades. The advantage of elliptic curve based cryptosystems, ECC, over other public-key cryptosystems is their short key size, high processing throughput, and low bandwidth. For example, the typical key size of ECC that guarantees the security comparable to that of 1024 bit key size with the RSA cryptosystems is considered to be just 160 bits. Therefore, several of the most efficient public-key encryption schemes and digital signatures are ECC such as EC-ElGamal (the elliptic curve version of ElGamal) and EC-DSA.

The reason why ECC has such short key lengths is that the *index calculus* technique is considered to be ineffective for computing the discrete logarithm (DL) of the elliptic curve group over finite fields, while it can effectively compute integer factoring and DL of the multiplicative group of a finite field.

However, the mathematical features that are specific to elliptic curve groups compared with multiplicative groups are not only the inapplicability of the index calculus. The most characteristic property of an elliptic curve group is its group structure, which is isomorphic to the product of *two cyclic groups*.

The pairing on an elliptic curve to be introduced in this paper employs this group structure, and is specific to elliptic curve groups (and the generalizations such as hyperelliptic curve groups). In this sense, two decades after we started applying elliptic curves to cryptography, we have finally reached the application of the pairing to cryptographic design, the most essential and natural application of elliptic curves to cryptography.

2 Elliptic Curve Cryptosystems

The application of elliptic curves to cryptography uses elliptic curves defined over finite fields.

P.Q. Nguyen (Ed.): VIETCRYPT 2006, LNCS 4341, pp. 50–66, 2006.

We now introduce some notations. $E(\mathbb{F}_q)$ is a set of $\mathbb{F}_q$-rational points of elliptic curve E over finite field $\mathbb{F}_q$. That is, $E(\mathbb{F}_q)$ is a set of points satisfying $y^2 = x^3 + ax + b$ (other equations are used for finite field $\mathbb{F}_q$ with characteristic 2 and 3) and the special point $\mathcal{O}$ called the infinity point.

A group operation is defined over $E(\mathbb{F}_q)$ and $\mathcal{O}$ is the identity. We now express the group operation by $+$. The discrete logarithm (DL) problem of $E(\mathbb{F}_q)$ is to compute $x \in \mathbb{N}$, given (G, Y), where G is a base point of $E(\mathbb{F}_q)$ and $Y = xG$, which is $G + \cdots + G$ (G is added x times). (After Section 6, we will use the multiplicative form for the group operations in place of the conventional additive form here.)

Elliptic curve cryptosystems (ECC) are constructed on the group of $E(\mathbb{F}_q)$. The security of ECC depends on the difficulty of computing the DL problem of $E(\mathbb{F}_q)$. An ECC scheme can be designed in a manner similar to that of a scheme based on the multiplicative DL problem. For example, EC-DH, EC-ElGamal and EC-DSA are constructed over $E(\mathbb{F}_q)$ in a manner analogous to that of DH, ElGamal and DSA.

Cryptosystems based on pairing (of elliptic curves) are a class of elliptic curve cryptosystems, but have very different features than the conventional ECC.

3   Pairings

The Weil pairing is defined over elliptic curves as follows: Let $E/\mathbb{F}_q$ be an elliptic curve defined over $\mathbb{F}_q$ and m be an integer coprime to q. Let $E[m]$ be the set of m torsion points of $E/\mathbb{F}_q$ (i.e., $E[m] = \{P \mid P \in \overline{\mathbb{F}}_q \wedge mP = \mathcal{O}\}$). $E[m]$ is isomorphic to $\mathbb{Z}/m\mathbb{Z} \times \mathbb{Z}/m\mathbb{Z}$. The Weil pairing, $e_m(P, Q) \in \overline{\mathbb{F}}_q^*$, is defined for two points, P and Q, in $E[m]$, and has the following properties:

(1) For any $P, Q \in E[m]$, $(e_m(P, Q))^m = 1$.
(2) For all $P \in E[m]$, $e_m(P, P) = 1$.
(3) Bilinear: for any $P, Q, P_1, P_2, Q_1, Q_2 \in E[m]$,

$$e_m(P_1 + P_2, Q) = e_m(P_1, Q)e_m(P_2, Q),$$

$$e_m(P, Q_1 + Q_2) = e_m(P, Q_1)e_m(P, Q_2).$$

(4) Alternating: for any $P, Q \in E[m]$, $e_m(P, Q) = e_m(Q, P)^{-1}$.
(5) Non-degenerate: if $e_m(P, Q) = 1$ for any $P \in E[m]$, then $Q = \mathcal{O}$.

That is, $e_m(P, Q)$ bilinearly maps two points, P and Q, in $E[m]$ to an m-th root of unity in $\overline{\mathbb{F}}_q^*$.

Note that there exists an extension field, $\mathbb{F}_{q^k}$, such that $E(\mathbb{F}_{q^k})$ includes $E[m]$. Then $e_m(P, Q)$ is an m-th root of unity in $\mathbb{F}_{q^k}^*$.

The Weil pairing can be efficiently calculated by Miller's algorithm. The Tate pairing also has similar properties, and is often employed in cryptographic applications, since it is faster to compute a Tate pairing than a Weil pairing in typical implementations.

Historically, pairing was first used to attack elliptic curve cryptosystems on supersingular curves in the early 1990's [34] (the attack is often called the MOV reduction). However, in the recent application of pairings to cryptography, they are used not for a negative purposes (i.e., attacking cryptographic schemes) but for positive purposes (i.e., designing cryptographic schemes).

4   Curves Suitable for Pairings

When we apply the Weil/Tate pairing to a general elliptic curve, we have to use an extension field $\mathbb{F}_{q^k}$ with huge extension degree k (in general k is exponentially large in $|q|$) [2]. One of the most suitable curves for the application of Weil/Tate pairing to cryptography is *supersingular curves*, since the extension degree is at most 6 for supersingular curves.

In some applications, however, a generic curve (not a supersingular curve) may be more suitable, since a generic curve offers more freedom in selecting the extension degree and characteristics of the underlying finite field. It is not however so easy to find a generic curve suitable for pairings. Some methods have been proposed to efficiently select a generic curve that has a low extension degree applicable to pairings (e.g., MNT curves) [33,36].

The security of a group with pairing over an elliptic curve should be investigated from the following two viewpoints:

1. The intractability of the discrete logarithm over the elliptic curve (ECDL) with the ground finite field, $\mathbb{F}_q$ (i.e., ECDL over $E(\mathbb{F}_q)$).
2. The intractability of the discrete logarithm over the multiplicative group (DL) of the extension finite field, $\mathbb{F}_{q^k}$ (i.e., DL over $\mathbb{F}_{q^k}^*$).

To guarantee a certain level of security, these two conditions of security should clear the level simultaneously. If we use a *supersingular* elliptic curve with extension degree $k = 6$, the characteristic is restricted to be only 3. We should then consider the security of ECDL over $E(\mathbb{F}_{3^n})$ and DL over $\mathbb{F}_{3^{6n}}^*$. Note that the best algorithm for solving the DL of $\mathbb{F}_{3^{6n}}^*$ is the function field sieve method whose running time is comparable to that of the special number field sieve method and is faster than that of the general number field sieve method. So, when $n > 100$, the security condition of DL over $\mathbb{F}_{3^{6n}}^*$ dominates over that of ECDL over $E(\mathbb{F}_{3^n})$.

Roughly speaking, almost all generic curves are suitable for (conventional) ECC (under some conditions), but are not suitable for pairing-based cryptography. Only a limited class of curves such as supersingular and MNT curves are suitable for pairing-based cryptography, but are not suitable for ECC.

5   Symmetric Pairings

There is another merit of supersingular curves when employing pairings for cryptography. That is, a supersingular curve has (efficiently computable) isomorphism, ϕ, called the *distortion map*.

Let E be a supersingular curve over $\mathbb{F}_q$ and the order of point $G_1 \in E(\mathbb{F}_q)$ be m. Then, there exists an extension degree $k(\leq 6)$ and $G_2 \in E(\mathbb{F}_{q^k})$ such that $E[m] \cong \langle G_1 \rangle \times \langle G_2 \rangle$, and ϕ is the isomorphism from $\langle G_1 \rangle$ to $\langle G_2 \rangle$, where $G_2 = \phi(G_1)$. We can then define a variant of the Weil pairing $\hat{e}_m$ over two points, P and Q, in $E(\mathbb{F}_q)$ as follows:

$$\hat{e}_m(P, Q) = e_m(P, \phi(Q)) \in \mathbb{F}_{q^k}^*.$$

Here note that $\hat{e}_m(P, P) \neq 1$ and $\hat{e}_m(P, Q) = \hat{e}_m(Q, P)$, while $e_m(P, P) = 1$ and $e_m(P, Q) = e_m(Q, P)^{-1}$. So, this variant of Weil pairing $\hat{e}_m$ is called a *symmetric* pairing, while the original Weil pairing e_m is called an *asymmetric* pairing.

The advantage of this Weil pairing variant $\hat{e}_m : \langle G_1 \rangle \times \langle G_1 \rangle \to \mathbb{F}_{q^k}^*$ is that it is defined over two points in $E(\mathbb{F}_q)$ (two elements in $\langle G_1 \rangle$), while e_m is defined over a point in $E(\mathbb{F}_q)$ and another point in $E(\mathbb{F}_{q^k})$. (For example, if the size of an element of $\mathbb{F}_q$ is 300 bits and extension degree k is 6, then the size of an element of $\mathbb{F}_{q^k}$ and the size of $\hat{e}_m(P, Q)$ and $e_m(P, Q')$ are 1800 bits.)

6   Pairing-Based Cryptography

6.1   Pairing Group

Hereafter, we will use only *symmetric* pairings (not asymmetric pairings) as pairings, but almost all schemes that we will introduce in this paper can be also realized with asymmetric pairings.

For simplicity of description, we express the symmetric pairing $\hat{e}_m : \langle G_1 \rangle \times \langle G_1 \rangle \to \mathbb{F}_{q^k}^*$ by bilinear map e from a multiplicative group, $\mathbb{G}$, to another multiplicative group, $\mathbb{G}_T$, i.e., $e : \mathbb{G} \times \mathbb{G} \to \mathbb{G}_T$ such that:

1. $\mathbb{G}$ is a cyclic group of prime order p,
2. g is a generator of $\mathbb{G}$,
3. e is a non-degenerate bilinear map $e : \mathbb{G} \times \mathbb{G} \to \mathbb{G}_T$, where $|\mathbb{G}| = |\mathbb{G}_T| = p$, and
4. e and the group action in $\mathbb{G}$ and $\mathbb{G}_T$ can be computed efficiently.

6.2   Brief Overview of Pairing-Based Cryptography

Around 2000, application of the pairings to cryptography was initiated by Verheul [41], Joux [31], and Sakai, Ohgishi and Kasahara [39]. Verheul introduced the above-mentioned symmetric pairings, and Joux proposed a key distribution system among three parties (three party version of the Diffie-Hellman key distribution) by using the symmetric pairings. Sakai, Ohgishi and Kasahara solved the problem on how to efficiently construct the identity-based encryption (IBE) that had been open since 1984 when Shamir proposed the concept of IBE.

Following these pioneer works, Boneh and others drastically exploited the possibility of applying pairings to cryptography. Boneh and Franklin [12] formalized the security of IBE as the IND-ID-CCA2 (indistinguishable against adaptively chosen-ciphertext attacks under chosen identity attacks) security and proposed an IND-ID-CCA2 secure IBE scheme in the random oracle model [5]. Boneh, Lynn and Shacham [14] proposed a new signature scheme whose signatures are short. The security proof is also based on the random oracle model.

Then, an enormous number of papers on pairing-based cryptography have been published for the last several years, and they cover very broad areas of cryptography [3].

One of the most interesting applications of pairings to cryptography is to construct practical encryption/signature schemes that are proven to be secure in the standard model (without the random oracle model). Previously only a few such schemes (e.g., Cramer-Shoup schemes [23,24,25]) were proposed.

Interestingly IBE plays a key role of constructing practical secure schemes in the standard model. That is, a secure IBE scheme in the standard model can be used to construct secure public-key encryption/signature schemes in the standard model. (In addition, hierarchical IBE (HIBE) [29] is used to construct forward-secure public-key encryption schemes and CCA2 secure IBE schemes [21,22].)

Hereafter we will introduce how pairings are applied to constructing secure IBE/encryption/signature schemes in the standard model.

7   Computational Assumptions

Let $\mathbb{G}$ be a bilinear group of prime order p and g be a generator of $\mathbb{G}$. Here we review several computational assumptions on the bilinear maps, which are assumed in the pairing-based cryptographic schemes to be introduced in this paper.

7.1   Bilinear Diffie-Hellman Assumption

The Bilinear Diffie-Hellman (BDH) problem [12,31] in $\mathbb{G}$ is as follows: given a tuple $g, g^a, g^b, g^c \in \mathbb{G}$ as input, output $e(g, g)^{abc} \in \mathbb{G}_T$. The advantage of adversary $\mathcal{A}$ for the BDH problem is

$$\Pr[\mathcal{A}(g, g^a, g^b, g^c) = e(g, g)^{abc}].$$

Similarly, the advantage of adversary $\mathcal{B}$ for the Decisional BDH (DBDH) problem is

$$| \Pr[\mathcal{B}(g, g^a, g^b, g^c, e(g, g)^{abc}) = 0] - \Pr[\mathcal{B}(g, g^a, g^b, g^c, T) = 0] |,$$

where T is randomly selected from $\mathbb{G}_T$.

Definition 1. *We say that the (Decisional) BDH assumption holds in $\mathbb{G}$ if any probabilistic polynomial time adversary has negligible advantage for the (Decisional) BDH problem.*

7.2 Bilinear Diffie-Hellman Inversion Assumption

The q Bilinear Diffie-Hellman Inversion (q-BDHI) problem [8] is defined as follows: given the $(q + 1)$-tuple $(g, g^x, g^{x^2}, \ldots, g^{x^q}) \in (\mathbb{G})^{q+1}$ as input, compute $e(g, g)^{1/x} \in \mathbb{G}_T$. The advantage of an adversary $\mathcal{A}$ for q-BDHI is

$$\Pr[\mathcal{A}(g, g^x, g^{x^2}, \ldots, g^{x^q}) = e(g, g)^{1/x}].$$

Similarly, the advantage of adversary $\mathcal{B}$ for the Decisional q-BDHI (q-DBDHI) problem is

$$|\Pr[\mathcal{B}(g, g^x, g^{x^2}, \ldots, g^{x^q}, e(g, g)^{1/x}) = 0] - \Pr[\mathcal{B}(g, g^x, g^{x^2}, \ldots, g^{x^q}, T) = 0]|,$$

where T is randomly selected from $\mathbb{G}_T$.

Definition 2. *We say that the (Decisional) q-BDHI assumption holds in $\mathbb{G}$ if any probabilistic polynomial time adversary has negligible advantage for the (Decisional) q-BDHI problem.*

It is not known if the q-BDHI assumption, for $q > 1$, is equivalent to BDH.

In this paper, we often drop the q and refer to the (Decisional) BDHI assumption.

7.3 Strong Diffie-Hellman Assumption

The q Strong Diffie-Hellman (q-SDH) problem [9] is defined as follows: given the $(q + 1)$-tuple $(g, g^x, g^{x^2}, \ldots, g^{x^q}) \in (\mathbb{G})^{q+1}$ as input, compute $(g^{1/(x+c)}, c) \in \mathbb{G} \times \mathbb{N}$. The advantage of an adversary $\mathcal{A}$ for q-SDH is

$$\Pr[\mathcal{A}(g, g^x, g^{x^2}, \ldots, g^{x^q}) = (g^{1/(x+c)}, c)].$$

Definition 3. *We say that the q-SDH assumption holds in $\mathbb{G}$ if any probabilistic polynomial time adversary has negligible advantage for the q-SDH problem.*

In this paper, similarly to the BDHI assumption, we often drop the q and refer to the SDH assumption.

Remark: Cheon shows some weakness of the q-SDH problem [18], and the security parameter of the problem should be a bit longer to avoid Cheon's attack.

8   Identity-Based Encryption (IBE)

Identity-based encryption (IBE) [40] is a variant of public-key encryption (PKE), where the identity of a user is employed in place of the user's public-key. In this concept,

Setup: A trusted party (authority) generates a pair of secret-key x (master secret key) and public-key y (system parameter).

Extract: The trusted party also generates A's secret decryption key, s_A, from the identity of A and securely sends s_A to A.

Encrypt: When B encrypts a message m to A, B utilizes A's identity, ID_A (in place of A's public-key in PKE). Let c_A be a ciphertext of m encrypted by ID_A.

Decrypt: A can decrypt ciphertext c_A by using A's decryption key s_A.

Although IBE itself is a very useful primitive in cryptography, here we will review IBE as a building block of designing practical secure PKE/signature schemes in the standard model.

8.1 Security of IBE

Boneh and Franklin [12] define the security, IND-ID-CCA2 (indistinguishable against adaptively chosen-ciphertext attacks under chosen identity attacks), for IBE systems. We now informally introduce the definition as follows:

Definition 4. *(Security of IBE: IND-ID-CCA2) Let us consider the following experiment between an adversary, $\mathcal{A}$, and the challenger, $\mathcal{C}$.*

1. *First, $\mathcal{C}$ generates a system parameter of IBE and sends it to $\mathcal{A}$.*
2. *$\mathcal{A}$ is allowed to ask two types of queries, extraction queries and decryption queries, to $\mathcal{C}$. Here, an extraction query is an identity, ID_i, to which $\mathcal{C}$ replies the corresponding decryption key, d_i, and a decryption query is a ciphertext, c_j, along with an identity, ID_j, to which $\mathcal{C}$ replies with the corresponding plaintext, m_j.*
3. *$\mathcal{A}$ is also allowed to adaptively choose an identity, ID^*, and two messages, m_0 and m_1, that $\mathcal{C}$ wishes to attack, then $\mathcal{C}$ replies with a ciphertext, c^*, of m_b (b is randomly chosen from $\{0,1\}$) with respect to identity ID^*.*
4. *Finally $\mathcal{A}$ outputs a bit, b^*. Let Advantage be $|2\Pr[b = b^*] - 1|$.*

An IBE scheme is IND-ID-CCA2 if, for any probabilistic polynomial-time $\mathcal{A}$, Advantage is negligibly small.

In the above-mentioned definition of IND-ID-CCA2, $\mathcal{A}$ is allowed to adaptively choose the challenge identity, ID^*, that it wishes to attack.

Canetti, Halevi, and Katz [21,22] define a weaker notion of security in which the adversary $\mathcal{A}$ commits ahead of time to the challenge identity ID^* it will attack. We refer to this notion as *selective identity* adaptively chosen-ciphertext secure IBE (IND-sID-CCA2). In addition, they also define a weaker security notion of IBE, *selective-identity* chosen-plaintext secure IBE (IND-sID-CPA).

8.2 Boneh-Franklin IBE Scheme

The Boneh-Franklin IBE scheme [12] is proven to be secure in the random oracle model (not in the standard model). We now introduce this scheme as a typical example of pairing-based secure cryptosystems in the random oracle model (and as a bench mark to evaluate the efficiency of secure IBE schemes in the standard model).

Setup: Given $(\mathbb{G}, \mathbb{G}_T, p, k \ (k = |p|)$, a trusted party randomly selects a generator g in $\mathbb{G}$ as well as four hash functions, $H_1, \ldots, H_4$. The trusted party also randomly selects $x \in (\mathbb{Z}/p\mathbb{Z})^*$, and computes $y = g^x$. The system parameter is $(g, y, H_1, \ldots, H_4)$ and the (secret) master key is x.

Extract: Given ID_A of user A, ID_A is mapped (through H_1) to an element of $\mathbb{G}$, h_A, and A's secret key, $s_A = h_A^x$ is computed.

Encrypt: To encrypt a message $m \in \{0,1\}^k$ under ID_A, randomly select $\sigma \in \{0,1\}^k$, and compute

$$C = (g^r, \sigma \oplus H_2(e(h_A, y)^r), m \oplus H_4(\sigma)),$$

where $r = H_3(\sigma, m)$.

Decrypt: Let $C = (C_1, C_2, C_3)$ be a ciphertext encrypted using ID_A. To decrypt C, compute

$$\sigma = C_2 \oplus H_2(e(s_A, C_1)), \quad \text{and} \quad m = C_3 \oplus H_4(\sigma).$$

Set $r = H_3(\sigma, m)$ and check whether $C_1 = g^r$ holds. If not, rejects the decryption. Otherwise, output m.

Security: The Boneh-Franklin IBE scheme is IND-ID-CCA2 in the random oracle model (i.e., assuming $H_1, \ldots, H_4$ are truly random functions) under the BDH assumption.

Remark: The Sakai-Kasahara IBE scheme [38], whose security is proven in the random oracle model [17], is more practical than the Boneh-Franklin IBE scheme, since mapping of a bit string to an element of $\mathbb{G}$ is not necessary in the Sakai-Kasahara IBE scheme.

8.3 Boneh-Boyen IBE Scheme

There are three Boneh-Boyen IBE schemes that are secure in the standard model (two are in [8] and one is in [10]).

One of the two schemes in [8] is IND-sID-CPA secure, and the other is IND-sID-CCA2 secure. The IND-sID-CCA2 secure scheme [8] is constructed by converting from the IND-sID-CPA secure basic scheme through the conversion technique of [22]. The scheme in [10] is fully secure (IND-ID-CCA2 secure) (through the conversion technique of [22]).

The IND-sID-CPA secure scheme in [8] is much more efficient than the others. Since an IND-sID-CPA secure IBE scheme is sufficient as a building block to construct an IND-CCA2 PKE (Section 9.1), we now introduce the IND-sID-CPA secure IBE scheme in [8] as follows (another reason why we introduce this scheme is that it is closely related to the Boneh-Boyen signature scheme [9] in Section 10.1):

Setup: Given $(\mathbb{G}, \mathbb{G}_T, p, k)$ $(k = |p|)$, a trusted party randomly selects a generator g in $\mathbb{G}$ and $x, y \in (\mathbb{Z}/p\mathbb{Z})^*$, and computes $X = g^x$ and $Y = g^y$. The system parameter is (g, X, Y) and the (secret) master key is (x, y).

Extract: Given $v \in (\mathbb{Z}/p\mathbb{Z})^*$ as ID_A of user A, pick a random $r \in \mathbb{Z}/p\mathbb{Z}$, compute $K = g^{1/(v+x+ry)} \in \mathbb{G}$, and set A's secret key $d_A = (r, K)$.

Encrypt: To encrypt a message $m \in \mathbb{G}_T$ under ID_A (i.e., v), pick a random $s \in \mathbb{Z}/p\mathbb{Z}$ and output the ciphertext

$$C = (g^{sv}X^s, Y^s, e(g,g)^s m).$$

Decrypt: Let $C = (C_1, C_2, C_3)$ be a ciphertext encrypted using ID_A. To decrypt C using $d_A = (r, K)$, compute

$$\frac{C_3}{e(C_1 C_2^r, K)},$$

which is m when C is valid.

For a valid ciphertext we have

$$\frac{C_3}{e(C_1 C_2^r, K)} = \frac{C_3}{e(g^{sv}X^s Y^{sr}, g^{1/(v+x+ry)})}$$
$$= \frac{C_3}{e(g^{s(v+x+ry)}, g^{1/(v+x+ry)})} = \frac{C_3}{e(g,g)^s} = m.$$

Security: The above-mentioned Boneh-Boyen IBE scheme is IND-sID-CPA (selective-identity chosen-plaintext secure) under the Decisional BDHI (DBDHI) assumption.

8.4   Waters IBE Scheme

The Waters IBE scheme [42] is an efficient IND-ID-CCA2 secure IBE in the standard model. Similarly to the Boneh-Boyen IBE scheme [10], the Waters IBE scheme is converted from the IND-ID-CPA secure basic scheme (the *Waters basic IBE scheme*) through the conversion technique of [22].

Efficient secure (IND-CCA2) PKE and secure (EUF-CMA) signatures in the standard model are constructed from the Waters basic IBE scheme (Sections 9.2 and 10.2). The Waters basic IBE scheme is as follows:

Setup: Given $(\mathbb{G}, \mathbb{G}_T, p, k)$ ($k = |p|$), a trusted party randomly selects generators, g and g_2, in $\mathbb{G}$ and $\alpha \in \mathbb{Z}/p\mathbb{Z}$, and computes $g_1 = g^\alpha$. Additionally the party randomly selects $u' \in \mathbb{G}$ and k-length vector $(u_1, \ldots, u_k) \in \mathbb{G}^k$, The public parameter is $(g, g_1, g_2, u', u_1, \ldots, u_k)$. The master secret key is g_2^α.

Extract: Let v be an k bit string representing an identity ID_A of user A, v_i denote the i-th bit of v, and $\mathcal{V} \subseteq \{1, \ldots, k\}$ be the set of all i for which $v_i = 1$. A's secret key, d_A, for identity v is generated as follows. First, a random $r \in \mathbb{Z}/p\mathbb{Z}$ is chosen. Then the secret key is constructed as:

$$d_A = (g_2^\alpha (u' \prod_{j \in \mathcal{V}} u_j)^r, g^r).$$

Encrypt: To encrypt a message $m \in \mathbb{G}_T$ under ID_A (i.e., v), pick a random $s \in (\mathbb{Z}/p\mathbb{Z})^*$ and output the ciphertext

$$C = ((u' \prod_{j \in \mathcal{V}} u_j)^s, g^s, e(g_1, g_2)^s m).$$

Decrypt: Let $C = (C_1, C_2, C_3)$ be a ciphertext encrypted using ID_A (i.e., v). To decrypt C using $d_A = (d_1, d_2)$, compute

$$C_3 \frac{e(d_2, C_1)}{e(d_1, C_2)}$$

which is m when C is valid.

For a valid ciphertext we have

$$C_3 \frac{e(d_2, C_1)}{e(d_1, C_2)} = (e(g, g)^s m) \frac{e(g^r, (u' \prod_{j \in \mathcal{V}} u_j)^s)}{e(g_2^\alpha (u' \prod_{j \in \mathcal{V}} u_j)^r, g^s)}$$

$$= (e(g, g)^s m) \frac{e(g, (u' \prod_{j \in \mathcal{V}} u_j))^{rs}}{e(g_1, g_2)^s e((u' \prod_{j \in \mathcal{V}} u_j), g)^{rs}} = m.$$

Security: The Waters basic IBE scheme is IND-ID-CPA under the Decisional BDH (DBDH) assumption.

Remark: Among IND-ID-CCA2 secure IBE schemes in the standard model, the Gentry IBE scheme [28] and the Kiltz IBE scheme [32] are more efficient than the Waters scheme.

9 Public-Key Encryption

The desirable security of a public-key encryption (PKE) scheme is formulated as semantic security against adaptively chosen-ciphertext attacks (IND-CCA2) [4]. Although there are several ways to construct practical IND-CCA2 secure PKE schemes in the *random oracle model* [5], only a few practical schemes such as the Cramer-Shoup PKE scheme [23,25] were proven to be secure in the *standard model*.

Pairings are exploiting a new methodology to design a practical IND-CCA2 secure PKE schemes in the standard model. The new methodology uses transformation from an IBE scheme to a PKE scheme.

9.1 Canetti-Halevi-Katz Construction

Canetti, Halevi and Katz [22] have shown how to construct an IND-CCA2 secure PKE scheme from any IND-sID-CPA secure IBE scheme. In the construction, a one-time signature scheme is also employed. Since this construction is efficient, we can construct an efficient IND-CCA2 secure PKE scheme in the standard model using the Boneh-Boyen IBE scheme [8].

We now show the construction of a PKE scheme as follows:

Key Generation: Run the setup process of IBE to obtain a pair of system parameter and master key. The public key, PK, is the system parameter and the secret key, SK, is the master key.

Encrypt: To encrypt message m using public key PK (IBE's system parameter), the sender first generates a pair of verification key vk and signing key sk of a one-time signature scheme. The sender then computes IBE's ciphertext C of message m with respect to identity vk, and signature σ of C by using signing key sk. The ciphertext is (vk, C, σ).

Decrypt: To decrypt ciphertext (vk, C, σ) using secret key SK (IBE's master key), the receiver first checks whether σ is a valid signature of C with respect verification key vk. If not, the receiver outputs $\bot$. Otherwise, the receiver computes IBE's decryption key d_{vk} for identity vk, and output m decrypted from C by d_{vk}.

Security: If the underlying IBE scheme is IND-sID-CPA and the one-time signature scheme is strongly unforgeable (see [9] for the definition of strong unforgeability) then the Canetti-Halevi-Katz construction of PKE is IND-CCA2.

If the underlying one-time signature scheme is efficient, the Canetti-Halevi-Katz PKE scheme from the Boneh-Boyen IBE scheme [8] is relatively as efficient as (but less efficient than) Cramer-Shoup. The major advantage of this construction over Cramer-Shoup is that the validity of a ciphertext can be verified publicly, while a ciphertext should be verified secretly (i.e., the verification requires the secret key) in Cramer-Shoup. This property is useful in constructing a threshold PKE scheme like [11].

Boneh and Katz [13] improved the Canetti-Halevi-Katz construction by using a message authentication code in place of a one-time signature. The Boneh-Katz construction however is not publicly verifiable.

9.2   Boyen-Mei-Waters PKE Scheme

Boyen, Mei and Waters [15] presented a new way (inspired by [22]) of constructing IND-CCA2 secure PKE schemes in the standard model. Their construction is based on two efficient IBE schemes, the Boneh-Boyen and Waters basic IBE schemes. Unlike the Canetti-Halevi-Katz and Boneh-Katz constructions that use IBE as a black box, the Boyen-Mei-Waters construction directly uses the underlying IBE structure, and requires no cryptographic primitive other than the IBE scheme itself. In addition, the validity of ciphertexts can be checked publicly.

We now introduce the Boyen-Mei-Waters PKE scheme based on the Waters basic IBE scheme.

Key Generation: A user's public/private key pair generation algorithm proceeds as follows. Given $(\mathbb{G}, \mathbb{G}_T, p, k)$ $(k = |p|)$, randomly select a generator g in $\mathbb{G}$ and $\alpha \in \mathbb{Z}/p\mathbb{Z}$, and computes $g_1 = g^{\alpha}$ and $z = e(g, g_1) = e(g, g)^{\alpha}$. Next, choose a random $y' \in \mathbb{Z}/p\mathbb{Z}$ and a random k-length vector $(y_1, \ldots, y_n)$, whose elements are chosen at random from $\mathbb{Z}/p\mathbb{Z}$. Then calculate $u' = g^{y'}$ and $u_i = g^{y_i}$ for $i = 1$ to k. Finally, a random seed δ for a collision resistant hash function family H is chosen. The published public key is

$$(z = e(g, g_1), u' = g^{y'}, u_1 = g^{y_1}, \ldots, u_k = g^{y_k}, \delta),$$

and the private key is

$$(g_1 = g^\alpha, y', y_1, \ldots, y_k).$$

Encrypt: A message $m \in \mathbb{G}_T$ is encrypted as follows. First, a value $s \in \mathbb{Z}/p\mathbb{Z}$ is randomly selected. Then compute $C_2 = g^s$ and $C_3 = z^s m = e(g, g_1)^s m = e(g, g)^{\alpha s} m$. Next, compute $w = H_\delta(C_2, C_3)$ and $w_1 w_2 \ldots w_k$ denote the binary expansion of w, where each bit $w_i \in \{0, 1\}$. Let $\mathcal{W} \subseteq \{1, \ldots, k\}$ be the set of all i for which $w_i = 1$. Finally compute $C_1 = (u' \prod_{i=1}^k u_i^{w_i})^s$. The ciphertext is

$$C = (C_1, C_2, C_3) = ((u' \prod_{j \in \mathcal{W}} u_j)^s, g^s, e(g, g_1)^s m).$$

Decrypt: Given ciphertext $C = (C_1, C_2, C_3)$, first compute $w = H_\delta(C_2, C_3)$, expressed in binary as $w_1 w_2 \ldots w_k$. Next, compute $w' = y' + \sum_{i=1}^k y_i w_i \bmod p$, and check whether $(C_2)^{w'} = C_1$. If not, output $\perp$. Otherwise, the ciphertext is valid, and decrypt the message as

$$\frac{C_3}{e(C_2, g_1)} = m.$$

Although the Boyen-Mei-Waters PKE scheme is less efficient than the Cramer-Shoup PKE scheme and the variants, the validity of a ciphertext is *publicly* verifiable in the Boyen-Mei-Waters PKE scheme, while it is *privately* verifiable in the Cramer-Shoup PKE scheme and the variants. Here, in the Boyen-Mei-Waters PKE scheme, the check of $(C_2)^{w'} = C_1$ using private information w' can be replaced by the equivalent check with using the pairing and public information. Due to the public verifiability, an efficient threshold PKE scheme in the standard model can be constructed on this PKE scheme [11].

Security: Let H be a collision resistant hash function family. Then the Boyen-Mei-Waters PKE scheme is IND-CCA2 under the Decisional BDH (DBDH) assumption.

10   Digital Signatures

The current status on designing secure digital signatures in the standard model is fairly similar to that on designing secure PKE schemes in the standard model.

The desirable security of a digital signature scheme is formulated as existential unforgeability against adaptively chosen-message attacks (EUF-CMA) [30]. Although there are several ways to construct practical EUF-CMA secure signature schemes in the random oracle models [6,7,14], only a few practical schemes were proven to be secure in the standard model (the Cramer-Shoup signature scheme etc. [19,24,27]).

Similarly to PKE, pairings are exploiting a new methodology to design practical EUF-CMA secure signature schemes in the standard model. There are two ways in the new methodology; one is to directly design (and prove the security of) a signature scheme from pairings (the Boneh-Boyen signature scheme etc. [9,37,43]), and the other is to convert an IND-ID-CPA secure IBE scheme to a signature scheme (e.g., the Waters signature scheme [42]).

The Boneh-Boyen signature scheme may be considered to be converted from the Boneh-Boyen IBE scheme [8] in Section 8.3, but it is a bit different from the case of the Waters signature scheme. Since the Waters basic IBE scheme is IND-ID-CPA, the converted signature scheme is EUF-CMA under the same assumption as that for the IBE scheme. On the other hand, since the Boneh-Boyen IBE scheme is IND-sID-CPA, the converted signature scheme is not guaranteed to be EUF-CMA under the same assumption. Actually, the assumption (SDH) for the Boneh-Boyen signature scheme is different from that (DBDHI) for the Boneh-Boyen IBE scheme.

10.1 Boneh-Boyen Signature Scheme

Boneh and Boyen presented a very practical signature scheme that is EUF-CMA secure in the standard model. Signatures in their scheme are much shorter and simpler than the previous secure signature schemes in the standard model.

The Boneh-Boyen signature scheme [9] is as follows:

Key Generation: Given $(\mathbb{G}, \mathbb{G}_T, p, k)$ $(k = |p|)$, randomly select a generator g in $\mathbb{G}$ and $x, y \in (\mathbb{Z}/p\mathbb{Z})^*$, and computes $u = g^x$ and $v = g^y$. The public key is (g, u, v). The secret key is (x, y).

Sign: Given a secret key (x, y) and a message $m \in (\mathbb{Z}/p\mathbb{Z})^*$, pick a random $r \in (\mathbb{Z}/p\mathbb{Z})^*$ and compute

$$\sigma = g^{1/(x+m+yr)}.$$

Here $1/(x + m + yr)$ is computed modulo p. The signature is (σ, r).

Verify: Given a public key (g, u, v), a message $m \in (\mathbb{Z}/p\mathbb{Z})^*$, and a signature (σ, r), verify that

$$e(\sigma, u g^m v^r) = e(g, g).$$

If the equality holds the result is valid; otherwise the result is invalid.

Security: The Bone-Boyen signature scheme is EUF-CMA under the strong DH (SDH) assumption.

10.2 Waters Signature Scheme

The Waters signature scheme is converted from the Waters basic IBE scheme.

Key Generation: Given $(\mathbb{G}, \mathbb{G}_T, p, k)$ $(k = |p|)$, randomly select generators, g and g_2, in $\mathbb{G}$ and $\alpha \in \mathbb{Z}/p\mathbb{Z}$, and compute $g_1 = g^\alpha$. Randomly select $u' \in \mathbb{G}$ and k-length vector $(u_1, \ldots, u_k) \in \mathbb{G}^k$. The public key is $(g, g_1, g_2, u', u_1, \ldots, u_k)$. The secret key is g_2^α.

Sign: Let m be an k-bit message to be signed and m_i denotes the ith bit of m, and $\mathcal{M} \subseteq \{1, \ldots, k\}$ be the set of i for which $m_i = 1$. A signature of m is generated as follows. First, a random r is chosen. Then the signature is constructed as:

$$\sigma = (g_2^\alpha (u' \prod_{j \in \mathcal{M}} u_j)^r, g^r).$$

Verify: Given a public-key $(g, g_1, g_2, u', u_1, \ldots, u_k)$, a message $m \in \{0, 1\}^k$, and a signature $\sigma = (\sigma_1, \sigma_2)$, check

$$\frac{e(\sigma_1, g)}{e(\sigma_2, u' \prod_{j \in \mathcal{M}} u_j)} = e(g_1, g_2).$$

If it holds, the verification result is valid; otherwise the result is invalid.

Security: The Waters signature scheme is EUF-CMA under the Decisional BDH (DBDH) assumption.

Remark: A signature scheme and its variant [37] are more suitable for many cryptographic protocol applications such as blind signatures.

11   Concluding Remarks

This paper introduced how the pairing technique is used to design efficient IBE/PKE/signatures that are provably secure in the standard model. The methodology of using pairings will be applied to more wide areas of secure cryptosystems and protocols. For example, it is applied to more protocol-oriented primitives like group signatures [1,16], blind signatures [37], threshold PKE [11], verifiable random functions [26] and broadcast encryption.

Acknowledgements

The author would like to thank Phong Nguyen, the Program Committee Chair of VietCrypt, for inviting him to the conference and proceedings. He would also like to thank Goichiro Hanaoka for his valuable comments on the preliminary manuscript.

References

1. Ateniese, G., Camenisch, J., de Medeiros, B. and Hohenberger, S., Practical Group Signatures without Random Oracles, IACR ePrint Archive, 2005/385, http://eprint.iacr.org/2005/385 (2005)
2. Balasubramanian, R. and Koblitz, N., The Improbability that an Elliptic Curve Has Subexponential Discrete Log Problem under the Menezes-Okamoto-Vanstone Algorithm, J. Cryptology, 11, pp.141-145 (1998).

3. Barreto, P., The Pairing-Based Crypto Lounge,
   http://paginas.terra.com.br/informatica/paulobarreto/pblounge.html.
4. Bellare, M., Desai, A., Pointcheval, D. and Rogaway, P., Relations Among Notions of Security for Public-Key Encryption Schemes, Adv. in Cryptology – Crypto 1998, LNCS 1462, Springer-Verlag, pp. 26-45 (1998).
5. Bellare, M. and Rogaway, P., Random Oracles are Practical: a Paradigm for Designing Efficient Protocols, Proceedings of the 1st ACM Conference on Computer and Communications Security, CCS 1993, ACM, pp. 62–73 (1993).
6. Bellare, M. and Rogaway, P., The Exact Security of Digital Signatures: How to Sign with RSA and Rabin, Adv. in Cryptology – Eurocrypt 1996, LNCS 1070, Springer-Verlag, pp. 399-416 (1996).
7. Boldyreva, A., Threshold Signature, Multisignature and Blind Signature Schemes Based on the Gap-Diffie-Hellman-Group Signature Scheme, Proceedings of PKC 2003, LNCS 2567, Springer-Verlag, pp.31-46 (2003).
8. Boneh, D. and Boyen, X., Efficient Selective-ID Secure Identity Based Encryption Without Random Oracles, Adv. in Cryptology – Eurocrypt 2004, LNCS 3027, Springer-Verlag, pp. 223-238 (2004).
9. Boneh, D. and Boyen, X., Short Signatures Without Random Oracles, Adv. in Cryptology – Eurocrypt 2004, LNCS 3027, Springer-Verlag, pp. 56–73 (2004).
10. Boneh, D. and Boyen, X., Secure Identity Based Encryption Without Random Oracles, Adv. In Cryptology – Crypto 2004, LNCS 3152, Springer-Verlag, pp. 443–459 (2004).
11. Boneh, D., Boyen, X. and Halevi, S., Chosen Ciphertext Secure Public Key Threshold Encryption Without Random Oracles, to appear in Proceedings of CT-RSA 2006, Springer-Verlag (2006).
    Available at http://crypto.stanford.edu/ dabo/abstracts/threshold.html.
12. Boneh, D. and Franklin, M., Identity-Based Encryption from the Weil Pairing, Adv. in Cryptology – Crypto 2001, LNCS 2139, Springer-Verlag, pp.213–229 (2001). Journal version in SIAM Journal of Computing, 32(3), pp. 586–615 (2003).
13. Boneh, D. and Katz, J., Improved Efficiency for CCA-Secure Cryptosystems Built Using Identity Based Encryption, Proceedings of CT-RSA 2005, LNCS 3376, Springer-Verlag, pp.87-103 (2005).
14. Boneh, D., Lynn, B. and Shacham, H., Short Signatures from the Weil Pairing, Adv. in Cryptology – Asiacrypt 2001, LNCS 2248, Springer-Verlag, pp.514–532 (2001).
15. Boyen, X., Mei, Q. and Waters, B., Direct Chosen Ciphertext Security from Identity-Based Techniques, Proceedings of the 12th ACM Conference on Computer and Communications Security, CCS 2005, ACM (2005).
    Full version available at http://www.cs.stanford.edu/ xb/ccs05/.
16. Boyen, X. and Waters, B., Compact Group Signatures Without Random Oracles, IACR ePrint Archive, 2005/381, http://eprint.iacr.org/2005/381 (2005)
17. Chen, L. and Cheng, Z., Security Proof of Sakai-Kasahara's Identity-Based Encryption Scheme, IMA International Conference 2005, LNCS 3796, Springer-Verlag, pp.442–459 (2005).
18. Cheon, J. H., Security Analysis of the Strong Diffie-Hellman Problem, Adv. in Cryptology – EUROCRYPT 2006, LNCS 4004, Springer-Verlag, pp.1–11 (2006).
19. Camenisch, J. and Lysyanskaya, A., A Signature Scheme with Efficient Protocols, Security in communication networks, LNCS 2576, Springer-Verlag, pp. 268-289 (2002).

20. Camenisch, J. and Lysyanskaya,A., Signature Schemes and Anonymous Credentials from Bilinear Maps, Adv. In Cryptology – Crypto 2004, LNCS 3152, Springer-Verlag, pp.56–72 (2004)

21. Canetti, R., Halevi, S. and Katz, J., A Forward-Secure Public-Key Encryption Scheme, Adv. in Cryptology – Eurocrypt 2003, LNCS, Springer-Verlag, pp.255-271 (2003).
Full version available at http://eprint.iacr.org/2003/083.

22. Canetti, R., Halevi, S. and Katz, J., Chosen-Ciphertext Security from Identity-Based Encryption, Adv. in Cryptology – Eurocrypt 2004, LNCS 3027, Springer-Verlag, pp. 207-222 (2004).
Full version available at http://eprint.iacr.org/2003/182.

23. Cramer, R. and Shoup, V., A Practical Public Key Cryptosystem Provably Secure Against Chosen Ciphertext Attack, Adv. in Cryptology – Crypto 1998, LNCS 1462, Springer-Verlag, pp. 13-25 (1998).

24. Cramer, R. and Shoup, V., Signature Schemes Based on the Strong RSA Assumption, ACM TISSEC, 3(3), pp.161–185 (2000). Extended abstract in Proc. 6th ACM CCS (1999).

25. Cramer, R. and Shoup, V., Universal Hash Proofs and a Paradigm for Adaptive Chosen Ciphertext Secure Public-Key Encryption, Adv. in Cryptology – Eurocrypt 2002, LNCS 2332, Springer-Verlag, pp. 45-64 (2002).

26. Dodis, Y. and Yampolskiy, A., A Verifiable Random Function with Short Proofs and Keys, Proceedings of PKC 2005, LNCS 3386, Springer-Verlag, pp.416–431 (2005).

27. Fischlin, M., The Cramer-Shoup Strong-RSA Signature Scheme Revisited, Proceedings of PKC 2003, LNCS 2567, Springer-Verlag, pp.116–129 (2003).

28. Gentry, C., Practical Identity-Based Encryption Without Random Oracles, Adv. in Cryptology – Eurocrypt 2006, LNCS 4004, Springer-Verlag, pp. 445-464 (2006).

29. Gentry, C. and Silverberg, A., Hierarchical Identity-Based Cryptography, Adv. in Cryptology – Asiacrypt 2002, LNCS 2501, Springer-Verlag, pp. 548-566 (2002).

30. Goldwasser, S., Micali, S. and Rivest, R., A Digital Signature Scheme Secure against Adaptive Chosen-Message Attacks, SIAM J. Computing 17(2): 281-308 (1988).

31. Joux, A., A One Round Protocol for Tripartite Diffie-Hellman, Proceedings of Algorithmic Number Theory Symposium IV, LNCS 1838, Springer-Verlag, pp.385–394 (2000).
Journal version in Journal of Cryptology, 17(4), pp.263–276 (2004).

32. Kiltz, E., Direct Chosen-Ciphertext Secure Identity-Based Encryption in the Standard Model with Short Ciphertexts, IACR ePrint Archive, 2006/122, http://eprint.iacr.org/2006/122 (2006)

33. Koblitz, N. and Menezes, A., Pairing-Based Cryptography at High Security Levels, IACR ePrint Archive, 2005/076, http://eprint.iacr.org/2005/076 (2005)

34. Menezes, A., Okamoto, T., and Vanstone, S., Reducing Elliptic Curve Logarithms to Logarithms in a Finite Field, IEEE Transactions on Information Theory 39, pp. 1639–1646 (1993).

35. Miller, V., The Weil Pairing, and its Efficient Calculation, Journal of Cryptology, 17(4) (2004).

36. Miyaji, A., Nakabayashi, M. and Takano, S., New Explicit Conditions of Elliptic Curve Traces for FR-reduction, IEICE Trans. Fundamentals, E84-A(5) (2001).

37. Okamoto,T., Efficient Blind and Partially Blind Signatures Witout Random Oracles, to appear in Proceedings of TCC 2006, LNCS, Springer-Verlag (2006).

38. Sakai, R. and Kasahara, M., ID Based Cryptosystems with Pairing on Elliptic Curve, IACR ePrint Archive, 2003/054, http://eprint.iacr.org/2003/054 (2003)
39. Sakai, R., Ohgishi, K. and Kasahara, M., Cryptosystems Based on Pairings, In Symposium on Cryptography and Information Security, SCIS 2000, Japan (2000).
40. Shamir, A., Identity-Based Cryptosystems and Signature Schemes, Adv. in Cryptology – Crypto 1984, LNCS 196, Springer-Verlag, pp. 47-53 (1984).
41. Verheul, E., Self-blindable Credential Certificates from the Weil Pairing, Adv. in Cryptology – Asiacrypt 2001, LNCS 2248, pp. 533–551, Springer-Verlag (2002).
42. Waters, B., Efficient Identity-Based Encryption Without Random Oracles, Adv. in Cryptology – Eurocrypt 2005, LNCS 3494, pp. 114-127, Springer-Verlag (2005). Available at http: //eprint.iacr.org/2004/180
43. Zhang, F., Chen, X., Susilo, W. and Mu, Y., A New Short Signature Scheme Without Random Oracles from Bilinear Pairings, IACR ePrint Archive, 2005/386, http://eprint.iacr.org/2005/386 (2005)

A New Signature Scheme Without Random Oracles from Bilinear Pairings

Fangguo Zhang[1,3], Xiaofeng Chen[2,3], Willy Susilo[4], and Yi Mu[4]

[1] Department of Electronics and Communication Engineering,
Sun Yat-Sen University, Guangzhou 510275, P.R. China
isszhfg@mail.sysu.edu.cn
[2] Department of Computer Science,
Sun Yat-Sen University, Guangzhou 510275, P.R. China
isschxf@mail.sysu.edu.cn
[3] Guangdong Key Laboratory of Information Security Technology
Guangzhou 510275, P.R. China
[4] School of IT and Computer Science
University of Wollongong, Wollongong, NSW 2522, Australia
{wsusilo,ymu}@uow.edu.au

Abstract. In this paper, we propose a new signature scheme that is existentially unforgeable under a chosen message attack *without* random oracle. The security of the proposed scheme depends on a new complexity assumption called the $k+1$ square roots assumption. Moreover, the $k+1$ square roots assumption can be used to construct shorter signatures under the random oracle model.

Keywords: Short signature, Bilinear pairings, Standard model, Random oracle.

1 Introduction

Digital signatures are important and fundamental cryptographic primitives, they not only provide basic signing functionality but also are building blocks in cryptographic protocol design.

Short digital signatures are always desirable. They are necessary in some situation where people need to enter the signature manually, such as using a PDA that is not equipped with a keyboard. Additionally, short digital signatures are essential to ensure the authenticity of messages in low-bandwidth communication channels. In general, short digital signatures are used to reduce the communication complexity of any transmission. As noted in [24], when one needs to sign a postcard, it is desirable to minimize the total length of the original message and the appended signature. In the early days, research in this area has been mainly focusing on how to minimize the total length of the message and the appended signature [25,1] and how to shorten the DSA signature scheme while preserving the same level of security [24]. From Hidden Field Equation (HFE)

P.Q. Nguyen (Ed.): VIETCRYPT 2006, LNCS 4341, pp. 67–80, 2006.

problem and Syndrome Decoding problem, a number of short signature schemes, such as Quartz [26,14], McEliece-based signature [15], have been proposed.

Boneh, Lynn and Shacham [9] used a totally new approach to design short digital signatures. The resulting signature scheme, referred to as the BLS signature scheme, is based on the Computational Diffie-Hellman (CDH) assumption on elliptic curves with low embedding degree. In BLS signature scheme, with a signature length $\ell = 160$ bits (which is approximately half the size of DSS signatures with the same security level), it provides a security level of approximately $\mathcal{O}(2^{80})$ in the random oracle model. In [28,5], a more efficient approach to produce a signature of the same length as BLS scheme was proposed. Nonetheless, its security is based on a stronger assumption.

Provable security is the basic requirement for signature schemes. Currently, most of the practical secure signature schemes were proven in the random oracle model [3]. Security in the random oracle model does *not* imply security in the real world. The first provably secure signature scheme in the standard model was proposed by Goldwasser *et al.* [21] in 1984. However, in this scheme, a signature is produced by signing the message bit-by-bit and hence, it is regarded as impractical for some applications. Independently, Gennaro, Halevi and Rabin [20] and Cramer and Shoup [16] proposed secure signature schemes under the so-called Strong RSA assumption in the standard model and the efficiency of which is suitable for practical use. Later, Camenisch and Lysyanskaya [11] and Fischlin [18] constructed two provably secure signature schemes under the strong RSA assumption in the standard model. In 2004, Boneh and Boyen [5] proposed a short signature scheme (BB04) from bilinear groups which is existentially unforgeable under a chosen message attack without using random oracles. The security of the scheme depends on a new complexity assumption, called *the Strong Diffie-Hellman assumption*. We note that Cheon [13] recently showed that SDH and related problems are slightly easier than discrete logarithm problem. However, his analysis is generic and does not violate the generic lower bounds on the hardness of SDH given in [5]. Nevertheless, it is worthwhile to design provably secure signature schemes using different hard problems.

In this paper, we construct a new, efficient and provably secure short signature scheme in the standard model from bilinear pairings. The signature size of the proposed scheme is the same as in the BB04 scheme. We note that our scheme is the second short signature scheme *without* random oracles. The security of our scheme depends on a new complexity assumption called the $k+1$ square roots assumption. In the random oracle model, we present a signature scheme that produces *even shorter signature length*. It produces a signature whose length is approximately 160 bits.

The rest of the paper is organized as follows. The next section contains some preliminaries required throughout the paper. We briefly review the bilinear pairings and secure signature schemes, and propose the $k+1$ square roots problem and $k+1$ square roots assumption. In Section 3, we propose our new short signature scheme and its security analysis *without* random oracles. In Section 4 we show that by employing random oracles, the $k+1$ square roots assumption can be used to build even shorter signatures. In this section, we also provide a security proof under the random oracle model. Section 5 concludes this paper.

2 Preliminaries

2.1 Bilinear Pairings

In recent years, the bilinear pairings have been found to be very useful in various applications in cryptography and have allowed us to construct new cryptographic primitives. We briefly review the bilinear pairings using the same notation as in [7,9]:

Let $\mathbb{G}$ be (mutiplicative) cyclic groups of prime order q. Let g be a generator of $\mathbb{G}$.

Definition 1. *A map $e : \mathbb{G} \times \mathbb{G} \rightarrow \mathbb{G}_T$ (here $\mathbb{G}_T$ is another mutiplicative cyclic group such that $|\mathbb{G}| = |\mathbb{G}_T| = q$) is called a bilinear pairing if it satisfies the following properties:*

1. **Bilinearity:** *For all $u, v \in \mathbb{G}$ and $a, b \in \mathbb{Z}_q$, we have $e(u^a, v^b) = e(u, v)^{ab}$.*
2. **Non-degeneracy:** *$e(g, g) \neq 1$. In other words, if g is a generator of $\mathbb{G}$, then $e(g, g)$ generates $\mathbb{G}_T$.*
3. **Computability:** *There is an efficient algorithm to compute $e(u, v)$ for all $u, v \in \mathbb{G}$.*

We say that $\mathbb{G}$ is a bilinear group if there exists a group $\mathbb{G}_T$, and a bilinear pairing $e : \mathbb{G} \times \mathbb{G} \rightarrow \mathbb{G}_T$ as above. Such groups can be found on supersingular elliptic curves or hyperelliptic curves over finite fields, and the bilinear parings can be derived from the Weil or Tate pairing.

2.2 The $k + 1$ Square Roots Assumption

In this subsection, we first introduce a new hard problem on which the new signature scheme in this paper is based.

Definition 2 ($k+1$-SRP). *The $k+1$ **Square Roots Problem** in $(\mathbb{G}, \mathbb{G}_T)$ is as follows: For an integer k, and $x \in_R \mathbb{Z}_q$, $g \in \mathbb{G}$, given*

$$\{g, \alpha = g^x, h_1, \ldots, h_k \in \mathbb{Z}_q, g^{(x+h_1)^{\frac{1}{2}}}, \ldots, g^{(x+h_k)^{\frac{1}{2}}}\},$$

compute $g^{(x+h)^{\frac{1}{2}}}$ for some $h \notin \{h_1, \ldots, h_k\}$.

We say that the $k + 1$-SRP is (t, ϵ)-hard if for any t-time adversary $\mathcal{A}$, we have

$$\Pr \left[\begin{array}{l} \mathcal{A}(g, \alpha = g^x, g^{(x+h_1)^{\frac{1}{2}}}, \ldots, g^{(x+h_k)^{\frac{1}{2}}} | x \in_R \mathbb{Z}_q, g \in \mathbb{G}, h_1, \ldots, h_k \in \mathbb{Z}_q) \\ = g^{(x+h)^{\frac{1}{2}}}, \ h \notin \{h_1, \ldots, h_k\} \end{array} \right] < \epsilon$$

where ϵ is negligible.

Definition 3 ($k + 1$-SR Assumption). *We say that the $(k + 1, t, \epsilon)$-SR assumption holds in $(\mathbb{G}, \mathbb{G}_T)$ if no t-time algorithm has advantage at least ϵ in solving the k+1-SRP in $(\mathbb{G}, \mathbb{G}_T)$, i.e., $k + 1$-SRP is (t, ϵ)-hard in $(\mathbb{G}, \mathbb{G}_T)$.*

Remarks. $k + 1$ Square Roots Problem is not a well studied problem and we are uncertain of its difficulty. A simple observation is that when we obtain enough values of h_i (about $\log q$) such that for each h_i, $x + h_i$ is a quadratic residue modulo q, then there exists a unique x that satisfies these equations. The explanation of this observation is as follows.

Given $h_1, h_2, ..., h_k$, for each h_i, there are many elements in G such that the sum of each of these elements and h_i is a quadratic residue. For convenience, we denote by S_i (wrt. h_i) those elements such that for any element $x \in S_i$, $x + h_i$ is a quadratic residue. The solution of $k + 1$ Square Roots Problem is in the intersection of $\{S_i\}$, $i = 1, \cdots, k$. Therefore, x is unique when k is large enough. However, when q is large, there exists no efficient algorithm to find $\{S_i\}$ for each h_i. The fact that x is unique given the above sets also precludes lower bounds on the hardness of our assumption in the generic group model. We note that this property does not degrade the security of our schemes.

2.3 Secure Signature Schemes

A signature scheme consists of the following four algorithms: a parameter generation algorithm ParamGen, a key generation algorithm KeyGen, a signature generation algorithm Sign and a signature verification algorithm Ver.

There are two types of attacks against signature schemes, namely the *no-message attack* and the *known-message attack*. In the first case, the attacker only knows the public key of the signer. In the second case, the attacker has access to a list of message-signature pairs. The strongest type of chosen-message attack is called the adaptively chosen-message attack, where the attacker has the knowledge of the public key of the signer, and he can ask the signer to sign *any* message that he wants. He can then adapt his queries according to the previous message-signature pairs. The strongest notion of security for signature schemes was defined by Goldwasser, Micali and Rivest [21,22] as follows:

Definition 4 (Secure signatures [21,22]). *A signature scheme $\mathcal{S} = \langle ParamGen, KeyGen, Sign, Ver \rangle$ is existentially unforgeable under an adaptive chosen message attack if it is infeasible for a forger who only knows the public key to produce a valid message-signature pair after obtaining polynomially many signatures on messages of its choice from the signer.*

Formally, for every probabilistic polynomial time forger algorithm $\mathcal{F}$ there exist no non-negligible probability ϵ such that

$$\mathbf{Adv}(\mathcal{F}) = \Pr \left[\begin{array}{l} \langle pk, sk \rangle \leftarrow \langle ParamGen, KeyGen \rangle (1^l); \\ for\ i = 1, 2, \dots, k; \\ m_i \leftarrow \mathcal{F}(pk, m_1, \sigma_1, \dots, m_{i-1}, \sigma_{i-1}), \sigma_i \leftarrow Sign(sk, m_i); \\ \langle m, \sigma \rangle \leftarrow \mathcal{F}(pk, m_1, \sigma_1, \dots, m_k, \sigma_k); \\ m \notin \{m_1, \dots, m_k\}\ and\ Ver(pk, m, \sigma) = accept \end{array} \right] \geq \epsilon.$$

Goldwasser *et al.* also constructed a signature scheme that satisfies the above security notion. Their scheme has an advantage that it does not use hash functions for message formatting. It is the first secure signature scheme under the standard model.

Here, we use the definition of [4] that takes into account the presence of an ideal hash function (the cryptographic hash function is seen as an oracle that produces a random value for each new query), and gives a concrete security analysis of digital signatures.

Definition 5 (Exact security of signatures [4]). *A forger $\mathcal{F}$ is said to (t, q_H, q_S, ϵ)-break the signature scheme $\mathcal{S} = < ParamGen, KeyGen, Sign, Ver >$ via an adaptive chosen message attack if after at most q_H queries to the hash oracle, q_S signatures queries and t processing time, it outputs a valid forgery with probability at least ϵ.*

A signature scheme $\mathcal{S}$ is (t, q_H, q_S, ϵ)-secure if there is no forger who (t, q_H, q_S, ϵ)-breaks the scheme.

3 New Short Signatures Without Random Oracles

3.1 Construction

We describe the new signature scheme as follows:

Let $e : \mathbb{G} \times \mathbb{G} \to \mathbb{G}_T$ be the bilinear pairing where $|\mathbb{G}| = |\mathbb{G}_T| = q$ for some prime q. We assume that $|q| \geq 160$. As for the message space, if the signature scheme is intended to be used directly for signing messages, then $|m| = 160$ is good enough, since given a suitable collision resistant hash function, one can first hash a message to 160 bits, and then sign the resulting value. Hence, the messages m to be signed can be regarded as an element in $\mathbb{Z}_q$.

In order to give an exact security proof with a good bound for the new signature scheme, we assume that $q \equiv 3 \bmod 4$ (so that -1 is a non-quadratic residue modulo q), and the message space is $\{1, ..., (q-1)/2\}$. For any message $m \in \{1, ..., (q-1)/2\}$, if m is not a quadratic residue modulo q, then $q - m$ or $-m$ will be a quadratic residue modulo q. The system parameters are $(\mathbb{G}, \mathbb{G}_T, e, q, g)$, where $g \in \mathbb{G}$ is a random generator.

Key Generation. Randomly select $x, y \in_R \mathbb{Z}_q^*$, and compute $u = g^x$, $v = g^y$. The public key is (u, v). The secret key is (x, y).

Signing: Given a secret key $x, y \in_R \mathbb{Z}_q^*$, and a message $m \in \{1, ..., (q-1)/2\}$, pick a random $r \in_R \mathbb{Z}_q^*$,

– If m is a quadratic residue modulo q, then compute

$$\sigma = g^{(x+my+r)^{\frac{1}{2}}} \in \mathbb{G}$$

– Otherwise, if m is a non-quadratic residue modulo q, then compute

$$\sigma = g^{(x+(-m)y+r)^{\frac{1}{2}}} \in \mathbb{G}$$

Here $(x + my + r)^{\frac{1}{2}}$ or $(x + (-m)y + r)^{\frac{1}{2}}$ is computed modulo q. When they are not quadratic residues modulo q, we try again with a different random r. The signature is (σ, r).

Verification: Given a public key $(\mathbb{G}, \mathbb{G}_T, q, g, u, v)$, a message $m \in \{1, ..., (q-1)/2\}$, and a signature (σ, r), verify that

$$e(\sigma, \ \sigma) = e(uv^m g^r, \ g)$$

or

$$e(\sigma, \ \sigma) = e(uv^{-m} g^r, \ g)$$

The verification is correct due to the following equations:

$$e(\sigma, \ \sigma) = e(g^{(x \pm my + r)^{\frac{1}{2}}}, \ g^{(x \pm my + r)^{\frac{1}{2}}})$$
$$= e(g, \ g)^{(x \pm my + r)^{\frac{1}{2}} \cdot (x \pm my + r)^{\frac{1}{2}}}$$
$$= e(g, \ g)^{x \pm my + r}$$
$$= e(uv^{\pm m} g^r, \ g)$$

3.2 Efficiency

To date, there exist three secure signature schemes without random oracles from the bilinear groups, namely BB04 scheme [5], BMS03 scheme [10] and CL04 scheme [12]. BMS03 signature scheme is based on a signature authentication tree with a large branching factor. Compared to BMS03 and CL04 schemes, our scheme has the obvious advantages in all parameters, such as the public key, signature lengths and performance.

The new signature scheme requires one computation of square root in $\mathbb{Z}_q^*$ and one exponentiation in $\mathbb{G}$ to sign. For the verification, it requires two or three pairings and two exponentiations in $\mathbb{G}$.

We note that the computation of the pairing is the most time-consuming in pairing based cryptosystems. Although there have been many papers discussing the complexity of pairings and how to speed up the pairing computation [2,17,19], the computation of the pairing still remains time-consuming. Similar to BB04 scheme, some pairings in the proposed signature scheme can be pre-computed and published as part of the signer's public key, such that there is only *one* pairing operation in the verification. We pre-compute $a = e(u, \ g)$, $b = e(v, \ g)$ and $c = e(g, \ g)$, and publish them as part of the signer's public key. Then, for a message $m \in \mathbb{Z}_q^*$, and a signature (σ, r), the verification can be done as follows:

$$e(\sigma, \ \sigma) \stackrel{?}{=} a \cdot b^{\pm m} \cdot c^r.$$

Hence, the verification requires only one pairing and two exponentiations in $\mathbb{G}_T$, and we note that the exponentiations in $\mathbb{G}_T$ are significantly faster than pairing operations.

Signature Length. A signature in the new scheme contains of two elements (σ, r), where one element is in $\mathbb{G}$ and the other element is in $\mathbb{Z}_q^*$. When using a supersingular elliptic curve over finite field F_{p^n} with embedding degree $k = 6$ and the modified Weil pairing or Tate pairing [9,23], the length of an element in $\mathbb{Z}_q^*$ and $\mathbb{G}$ can be approximately $\log_2 q$ bits, and therefore the total signature length is approximately $2 \log_2 q$ bits. To be more precisely, let $P \in E(F_{p^n})$, $ord(P) = q$, $\mathbb{G} = < P > \subset E[q]$ ($E[q]$ is the group of q-torsion points of E). Let ϕ be a distortion map, *i.e.*, an efficiently computable automorphism of $E[q] \cong \mathbb{Z}_q \times \mathbb{Z}_q$ such that $\phi(P) \notin < P > = \mathbb{G}$. Actually, the map ϕ maps q-torsion points defined over F_{p^n} to q-torsion points defined over the extension field $F_{p^{nk}}$ (For supersingular elliptic curve, such distortion map always exists). Consider the bilinear pairing

$$\hat{e} : \mathbb{G} \times \mathbb{G} \to \mu_q,$$

defined by

$$\hat{e}(P, Q) := e_w(P, \phi(Q)),$$

here e_w denotes the Weil pairing and μ_q is the subgroup of order q in $F_{p^{nk}}^*$.

We can select the parameter such that the elements in $\mathbb{G}$ are 171-bits strings. A possible choice of these parameters can be from Boneh *et al.*'s short signature scheme [9] : $\mathbb{G}$ is derived from the curve $E/GF(3^{97})$ defined by $y^2 = x^3 - x + 1$, which has 923-bit discrete-log security. Therefore, at the current security requirement, we can obtain a signature whose length is approximately the same as a DSA signature with the same level of security, but which is provably secure and existentially unforgeable under a chosen message attack without the random oracle model, which is the same as BB04. Hence, this is the second short signature scheme without random oracles.

However, the proposed signature scheme has a drawback, that is the scheme requires a symmetric bilinear map, whereas BLS and BB04 can work with a symmetric or an asymmetric map. Currently, the symmetric bilinear map with short representation of group element can only be found on supersingular curves. Since these curves have an embedding degree of at most 6, this will make the new signatures bigger and harder to scale, compared to BB04 and BLS, at higher security levels.

3.3   Proof of Security

The following theorem shows that the scheme above is existentially unforgeable in the strong sense under chosen message attacks, provided that the $k + 1$-SR assumption holds in $(\mathbb{G}, \mathbb{G}_T)$.

Theorem 1. *Suppose the $(k + 1, t', \epsilon')$-SR assumption holds in $(\mathbb{G}, \mathbb{G}_T)$. Then the signature scheme above is (t, q_S, ϵ)-secure against existential forgery under an adaptive chosen message attack provided that*

$$q_S < k+1, \ \epsilon = 2\epsilon' + 4\frac{q_S}{q} \approx 2\epsilon', \ t \le t' - \Theta(q_S T).$$

where T is the maximum time for computing a square root in $\mathbb{Z}_q^*$ and an exponentiation in $\mathbb{G}$.

Proof. To prove the theorem, we will prove the following: "If there exists a (t, q_S, ϵ)-forger $\mathcal{F}$ using adaptive chosen message attack for the proposed signature scheme, then there exists a (t', ϵ')-algorithm $\mathcal{A}$ solving q_S-SRP (also $k+1$-SRP, if $k+1 > q_S$), where $t' \ge t + \Theta(q_S T)$, $\epsilon' = \frac{\epsilon}{2} - 2\frac{q_S}{q}$."

Assume $\mathcal{F}$ is a forger that (t, q_S, ϵ)-breaks the signature scheme. We construct an algorithm $\mathcal{A}$ that, by interacting with $\mathcal{F}$, solves the q_S-SRP in time t' with advantage ϵ'.

Suppose $\mathcal{A}$ is given a challenge – a random instance of q_S-SRP:

" *For an integer q_S, and $x \in_R \mathbb{Z}_q$, $g \in \mathbb{G}$, given*

$$\{g, \ \alpha = g^x, \ h_1, \ldots, h_{q_S} \in \mathbb{Z}_q, \ g^{(x+h_1)^{\frac{1}{2}}}, \ldots, g^{(x+h_{q_S})^{\frac{1}{2}}}\},$$

to compute $g^{(x+h)^{\frac{1}{2}}}$ for some $h \notin \{h_1, \ldots, h_{q_S}\}$. "

Next, we describe how the algorithm $\mathcal{A}$ to solve the q_S-SRP by interacting with $\mathcal{F}$. The approach is similar to BB04 [5]. We distinguish between two types of forgers that $\mathcal{F}$ can emulate. Let $(\mathbb{G}, \ \mathbb{G}_T, \ q, \ g, u, \ v)$ be the public key given to forger $\mathcal{F}$ where $u = g^x$ and $v = g^y$. Suppose $\mathcal{F}$ asks for signatures on messages $m_1, m_2, \cdots, m_{q_S} \in \mathbb{Z}_q^*$ and is given signatures $(r_i, \ \sigma_i)$ on these messages for $i = 1, \cdots, q_S$. Let $h_i = m_i y + r_i$ and let (m, r, σ) be the forgery produced by $\mathcal{F}$. Denote two types of forger $\mathcal{F}$ as:

Type-1 Forger which either makes query for $m_i = -x$, or outputs a forgery where $my + r \notin \{h_1, h_2, \cdots, h_{q_S}\}$.

Type-2 Forger which never makes any query for a message $m = -x$, and outputs a forgery where $my + r \in \{h_1, h_2, \cdots, h_{q_S}\}$.

$\mathcal{A}$ plays the role of the signer, it produces a forgery for the signature scheme as follows:

Setup: $\mathcal{A}$ is given g, $\alpha = g^x$, with q_S known solutions $(h_i \in \mathbb{Z}_q, \ s_i = g^{(x+h_i)^{\frac{1}{2}}} \in \mathbb{G})$ for random h_i $(i = 1, \cdots, q_S)$. $\mathcal{A}$ picks random $y \in \mathbb{Z}_q$ and a bit $b_{mode} \in \{1, 2\}$ randomly. If $b_{mode} = 1$, $\mathcal{A}$ publishes the public key $PK_1 = (\mathbb{G}, \ \mathbb{G}_T, \ q, \ g, u, \ v)$, here $u = \alpha$, $v = g^y$. If $b_{mode} = 2$, $\mathcal{A}$ publishes the public key $PK_2 = (\mathbb{G}, \ \mathbb{G}_T, \ q, \ g, \ u, \ v)$, here $u = g^y$, $v = \alpha$. In $\mathcal{F}$'s view, both PK_1 and PK_2 are valid public keys for the signature scheme.

Simulation: The forger $\mathcal{F}$ can issue up to q_S signature queries in an adaptive fashion. To respond these signature queries, $\mathcal{A}$ maintains a list H-list of tuples (m_i, r_i, h_i) and a query counter l which is initially set to 0.

Upon receiving a signature query for m_i, $\mathcal{A}$ increments l by one, and checks if $l > q_S$. If $l > q_S$, it neglects further queries by $\mathcal{F}$ and terminates $\mathcal{F}$. Otherwise, it checks if $g^{-m_i} = u$. If so, then $\mathcal{A}$ just obtained the private key for the public key $PK = (\mathbb{G}, \mathbb{G}_T, q, g, u, v)$ it was given, which allows it to forge the signature on any message of its choice. At this point $\mathcal{A}$ successfully terminates the simulation.

Otherwise, if $b_{mode} = 1$, set $r_i = h_i - m_i y \in \mathbb{Z}_q$. In the very unlikely event that $r_i = 0$, $\mathcal{A}$ reports failure and aborts. Otherwise, $\mathcal{A}$ gives $\mathcal{F}$ the signature $(r_i, \ \sigma_i = s_i)$. This is a valid signature on m_i under the public key $PK_1 = (\mathbb{G}, \mathbb{G}_T, q, g, u, v)$ since r_i is uniform in $\mathbb{Z}_q$ and

$$e(\sigma_i, \ \sigma_i) = e(g^{(x+h_i)^{\frac{1}{2}}}, \ g^{(x+h_i)^{\frac{1}{2}}}) = e(ug^{h_i}, \ g) = e(ug^{r_i + m_i y}, \ g) = e(uv^{m_i}g^{r_i}, \ g).$$

If $b_{mode} = 2$, set $r_i = m_i h_i - y \in \mathbb{Z}_q$. If $r_i = 0$, $\mathcal{A}$ reports failure and aborts. Otherwise, $\mathcal{A}$ returns $(r_i, \ \sigma_i = s_i^{\sqrt{m_i}})$ (If m_i is a quadratic residue modulo q) or $(r_i, \ \sigma_i = s_i^{\sqrt{-m_i}})$ (If m_i is a non-quadratic residue modulo q) as answer. This is a valid signature on m_i for PK_2 because r_i is uniform in $\mathbb{Z}_q$ and

$$\begin{aligned}
e(\sigma_i, \ \sigma_i) &= e(g^{(x+h_i)^{\frac{1}{2}}\sqrt{m_i}}, \ g^{(x+h_i)^{\frac{1}{2}}\sqrt{m_i}}) \\
&= e(g^{m_i h_i} v^{m_i}, \ g) \\
&= e(g^{y+r_i} v^{m_i}, \ g) \\
&= e(uv^{m_i} g^{r_i}, \ g)
\end{aligned}$$

$\mathcal{A}$ adds the tuple $(m_i, \ r_i, \ v^{m_i} g^{r_i})$ to H-list.

Reduction: Eventually, the forger $\mathcal{F}$ returns a forgery (m, r, σ), where (r, σ) is a valid forgery distinct from any previously given signature on message m. Note that by adding dummy queries as required, we may assume that $\mathcal{F}$ made exactly q_S signature queries. Let $W \leftarrow v^m g^r$. Algorithm $\mathcal{A}$ searches the H-list for a tuple whose rightmost component is equal to W. Then according to two types of forger $\mathcal{F}$, we denote the following events as:

F1: **(Type-1 forgery:)** No tuple of the form $(\cdot, \ \cdot, \ W)$ appears on the H-list.

F2: **(Type-2 forgery:)** The H-list contains at least one tuple (m_j, r_j, W_j) such that $W_j = W$.

Denote E1 to be the event $b_{mode} = 1$ (i.e., $\mathcal{F}$ produced a type-1 forgery, or $\mathcal{F}$ made a signature query for a message m_i such that $g^{-m_i} = u$.) and denote E2 to be the event $b_{mode} = 2$. We claim that $\mathcal{A}$ can succeed in breaking the signature scheme if $(E1 \wedge F1) \vee (E2 \wedge F2)$ happens.

Case 1. If $u = g^{-m_i}$, then $\mathcal{A}$ has already recovered the secret key of its challenger, $\mathcal{A}$ can forge a signature on any message of his choice. We assume

that $\mathcal{F}$ produced a type-1 forgery (m, r, σ). Since the forgery is valid, we have

$$e(\sigma, \; \sigma) = e(uv^m g^r, \; g) = e(ug^{my+r}, \; g).$$

Let $h = my + r$. So, the forgery (m, r, σ) provides a new $q_S - SRP$ solution (h, σ).

Case 2. Since $v = \alpha = g^x$, then we know that there exists a pair $v^{m_j} g^{r_j} = v^m g^r$. Since $(m, r) \neq (m_j, r_j)$, otherwise it is not regarded as a forgery, so, $m \neq m_j$, $r \neq r_j$. Therefore, $\mathcal{A}$ can compute $x = \frac{r_j - r}{m - m_j}$ which also enables $\mathcal{A}$ to recover the secret key of its challenger. He can now forge a signature on any message of its choice.

Any valid forgery (m, r, σ) will give a new $q_S - SRP$ solution under at least one of the 2 above reductions.

This completes the description of Algorithm $\mathcal{A}$. A standard argument shows that if $\mathcal{A}$ does not abort, then, from the viewpoint of $\mathcal{F}$, the simulation provided by $\mathcal{A}$ is indistinguishable from a real attack scenario. Since the simulations are perfect, $\mathcal{F}$ cannot guess which reduction the simulator is using. Therefore, $\mathcal{F}$ produces a valid forgery in time t with probability at least ϵ.

Since E1 and F1 are independent with uniform distribution, $Pr[E1 \vee E2] = 1$ and $Pr[F1 \vee F2] = 1$, the probability that $\mathcal{A}$ succeeds is $Pr[(E1 \wedge F1) \vee (E2 \wedge F2)] = \frac{1}{2}$.

Next we bound the probability that $\mathcal{A}$ dos not abort. ¿From above description of $\mathcal{A}$ we know that $\mathcal{A}$ aborts if

- At $E1 \wedge F1$, only if $r_i = 0$, i.e., $m_i y = h_i$. For given y, this happens with probability at most $\frac{q_S}{q}$.
- or at $E2 \wedge F2$, only if $r_i = 0$, i.e., $m_i h_i = y$. For given y, this happens with probability at most $\frac{q_S}{q}$.

So, $\mathcal{A}$ succeeds with probability at least $\frac{\epsilon}{2} - 2\frac{q_S}{q}$.

Let T be the maximum time for a computing square root in $\mathbb{Z}_q^*$ and an exponentiation in $\mathbb{G}$. The running time of $\mathcal{A}$ is $t' \geq t + \Theta(q_S T)$. This complete the proof.   □

4   Shorter Signature with Random Oracles

In this section, we present a more efficient short signature scheme based on $q_S - SRP$ in the random oracle model. The proposed new short signature scheme with random oracle is described as follows:

The system parameters are $(\mathbb{G}, \mathbb{G}_T, e, q, g, H)$, here $g \in \mathbb{G}$ is a random generator and $H : \{0, 1\}^* \rightarrow \mathbb{Z}_q^*$ is a cryptographic hash function. We assume that $q \equiv 3 \bmod 4$ (so that -1 is a non-quadratic residue modulo q).

Key Generation. Randomly select $x \in_R \mathbb{Z}_q^*$, and compute $u = g^x$. The public key is u. The secret key is x.

Signing: Given a secret key x, and a message m, computes $\sigma = g^{(H(m)+x)^{\frac{1}{2}}}$. If $(H(m) + x)$ is a non-quadratic residue modulo q, compute $\sigma = g^{(-(H(m)+x))^{\frac{1}{2}}}$.

Verification: Given a public key $(\mathbb{G}, \mathbb{G}_T, e, q, g, u, H)$, a message m, and a signature σ, verify that

$$e(\sigma, \sigma) = e(g^{H(m)}u, g)$$

or

$$e(\sigma, \sigma) = e(g^{H(m)}u, g)^{-1}.$$

This signature scheme can provide the same signature length as BLS scheme. We compare this signature scheme with the BLS scheme from the view point of computation overhead. The key and signature generation times are comparable to BLS signatures. The verification time is faster, since the verification requires only one pairing and one exponentiation due to the pre-computation of $a = e(u, g)$ and $c = e(g, g)$. This is comparable to the random-oracle version of the BB signature, which also uses a single pairing. By contrast, the BLS signature requires two pairings.

About the security of proposed signature scheme against an adaptive chosen message attack, we obtain the following theorem:

Theorem 2. *If there exists a (t, q_H, q_S, ϵ)-forger $\mathcal{F}$ using adaptive chosen message attack for the proposed signature scheme, then there exists a (t', ϵ')-algorithm $\mathcal{A}$ solving $q_H - k$-SRP (for a constant $k \in \mathbb{Z}^+$), where*

$$t = t', \ \epsilon' \geq \prod_{j=0}^{q_S-1} \frac{q_H - k - j}{q_H - j} \cdot \frac{k}{q_H} \cdot \epsilon.$$

Especially, there exists a $(t' = t, \ \epsilon' \geq \frac{q_S}{q_H^2} \cdot \epsilon)$-algorithm $\mathcal{A}$ solving $q_H - 1$-SRP.

Proof. In the proposed signature scheme, before signing a message m, we need to make a query $H(m)$. Our proof is in the random oracle model (the hash function is seen as a random oracle, *i.e.*, the output of the hash function is uniformly distributed).

Suppose that a forger $\mathcal{F}$ (t, q_H, q_S, ϵ)-break the signature scheme using an adaptive chosen message attack. We will use $\mathcal{F}$ to construct an algorithm $\mathcal{A}$ to solve $q_H - 1$-SRP.

Suppose $\mathcal{A}$ is given a challenge:

" *For integer q_H and k, and $x \in_R \mathbb{Z}_q$, $g \in \mathbb{G}$, given*

$$\{g, \ \alpha = g^x, \ h_1, \ldots, h_{q_H-k} \in \mathbb{Z}_q, \ g^{(x+h_1)^{\frac{1}{2}}}, \ldots, g^{(x+h_{q_H-k})^{\frac{1}{2}}}\},$$

to compute $g^{(x+h)^{\frac{1}{2}}}$ for some $h \notin \{h_1, \ldots, h_{q_H-k}\}$. "

Now $\mathcal{A}$ plays the role of the signer and sets the public key be $u = \alpha$. $\mathcal{A}$ will answer hash oracle queries and signing queries itself. We assume that $\mathcal{F}$ never repeats a hash query or a signature query.

S1. $\mathcal{A}$ prepares q_H responses $\{w_1, w_2, \ldots, w_{q_H}\}$ of the hash oracle queries, $h_1, \ldots,$ h_{q_H-k} are distributed randomly in this response set.

S2. $\mathcal{F}$ makes a hash oracle query on m_j for $1 \le j \le q_H$. $\mathcal{A}$ sends w_j to $\mathcal{F}$ as the response of the hash oracle query on m_j.

S3. $\mathcal{F}$ makes a signature oracle query for w_j. If $w_i = h_j$, $\mathcal{A}$ returns $g^{(x+h_j)^{\frac{1}{2}}}$ to $\mathcal{F}$ as the response. Otherwise, $\mathcal{A}$ reports failure and aborts.

S4. Eventually, $\mathcal{F}$ halts and outputs a message-signature pair (m, σ). Here the hash value of m is some w_l and $w_l \notin \{h_1, \ldots, h_{q_H-k}\}$. Since (m, σ) is a valid forgery and $H(m) = w_l$, it satisfies:

$$e(\sigma,\ \sigma) = e(g^{H(m)}u,\ g).$$

So, $\sigma = g^{(x+w_l)^{\frac{1}{2}}}$. $\mathcal{A}$ outputs (w_l, σ) as a solution to $\mathcal{A}$'s challenge.

Algorithm $\mathcal{A}$ simulates the random oracles and signature oracle perfectly for $\mathcal{F}$. $\mathcal{F}$ cannot distinguish between $\mathcal{A}$'s simulation and real life because the hash function behaves as a random oracle. Therefore $\mathcal{F}$ produces a valid forgery for the signature scheme with probability at least ϵ.

Now, we bound the probability $\mathcal{A}$ dos not abort. In step S3, the success probability of $\mathcal{A}$ is $\frac{q_H-k}{q_H}$, and hence, for all signature oracle queries, $\mathcal{A}$ will not fail with probability

$$\rho \ge \prod_{j=0}^{q_S-1} \frac{q_H - k - j}{q_H - j}$$

(if $\mathcal{F}$ only makes $s(\le q_S)$ signature oracle queries, the success probability of $\mathcal{A}$ is $\prod_{j=0}^{s-1} \frac{q_H-k-j}{q_H-j}$). Hence, after the algorithm $\mathcal{A}$ finished the step S4, the success probability of $\mathcal{A}$ is:

$$\epsilon' \ge \prod_{j=0}^{q_S-1} \frac{q_H - k - j}{q_H - j} \cdot \frac{k}{q_H} \cdot \epsilon.$$

In particular, if we let $k = 1$, then the success probability of $\mathcal{A}$ is:

$$\epsilon' \ge \frac{q_S}{q_H^2} \cdot \epsilon.$$

The running time of $\mathcal{A}$ is equal to the running time of $\mathcal{F}$, where $t' = t$. $\square$

5   Conclusion and Further Works

In this paper, we proposed the second short signature scheme from bilinear pairing which is existentially unforgeable under a chosen message attack without using random oracles. The security of our scheme depends on a new complexity assumption called the $k+1$ square roots assumption. Furthermore, the $k+1$ square roots assumption gives even shorter signatures in the random oracle model, where a signature is only one element in a bilinear group.

As for applications of our signature schemes, we present a new chameleon hash signature scheme, an on-line/off-line signature scheme and a new efficient anonymous credential scheme based on the proposed signature scheme in the earlier version of this paper [27]. These applications are omitted here due to the page limitation. BLS[9], BB04 [5] and ZSS [28] short signature schemes play an important role in many pairing-based cryptographic systems. The proposed signature scheme is comparable to them and we expect to see many other schemes based on it, such as group signatures [6], aggregate signatures [8] and others.

Acknowledgements

We would like to thank Xavier Boyen and the anonymous reviewers of VietCrypt 2006 for their helpful comments and suggestions. We would also like to thank Serge Vaudenay for a constructive suggestion during the conference.

This work has been supported by the National Natural Science Foundation of China (No. 60403007 and No. 60503006) and ARC Discovery Grant DP0557493 and the Project-sponsored by SRF for ROCS, SEM.

References

1. M. Abe and T. Okamoto. *A signature scheme with message recovery as secure as discrete logarithm.* Advances in Cryptology -Asiacrypt 1999, LNCS 1716, pp.378-389, Springer-Verlag, 1999.
2. P.S.L.M. Barreto, H.Y. Kim, B.Lynn, and M.Scott, *Efficient algorithms for pairing-based cryptosystems*, Advances in Cryptology-Crypto 2002, LNCS 2442, pp.354-368, Springer-Verlag, 2002.
3. M. Bellare and P. Rogaway, *Random oracles are practical: a paradigm for designing effiient protocols*, Proceedings of the 1st ACM Conference on Computer and Communications Security, pp.62-73, ACM press, 1993.
4. M. Bellare and P. Rogaway, *The exact security of digital signatures - How to sign with RSA and Rabin*, Advances in Cryptology-Eurocrypt 1996, LNCS 1070, pp. 399-416, Springer- Verlag, 1996.
5. D. Boneh and X. Boyen, *Short signatures without random oracles*, Advances in Cryptology-Eurocrypt 2004, LNCS 3027, pp.56-73, Springer-Verlag, 2004.
6. D. Boneh, X. Boyen and H. Shacham, *Short group signatures*, Advances in Cryptology-Crypto 2004, LNCS 3152, pp.41-55, Springer-Verlag, 2004.
7. D. Boneh and M. Franklin, *Identity-based encryption from the Weil pairing*, Advances in Cryptology-Crypto 2001, LNCS 2139, pp.213-229, Springer-Verlag, 2001.
8. D. Boneh, C. Gentry, B. Lynn and H. Shacham, *Aggregate and verifiably encrypted signatures from bilinear maps*, Advances in Cryptology-Eurocrypt 2003, LNCS 2656, pp.272-293, Springer-Verlag, 2003.
9. D. Boneh, B. Lynn, and H. Shacham, *Short signatures from the Weil pairing*, Advances in Cryptology-Asiacrypt 2001, LNCS 2248, pp.514-532, Springer-Verlag, 2001.
10. D. Boneh, I. Mironov and V. Shoup, *A secure signature scheme from bilinear maps*, CT-RSA 2003, LNCS 2612, pp.98-110, Springer-Verlag, 2003.

11. J. Camenisch and A. Lysyanskaya, *A signature scheme with efficient protocols*, SCN 2002, LNCS 2576, pp.274-295, Springer- Verlag, 2003.

12. J. Camenisch and A. Lysyanskaya, *Signature schemes and anonymous credentials from bilinear maps*, Advances in Cryptology-Crypto 2004, LNCS 3152, pp.56-72, Springer- Verlag, 2004.

13. J.H. Cheon, *Security analysis of the strong Diffie-Hellman problem*, Advances in Cryptology-Eurocrypt 2006, LNCS 4004, pp.1-11, Springer-Verlag, 2006.

14. N. Courtois, M. Daum and P. Felke, *On the security of HFE, HFEv- and Quartz*, PKC 2003, LNCS 2567, pp.337-350. Springer- Verlag, 2003.

15. N.T. Courtois, M. Finiasz and N. Sendrier, *How to achieve a McEliece-based Digital Signature Schem*, Advances in Cryptology-Asiacrypt 2001, LNCS 2248, pp.157-174, Springer-Verlag, 2001.

16. R. Cramer and V. Shoup, *Signature schemes based on the strong RSA assumption*, Proceedings of the 6th ACM Conference on Computer and Communications Security, pp.46-52, ACM press, 1999.

17. I. M. Duursma and H.-S. Lee, *Tate pairing implementation for hyperelliptic curves $y^2 = x^p - x + d$*, Advances in Cryptology -Asiacrypt 2003, LNCS 2894, pp.111-123, Springer-Verlag, 2003.

18. M. Fischlin, *The Cramer-Shoup strong-RSA signature scheme revisited*, PKC 2003, LNCS 2567, pp.116-129, Springer-Verlag, 2003.

19. S. D. Galbraith, K. Harrison, and D. Soldera, *Implementing the Tate pairing*, ANTS 2002, LNCS 2369, pp.324-337, Springer-Verlag, 2002.

20. R. Gennaro, S. Halevi and T. Rabin, *Secure hash-and-sign signature without the random oracle*, Advances in Cryptology-Eurocrypt 1999, LNCS 1592, pp.123-139, Springer-Verlag, 1999.

21. S. Goldwasser, S. Micali and R. Rivest, *A 'paradoxical' solution to the signature problem (extended abstract)*, Proc. of FOCS'84, pp.441-448, 1984.

22. S. Goldwasser, S. Micali and R. Rivest, *A digital signature scheme secure against adaptive chosen-message attacks*, SIAM Journal of Computing, 17(2), pp. 281-308, 1988.

23. A. Joux, *The Weil and Tate pairings as building blocks for public key cryptosystems*, ANTS 2002, LNCS 2369, pp. 20-32, Springer-Verlag, 2002

24. D. Naccache and J. Stern, *Signing on a postcard*, Financial Cryptography and Data Security 2000, LNCS 1962, pp.121-135, Springer-Verlag, 2000.

25. K. Nyberg and R. Rueppel, *A new signature scheme based on the DSA, giving message recovery*, Proceedings of the 1st ACM Conference on Communications and Computer Security, pp. 58-61, 1993.

26. J. Patarin, N. Courtois and L. Goubin, *QUARTZ, 128-bit long digital signatures*, CT-RSA 2001, LNCS 2020, pp. 282-297, Springer-Verlag, 2001.

27. F. Zhang, X. Chen, W. Susilo and Y. Mu, *A New Signature Scheme without Random Oracles and Its Applications*, Cryptology ePrint Archive: Report 2005/386.

28. F. Zhang, R. Safavi-Naini and W. Susilo, *An efficient signature scheme from bilinear pairings and its applications*, PKC 2004, LNCS 2947, pp.277-290, Springer-Verlag, 2004.

Efficient Dynamic k-Times Anonymous Authentication

Lan Nguyen

CSIRO ICT Centre, Australia
WinMagic, Canada
Lan.Nguyen@winmagic.com

Abstract. In *k-times anonymous authentication* (k-TAA) schemes, members of a group can be anonymously authenticated to access applications for a bounded number of times determined by application providers. *Dynamic* k-TAA allows application providers to independently grant or revoke group members from accessing their applications. Dynamic k-TAA can be applied in several scenarios, such as k-show anonymous credentials, digital rights management, anonymous trial of Internet services, e-voting, e-coupons etc. This paper proposes the first provably secure dynamic k-TAA scheme, where authentication costs do not depend on k. This efficiency is achieved by using a technique called "efficient provable e-tag", which could be applicable to other e-tag systems.

Keywords: privacy, anonymity, dynamic k-times anonymous authentication, k-show anonymous credentials, e-tag.

1 Introduction

In a k-times anonymous authentication system [13], participants include a group manager (GM), some application providers (AP) and many users. The GM registers users into the group and each AP independently announces the number of times a group member can access her application. A group member can be anonymously authenticated by APs within their allowed numbers of times and without contacting the GM. No one, even the GM or APs, is able to identify honest users or link two authentication executions of the same user while anyone can trace dishonest users. No party, even the GM, can successfully impersonate an honest user in an authentication execution.

However, k-TAA schemes are inflexible in the sense that the GM decides on the group membership and APs do not have any control over giving users access permission to their services. APs are passive and their role is limited to announcing the number of times a user can access their applications. In practice, APs want to select their user groups and grant or revoke access to users independently. For example, the AP may prefer to give access to users with good profile, or the AP may need to put an expiry date on users' access. *Dynamic* k-TAA [12] was introduced to provide these properties. In dynamic k-TAA, APs have more control over granting and revoking access to their services, and less

P.Q. Nguyen (Ed.): VIETCRYPT 2006, LNCS 4341, pp. 81–98, 2006.

trust and computation from the GM is required. Dynamic k-TAA allows APs to restrict access to their services based on not only the number of times but also other factors such as expiry date and so can be used in much wider range of realistic scenarios.

A primitive close to k-TAA is Privacy-Protecting Coupon (PPC) system [6,11], which consists of an *Initialisation* algorithm and 2 protocols, *Issue* and *Redeem*. There is a *vendor* and many *users*. The vendor can *issue* a k-redeemable coupon to a user such that the user can unlinkably *redeem* the coupon for exactly k times. There could be another algorithm, *Terminate*, which allows the vendor to *terminate* coupons. Compared to k-TAA, PPC does not allow traceability of malicious users and the vendor acts as the group manager and a single application provider.

Applications of k-TAA can be found in digital rights management (DRM). For example, k-TAA can be used to provide pay-per-use anonymous access to online digital content, such as music, movies, interactive games, betting and gambling, that are supplied by different application providers. A user can buy credits to download hundreds of songs or movies over a year at a discount price. Another example is trial browsing [13], where each provider allows members of a group, such as XXX community, to anonymously and freely browse content such as movies or music on trial. The provider also wants to limit the number of times that a user can access the service on trial and users, who try to go over the prescribed quota, must be identified. k-TAA can also be used to construct k-*show anonymous credential* systems [15], where credential-issuing organizations can limit the number of times a user can show her credentials.

In previous k-TAA schemes, the authentication procedure has computation and communication costs linearly depending on the bound k. If an application provider sets k to be a large number, the authentication procedure becomes expensive. For example, a music web site may sell e-vouchers each of which can be used to anonymously download 10000 songs within a year. Then each user has to run the same expensive authentication protocol for each downloaded song. If there are many users in the group, the authentication cost multiplies by the number of users. So, the open problem is to construct k-TAA schemes where the computation and communication costs in the authentication procedure do not depend on k.

1.1   Our Contribution

We propose the first dynamic k-TAA scheme with constant authentication costs, extended from the NS05 scheme [12], and prove its security. It can be used to construct the first k-show anonymous credential system with constant costs. It can be converted to a k-TAA scheme using the approach in [12]. It is also possible to construct a combined scheme, where some of the APs have the dynamic property and other APs do not. Section 4.3 details efficiency comparison with previous k-TAA schemes [13,12].

Our scheme still uses tag as in the TFS04 [13] and NS05 [12] schemes. In these schemes, the GM issues some secret key to each user. An AP with bound

k provides a set of k tag bases. For each authentication, the user uses his secret key and a tag base to computes a value, called a tag, and sends it to the verifier with a zero-knowledge proof that the tag is correctly computed and the user is a group member. If the user attempts to access more than k times, he has to use a tag base twice and his identity will be revealed. The problem with these constructions is that the proof that the tag is correctly computed from one of the k tag bases requires a proof of knowledge of one of k elements and its cost linearly depends on k. Our objective is to remove this dependency.

We use a methodology, called "efficient provable e-tag", which was first proposed in [11] for a PPC system. An ordinary k-TAA scheme with constant costs [14] also uses this method.

In this method, each AP with bound k uses its secret key to issue k signatures on k random messages and these message-signature tuples are used as tag bases. Then the proof of knowledge of one of k elements is replaced by a proof of knowledge of a message-signature tuple. However, using our message-signature tuples with the function to compute tags from tag bases as in [13,12] will result in a "cut and choose" zero-knowledge proof. So we use another function similar to the verifiable random function proposed in [7] that is used for the efficient compact e-cash scheme in [5]. We also need a different way for the GM to issue member secret and public keys to users.

The organization of the paper is as follows. We give the background in section 2 and present the model of dynamic k-TAA in section 3. Section 4 provides technical description of the proposed dynamic k-TAA scheme.

2 Preliminaries

We follow notation in [12,13] and use some complexity assumptions, including Strong Diffie-Hellman (SDH), Decisional Bilinear Diffie-Hellman Inversion (DBDHI) and Computational Bilinear Diffie-Hellman Inversion 2 (CBDHI2). The notation and assumptions are provided in Appendix A.

2.1 Bilinear Groups

Let $\mathbb{G}_1$, $\mathbb{G}_2$ and $\mathbb{G}_T$ be multiplicative cyclic groups of prime order p. Suppose P_1 and P_2 are generators of $\mathbb{G}_1$ and $\mathbb{G}_2$ respectively, and there is an isomorphism $\psi : \mathbb{G}_2 \to \mathbb{G}_1$ such that $\psi(P_2) = P_1$. A function $e : \mathbb{G}_1 \times \mathbb{G}_2 \to \mathbb{G}_T$ is said to be a bilinear pairing if it satisfies the following properties:

1. **Bilinearity:** $e(P^a, Q^b) = e(P, Q)^{ab}$ for all $P \in \mathbb{G}_1$, $Q \in \mathbb{G}_2$ and $a, b \in \mathbb{Z}_p$.
2. **Non-degeneracy:** $e(P_1, P_2) \neq 1$.
3. **Computability:** $e(P, Q)$ is efficiently computed, $\forall P \in \mathbb{G}_1, Q \in \mathbb{G}_2$.

For simplicity, hereafter, we set $\mathbb{G}_1 = \mathbb{G}_2 = \mathbb{G}$ and $P_1 = P_2$ but the proposed scheme can be easily modified for $\mathbb{G}_1 \neq \mathbb{G}_2$. We define a Bilinear Pairing Instance Generator as a PPT algorithm $\mathcal{G}$ that takes 1^κ and returns a random tuple $\mathbf{t} = (p, \mathbb{G}, \mathbb{G}_T, e, P)$ of bilinear pairing parameters where p is of size κ.

2.2   General BB Signatures

This is a generalization of the Boneh-Boyen signature scheme [1], which is unforgeable under a weak chosen message attack if the SDH assumption holds. It allows generation of a single signature for two random messages and an efficient knowledge proof of the signature and messages without revealing anything about the signature and messages.

Key Generating. Suppose $(p, \mathbb{G}, \mathbb{G}_T, e, Q)$ is a bilinear pairing tuple. Generate random $H' \leftarrow \mathbb{G}$ and $s' \leftarrow \mathbb{Z}_p^*$ and obtain $Q'_{pub} = Q^{s'}$. The public key is (Q, H', Q'_{pub}) and the secret key is s'.

Signing. For messages $t \in \mathbb{Z}_p^*$ and $\check{t} \in \mathbb{Z}_p \setminus \{-s'\}$, output the signature $R = (Q^t H')^{1/(s'+\check{t})}$.

Verifying. For a public key (Q, H', Q'_{pub}), messages $t \in \mathbb{Z}_p^*$ and $\check{t} \in \mathbb{Z}_p \setminus \{-s'\}$, and a signature $R \in \mathbb{G}$, verify that $e(R, Q^{\check{t}} Q'_{pub}) = e(Q^t H', Q)$.

2.3   CL-SDH Signatures

This is a variant of the Camenisch-Lysyanskaya signature scheme [4] using the SDH assumption. Note that, as shown in [11], there is an efficient protocol between a user and a signer to generate a CL-SDH signature for the user's message without the signer learning anything about the message; and there is an efficient zero-knowledge proof of knowledge of a CL-SDH message-signature pair.

Key Generating. Suppose $(p, \mathbb{G}, \mathbb{G}_T, e, P)$ is a bilinear pairing tuple. Generate random $P_0, H' \leftarrow \mathbb{G}$ and $\gamma \leftarrow \mathbb{Z}_p^*$ and obtain $P_{pub} = P^\gamma$. The public key is (P, P_0, H', P_{pub}) and the secret key is γ.

Signing. For message $x \in \mathbb{Z}_p^*$, generate random $v \leftarrow \mathbb{Z}_p$ and $a \leftarrow \mathbb{Z}_p \setminus \{-\gamma\}$ and compute $S = (P^x H'^v P_0)^{1/(\gamma+a)}$. The signature is (a, S, v).

Verifying. For a public key (P, P_0, H', P_{pub}), a message $x \in \mathbb{Z}_p^*$, and a signature (a, S, v), verify that $e(S, P^a P_{pub}) = e(P^x H'^v P_0, P)$.

3   Model

This section revises the formal model for dynamic k-TAA [13,12].

3.1   Procedures

A dynamic k-TAA system is specified as a tuple of PT algorithms (GKg, AKg, Join^U, Join^M, Bound, Grant, Revoke, Authen^U, Authen^P, Trace), operated by a *group manager* (GM), *application providers* (AP) and *users*. Each AP $\mathcal{V}$ has a public *authentication log* LOG$_\mathcal{V}$, an *access group* $AG_\mathcal{V}$ of users who are allowed to access its application, and some *public information* $PI_\mathcal{V}$. The algorithms are described as follows.

GKg: The GM runs this *setup* PPT algorithm on input 1^l to obtain a group public key *gpk* and the GM's secret key *gsk*.

AKg: An AP $\mathcal{V}$ runs this PPT algorithm on input a group public key gpk to obtain a pair of AP public key and secret key $(apk_{\mathcal{V}}, ask_{\mathcal{V}})$.

Join^U, Join^M: This *joining* protocol allows the GM to register a user into the group. Both of the interactive algorithms Join^U (the user) and Join^M (the GM) take as input the group public key gpk and Join^M is also given the GM's secret key gsk. Join^M returns either accept or reject. If it is accept, Join^U outputs a pair of member public key and secret key (mpk_i, msk_i).

Bound: An AP $\mathcal{V}$ uses this *bound announcement* PPT algorithm to announce the number of times a user in its access group can use its application. It takes as input gpk, $apk_{\mathcal{V}}$ and $ask_{\mathcal{V}}$ and outputs the upper bound k and some information which is published with the AP's identity $ID_{\mathcal{V}}$.

Grant: An AP $\mathcal{V}$ runs this algorithm to grant a group member access to its application. The AP adds the member to its access group $AG_{\mathcal{V}}$ and updates his public information $PI_{\mathcal{V}}$. From $PI_{\mathcal{V}}$, the member can obtain an access key mak.

Revoke: This algorithm allows an AP to revoke a group member from accessing its application. It removes the member from the AP's access group and updates its public information.

Authen^U, Authen^P: This *authentication* protocol, between a user (Authen^U) and an AP $\mathcal{V}$ (Authen^P), allows the AP to authenticate the user for accessing its application. The protocol input is all of the AP and group's public information, and Authen^U's private input includes the user's keys mpk, msk and mak. Authen^P returns accept, if the user is in the AP's access group and has been authenticated by the AP less than k times, or reject otherwise. The authentication transcript is added to the log LOG$_{\mathcal{V}}$.

Trace: Anyone can run this *public tracing* PPT algorithm to trace a malicious user. It takes as input all group public information and an authentication log and outputs a user identity, GM or NONE, which respectively mean "the user attempts to access more than the announced bound", "the GM published information maliciously", and "there is no malicious entity recorded in this log".

3.2   Correctness and Security Requirements

The adversary has access to a number of oracles and can query them according to the brief description below, to learn about the system and increase his success chance in the attacks. The oracles include $\mathcal{O}_{LIST}$, $\mathcal{O}_{QUERY}$, $\mathcal{O}_{JOIN-U}$, $\mathcal{O}_{AUTH-U}$, $\mathcal{O}_{JOIN-GM}$, $\mathcal{O}_{AUTH-AP}$, $\mathcal{O}_{GRAN-AP}$, $\mathcal{O}_{REVO-AP}$ and $\mathcal{O}_{CORR-AP}$ whose descriptions are provided in the full version. The correctness condition and security requirements for dynamic k-TAA are summarized as follows. Formal definitions of oracles and requirements can be found in [13,12].

Correctness: It requires that an honest member who is in the access group of an honest AP and has performed the authentication protocol with the AP for less than the allowed number of times, is successfully authenticated by the AP.

Anonymity: Intuitively, it means that given two honest group members i_0 and i_1, who are in the access group of an AP, it is computationally hard to distinguish between authentication executions, which are performed by the AP and one of the two members. In the experiment, the adversary is allowed to collude with the GM, all APs, and all users except target users i_0 and i_1, and to query oracles $\mathcal{O}_{LIST}$, $\mathcal{O}_{JOIN-U}$, $\mathcal{O}_{AUTH-U}$ and $\mathcal{O}_{QUERY}$. The adversary is allowed to make only one query to $\mathcal{O}_{QUERY}$ on input i_0, i_1 and an AP whose access group contains i_0 and i_1. On receiving such a query, $\mathcal{O}_{QUERY}$ makes either i_0 or i_1 to execute the authentication protocol with the AP and outputs the protocol transcript. Each of the users i_0 and i_1 must be authenticated by the AP within k times. The anonymity condition holds if the probability that the adversary can correctly guess the user identity used in $\mathcal{O}_{QUERY}$'s authentication execution is negligibly better than a random guess.

This anonymity definition is general enough to capture desirable privacy properties. For example, if the adversary can link authentication executions of the same user with different APs with non-negligible probability, then the adversary can break the anonymity experiment with non-negligible probability. In the experiment, the adversary can use $\mathcal{O}_{AUTH-U}$ to trigger authentication executions between i_0 or i_1 with different APs. When $\mathcal{O}_{QUERY}$ generates a challenged authentication execution, the adversary can link it to the executions generated by $\mathcal{O}_{AUTH-U}$ with non-negligible probability. As the adversary knows the user identity of each execution generated by $\mathcal{O}_{AUTH-U}$, it can tell the user identity of the challenged authentication execution with non-negligible probability.

Detectability: It loosely means that if a subgroup of corrupted members have performed the authentication procedure with the same honest AP for more than the total allowed number of times, then the public tracing algorithm using the AP's authentication log outputs NONE with negligible probability. The experiment has two stages and the adversary is allowed to corrupt all users. In the first stage, the adversary can query $\mathcal{O}_{LIST}$, $\mathcal{O}_{JOIN-GM}$, $\mathcal{O}_{AUTH-AP}$, $\mathcal{O}_{GRAN-AP}$, $\mathcal{O}_{REVO-AP}$ and $\mathcal{O}_{CORR-AP}$. Then all authentication logs of all APs are emptied. In the second stage, the adversary continues the experiment, but without access to the revoking oracle $\mathcal{O}_{REVO-AP}$. The adversary wins if he can be successfully authenticated by an honest AP $\mathcal{V}$ with access bound k for more than $k \times \#AG_\mathcal{V}$ times, where $\#AG_\mathcal{V}$ is the number of members in the AP's access group. The detectability condition requires that the probability that the adversary wins is negligible.

Exculpability for users: It intuitively means that the tracing algorithm does not output the identity of an honest user even if other users, the GM and all APs are corrupted. In the experiment, the adversary, who wants to frame an honest user i, is allowed to corrupt all entities except the user i and can access $\mathcal{O}_{LIST}$, $\mathcal{O}_{JOIN-U}$, and $\mathcal{O}_{AUTH-U}$. The adversary must authenticate user i using $\mathcal{O}_{AUTH-U}$ within the allowable numbers of times set by the APs. If the adversary succeeds in computing an authentication log, with which the public tracing algorithm outputs i, the adversary wins. The exculpability condition for users requires that the probability that the adversary wins is negligible.

Exculpability for the GM: Loosely speaking, it means that the tracing algorithm does not output the honest GM even if all users and all APs are corrupted. In the experiment, the adversary wants to frame the honest GM and he is allowed to corrupt all users and all APs and access $\mathcal{O}_{LIST}$ and $\mathcal{O}_{JOIN-GM}$. If the adversary succeeds in computing an authentication log, with which the public tracing algorithm outputs GM, the adversary wins. The exculpability condition for the GM requires that the probability that the adversary wins is negligible.

4 Dynamic k-TAA with Constant Authentication Costs

4.1 Overview

Section 1.1 has already given the general intuition of the approach "efficient provable e-tag", which substantially improves efficiency of our scheme over the NS05 and TFS04 schemes [12,13]. We now provide an outline of this scheme and note where this scheme is similar to NS05. In the GKg algorithm, a bilinear pairing tuple $(p, \mathbb{G}, \mathbb{G}_T, e, P)$ is generated, the GM's secret key is a CL-SDH secret key $\gamma \leftarrow \mathbb{Z}_p^*$ and the group public key includes the corresponding CL-SDH public key (P, P_0, H', P_{pub}) and a value $\Phi \leftarrow \mathbb{G}_T$.

As noted in section 2.3, there is an efficient protocol between a user and a signer to generate a CL-SDH signature for the user's secret message x without the signer learning anything about the message. This protocol underlies the joining protocol $(\text{Join}^U, \text{Join}^M)$, where the user also has to publish his identity and $\beta = \Phi^{1/x}$ in the identification list LIST that allows tracing of malicious users in the Trace algorithm. At the end of the joining protocol, the user obtains a CL-SDH signature (a, S, v) for a message x, where v is also the user's random secret. The user's member secret key is (x, v) and his member public key is (a, S, β). As also noted in section 2.3, there is an efficient zero-knowledge proof of knowledge of a CL-SDH message-signature pair (by proving the knowledge of (a, S, v) and x such that $e(S, P^a P_{pub}) = e(P^x H'^v P_0, P)$). The user can be anonymously authenticated as a group member by using this proof, as shown in the authentication protocol.

In the AKg algorithm, an AP's public-secret key pair includes a general BB public key (Q, H', Q'_{pub}) and the corresponding BB secret key is s'. The Bound algorithm, for a bound k, generates k random message couples $(t_1, \check{t}_1), ..., (t_k, \check{t}_k)$ and k corresponding general BB signatures $R_1, ..., R_k$. The AP publishes k *tag bases* $(t_1, \check{t}_1, R_1), ..., (t_k, \check{t}_k, R_k)$ to be used for up to k times user access to the AP's service (each tag base is a general BB message-signature triplet).

In the authentication protocol between the AP and a group member with key pair $((x, v), (a, S, \beta))$, the user obtains a random l from the AP, chooses a tag base $(t_i, \check{t}_i, R_i)$ and sends back a tag $(\Gamma, \check{\Gamma}) = (F(x, t_i), \check{F}(x, \check{t}_i, l))$, where F and $\check{F}$ are two functions. The user also shows the AP a zero-knowledge proof $Proof_2$ which proves four properties: (i) the user is a group member (by proving knowledge of a CL-SDH message-signature pair $(x, (a, S, v))$); (ii) the user knows a general BB message-signature triplet $(t_i, \check{t}_i, R_i)$ (without revealing the triplet); (iii) $(\Gamma, \check{\Gamma})$ is correctly computed from $l, x, (t_i, \check{t}_i), F$ and $\check{F}$ (that means $(\Gamma, \check{\Gamma}) = (F(x, t_i), \check{F}(x, \check{t}_i, l))$); and (iv) the AP has granted access to the user. Part (iv) is the same as in NS05 and we will talk about it afterwards. This protocol differs from NS05's authentication protocol with the construction of F and $\check{F}$ and parts (i), (ii) and (iii).

In the authentication protocols of TFS04 and NS05, the proof that one of the k announced tag bases has been used to compute the tag requires a proof of knowledge of one of k elements and its cost linearly depends on k. In our authentication protocol, that proof of knowledge of one of k tag bases is replaced by the proof of knowledge of a general BB message-signature triplet. Therefore, our authentication cost does not depend on k. The general BB signatures prevents the user from forging a new tag base without colluding the AP.

Similar to NS05, if the user uses the same tag base to compute another tag $(\Gamma', \check{\Gamma}')$, anyone can find these from the AP's authentication log (since $\Gamma = \Gamma'$) and use it to compute $\beta = (\check{\Gamma}/\check{\Gamma}')^{1/(l-l')}$, which is part of the user's public key ($\check{F}$ must be designed to allow this computation). However, if the member does not use the same tag base twice, his anonymity is protected (F and $\check{F}$ must be designed to allow this anonymity). The cost of checking if Γ has already appeared in the AP's authentication log is the same as in TFS04 and NS05, and is trivial if tags are orderly indexed by Γ, so we ignore that cost in claiming the 'constant' property.

F and $\check{F}$ must be designed so that: tags are not linkable; the property (iii) can be efficiently proved; and if a user uses the same tag base twice, his public key is computable from the two tags ($\beta = (\check{\Gamma}/\check{\Gamma}')^{1/(l-l')}$). We construct these two functions as $(\Gamma, \check{\Gamma}) = (\Phi^{1/(x+t_i)}, \Phi^{(lx+l\check{t}_i+x)/(x^2+x\check{t}_i)})$. This tag construction is different from [13,12] and developed from a recently proposed verifiable random function [7] using bilinear pairings. It possesses a precious feature of having both key x and tag base $t_i, \check{t}_i$ in the exponents of $\Phi^{1/(x+t_i)}$ and $\Phi^{(lx+l\check{t}_i+x)/(x^2+x\check{t}_i)}$. This feature allows the user's zero-knowledge proof $Proof_2$ in the authentication protocol to avoid the cut-and-choose method.

Now, we talk about the property (iv) and the Grant and Revoke algorithms, which are quite the same as in NS05. We also use dynamic accumulators to provide the dynamic property, which means the AP grants access to or revokes access from users. Each AP has a public key/secret key pair $((Q, Q_{pub}), s)$, where $Q_{pub} = Q^s$. To grant access to a member with a public key (a, S, β), the AP accumulates the value a of the public key into an *accumulated value* $V \leftarrow V^{s+a}$, and the member obtains the old accumulated value as the witness W. The member shows that the AP has granted access to him by proving the knowledge of (a, W) such that $e(W, Q^a Q_{pub}) = e(V, Q)$. To revoke access from the member, the AP computes a new *accumulated value* $V \leftarrow V^{1/(s+a)}$.

As in NS05, there is a Public Inspection algorithm (presented in the full version) executable by anyone to check if the APs perform the Bound, Grant and Revoke algorithms correctly.

4.2   Description

GKg
On input 1^κ, the Bilinear Pairing Instance Generator returns $(p, \mathbb{G}, \mathbb{G}_T, e, P)$. Generate $P_0, P_1, P_2, H, H' \leftarrow \mathbb{G}$, $\gamma \leftarrow \mathbb{Z}_p^*$ and $\Phi \leftarrow \mathbb{G}_T$, and let $P_{pub} = P^\gamma$.

The GM's secret key is a CL-SDH secret key $gsk = \gamma$. The group public key gpk consists of the corresponding CL-SDH group public key (P, P_0, H', P_{pub}) and values Φ, H, P_1, P_2. The identification list LIST of group members is initially empty.

AKg

An AP $\mathcal{V}$ generates $Q \leftarrow \mathbb{G}$, $s, s' \leftarrow \mathbb{Z}_p^*$ and computes $Q_{pub} = Q^s$, $Q'_{pub} = Q^{s'}$. The public and secret keys for the AP are $apk = (Q, Q_{pub}, Q'_{pub})$ and $ask = (s, s')$, respectively. They form a general BB key pair $((Q, H', Q'_{pub}), s')$. Then, same as NS05 [12], AP maintains an authentication log LOG, an *accumulated value*, which is published and updated after granting or revoking a member, and a public archive ARC (as the other public information PI in the formal model), which is a list of 3-tuples. The first component of the tuple is an element in the public key of a member, who was granted or revoked from accessing the AP. The second component is a single bit indicating whether the member was granted (1) or revoked (0). The third component is the accumulated value after granting or revoking the member. Initially, the accumulated value is set to $V_0 \leftarrow \mathbb{G}$ and LOG and ARC are empty.

$\mathbf{Join}^U$, $\mathbf{Join}^M$

This protocol allows the GM to generate a CL-SDH signature (a, S, v) for the user's secret x without learning anything about (x, v). The user also publishes $\beta = \Phi^{1/x}$. A user U_i can join the group as follows.

1. User U_i chooses $x, v' \leftarrow \mathbb{Z}_p^*$, computes $\beta = \Phi^{1/x}$ and a commitment $C = P^x H'^{v'}$ of x and adds (i, β) to the identification list LIST. The user then sends β and C to the GM with a standard non-interactive zero-knowledge proof $Proof_1 = PK\{(x, v') : C = P^x H'^{v'} \wedge \Phi = \beta^x\}$.
2. The GM verifies that (i, β) is an element of LIST and the proof is valid. The GM then generates $a \leftarrow \mathbb{Z}_p$ different from all corresponding previously generated values and $\tilde{v} \leftarrow \mathbb{Z}_p^*$, computes $S = (CH'^{\tilde{v}} P_0)^{1/(\gamma + a)}$, and sends $(S, a, \tilde{v})$ to user U_i.
3. User U_i computes $v = v' + \tilde{v}$ and confirms that equation $e(S, P^a P_{pub}) = e(P^x H'^v P_0, P)$ is satisfied. The new member U_i's secret key is $msk = (x, v)$, and his public key is $mpk = (a, S, \beta)$.

Bound

An AP publishes his identity ID and a number k as the bound. Let $(t_j, \check{t}_j) = \mathcal{H}_{\mathbb{Z}_p^* \times \mathbb{Z}_p^*}(ID, k, j)$ for $j = 1, ..., k$. The AP computes general BB signatures $R_j = (Q^{t_j} H')^{1/(s' + \check{t}_j)}$ for $j = 1, ..., k$ and publishes $(t_1, \check{t}_1, R_1), ..., (t_k, \check{t}_k, R_k)$. We call $(t_j, \check{t}_j, R_j)$ the jth *tag base* of the AP.

The Public Inspection algorithm (in the full version) can be run by anyone to check if the APs perform the Bound, Grant and Revoke algorithms correctly. So it is negligible that the APs can generate tag bases maliciously, for example, two APs setting the same t_j.

Grant

This is the same as in NS05. An AP grants access to a user U_i with public key $mpk = (a, \cdot, \cdot)$ as follows. Suppose there are j tuples in the AP's ARC and the AP's current accumulated value is V_j. The AP computes a new accumulated value $V_{j+1} = V_j^{s+a}$ and adds $(a, 1, V_{j+1})$ to his ARC. From the AP's ARC, the user U_i forms his access key $mak = (j + 1, W)$, where $W = V_j$, and keeps a counter d, which is initially set to 0.

Revoke

This is the same as in NS05. An AP revokes access from a user U_i with public key $mpk = (a, \cdot, \cdot)$ as follows. Suppose there are j tuples in the AP's ARC and the AP's current accumulated value is V_j. The AP computes a new accumulated value $V_{j+1} = V_j^{1/(s+a)}$, and adds $(a, 0, V_{j+1})$ to ARC.

Authen^U, Authen^P

The difference from NS05's authentication protocol lies in the second step, which is also the most important step of the protocol. In this step, the tag computation and $Proof_2$ are completely different from those in NS05. An AP (ID, k), whose public key and current accumulated value are $apk = (Q, Q_{pub}, Q'_{pub})$ and V respectively, authenticates a user U with public and secret keys $mpk = (a, S, \beta)$ and $msk = (x, v)$, respectively, as follows.

1. U increases counter d. If $d > k$, then U sends $\perp$ to the AP and stops. Otherwise, U runs the algorithm Update (as in [12] and in the full version) to update his access key $mak = (j, W)$. The AP then sends a random integer $l \leftarrow \mathbb{Z}_p^*$ to U.

2. U chooses an unused tag base $(t_\iota, \check{t}_\iota, R_\iota)$, computes tag $(\Gamma, \check{\Gamma}) = (\varPhi^{1/(x+t_\iota)}, \varPhi^{(lx+l\check{t}_\iota+x)/(x^2+x\check{t}_\iota)})$, and sends $(\Gamma, \check{\Gamma})$ to the AP with a proof
   $Proof_2 = PK\{(t_\iota, \check{t}_\iota, R_\iota, a, S, x, v, W) : \Gamma = \varPhi^{1/(x+t_\iota)} \wedge$
   $\check{\Gamma} = \varPhi^{(lx+l\check{t}_\iota+x)/(x^2+x\check{t}_\iota)} \wedge e(S, P^a P_{pub}) = e(P^x H'^v P_0, P) \wedge e(W, Q^a Q_{pub}) = e(V, Q) \wedge e(R_\iota, Q^{\check{t}_\iota} Q'_{pub}) = e(Q^{t_\iota} H', Q)\}$ ($Proof_2$ is described below).

3. If the proof is valid and if Γ is different from all corresponding tags in the AP's LOG, the AP adds tuple $(\Gamma, \check{\Gamma}, l)$ and the proof to LOG, and outputs accept. If the proof is valid and Γ is already written in LOG, the AP adds tuple $(\Gamma, \check{\Gamma}, l)$ and the proof to the LOG, outputs (detect,LOG) and stops. If the proof is invalid, the AP outputs reject and stops.

$Proof_2$

Let $U_1 = SH^{r_1}$; $U_2 = WH^{r_2}$; $U_3 = R_\iota H^{r_3}$ where $r_1, r_2, r_3 \leftarrow \mathbb{Z}_p$, then $Proof_2$ is equivalent to a proof of knowledge of $(t_\iota, \check{t}_\iota, a, x, v, r_1, r_2, r_3)$ such that

$$\Gamma^{x+t_\iota} = \varPhi; \quad \check{\Gamma}^{(x+\check{t}_\iota)x} \varPhi^{-lx-l\check{t}_\iota-x} = 1;$$

$$e(U_1, P)^a e(H, P)^{-r_1 a} e(H, P_{pub})^{-r_1} e(P, P)^{-x} e(H', P)^{-v}$$
$$= e(U_1, P_{pub})^{-1} e(P_0, P);$$

$$e(U_2, Q)^a e(H, Q)^{-r_2 a} e(H, Q_{pub})^{-r_2} = e(U_2, Q_{pub})^{-1} e(V, Q);$$

$$e(U_3, Q)^{\check{t}_\iota} e(H, Q)^{-r_3 \check{t}_\iota} e(H, Q'_{pub})^{-r_3} e(Q, Q)^{-t_\iota} = e(U_3, Q'_{pub})^{-1} e(H', Q)$$

Most of the pairing operations in this proof can be pre-computed. The member M computes the proof as follows.

1. Generate $r_1, r_2, r_3, k_1, ..., k_{18} \leftarrow \mathbb{Z}_p$ and compute
   $U_1 = SH^{r_1}; U_2 = WH^{r_2}; U_3 = R_\iota H^{r_3};$
   $U_4 = P_1^{r_1} P_2^{r_2} H^{r_4}; U_5 = P_1^{r_3} H^{r_5}; U_6 = P_1^{x+\tilde{t}_\iota} H^{r_6};$
   $T_1 = P_1^{k_1} P_2^{k_2} H^{k_4}; T_2 = P_1^{k_7} P_2^{k_8} H^{k_9} U_4^{-k_{10}}; T_3 = P_1^{k_3} H^{k_5};$
   $T_4 = P_1^{k_{11}} H^{k_{12}} U_5^{-k_{13}}; T_5 = P_1^{k_{14}+k_{13}} H^{k_6}; T_6 = P_1^{k_{15}} H^{k_{16}} U_6^{-k_{14}};$
   $\Pi_1 = \Gamma^{k_{14}+k_{17}}; \Pi_2 = \check{\Gamma}^{k_{15}} \Phi^{-lk_{14}-lk_{13}-k_{14}};$
   $\Pi_3 = e(U_1, P)^{k_{10}} e(H, P)^{-k_7} e(H, P_{pub})^{-k_1} e(P, P)^{-k_{14}} e(H', P)^{-k_{18}};$
   $\Pi_4 = e(U_2, Q)^{k_{10}} e(H, Q)^{-k_8} e(H, Q_{pub})^{-k_2};$
   $\Pi_5 = e(U_3, Q)^{k_{13}} e(H, Q)^{-k_{11}} e(H, Q'_{pub})^{-k_3} e(Q, Q)^{-k_{17}};$
2. Compute $c = \mathcal{H}_{\mathbb{Z}_p}(P||P_{pub}||P_0||H||H'||P_1||P_2||\Phi||Q||Q_{pub}||Q'_{pub}||ID||k||l||$
   $V||U_1||...||U_6||T_1||...||T_6||\Pi_1||...||\Pi_5)$
3. Compute in $\mathbb{Z}_p$: $s_1 = k_1 + cr_1; s_2 = k_2 + cr_2; s_3 = k_3 + cr_3; s_4 = k_4 + cr_4;$
   $s_5 = k_5 + cr_5; s_6 = k_6 + cr_6; s_7 = k_7 + cr_1 a; s_8 = k_8 + cr_2 a; s_9 = k_9 + cr_4 a;$
   $s_{10} = k_{10} + ca; s_{11} = k_{11} + cr_3 \tilde{t}_\iota; s_{12} = k_{12} + cr_5 \tilde{t}_\iota; s_{13} = k_{13} + c\tilde{t}_\iota; s_{14} = k_{14} + cx; s_{15} = k_{15} + c(x + \tilde{t}_\iota)x; s_{16} = k_{16} + cr_6 x; s_{17} = k_{17} + ct_\iota; s_{18} = k_{18} + cv$
4. Output $(U_1, ..., U_6, c, s_1, ..., s_{18})$

Verification of $Proof_{2b}$. Checking the following equation

$c \overset{?}{=} \mathcal{H}_{\mathbb{Z}_p}(P||P_{pub}||P_0||H||H'||P_1||P_2||\Phi||Q||Q_{pub}||Q'_{pub}||ID||k||l||V||U_1||...||U_6||$
$P_1^{s_1} P_2^{s_2} H^{s_4} U_4^{-c} || P_1^{s_7} P_2^{s_8} H^{s_9} U_4^{-s_{10}} || P_1^{s_3} H^{s_5} U_5^{-c} || P_1^{s_{11}} H^{s_{12}} U_5^{-s_{13}} ||$
$P_1^{s_{14}+s_{13}} H^{s_6} U_6^{-c} || P_1^{s_{15}} H^{s_{16}} U_6^{-s_{14}} || \Gamma^{s_{14}+s_{17}} \Phi^{-c} || \check{\Gamma}^{s_{15}} \Phi^{-ls_{14}-ls_{13}-s_{14}} ||$
$e(U_1, P)^{s_{10}} e(H, P)^{-s_7} e(H, P_{pub})^{-s_1} e(P, P)^{-s_{14}} e(H', P)^{-s_{18}} e(U_1, P_{pub})^c$
$e(P_0, P)^{-c} || e(U_2, Q)^{s_{10}} e(H, Q)^{-s_8} e(H, Q_{pub})^{-s_2} e(U_2, Q_{pub})^c e(V, Q)^{-c} ||$
$e(U_3, Q)^{s_{13}} e(H, Q)^{-s_{11}} e(H, Q'_{pub})^{-s_3} e(Q, Q)^{-s_{17}} e(U_3, Q'_{pub})^c e(H', Q)^{-c}.$

Trace

This algorithm is almost the same as in NS05. The identity of a malicious user can be traced from an AP's LOG as follows.

1. Look for two entries $(\Gamma, \check{\Gamma}, l, Proof)$ and $(\Gamma', \check{\Gamma}', l', Proof')$ in the LOG, such that $\Gamma = \Gamma'$ and $l \neq l'$, and that $Proof$ and $Proof'$ are valid. If no such entry can be found, output NONE.
2. Compute $\beta = (\check{\Gamma}/\check{\Gamma}')^{1/(l-l')} = (\Phi^{(lx+l\tilde{t}_\iota+x)/(x^2+x\tilde{t}_\iota)} / \Phi^{(l'x+l'\tilde{t}_\iota+x)/(x^2+x\tilde{t}_\iota)})^{1/(l-l')} = \Phi^{1/x}$, and look for a pair (i, β) from the LIST. Output member identity i, or if no such (i, β) can be found conclude that the GM has deleted some data from LIST, and output GM.

4.3 Comparison

Apart from providing the same desirable properties of the NS05 and TFS04 schemes, a significant advantage of our scheme is that its authentication costs do not depend on k or any parameter. Its only tradeoff is that the Bound algorithm needs to compute $\{R_1, ..., R_k\}$ for the tag bases. However, each AP needs to run the Bound algorithm only once whereas the authentication protocol is executed by all granted members for k times. So the tradeoff is very trivial compared to the advantage.

We have the following comparison on the number of exponentiations (EX), scalar multiplications (SM), pairings (PA) and transmitted bytes in the authentication protocol. For the communication comparison, we use the parameters in [12]. The TFS04 scheme has $\nu = 1024$, $\varepsilon = \mu = \kappa = 160$. For other schemes, p is a 160-bit prime, $\mathbb{G}_T$ is a subgroup of a finite field of size approximately 2^{1024} and $\mathbb{G}_T$ elements can be compressed by a factor of three using techniques in [9]. Most of the pairings can be pre-computed. The user can compute $e(U_1, P)^{k_{10}}$ by pre-computing $e(S, P)$ and $e(H, P)$ and computing $e(S, P)^{k_{10}} e(H, P)^{k_{10} r_1}$ (this way removes pairing computation but increases the number of exponentiations). It is similar for $e(U_2, Q)^{k_{10}}$ and $e(U_3, Q)^{k_{13}}$. Note that the TFS04 scheme does not provide the dynamic property and does not have the Update algorithm. That algorithm is the same for NS05 and our scheme. It is not needed if NS05 and our scheme are modified to remove the dynamic property. So we do not count the cost of the Update algorithm in the comparison table. Besides, the number of bytes sent by a user in the NS05 scheme we computed ($60\,k + 408$) is different from that in [12] ($60\,k + 304$).

	TFS04	NS05	Our scheme
Computation by AP	$(17+8k)$EXs	$(15+8k)$EXs $+8$SMs$+4$PAs	21EXs$+$ 20SMs$+6$PAs
Computation by User	$(28+8k)$EXs	$(21+8k)$EXs $+12$SMs	22EXs$+$ 27SMs
Bytes sent by AP	40	20	20
Bytes sent by User	$60\,k + 1617$	$60\,k + 408$	585
Dynamic	No	Yes	Yes

4.4 Security

Security of our scheme is stated in Theorem 1, which is proved in Appendix B.

Theorem 1. *In the random oracle model, the dynamic k-TAA scheme provides: (i) Correctness; (ii) Anonymity under the Decisional Bilinear Diffie-Hellman Inversion assumption; (iii) Detectability under the Strong Diffie-Hellman assumption; (iv) Exculpability for users under the Computational Bilinear Diffie-Hellman Inversion 2 assumption; (v) Exculpability for the GM under the Strong Diffie-Hellman assumption.*

References

1. D. Boneh and X. Boyen. Short Signatures Without Random Oracles. EURO-CRYPT 2004, Springer-Verlag, LNCS 3027, pp. 56-73.
2. D. Boneh and X. Boyen. Efficient Selective-ID Secure Identity-Based Encryption Without Random Oracles. EUROCRYPT 2004, Springer-Verlag, LNCS 3027, pp. 223-238.

3. J. Camenisch, and A. Lysyanskaya. Dynamic Accumulators and Application to Efficient Revocation of Anonymous Credentials. CRYPTO 2002, Springer-Verlag, LNCS 2442, pp. 61-76.
4. J. Camenisch and A. Lysyanskaya. A Signature Scheme with Efficient Protocols. SCN 2002, Springer-Verlag, LNCS 2576.
5. J. Camenisch, S. Hohenberger, and A. Lysyanskaya. Compact E-Cash. EURO-CRYPT 2005, Springer-Verlag, LNCS 3494, pp. 302-321, 2005.
6. L. Chen, M. Enzmann, A. Sadeghi, M. Schneider, and M. Steiner. A Privacy-Protecting Coupon System. Financial Cryptography 2005, Springer-Verlag, LNCS 3570, pp. 93-109.
7. Y. Dodis and A. Yampolskiy. A Verifiable Random Function with Short Proofs and Keys. Public Key Cryptography 2005, Springer-Verlag, LNCS 3386, pp. 416-431.
8. A. Fiat, and A. Shamir. How to prove yourself: practical solutions to identification and signature problems. CRYPTO 1986, Springer-Verlag, LNCS 263, pp. 186-194.
9. R. Granger, D. Page, and M. Stam. A Comparison of CEILIDH and XTR. Algorithmic Number Theory, 6th International Symposium, ANTS-VI, pages 235-249. Springer, June 2004.
10. A. Kiayias, and Moti Yung. Group Signatures: Provable Security, Efficient Constructions and Anonymity from Trapdoor-Holders. Cryptology ePrint Archive: Report 2004/076.
11. Lan Nguyen. Privacy-Protecting Coupon System Revisited. Financial Cryptography Conference (FC) 2006, LNCS, Springer, 2006.
12. L. Nguyen and R. Safavi-Naini. Dynamic k-Times Anonymous Authentication. Applied Cryptography and Network Security (ACNS) 2005, Springer-Verlag, LNCS 3531, 2005.
13. I. Teranisi, J. Furukawa, and K. Sako. k-Times Anonymous Authentication. ASIACRYPT 2004, Springer-Verlag, LNCS 3329, pp. 308-322, 2004.
14. I. Teranishi and K. Sako. k-Times Anonymous Authentication with a Constant Proving Cost. Public Key Cryptography 2006, Springer-Verlag, LNCS 3958, pp. 525-542, 2006.
15. V. Wei. Tracing-by-Linking Group Signatures. Information Security Conference (ISC) 2005, Springer-Verlag, LNCS 3650, pp. 149-163, 2005.

A Preliminaries

A.1 Notation

For a function $f : \mathbb{N} \to \mathbb{R}^{+}$, if for every positive number α, there exists a positive integer κ_0 such that for every integer $\kappa > \kappa_0$, it holds that $f(\kappa) < \kappa^{-\alpha}$, then f is said to be *negligible*. Let PT denote *polynomial-time*, PPT denote *probabilistic* PT and DPT denote *deterministic* PT. For a PT algorithm $\mathcal{A}(\cdot)$, "$x \leftarrow A(\cdot)$" denotes an output from the algorithm. For a set $\mathbf{X}$, "$x \leftarrow \mathbf{X}$" denotes an element uniformly chosen from $\mathbf{X}$, and $\#\mathbf{X}$ denotes the number of elements in $\mathbf{X}$. Let "$\Pr[Procedures|Predicate]$" denote the probability that *Predicate* is true after executing the *Procedures*, $\mathcal{H}_{\mathbf{X}}$ denote a hash function from the set of all finite binary strings $\{0,1\}^*$ onto the set $\mathbf{X}$, and $PK\{x : R(x)\}$ denote a proof of knowledge of x that satisfies the relation $R(x)$. An *adversary* is modelled by an interactive Turing machine, which interacts with some oracles.

Each oracle performs operations and produces outputs required by queries from the adversary. An entity is *corrupted* if the adversary has the entity's secret keys and completely controls the entity's actions. We define $1/0$ to be 0.

A.2   Complexity Assumptions

q-Strong Diffie-Hellman (q-SDH) Assumption [1]. *For every PPT algorithm $\mathcal{A}$, the following function $Adv_{\mathcal{A}}^{q\text{-}SDH}(\kappa)$ is negligible.*

$$Adv_{\mathcal{A}}^{q\text{-}SDH}(\kappa) = Pr[(\mathcal{A}(\mathbf{t}, P, P^s, \ldots, P^{(s^q)}) = (c, P^{1/(s+c)})) \wedge (c \in \mathbb{Z}_p)]$$

where $\mathbf{t} = (p, \mathbb{G}, \mathbb{G}_T, e, P) \leftarrow \mathcal{G}(1^\kappa)$ and $s \leftarrow \mathbb{Z}_p^*$.
The assumption informally means that there is no PPT algorithm that can compute a pair $(c, P^{1/(s+c)})$, where $c \in \mathbb{Z}_p$, from a tuple $(P, P^s, \ldots, P^{(s^q)})$, where $s \leftarrow \mathbb{Z}_p^*$.

Decisional Bilinear Diffie-Hellman Inversion (DBDHI) Assumption [2]. *For every PPT algorithm $\mathcal{A}$, the following function $Adv_{\mathcal{A}}^{DBDHI}(\kappa)$ is negligible.*

$$\begin{aligned} Adv_{\mathcal{A}}^{DBDHI}(\kappa) = |&Pr[\mathcal{A}(\mathbf{t}, P, P^s, \ldots, P^{(s^q)}, e(P, P)^{1/s}) = 1] \\ -&Pr[\mathcal{A}(\mathbf{t}, P, P^s, \ldots, P^{(s^q)}, \Gamma) = 1]| \end{aligned}$$

where $\mathbf{t} = (p, \mathbb{G}, \mathbb{G}_T, e, P) \leftarrow \mathcal{G}(1^\kappa)$, $\Gamma \leftarrow \mathbb{G}_T^*$ and $s \leftarrow \mathbb{Z}_p^*$.
Intuitively the DBDHI assumption [2] states that there is no PPT algorithm that can distinguish between a tuple $(P, P^s, \ldots, P^{(s^q)}, e(P, P)^{1/s})$ and a tuple $(P, P^s, \ldots, P^{(s^q)}, \Gamma)$, where $\Gamma \leftarrow \mathbb{G}_T^*$ and $s \leftarrow \mathbb{Z}_p^*$. We define the Computational Bilinear Diffie-Hellman Inversion 2 assumption, which holds if either DBDHI or SDH holds.

Computational Bilinear Diffie-Hellman Inversion 2 (CBDHI2) Assumption. *For every PPT algorithm $\mathcal{A}$, the following function $Adv_{\mathcal{A}}^{CBDHI2}(\kappa)$ is negligible.*

$$Adv_{\mathcal{A}}^{CBDHI2}(\kappa) = Pr[\mathcal{A}(\mathbf{t}, P, P^s, \ldots, P^{(s^q)}, e(P, P)^{1/s}) = s]$$

where $\mathbf{t} = (p, \mathbb{G}, \mathbb{G}_T, e, P) \leftarrow \mathcal{G}(1^\kappa)$ and $s \leftarrow \mathbb{Z}_p^*$.

B   Security Proofs

For Theorem 1, as part (i) can easily be proved by checking equations, we only provide proofs for parts (ii), (iii), (iv) and (v). Due to space limitation, we omit the proof that $Proof_2$ is non-interactive zero-knowledge, which is standard.

B.1   Proof of Theorem 1 (ii)

Suppose there exists a PPT adversary $\mathcal{A}$ breaking the Anonymity property of our scheme, we show a PPT adversary $\mathcal{B}$ that can break the DBDHI assumption.

Let $\mathbf{t} = (p, \mathbb{G}, \mathbb{G}_T, e, P') \leftarrow \mathcal{G}(1^\kappa)$ and a tuple $\alpha = (P', P'^w, \ldots, P'^{(w^q)}, \Lambda)$ be uniformly chosen from either $S_0 = \{(P', P'^w, \ldots, P'^{(w^q)}, e(P', P')^{1/w}) | w \leftarrow \mathbb{Z}_p^*\}$ or $S_1 = \{(P', P'^w, \ldots, P'^{(w^q)}, \Lambda) | w \leftarrow \mathbb{Z}_p^*, \Lambda \leftarrow \mathbb{G}_T^*\}$. $\mathcal{B}$'s challenge is to guess whether α is chosen from S_0 or S_1. $\mathcal{B}$ interacts with $\mathcal{A}$ as follows.

$\mathcal{B}$ randomly chooses a bit $b \leftarrow \{0, 1\}$ and let b' be the other bit. $\mathcal{B}$ generates different $\delta_0, \check{\delta}_0, \delta_1, \check{\delta}_1, \ldots, \delta_{q-1}, \check{\delta}_{q-1} \leftarrow \mathbb{Z}_p^*$ and sets $x_b = w - \delta_0$ (without knowing x_b). Let $F = x_b(x_b + \check{\delta}_0) \prod_{i=1}^{q-1}(x_b + \delta_i)(x_b + \check{\delta}_i)$, then it can be presented as a polynomial $F = \sum_{i=0}^{2q} A_i w^i$, where $A_0, \ldots, A_{2q}$ are computable from $\delta_0, \check{\delta}_0, \delta_1, \check{\delta}_1, \ldots, \delta_{q-1}, \check{\delta}_{q-1}$ and $A_0 \neq 0$. Therefore, $\mathcal{B}$ can compute $\Phi = e(P', P')^F$, $\beta_b = \Phi^{1/x_b}$, $\check{\Theta}_0 = \Phi^{1/(x_b + \check{\delta}_0)}$, $\Theta_i = \Phi^{1/(x_b + \delta_i)}$ and $\check{\Theta}_i = \Phi^{1/(x_b + \check{\delta}_i)}$, $i = 1, \ldots, q - 1$ from $(P', P'^w, \ldots, P'^{(w^q)})$. Given l, $\mathcal{B}$ can also compute $\Phi^{(lx_b + l\check{\delta}_i + x_b)/(x_b^2 + x_b\check{\delta}_i)} = \beta_b^l \check{\Theta}_i$ for $i = 0, \ldots, q - 1$. Let $\Theta_0 = e(P', P')^{\sum_{i=1}^{2q} A_i w^{i-1}}$. Λ^{A_0}, if $\Lambda = e(P', P')^{1/w}$ then $\Theta_0 = \Phi^{1/(x_b + \delta_0)}$.

$\mathcal{B}$ selects $P, P_0, P_1, P_2, H, H' \leftarrow \mathbb{G}$, $\gamma \leftarrow \mathbb{Z}_p^*$, and computes $P_{pub} = P^\gamma$. $\mathcal{B}$ provides $\mathcal{A}$ the group public key $gpk = (P, P_0, H', P_{pub}, \Phi, H, P_1, P_2)$ and the group secret key $gsk = \gamma$. $\mathcal{B}$ creates a number of users including two target users i_0 and i_1 that will be sent to $\mathcal{O}_{QUERY}$ later.

At any time, $\mathcal{A}$ can create a new AP by generating apk, ask, an initial accumulated value, LOG and ARC as described in the AKg algorithm. Because $\mathcal{A}$ determines the AP's identity to be sent to $\mathcal{O}_{QUERY}$, it can create more APs without detriment to its attack. Therefore, let ζ be the upper bound on the number of APs, we can assume $\mathcal{A}$ always creates ζ APs. $\mathcal{B}$ randomly picks $m \leftarrow \mathbb{Z}_\zeta^*$. Suppose the mth AP ID_m has bound k_m, $\mathcal{B}$ randomly picks $j_m \leftarrow \mathbb{Z}_{k_m}^*$.

$\mathcal{B}$ simulates oracles accessible by $\mathcal{A}$ as follows.

- Random oracle $\mathcal{H}_{\mathbb{Z}_p^* \times \mathbb{Z}_p^*}$: This oracle is queried in the Bound algorithm. If the query is (ID_m, k_m, j_m), the oracles returns $(t = \delta_0, \check{t} = \check{\delta}_0)$. Otherwise, on the ith query, the oracle returns $(t = \delta_i, \check{t} = \check{\delta}_i)$.

- $\mathcal{O}_{LIST}$: This oracle operates as in the definition of $\mathcal{O}_{LIST}$, with regard to an *identification list* LIST of user identity/public-key pairs. $\mathcal{A}$ can query the oracle to view a user's public key. $\mathcal{A}$ can request the oracle to record the identity and public key of a user, who is not i_0 or i_1, to LIST. $\mathcal{A}$ can request the oracle to delete data from LIST.

- $\mathcal{O}_{JOIN-U}$: $\mathcal{A}$ just needs to query this oracle to register i_0 and i_1 to the group, as $\mathcal{A}$ can collude other users and the GM.

  If $\mathcal{A}$ asks the oracle to register i_b, $\mathcal{B}$ chooses $C_b \leftarrow \mathbb{G}$, computes $\beta_b = \Phi^{1/x_b}$ and adds (i_b, β_b) to LIST. $\mathcal{B}$ (the oracle) then returns β_b and C_b to $\mathcal{A}$ (the GM) with a simulation of the standard non-interactive zero-knowledge proof $Proof_1 = PK\{(x_b, v_b') : C_b = P^{x_b}H'^{v_b'} \wedge \Phi = \beta_b^{x_b}\}$. The GM follows the Join protocol's description and sends back $(S_b, a_b, \tilde{v}_b)$. $\mathcal{B}$ then checks that $e(S_b, P^{a_b}P_{pub}) = e(C_bH'^{\tilde{v}_b}P_0, P)$ and sets i_b's public key as (a_b, S_b, β_b) (i_b's secret key $(x_b, v_b = v_b' + \tilde{v}_b)$ is unknown).

  If $\mathcal{A}$ asks the oracle to register $i_{b'}$, $\mathcal{B}$ chooses $x_{b'}, v_{b'}' \leftarrow \mathbb{Z}_p^*$, computes $\beta_{b'} = \Phi^{1/x_{b'}}$ and follows the Join protocol's description so that $(i_{b'}, \beta_{b'})$ is added to LIST, $i_{b'}$'s public key is $(a_{b'}, S_{b'}, \beta_{b'})$ and $i_{b'}$'s secret key is $(x_{b'}, v_{b'})$.

- $\mathcal{O}_{AUTH-U}$: $\mathcal{A}$ just needs to query this oracle to authenticate i_0 and i_1, as $\mathcal{A}$ can collude other users, the APs and the GM.

  If i_b is queried to be authenticated by an AP (ID, k), whose public key and current accumulated value are $apk = (Q, Q_{pub}, Q'_{pub})$ and V respectively, and i_b's counter d for this AP is not greater than k, $\mathcal{B}$ runs the algorithm Update to update his access key $mak = (j, W_b)$. On receiving a random integer $l \leftarrow \mathbb{Z}_p^*$ from the AP, $\mathcal{B}$ chooses a unused tag base $(t_\iota, \check{t}_\iota, R_\iota)$, where $(t_\iota, \check{t}_\iota)$ is different from $(\delta_0, \check{\delta}_0)$, computes tag $(\Gamma, \check{\Gamma}) = (\Phi^{1/(x_b+t_\iota)}, \Phi^{(lx_b+l\check{t}_\iota+x_b)/(x_b^2+x_b\check{t}_\iota)})$, and sends $(\Gamma, \check{\Gamma})$ to the AP with a simulation of the proof $Proof_2$, which can be done by using the simulator in the proof for $Proof_2$'s zero-knowledge property and resetting the random oracle. $\mathcal{A}$ and $\mathcal{B}$ perform the rest of the authentication protocol as specified in Section 4.2.

  If $i_{b'}$ is queried to be authenticated by an AP, as $\mathcal{B}$ knows $i_{b'}$'s secret key, $\mathcal{A}$ and $\mathcal{B}$ can simulate the authentication protocol as specified in Section 4.2.

- $\mathcal{O}_{QUERY}$: If the queried AP is not ID_m, $\mathcal{B}$ fails and exits. Otherwise, as m is randomly chosen, the probability that the queried AP is ID_m is at least $1/\zeta$. In this case, suppose the AP ID_m has public key $apk = (Q, Q_{pub}, Q'_{pub})$ and current accumulated value V, and i_b's counter d for this AP is not greater than k. $\mathcal{B}$ runs the algorithm Update to update his access key $mak = (j, W_b)$. On receiving a random integer $l \leftarrow \mathbb{Z}_p^*$ from the AP, $\mathcal{B}$ chooses the tag base $(t_\iota = \delta_0, \check{t}_\iota = \check{\delta}_0, R_\iota)$, computes tag $(\Gamma, \check{\Gamma}) = (\Theta_0, \Phi^{(lx_b+l\check{t}_\iota+x_b)/(x_b^2+x_b\check{t}_\iota)})$, and sends $(\Gamma, \check{\Gamma})$ to the AP with a simulation of the proof $Proof_2$, which can be done by using the simulator in the proof for $Proof_2$'s zero-knowledge property and resetting the random oracle. The AP ID_m and $\mathcal{B}$ perform the rest of the authentication protocol as specified in Section 4.2. $\mathcal{B}$ then outputs the transcript of the protocol.

From the transcript outputted by $\mathcal{O}_{QUERY}$, if $\mathcal{A}$ returns the bit b, then $\mathcal{B}$ decides that the tuple α is chosen from S_0. Otherwise, $\mathcal{B}$ decides that the tuple α is chosen from S_1. Then if $\mathcal{A}$ can break the Anonymity property of the k-TAA scheme, then $\mathcal{B}$ can break the DBDHI assumption.

B.2   Proof of Theorem 1 (iii)

Suppose there exists a PPT adversary $\mathcal{A}$ breaking the Detectability property of our scheme, we show a PPT adversary $\mathcal{B}$ that can break the SDH assumption. Let $challenge = (R, R^z, \ldots, R^{z^q})$ be a tuple of the SDH assumption, where $z \leftarrow \mathbb{Z}_p^*$, $\mathcal{B}$'s challenge is to compute $(c, R^{1/(z+c)})$, where $c \in \mathbb{Z}_p$.

As $\mathcal{A}$ can break Detectability, at the end of the experiment with non negligible probability, $\mathcal{A}$ can be successfully authenticated by an honest AP $\mathcal{V}$ with access bound k for more than $k \times \#AG_\mathcal{V}$ times, where $\#AG_\mathcal{V}$ is the number of members in the AP's access group. As $Proof_2$ is zero-knowledge, for each of these successful authentication runs, $\mathcal{A}$ must have the knowledge of a tag base $(t_\iota, \check{t}_\iota, R_\iota)$, a member public key (a, S), a member secret key (x, v) and a member access key W. There are 3 possible cases:

- A member secret key (x, v) in $\mathcal{V}$'s access group is used for authentication for more than k times. As $\mathcal{V}$ provides only k tag bases, $\mathcal{A}$ must generate a new valid tag base $(t, \check{t}, R)$ to use with (x, v). Following arguments (which can't be shown due to space limitation) similar to the proof of Lemma 2 in [12], if $\mathcal{A}$ can generate a new valid tag base $(t, \check{t}, R)$, then the SDH assumption does not hold.
- No member secret key in $\mathcal{V}$'s access group is used for authentication for more than k times and $\mathcal{A}$ can generate a new member key pair $((a, S, \beta), (x, v))$ different from any member key pair of the whole group. Following arguments (which can't be shown due to space limitation) similar to the proof of Lemma 2 in [12], if this can be done, then the SDH assumption does not hold.
- No member secret key in $\mathcal{V}$'s access group is used for authentication for more than k times and $\mathcal{A}$ can generate a new member access key W for a group member, who is not in $\mathcal{V}$'s access group and has a member key pair $((a, S, \beta), (x, v))$. In this case, $\mathcal{B}$ simulates the GM, the users, the APs and randomly chooses an AP $\mathcal{V}$ with bound k and provides them to $\mathcal{A}$. $\mathcal{B}$ then runs the GKg algorithm to generate $gpk = (P, P_0, H', P_{pub}, \Phi, H, P_1, P_2)$ and $gsk = \gamma$ and runs the AKg algorithm for all APs, except $\mathcal{V}$. For $\mathcal{V}$, $\mathcal{B}$ selects $f, s' \leftarrow \mathbb{Z}_p^*$, and set $Q = R$, $Q_{pub} = R^z$, $Q'_{pub} = Q^{s'}$ and $V_0 = R^f$. The initial accumulated value is V_0 and $\mathcal{V}$'s keys are $((Q, Q_{pub}, Q'_{pub}), (z, s'))$, where $\mathcal{B}$ does not know z. With these capabilities, $\mathcal{B}$ can easily provide $\mathcal{A}$ access to simulations of the oracles $\mathcal{O}_{JOIN-GM}$, $\mathcal{O}_{AUTH-AP}$, $\mathcal{O}_{GRAN-AP}$, $\mathcal{O}_{REVO-AP}$ and $\mathcal{O}_{CORR-AP}$, except when $\mathcal{A}$ uses $\mathcal{O}_{CORR-AP}$ to corrupt $\mathcal{V}$, $\mathcal{B}$ fails and stops. Note that when $\mathcal{A}$ uses $\mathcal{O}_{GRAN-AP}$ or $\mathcal{O}_{REVO-AP}$ to ask $\mathcal{V}$ to grant access to or revoke access from a user, $\mathcal{B}$ can always use the tuple *challenge* to compute the new accumulated value, as long as the number of users is less than q. As $\mathcal{B}$ randomly chooses $\mathcal{V}$, with non-negligible probability, $\mathcal{A}$ can be successfully authenticated by $\mathcal{V}$ for more than $k \times \#AG_{\mathcal{V}}$ times and generate a new member access key W for a group member, who is not in $\mathcal{V}$'s access group and has a member key pair $((a, S, \beta), (x, v))$. Suppose the public keys of all members in AG_{ID} are $\{(a_i, \cdot, \cdot)\}_{i=1}^m$, then the current accumulated value of the AP is $V = R^{f \prod_{i=1}^m (a_i + z)}$, therefore $W = R^{f \prod_{i=1}^m (a_i + z)/(a+z)}$. From W and the tuple *challenge*, $\mathcal{B}$ can compute $R^{1/(a+z)}$ and thereby break the SDH assumption.

B.3   Proof of Theorem 1 (iv)

We show that if there exists a PPT adversary $\mathcal{A}$ breaking Exculpability for users in our scheme, then there exists a PPT adversary $\mathcal{B}$ breaking the Computational Bilinear Diffie-Hellman Inversion 2 assumption. Let $\mathbf{t} = (p, \mathbb{G}, \mathbb{G}_T, e, P') \leftarrow \mathcal{G}(1^\kappa)$ and suppose that $\mathcal{B}$ is given a challenge $\alpha = (P', P'^w, \ldots, P'^{(w^q)}, e(P', P')^{1/w})$ and $\mathcal{B}$ needs to compute w. $\mathcal{B}$ generates different $\delta_0, \check{\delta}_0, \delta_1, \check{\delta}_1, \ldots,$ $\delta_{q-1}, \check{\delta}_{q-1} \leftarrow \mathbb{Z}_p^*$ and sets $x = w - \delta_0$. $\mathcal{B}$ simulates an instance of the dynamic k-TAA scheme and the oracles in the same way as simulations in the experiment of Anonymity proof, except that there is only one target user i and $e(P', P')^{1/w}$ is used instead of Λ. So we omit the description of simulations.

If $\mathcal{A}$ can break Exculpability for users, then Trace outputs i at the end of the experiment. That means there exist $(\Gamma_1, \check{\Gamma}_1, l_1, Proof)$ and $(\Gamma_2, \check{\Gamma}_2, l_2, Proof')$ in the log of an AP such that $\Gamma_1 = \Gamma_2$, $(\check{\Gamma}_1/\check{\Gamma}_2)^{1/(l_1-l_2)} = \beta(= \Phi^{1/x})$ and $Proof$ and $Proof'$ are valid. As $\mathcal{A}$ can only use $\mathcal{O}_{AUTH-U}$ within the allowable numbers of times, not both $(\Gamma_1, \check{\Gamma}_1, l_1, Proof)$ and $(\Gamma_2, \check{\Gamma}_2, l_2, Proof')$ is created by $\mathcal{B}$ using the oracle.

In the case neither of them was created by $\mathcal{B}$ using the oracle, as $Proof_2$ is zero-knowledge, $\mathcal{A}$ must have the knowledge of $(x_1, t_1, \check{t}_1)$ and $(x_2, t_2, \check{t}_2)$ such that $(\Gamma_1, \check{\Gamma}_1) = (\Phi^{1/(x_1+t_1)}, \Phi^{(l_1 x_1 + l_1 \check{t}_1 + x_1)/(x_1^2 + x_1 \check{t}_1)})$; $(\Gamma_2, \check{\Gamma}_2) = (\Phi^{1/(x_2+t_2)}, \Phi^{(l_2 x_2 + l_2 \check{t}_2 + x_2)/(x_2^2 + x_2 \check{t}_2)})$; $\Gamma_1 = \Gamma_2$ and $\check{\Gamma}_1/\check{\Gamma}_2 = \Phi^{(l_1-l_2)/x}$. By converting all elements into exponents of Φ, one can compute x from $x_1, t_1, \check{t}_1, l_1, x_2, t_2, \check{t}_2, l_2$. Therefore, w is computable. By similar arguments for the case when one of $(\Gamma_1, \check{\Gamma}_1, l_1, Proof)$ or $(\Gamma_2, \check{\Gamma}_2, l_2, Proof')$ was created by $\mathcal{B}$ using the oracle, one can also find w.

B.4   Proof of Theorem 1 (v)

Suppose a PPT adversary $\mathcal{A}$ can break Exculpability for the GM in our scheme, we show that the SDH assumption does not hold. If Trace outputs GM at the end of the experiment, there exist $(\Gamma, \check{\Gamma}, l, Proof)$ and $(\Gamma', \check{\Gamma}', l', Proof')$ in the log of an AP such that $\Gamma = \Gamma'$, $(\check{\Gamma}/\check{\Gamma}')^{1/(l-l')} \notin LIST$ and $Proof$ and $Proof'$ are valid. As $Proof_2$ is zero-knowledge, $\mathcal{A}$ must have the knowledge of $(t, \check{t}, a, S, x, v)$ and $(t', \check{t}', a', S', x', v')$ such that $x+t = x'+t'$. If $t \neq t'$, then with non-negligible probability, either x or x' is not issued in the Join protocol with the GM; so a new valid member public key/secret key pair has been created without the GM. If $t = t'$, then $x = x'$. But $\Phi^{1/x} \notin LIST$, so x is not issued in the Join protocol with the GM; so a new valid member public key/secret key pair has also been created without the GM. Following arguments (which can't be shown due to space limitation) similar to the proof of Lemma 2 in [12], if a new valid member public key/secret key pair can be created without the GM, then the SDH assumption does not hold.

Side Channel Analysis of Practical Pairing Implementations: Which Path Is More Secure?

Claire Whelan* and Mike Scott

School of Computing, Dublin City University
Ballymun, Dublin 9. Ireland
{cwhelan, mike}@computing.dcu.ie

Abstract. We present an investigation into the security of three practical pairing algorithms; the Tate, truncated Eta (η_T) and Ate pairing, in terms of side channel vulnerability. These three algorithms have recently shown to be efficiently computable on the resource constrained smart card, however no in depth side channel analysis of these specific pairing implementations has yet appeared in the literature. We assess these algorithms based on two main avenues of attack since the secret parameter input to the pairing can potentially be entered in two possible positions, i.e. $e(P, Q)$ or $e(Q, P)$ where P is public and Q is private. We analyse the core operations fundamental to pairings and propose how they can be attacked in a computationally efficient way. Building on this we show how each implementation may potentially succumb to a side channel attack and demonstrate how one path is more susceptible than the other in Tate and Ate. For those who wish to deploy pairing based systems we make a simple suggestion to improve resistance to side channel attacks.

Keywords: Side Channel Analysis (SCA), Pairing Based Cryptography, Correlation Power Analysis (CPA), Tate Pairing, Ate Pairing, η_T Pairing.

1 Introduction

Pairings are a relatively new primitive in the world of cryptography. Pairings are bilinear maps, which make them attractive for cryptographic constructions. Since their introduction in the constructive sense[1], a multitude of pairing based protocols have been suggested and a handful of efficient pairing implementations have been developed. We refer the reader to [1] for a comprehensive listing of such papers.

Side Channel Analysis (SCA) has advanced immeasurably since its breakthrough into the security community almost a decade ago [10]. Almost every cryptographic construction, especially those intended for use in the smart card, have been subject to some form of SCA or another. These powerful attacks,

---

* Supported by the Irish Research Council for Science, Engineering and Technology (IRCSET).

[1] Initially pairings were suggested for cryptanalytic purposes [12].

P.Q. Nguyen (Ed.): VIETCRYPT 2006, LNCS 4341, pp. 99–114, 2006.

which do not play by the rules of traditional cryptanalysis, have proven successful against many algorithms.

In this paper we perform passive differential side channel analysis (in the form of correlation power analysis (CPA)) of three pairing algorithms, namely the Tate [3], truncated Eta (η_T) [2] and Ate pairing [9].

1.1   Related Work

The first mention of side channel analysis of pairings was in 2004 when Page and Vercauteren [14] described a fault attack of Duursma-Lee algorithm [7] for characteristic three and how the multiplication operation in general pairings could be attacked using Simple Power Analysis (SPA) and a Messerges style Differential Power Analysis (DPA) [13]. While this paper identified the vulnerable operation in pairings (i.e. finite field multiplication), the method described to attack it was computationally infeasible. The method extracted one bit at a time of one of the coordinates of the secret parameter (for example, extracted one bit of x of $Q(x, y)$ at a time). Given that each coordinate is a element of the underlying finite field and potentially n bits (where n is at least 160 bits), extracting one bit at a time is unrealistic. This is without even considering the additional task of data acquisition and data processing required for DPA to extract one bit.

1.2   Motivation

We choose three specific pairing algorithms to assess, namely the BKLS algorithm for the Tate pairing [3], the Ate pairing [9], and the BGOhES algorithm for a truncated version of the Eta pairing η_T [2]. Our reasoning for choosing these implementations is that recently Scott et al. [16] presented the first timings for the computation of these pairings which was comparable with contemporary alternative cryptosystems on a 32 bit smart card. Since this contentious aspect that has previously hindered the widespread adoption of pairings from a commercial perspective is no longer an issue, the door is open for adoption of pairings on these potentially side channel attackable devices. Therefore thorough side channel evaluation of employable pairing implementations is necessary and vital.

1.3   Contributions of This Work

In this paper we build on Page and Vercauterens work by describing a more in depth approach to performing side channel analysis of three specific pairing implementations. We solely concentrate on passive side channel attacks which monitor the natural inescapable emanations of a device such as power analysis [11], as opposed to determining the effects of purposely induced faults. We provide a computationally feasible method of attacking the finite field operations fundamental to pairings. We describe this attack in terms of both finite field multiplication and square root, since the square root operation is a potentially attackable operation in the η_T pairing. However we approach our analysis from a different perspective. Instead of focusing on specific algorithms for specific operations, we focus on how operations are computed from a structural perspective. We assess each candidate pairing algorithm based on the prospect of the secret parameter being entered in either parameter position and show how some pairing implementations are more susceptible to attack than others.

The paper is organised as follows. A brief overview of the candidate pairing algorithms and correlation power analysis (CPA) is presented in section 2. In section 3 we analyse the core pairing operations in terms of how they may be attacked using side channels from a structural sense. We define a strategy of attack for each pairing algorithm and consequently compare Tate, η_T and Ate in section 4. We present possible countermeasures and address their effectiveness in deterring SCA in section 5. Finally we conclude and summarise our findings in section 6. Note that the specific pairing algorithms themselves can be found in appendix A.

2   Background

We briefly review relevant details on pairings and CPA.

2.1   Overview of Practical Pairings

Let E be an elliptic curve over a finite field $\mathbb{F}_q$. Pairings are functions which map a pair of elliptic curve points $P, Q \in E(\mathbb{F}_q)$, to an element of a multiplicative group of an underlying finite field $\mu \in \mathbb{F}_q$. Algorithms A.1, A.2 and A.3 describe implementations of Tate, Ate and η_T respectively. Each of these algorithms are efficiently computable on a 32 bit smart card, executing in under half a second [16]. Each of the algorithms we consider are optimised pairing algorithms.

The BKLS [3] algorithm is a particularly fast method for computing the Tate Pairing $e(P, Q)$, where $P \in E(\mathbb{F}_p)$ is a point on the base curve and $Q \in E'(\mathbb{F}_{p^{k/d}})$ is a point on the d-th order twist with embedding degree k, where d is at least 2 when k is even [4]. BKLS can be calculated over supersingular or non-supersingular curves over finite fields of arbitrary characteristic.

The Ate pairing [9] $a(P, Q)$ is the most recently discovered pairing algorithm, and is potentially faster than BKLS for non-supersingular curves. Ate cleverly observes that it is more efficient to make the first parameter $P \in E'(\mathbb{F}_{p^{k/d}})$ and the second parameter $Q \in E(\mathbb{F}_p)$.

The BGOhES algorithm for the η_T pairing, is a generalisation of Duursma-Lee pairing algorithm for the Tate pairing with a truncated loop. The pairing $\eta_T(P, Q)$ is calculated on a supersingular curve over small characteristic, where both parameters P and Q are elements in $E(\mathbb{F}_{p^m})$ where $p = 2$ or 3.

Each pairing algorithm ultimately consists of an application of Millers algorithm followed by a final exponentiation. The notable difference between the three pairing is that Ate and Eta both have half length loops compared to Tate. From a specific implementation standpoint, we will address the cases where the Tate and Ate pairing is calculated over the large prime field $\mathbb{F}_{p^k}$ and the η_T pairing is calculated over the binary field $\mathbb{F}_{2^{km}}$.

2.2   Correlation Power Analysis

Our reasoning for using CPA is that it focuses on words of data at a time instead of selection functions, and it overcomes some of the shortcomings of differential power analysis (DPA) such as ghost peaks [6].

The basis for correlation power analysis (CPA) [6] and other forms of passive differential SCA is that there exists a relationship between the data being processed during a computation and detectable physical manifestations such as power consumption. This dependance is magnified by capturing numerous acquisitions of the target in operation and then applying statistical analysis techniques to differentiate the signal of interest from noise.

Specifically, CPA builds a hypothetical model based on assumptions made about what constitutes energy dissipation. Then for key guesses, the correctness of a guess is established by estimating what the consumption of such data would be (based on the model) and then comparing it to actual data. This is generally performed using a correlation test such as Pearson's correlation coefficient:

$$\rho_{X,Y} = \frac{E(XY) - E(X)E(Y)}{\sqrt{E(X^2) - E^2(X)}\sqrt{E(Y^2) - E^2(Y)}} \tag{1}$$

where X relates to the actual data acquired from the attack such as the power consumption and Y relates to the estimated power consumption derived from the power model adopted (typical choices are the hamming weight or hamming distance model).

CPA reveals words (or partial words) of data at a time. We aim to employ CPA and recover the secret by iteratively extracting feasible portions of the secret.

3   Side Channel Analysis of Naive Pairings

In a number of pairing based protocols, either the P or Q parameter is secret. For example, in Boneh and Franklin's identity based encryption [5] the critical operation involving the secret key in a pairing is the decryption operation. Although we are analysing pairings in isolation, the associated side channel security of pairings have implications in the bigger picture.

In order to perform critical analysis of the candidate pairing algorithms, it is necessary to analyse the core pairing operations in terms of how much information they can potentially leak. In this section we will analyse the finite field calculations central to pairings. Before we address these operations individually, we make some observations about pairings.

3.1   Pairing Observations

We note the following possible opportunistic observations about pairings:

1. The secret parameter can potentially be entered as the first or second parameter in the pairing. If the curve is supersingular and a distortion map

$\psi(.)$ is used, as is the case with the Eta pairing, the parameters to the pairing $e(P, Q)$ can be switched, i.e. $e(Q, P)$ will yield the same result. In the case of the Tate and Ate pairing, while the parameter can take either path, it must hold for the entire protocol. Therefore this presents us with two avenues of attack; the P path and the Q path. We note that depending on which path is most vulnerable to SCA, such implications may lead to a simple method of defence.

2. Due to point compression we only need to extract the x coordinate of the secret point. Once this is found there are only two possibilities for y. Therefore we restrict our attention to the secret x coordinate.

3. We will try to focus on operations which involve elements from the base field $\mathbb{F}_q$, where $q = p$ or 2^m since extension field elements $\mathbb{F}_{q^k}$ are k bit times larger than that of base field elements.

3.2 Structural Analysis of Core Pairing Operations

One of the key requirements in performing a differential side channel attack is to identify an exploitable operation in the algorithm which involves some known (or computable) data and the secret key. Since elliptic curve arithmetic ultimately relies on the underlying finite field, we will restrict our analysis to multiplication, squaring, square root and reduction over the binary field and multiplication and reduction over the prime field. We refer the reader to appendix A to see when and where such operations are used in the candidate pairings.

We briefly recap on binary field and prime field arithmetic, since the candidate pairing implementations are over $\mathbb{F}_{2^m}$ and $\mathbb{F}_p$.

Characteristic two finite fields $\mathbb{F}_{2^m}$ are constructed using polynomial basis representation: $a(z) = \{a_i z^{m-1} + a_{i-1} z^{m-2} + \ldots + a_2 z^2 + a_1 z + a_0 \mid a_i \in \{0, 1\}\}$ where $a(z) \in \mathbb{F}_{2^m}$ has degree at most $m - 1$. Arithmetic over $\mathbb{F}_{2^m}$ is modulo the irreducible polynomial $f(z)$. We will represent $a(z) \in \mathbb{F}_{2^m}$ as the concatenation of w bit blocks: $a(z) = a_{(m/w)-1} | a_{(m/w)-2} | \ldots | a_0$, where w is the underlying processor's word length.

Characteristic p finite fields, $\mathbb{F}_p$, where p is a large prime, consist of the integers $0, 1, 2, \ldots, p - 1$ with arithmetic modulo p. Let $n = \lceil \log_2 p \rceil$ be the bit length of p. We will represent the elements $a \in \mathbb{F}_p$ as the concatenation of w bit blocks: $a = a_{(n/w)-1} | a_{(n/w)-2} | \ldots | a_0$.

Since we only need to deterministically calculate partial output of target operations, we revert back to the most basic methods for insight.

Multiplication. The most straightforward method for multiplication is the shift and xor method for $\mathbb{F}_{2^m}$ and the operand scanning method for $\mathbb{F}_p$. These methods are very similar, and so will only describe the former.

The multiplication[2] of two $\mathbb{F}_{2^m}$ elements $a(z) = \sum_{i=0}^{m-1} a_i z^i$ and $b(z) = \sum_{i=0}^{m-1} b_i z^i$ will produce the binary polynomial $c(z) = \sum_{i=0}^{2m-1} c_i z^i$, with degree

---

[2] In the context of binary fields by multiplication we mean carry-free binary polynomial multiplication.

$2m - 1$. The shift and xor method involves multiplying words of $b(z)$ by words of $a(z)$ at a time. This process is depicted in figure 1. In the smart card system used by Scott et al. [16] they used a special binary polynomial multiplication instruction. The main distinction between this method and the multiplication of two $\mathbb{F}_p$ elements is that instead of xor-ing, addition (with carry bits) is performed.

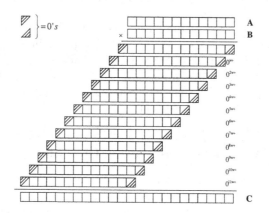

Fig. 1. Multiplication of $\mathbb{F}(2^m)$ elements: the shift and xor method

A simple power analysis (SPA) attack on the bitwise shift and xor method was suggested by Page et al. in [14], which could easily be extended to apply to the operand scanning method. However, since it is unlikely that this basic algorithm will be favoured in a constrained embedded device, it is doubtful that SPA will work. Other attacks on modular multiplication have also been suggested. Walter [17] demonstrates how Montgomery multiplication can be attacked with SPA if an extra reduction is included.

In reality a number of multiplication algorithms can be implemented. We suggest that instead of focusing on the multiplication algorithm itself, we focus on the result of the multiplication. Due to the structural evolution of multiplication, which the basic algorithms allow us to easily see (as in figure 1), we can easily identify which data portions effect the resulting product value (or partial product). A possible side channel attack of the multiplication operation is as follows:

Let the target operation for a CPA attack be the multiplication of two finite field elements. Note that we will deal with the act of multiplication and reduction separately for the moment. Let x be an n-bit known (or computable) value by the adversary and k be an n-bit unknown secret value. Let $y = x \cdot k$ be the resulting $2n$-bit product. We represent x, k and y as the concatenation of w bit blocks: $x = x_{(n/w)-1}|x_{(n/w)-2}|\dots|x_0$, $k = k_{(n/w)-1}|k_{(n/w)-2}|\dots|k_0$ and $y = y_{(2n/w)-1}|y_{(2n/w)-2}|\dots|y_0$ accordingly.

Since multiplication is not a suitable selection or partition function, and CPA is the attack of choice, w-bit portions of k will be extracted at a time[3]. To identify the target input block we denote x_l and k_l to be the lth w-bit block of x and k respectively, where $0 \leq l \leq n/w - 1$. To identify the hypothetical output block we denote y_r to be the rth w-bit block of y, where $0 \leq r \leq (2n/w) - 1$.

If we are dealing with implementations over the binary field, there are two possible positions from which the attack can commence, either the most or least significant word of k, since all middle words of the product $x \cdot k$ are polluted by the outermost words. If the implementation is over the prime field, we are restricted to commencing from the least significant word only since carry propagation will significantly effect all other words. We will describe the case where we begin searching the least significant word of k, k_0. First all the data for the correlation test is produced.

Algorithm 1. Generate hypothetical output of the multiplication $x_0 \cdot k_0$ for all possible k_0, where x_0 is known. N is the number of times the algorithm is executed and consequently relates to the number of acquisitions captured.

INPUT: x_0
OUTPUT: H_0
1: **for** $0 \leq j < 2^w$ **do**
2:     **for** $1 \leq i \leq N$ **do**
3:         $k_0 = j$
4:         $y_0 = k_0 \cdot x_0$
5:         $H_0(i,j) = y_0$
6:     **end for**
7: **end for**
8: **return** H_0

This will produce a $N \times 2^w$ matrix H_0 detailing the hypothetical product of all 2^w possible k_0's and the N known x_0's. Note that the actual multiplication of $k_0 \cdot x_0$ will produce a $2w$ bit product y_0, however only the least significant word of this is required as entry in H_0. The most significant word of y_0 contributes to the subsequent product word y_1.

To identify which is the correct least significant word k_0, the correlation is calculated between the estimated power consumption of each row in H_0 (this contributes to Y in equation (1)) and a discrete time interval in the acquired physical traces where the target operation is being executed (this contributes to X in equation (1)). The hypothesis with the highest correlation, is identified as the correct least significant word k_0.

To extract the remaining interior words of k, the attack proceeds similar to algorithm 1. It can be seen from figure 1 that in the $2n$-bit product y, all middle

---

[3] In the case of Scott et al. implementation, where $w = 32$ it will be a computationally intensive task to extract one word. However it is possible to calculate the partial correlation by just focusing on practical portions of w at a time.

words are influenced by more than one word in k. Therefore k_1 cannot be found unless k_0 is known, and k_2 cannot be found unless k_1 and k_0 is known, etc. Therefore, line 5 in algorithm 1 is replaced by $y_r = (k_l \cdot x_0) + ($ auxiliary words $)$ for $1 \leq l \leq n/w - 2$. For instance $y_1 = (k_1 \cdot x_0) + (k_0 \cdot x_1) + (k_0 \cdot x_0)$ and $y_2 = (k_2 \cdot x_0) + (k_1 \cdot x_1) + (k_0 \cdot x_2) + (k_1 \cdot x_0)$.

The computational cost of such an attack is $l \times 2^w$. Note that we can improve on this slightly when analysing binary field implementations. By observing the fact that the most and least significant words of k can be independently calculated (i.e. no middle words of k influence the multiplied output), we can simultaneously calculate hypotheses for k_0 and $k_{n/w-1}$. Once both of these terms have been extracted, then the search can step inwards, i.e. calculate hypotheses for k_1 and $k_{n/w-2}$ simultaneously, etc. This reduces the cost of extracting k to $\frac{l}{2} \times 2^w$.

Squaring. A variety of fast multiplication algorithms exist for squaring finite field elements. Here we will view squaring in its simplest form which is the multiplication of $x \cdot k$, where $x = k$, and so the attack just described can be applied in the same way to the squaring operation.

Square Root. The square root method is only called on in the Eta pairing implementation, and so only square root calculation over the binary field will be discussed. An efficient method for calculating the square root can be obtained from the observation that $\sqrt{a}$ can be expressed in terms of the square root of the element z [8]. Basically the value a is split into it's odd and even coefficients, as depicted in figure 2, and then the odd portion is multiplied by $\sqrt{z}$ and subsequently added to the even portion. If the irreducible polynomial $f(z)$ is a

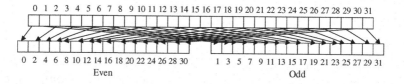

Fig. 2. Square Root of $\mathbb{F}(2^m)$ elements where $w = 32$

trinomial or pentanomial, then an efficient formula for calculating $\sqrt{z}$ can be used. For example, $\sqrt{z} = x^{\frac{m+1}{2}} + z^{\frac{n+1}{2}} \pmod{f(z)}$ when $f(z)$ is a trinomial and $\sqrt{z} = z^{\frac{m+1}{2}} + z^{\frac{n+1}{2}} + z^{\frac{q+1}{2}} + z^{\frac{r+1}{2}} \pmod{f(z)}$ when $f(z)$ is a pentanomial.

Since the act of square root is a single operand operation, the only way the act of calculating the square root of a secret value can be used in a side channel attack is if, after a prediction about a (and the resulting $\sqrt{a}$) has been made, this value is later used in an operation involving known data. The hypothetical output of this following operation is then used to verify the hypotheses. This is a form of second order attack and will be described later in more detail. For now we will describe how a value entered into the square root function changes in a structural sense, so that by knowing a portion of a we can deterministically calculate what a portion of $\sqrt{a}$ will be.

Let k equal the unknown secret value as before. The $\sqrt{k}$ is calculated as follows: First k is split into left and right chunks. Given that we will be predicting w bits of k at a time, this means that we will know $\frac{w}{2}$ bits of odd, and $\frac{w}{2}$ bits of even. We will denote these half words by k_o and k_e respectively.

In the multiplication step, two $\frac{m}{2}$ bit quantities, $k_o \cdot \sqrt{z}$, will be multiplied to produce an m bit product. Since we know one of the multipliers $\sqrt{z}$ we can predict what a portion of the product will be as before. This portion will be $\frac{w}{2}$ bits. Since there will not be multiple reduction steps (there might not even be one), we can calculate what the final output of the reduction will be.

The final step to the square root operation is the addition (xor) of the even values from before k_e and the product $k_o \cdot \sqrt{z}$ (mod $f(z)$). Since we know the $\frac{w}{2}$ bits of even, and $\frac{w}{2}$ bits of the product, we can calculate $\frac{w}{2}$ of the value $\sqrt{k}$. This can be carried on to the next step in the algorithm, where we can determine how these $\frac{w}{2}$ bits effect the result of the next operation. As before, we have two possible positions from which this attack can commence; the most significant word and the least significant word. Once either of these has been established the middle neighbour words can be searched for.

Even though $\frac{w}{2}$ bit portions are used to verify w bit hypotheses of k, the computational costs of extracting k is still $l \times 2^w$ or $\frac{l}{2} \times 2^w$ if k is attacked simultaneously from both ends.

Reduction. Almost all operations over finite fields are coupled with reduction. The protocol for modulo operations depends on the implementation, i.e. reduction can be performed either concurrently or consecutively. If reduction is performed consecutively, the attacks of the preceding operations can be applied as described. If reduction is performed concurrently, we will have to revise our attack strategy.

Over the binary field, the moduli chosen is of special form such that it permits fast reduction, for example irreducible trinomials or pentanomials are preferred.

Straightforward reduction can be performed using the shift and subtract method, where subtract over $\mathbb{F}_{2^m}$ is xor and subtract over $\mathbb{F}_p$ involves borrow bits. $a(z) \equiv b(z)$ (mod $f(z)$) or $a \equiv b$ (mod p) basically involves lining the modulus up with the most significant bit of a (or $a(z)$) and subtracting to produce an intermediate value t. The modulus is then repeatedly lined up with intermediate t's until the the bit length (or degree) of t is less than the bit length of p (or degree of $f(z)$).

If repeated reduce is implemented, then it is more difficult to definitively calculate the hypothetical output of interest. For example in the case of multiplication, if we are to predict partial output of $c \equiv a \cdot b$ (mod p), we must be able to calculate all of the product $a \cdot b$. Knowing only portions of $a \cdot b$ is not sufficient since the waterfall effect of reduction will lose these portions in a manner unpredictable by the adversary.

The implication of repeated reduce is that intermediate output of the calculation of $c \equiv a \cdot b$ (mod p) must now be used for the hypothesis testing. A possible attack might proceed as follows: We will describe this attack for the case where modular multiplication is the operation of interest, this technique may be applied similarly to other methods. Let x equal the known data, and k equal the unknown secret data as before. To extract k_0, we can hypothetically calculate the intermediate output $t \equiv x \cdot k_0$, where t is the intermediate $(m + w)$-bit result. t will then be reduced by p to once again produce a m-bit value. Since the modulus is public, the resultant m-bit value (or even portions of it) will be used to verify the correct k_0. To extract k_1, partial hypothetical output of $(t \equiv x \cdot k_1 \pmod p) + (t \equiv x \cdot k_0 \pmod p)$, where we assume k_0 has been found, is calculated. This process is repeated until no words of k remain unknown.

4   Possible Attacks

So far we have described how individual operations may be attacked using side channels. Now we will put these attacks into context as we describe when and where in the three candidate pairing algorithms these operations are performed and how they can be exploited to extract secret data. For each algorithm we will assess both paths where the first scenario details the situation where Q is secret (*Case 1*) and the second where P is secret (*Case 2*). Note that each case will be addressed relating to the specific implementation details given in [16].

4.1   The Tate Pairing: BKLS

BKLS [3] (algorithm 2, A.1) implements the Tate pairing $e(P, Q)$, where the first input parameter P is a point of order r on the base curve $E(\mathbb{F}_p)$, and the second parameter Q, is a point on the twist $E'(\mathbb{F}_{p^{k/2}})$. $e(P, Q)$ evaluates to an element in the finite field $\mathbb{F}_{p^k}$, where p is a large prime and k is the embedding degree. In the specific implementation described in [16] $k = 2$ and so the points P and Q have coordinates in $\mathbb{F}_p$. This means the coordinates (x, y) will be approximately the same length as the bit length of p.

Case 1. When P is public, since the order r will be a published parameter we can generate all intermediate jP values where $1 \leq j \leq r$ (for the calculation of rP). Q on the other hand will remain static throughout the pairing computation. The target operation in the algorithm involving the secret Q is: $m_j = y_j - \lambda_j(x_Q + x_j) - y_Q i$ where x_j, y_j and λ_j are known, $i = \sqrt{-1}$ and x_Q and y_Q are secret.

Since we only need extract x_Q, we can focus on the operation $\lambda_j(x_Q + x_j)$. As we will know x_j and λ_j, we can employ the attack of the multiplication operation described in 3.2 and extract words of x_Q at a time.

High order SCA can be applied here since we will know x_j, y_j, λ_j for all j, and so can calculate the hypothetical output of $\lambda_j(x_Q + x_j)$ at multiple points.

Case 2. Conversely in the scenario where P is secret and Q is public, our known value remains static through the attack, and the secret parameter is constantly changing. Intuitively this makes this path more difficult to attack, as even if intermediate values of P are recovered, the original P must be extracted requiring point subtraction and knowledge of the number of previous additions that have already been performed. On further inspection, based on our analysis of finite field operations, this avenue of attack is actually not possible for the following reason; In order to make an hypothesis, there must exist an operation where the adversary can deterministically calculate how the input effects the output (even partially). However given a section of x_j it is impossible to deterministically calculate any of the subsequent x_{j+1}, where x_{j+1} is the result from either the doubling or addition of the previous jP, since calculation of x_{j+1} requires knowledge of y_j. So even if we make predictions for the value of x_{j+1} in $\lambda_{j+1}(x_Q + x_{j+1})$, we have no way of determining what λ_{j+1} is and more importantly what the original x_j is.

This natural property of the BKLS algorithm can actually act as a deterrent by enforcing the secret parameter to be entered as the first parameter to the pairing.

4.2   The Ate Pairing

The Ate pairing $a(P,Q)$ [9] (algorithm 3, A.2) is also computed over the prime field, where P is chosen as a point of order r over the twisted curve $E'(\mathbb{F}_{p^{k/d}})$ with embedding degree k, and Q is chosen over the base field $E(\mathbb{F}_p)$. Here point scalar multiplication (the accumulation of P to rP) is calculated over the extension field $E'(\mathbb{F}_{p^{k/d}})$ (see [9] for details). This means that the underlying finite field arithmetic fundamental to point addition and doubling is performed over $\mathbb{F}_{p^{k/d}}$. However, the coordinates of Q will be over $\mathbb{F}_p$. In the specific implementation described in [16] $k = 4$ and $d = 2$.

Case 1. As with BKLS when P is public and Q is private the target operation is $m_j = i^2 y_Q - i(i^2 y_j/2 + \lambda_j(i^2 x_j/2 + x_Q))$ where elements from the twist jP are untwisted (hence the division by 2) and combined with Q to construct the Miller variable $m_j \in \mathbb{F}_{p^k}$. Note in this case $i = \sqrt{-2}$.

Isolating the operation $\lambda_j(i^2 x_j/2 + x_Q)$ involving the secret coordinate x_Q involves addition of an element in $\mathbb{F}_p$ to an element in $\mathbb{F}_{p^2}$ and multiplication over $\mathbb{F}_{p^2}$. Note $a+b$ where $a = x_1+y_1 i \in \mathbb{F}_p$ where $y_1 = 0$, and $b = x_2+y_2 i \in \mathbb{F}_{p^2}$, simply involves adding the real coefficient of a to the real coefficient of b.

To attack this operation, we once again utilise the observation about multiplication. Even though multiplication is now being performed over $\mathbb{F}_{p^2}$, we can still think in terms of multiplication over $\mathbb{F}_p$. So say $a = x_1 + y_1 i$ and $b = x_2 + y_2 i \in \mathbb{F}_{p^2}$, $a \cdot b$ is simply $(x_1 \cdot x_2 - 2y_1 \cdot y_2) + (x_1 \cdot y_2 + y_1 \cdot x_2)i$ where the internal multiplications are over $\mathbb{F}_p$. Relating to the our attack of $\lambda(i^2 x_j/2 + x_Q)$, this means we will be able to calculate the partial output of $x_1 \cdot x_2$ and $y_1 \cdot x_2$ where x_2 relates to the real coefficient of x_j which is added to x_Q. Note that we will be able to calculate all other portions of $a \cdot b$. Once again the attack of the multiplication operation as described in 3.2 can be employed to extract x_Q.

Case 2. The case where P is secret is almost analogous to the BKLS case, and hence appears to be impossible to attack. In Ate an attack would be even more complex since the calculation of jP involving point addition and doubling is over the extension field $\mathbb{F}_{p^2}$ and thus involves more complex arithmetic.

4.3 The η_T Pairing

The η_T algorithm (algorithm 4, A.3) is quite different from Tate and Ate. The implementation [2] for consideration is applicable to supersingular elliptic curves over the binary field $E(\mathbb{F}_{2^m})$. The pairing $\eta_T(P,Q)$ evaluates to an element in $\mathbb{F}_{2^{km}}$. Both parameters are points on the curve $E(\mathbb{F}_{2^m})$. No distortion map is explicitly used, as the map to the extension field is integrated into the algorithm.

Unlike BKLS and Ate, some preliminary computation takes place outside the loop. Operations on points are also completely avoided. The paths for P and Q are almost symmetric and so attack strategies for both paths are almost equivalent. The only difference is that where the square root of x_P and y_P is calculated, the squaring of x_Q and y_Q is performed.

Case 1. Here there are two main points of attack. The first point of attack is outside the loop; $f \leftarrow u \cdot (x_P + x_Q + 1) + y_P + y_Q + b + 1 + (u + x_Q)s + t$ where the first value in $\mathbb{F}_{2^{km}}$ is constructed. This is enabled by the incorporation of the public elements s and $t \in \mathbb{F}_{2^{km}}$.

Assuming that x_Q is secret and x_P is known, the operation $u \cdot (x_P + x_Q + 1)$ where $u = x_P + 1$, can be focused on. This operation basically involves addition (xor) and multiplication modulo the known irreducible polynomial $f(z)$. Hypothetical partial output of $u \cdot (x_P + x_Q + 1)$ can be calculated by guessing w-bit portions of x_Q.

The second point of attack is the squaring operation, i.e. $x_Q \leftarrow x_Q^2$. This squared value is subsequently used in the next round of the loop in $g \leftarrow u \cdot (x_P + x_Q) + y_P + y_Q + x_P + (u + x_Q)s + t$ where u in this calculation is x_P. By purely calculating what the hypothetical output of a portion of $x_Q \leftarrow x_Q^2$ is, we can analyse how this portion affects subsequent operations.

For example, assuming we have guessed what the least significant word of x_Q is, and calculated the least significant word of the resulting x_Q^2. This means that we can hypothetically calculate $u \cdot (x_P + x_Q + 1)$ after the first and second round of the for loop and still be able to easily trace back to the original Q.

Case 2. η_T is unique to Tate and Ate in that the two paths in the pairing are almost symmetric and so are equally as vulnerable. Similar to the attack of Q there are two main points of attack. The first point of attack is again outside the loop, where the operation $u \cdot (x_P + x_Q + 1)$ can be attacked.

The second point of attack is the square root function inside the loop. Given that we can deterministically calculate $\frac{w}{2}$ bits of hypothetical w bits of $\sqrt{x_P}$, this means that we can test how this predicted output effects the output of a number of subsequent operations to perform high order SCA. For example we can calculate the hypothetical output $u \cdot (x_P + x_Q)$ and $x_P + (u + x_Q)s + t$.

The only real obstacle that η_T provides is that regardless of whether the secret takes the either path, it is dynamic. Therefore if the adversary extracts an intermediate secret value they must work back to get the original point. This is in contrast to Ate and Tate which can be attacked at any point in the algorithm.

5   Possible Countermeasures and Their Implications

A number of countermeasures have already been anticipated to protect pairings against SCA [15], [14]. Taking advantage of bilinearity, the secret point can simply be blinded. A pairing can be calculated as $e(P,Q) = e(aP, bQ)^{1/ab}$ where a and b are random values or $e(P,Q) = e(P, Q + R)/e(P,R)$ where R is a random point. While these may be effective in deterring SCA since a new random value will be used every time the pairing is called, they are expensive, ultimately requiring point scalar multiplication and calculation of two pairings respectively.

Another more subtle countermeasure proposed by [15] observes that repeated multiplication of the Miller variable m in BKLS and Ate (or f in η_T) by a random element in $\mathbb{F}_p$ (or $\mathbb{F}_{2^m}$) will have no effect on the final pairing value since they will be eliminated in the final exponentiation. This is a less expensive deterrent only requiring a field multiplication per iteration of the Miller loop.

In order for this countermeasure to be effective, we recommend that the random value must not only be multiplied by the Miller variable, but must be multiplied by all intermediate values that make up the Miller variable. For example in the case of Tate; $m_j = r \cdot y_j - \lambda_j(r \cdot x_Q + r \cdot x_j) - r \cdot y_Q i$ where $r \in \mathbb{F}_p$. If a new random value is multiplied at every iteration of the loop, the attacks we have presented would no longer be possible.

6   Conclusion and Recommendations

We have presented the first passive differential side channel analysis of the Tate, Ate and η_T pairing. We performed this investigation in an analytical sense, where empirical knowledge of side channel attacks was used to determine where and how operations in the candidate algorithms could be exploited. We presented an attack of the multiplication, square root and reduction operations over finite fields, from a slightly different perspective. Instead of focusing on how these operations could be performed, we simply focus on trying to deterministically calculate partial output based on the structural expansion of basic algorithms.

We assessed the three candidate pairing algorithms based on the attack on both paths that a secret can take. From this we found that although none of the algorithms assessed proved to be resistant to SCA, Tate and Ate if implemented with the secret being stationed in the first parameter could withstand such attacks. η_T however, which is the most efficient algorithm computationally, is open to attack from either path proving that speed may not be the main consideration when choosing the best implementation.

From our findings we recommend two straightforward deterrents to protocol designers implementing pairing based protocols to protect against SCA: 1. If implementing the Tate or Ate pairing, ensure that the secret parameter is positioned in the first parameter (i.e. the secret takes the P path). 2. If implementing any of the pairings (but more specifically η_T), we recommend the adoption of the simple countermeasure proposed by [15] where intermediate finite field multiplication by a random value will successfully mask sensitive data.

References

1. P. Barreto. Pairing based crypto lounge. URL: http://paginas.terra.com.br/informatica/paulobarreto/pblounge.html.
2. P. Barreto, S. Galbraith, C. O'hEigeartaigh, and M. Scott. Efficient pairing computation on supersingular abelian varieties. Cryptology ePrint Archive: Report 2004/375. URL: http://eprint.iacr.org/2004/375.
3. P. Barreto, H. Kim, B. Lynn, and M. Scott. Efficient algorithms for pairing based cryptosystems. In *Advances in Cryptology - CRYPTO 02*, volume 2442 of *Lecture Notes in Computer Science*, pages 354–368. Springer Verlag, 2002.
4. P. S. L. M. Barreto, B. Lynn, and M. Scott. On the selection of pairing friendly groups. In *Selected Areas in Cryptography - SAC 2003*, Lecture Notes in Computer Science, Ottawa, Canada, 2003.
5. Dan Boneh and Matthew Franklin. Identity-based encryption from the weil pairing. In *Advances in Cryptology - CRYPTO 01*, volume 2139 of *Lecture Notes in Computer Science*, pages 213–229. Springer Verlag, 2001.
6. E. Brier, C. Clavier, and F. Olivier. Correlation power analysis with a leakage model. In M. Joye and J.J. Quisquater, editors, *Cryptographic Hardware and Embedded Systems - CHES 04*, volume 3156 of *Lecture Notes in Computer Science*, pages 16–29, 2004.
7. I. M. Duursma and H. S. Lee. Tate pairing implementation for hyperelliptic curves $y^2 = x^p - x + d$. In *Advances in Cryptology - Asiacrypt 2003*, volume 2894 of *Lecture Notes in Computer Science*, pages 111–123. Springer Verlag, 2003.
8. K. Fong, D. Hankerson, J. Lopez, and A. Menezes. Field inversion and point halving revisited. CACR Technical Report, CORR 2003-18. URL: http://www.cacr.math.waterloo.ca/, 2003.
9. F. Hess, N. Smart, and F. Vercauteren. The eta pairing revisited. Cryptology ePrint Archive: Report 2006/110. URL: http://eprint.iacr.org/2006/110.
10. P. Kocher. Timing attacks on implementations of diffie-hellman, rsa, dss and other systems. In *Advances in Cryptology - CRYPTO 96*, volume 1109 of *Lecture Notes in Computer Science*, pages 104–113. Springer Verlag, 1996.
11. P. Kocher, J. Jaffe, and B. Jun. Differential power analysis. In Michael J. Wiener, editor, *Advances in Cryptology - CRYPTO 99*, volume 1666 of *Lecture Notes in Computer Science*, pages 388–397. Springer Verlag, 1999.
12. A.J. Menezes, T. Okamoto, , and S. Vanstone. Reducing elliptic curve logarithms to logarithms in a finite field. In *IEEE Transactions on Information Theory*, volume 39, pages 1639–1646, 1993.
13. T. Messerges, E. Dabbish, and R. Sloan. Examining smart-card security under the threat of power analysis attacks. In *IEEE Transactions on Computers*, volume 51 of *4*, April 2002.

14. D. Page and F. Vercauteren. Fault and side-channel attacks on pairing based cryptography. Cryptology ePrint Archive, Report 2004/283. URL: http://eprint.iacr.org/2004/283.

15. M. Scott. Computing the tate pairing. In *CT-RSA*, volume 3376 of *Lecture Notes in Computer Science*, pages 293–304, 2005.

16. M. Scott, N. Costigan, and W. Abdulwahab. Implementing cryptographic pairings on smart cards. Cryptology ePrint Archive: Report 2006/144, URL: http://eprint.iacr.org/2006/144.

17. C. Walter. Simple power analysis of unified code for ecc double and add. In M. Joye and J. J. Quisquater, editors, *Cryptographic Hardware and Embedded Systems - CHES 04*, volume 3156 of *Lecture Notes in Computer Science*, pages 191–204, 2004.

A   Practical Pairing Implementations

A.1   The Tate Pairing

Algorithm 2. Computation of $e(P, Q)$ on $E(\mathbb{F}_p) : y^2 = x^3 + Ax + B$, where P is a point of prime order r on $E(\mathbb{F}_p)$ and Q is a point on the twisted curve $E'(\mathbb{F}_p)$

INPUT: $P = (x_P, y_P), Q = (x_Q, y_Q)$
OUTPUT: $m \in \mathbb{F}_p$

1: $m = 1$
2: $x_A, y_A \leftarrow x_P, y_P$
3: $n = r - 1$
4: **for** $i \leftarrow \lfloor \lg(r) \rfloor - 2$ **to** 0 **do**
5:      **if** $n_i = 1$ **then**
6:          $(\lambda, x_T, y_T) \leftarrow \text{double}(A, A)$
7:          $g = y_A - \lambda(x_Q + x_A) - i.y_Q$
8:          $x_A, y_A \leftarrow x_T, y_T$
9:          $m = m^2 \cdot g$
10:         $(\lambda, x_T, y_T) \leftarrow \text{add}(T, P)$
11:         $g = y_A - \lambda(x_Q + x_A) - i.y_Q$
12:         $x_A, y_A \leftarrow x_T, y_T$
13:         $m = m \cdot g$
14:     **else**
15:         $(\lambda, x_T, y_T) \leftarrow \text{double}(A, A)$
16:         $g = y_A - \lambda(x_Q + x_A) - i.y_Q$
17:         $x_A, y_A \leftarrow x_T, y_T$
18:         $m = m^2 \cdot g$
19:     **end if**
20: **end for**
21: $m = \frac{\bar{m}}{m}$
22: **return** $m^{(p+1)/r}$

The notation $\bar{m}$ denotes the conjugate of m.

A.2   The Ate Pairing

Algorithm 3. Computation of $a(P,Q)$ on $E(\mathbb{F}_p) : y^2 = x^3 + Ax + B$, where P is a point of prime order r on the twisted curve $E'(\mathbb{F}_{p^2})$ and Q is a point on the base curve $E(\mathbb{F}_p)$

INPUT: $P = (x_P, y_P), Q = (x_Q, y_Q)$
OUTPUT: $m \in \mathbb{F}_{p^2}$
  1: $m = 1$
  2: $x_A, y_A \leftarrow x_P, y_P$
  3: $n = t - 1$
  4: **for** $i \leftarrow \lfloor \lg(n) \rfloor - 2$ **to** 0 **do**
  5:    **if** $n_i = 1$ **then**
  6:       $(\lambda, x_T, y_T) \leftarrow \text{double}(A, A)$
  7:       $g = i^2 y_Q - i(i^2 y_A/2 + \lambda(i^2 x_A/2 + x_Q))$
  8:       $x_A, y_A \leftarrow x_T, y_T$
  9:       $m = m^2 \cdot g$
 10:       $(\lambda, x_T, y_T) \leftarrow \text{add}(A, P)$
 11:       $g = i^2 y_Q - i(i^2 y_A/2 + \lambda(i^2 x_A/2 + x_Q))$
 12:       $x_A, y_A \leftarrow x_T, y_T$
 13:       $m = m \cdot g$
 14:    **else**
 15:       $(\lambda, x_T, y_T) \leftarrow \text{double}(A, A)$
 16:       $g = i^2 y_Q - i(i^2 y_A/2 + \lambda(i^2 x_A/2 + x_Q))$
 17:       $x_A, y_A \leftarrow x_T, y_T$
 18:       $m = m^2 \cdot g$
 19:    **end if**
 20: **end for**
 21: $m = \frac{\bar{m}}{m}$
 22: **return** $m^{(p^2+1)/r}$

A.3   The η_T Pairing

Algorithm 4. Computation of $\eta_T(P,Q)$ on $E(\mathbb{F}_{2^m}) : y^2 + y = x^3 + x + b$

INPUT: $P = (x_P, y_P), Q = (x_Q, y_Q)$
OUTPUT: $f \in \mathbb{F}_{2^{km}}$
  1: $u \leftarrow x_P + 1$
  2: $f \leftarrow u \cdot (x_P + x_Q + 1) + y_P + y_Q + b + 1 + (u + x_Q)s + t$
  3: **for** $i \leftarrow 1$ **to** $(m+1)/2$ **do**
  4:    $u \leftarrow x_P, x_P \leftarrow \sqrt{x_P}, y_P \leftarrow \sqrt{y_P}$
  5:    $g \leftarrow u \cdot (x_P + x_Q) + y_P + y_Q + x_P + (u + x_Q)s + t$
  6:    $f \leftarrow f \cdot g$
  7:    $x_Q \leftarrow x_Q^2, y_Q \leftarrow y_Q^2$
  8: **end for**
  9: **return** $f^{(2^{2m}-1)(2^m - 2^{(m+1)/2}+1)(2^{(m+1/2)}+1)}$

Factorization of Square-Free Integers with High Bits Known

Bagus Santoso[1], Noboru Kunihiro[1], Naoki Kanayama[2,*],
and Kazuo Ohta[1]

[1] The University of Electro-Communications
1-5-1 Chofugaoka Chofu-shi, Tokyo 182-8585, Japan
[2] University of Tsukuba
1-1-1 Tennohdai Tsukuba-shi, Ibaraki 305-8573, Japan

Abstract. In this paper we propose an algorithm of factoring any integer N which has k different prime factors with the same bit-length, when $(\frac{1}{k+2} + \frac{\epsilon}{k(k-1)}) \log N$ high-order bits of each prime factor are given. For a fixed ϵ, the running time of our algorithm is heuristic polynomial in $(\log N)$. Our factoring algorithm is based on a new lattice-based algorithm of solving any k-variate polynomial equation over $\mathbb{Z}$, which might be an independent interest.

1 Introduction

In order to speed-up the exponent modulus computation in a cryptosystem which its security is based on the intractability of integer factorization of the modulus, e.g., RSA, the variants which use a multi-prime composite integer, combined with Chinese Remainder Theorem (CRT) have been proposed[4,8,13,14,16]. We call these variants as multi-prime variants. As an illustration, let N denote the public modulus of a cryptosystem. If the original cryptosystem uses N which is the product of two secret prime factors p_1, p_2, then the multi-prime variants use a composite integer N which is the product of k secret prime factors $p_1, \ldots, p_k$ for $k \geq 3$.

1.1 Motivation and Strategy

Several attacks on multi-prime RSA have been proposed [3,9]. However, as far as our knowledge, there has been no investigation for the case of multi-prime variant where N is the product of k *different secret prime factors* $p_1, \ldots, p_k$ for $k \geq 3$, and several high order bits of the prime-factors are known to the attacker. In this paper we propose a method of factoring N of the following case:

(1) N is the product of k different secret prime factors $p_1, \ldots, p_k$, and
(2) we are given several high-order bits of the prime-factors $p_1, \ldots, p_k$,
where $k \geq 2$.

---

* This work was done when the third author was at the University of Electro-Communications as JSPS Research Fellow.

P.Q. Nguyen (Ed.): VIETCRYPT 2006, LNCS 4341, pp. 115–130, 2006.

Our strategy is as follows. We extend the work of Coron[7], which originally proposed a method of solving bivariate polynomial equations over integer, into a new method of solving k-variate polynomial equations over integer for any $k \geq 2$. Then, we apply our new method to solve polynomial $p(x_1, \ldots, x_k) = (\tilde{p_1} + x_1)(\tilde{p_2} + x_2) \cdots (\tilde{p_k} + x_k) - N$ where $\tilde{p_i}$ is the given part of p_i.

1.2   Results and Contributions

This work gives some theoretical results as well as practical meaning.

1.2.1   Theoretical Results

First, our new method of solving k-variate polynomial equations is general and usable for any $k \geq 2$. This method is featured through the following theorem.

Theorem 1. *Let p be an irreducible k-variate integer polynomial of independent degree δ for $k \geq 2$. Let $X_1, \ldots, X_k$ be upper-bounds of $x_{0_1}, \ldots x_{0_k}$ such that $p(x_{0_1}, \ldots x_{0_k}) = 0$. Also let define $W := ||p(x_1 X_1, \ldots, x_k X_k)||_\infty$. If for some $\epsilon > 0$ such that*

$$\prod_{i=1}^{k} X_i < W^{\frac{2}{\delta(k+1)} - \epsilon}, \tag{1}$$

*holds, then there exists an algorithm such that within time **heuristic** polynomial in $(\log W, 2^{\delta^{k-1}})$, can find all integer pairs $x_{0_1}, \ldots x_{0_k}$ such that $p(x_{0_1}, \ldots x_{0_k}) = 0$, $|x_{0_i}| \leq X_i$ for $i \in [1, k]$.*

The "heuristic" part in Theorem 1 is due to the possibility that a resultant computation between two polynomials vanished to zero, while our algorithm works only if all required resultant computations are not vanished. The algorithm featured in Theorem 1 can be guaranteed to run in polynomial time if we assume that the following assumption holds.

Assumption 2. *The resultant computations of any two polynomials where neither is p, are not vanished.*

Assumption 2 enables us to prove that the algorithm featured in Theorem 1 works within polynomial time. Note that this assumption is *slightly weaker* than a more general assumption which might state: "*the resultant computations for any multivariate polynomials constructed yield non-zero polynomials*". We are succeeded in weakening our assumption by extending Lemma 3 of Coron[7]. Originally, Lemma 3 in [7] only concerned about the bound of the norm of two polynomials where one is a multiple of another in *bivariate* polynomial case. We extend it into a new lemma (Lemma 2 in this paper) which concerns about that in *k-variate* polynomial case. Additionally, we also prove an upper-bound of the norm of basis in LLL-reduced basis (Lemma 3) which is tighter than the one shown in [2].

Finally, as application of Theorem 1, we construct a factoring algorithm of a composite integer N which has k different square-free prime factors $p_1, \ldots, p_k$, given several bits of high-order bits of each $p_1, \ldots, p_k$s. The result is illustrated informally as the following theorem.

Theorem 3. *Let N be a composite integer with k different prime factors with the same bit length. If for each prime factor, we are given the at least $\left(\frac{1}{k+2} + \frac{\epsilon}{k(k-1)}\right) \log_2 N$ high order bits for some $\epsilon > 0$, then we have a (heuristic) polynomial time algorithm to factor N.*

1.2.2   Practical Meaning

Recently, some side-channel attacks on implementations of RSA where CRT and Montgomery Reduction are used, have been proposed[1,5,15]. It was shown that these attacks can reveal several high order bits of the prime factor of N where N is the public modulus of RSA. Note that almost all the implementation of multi-prime variants [4,8,13,14,16] also use CRT and Montgomery Reduction. Although up to this moment these attacks have only been proven to work on RSA where N is a product of only two primes yet, this kind of side-channel attacks might also apply to these multi-prime variants. Hence, if such side-channel attacks can reveal some high-order bits of the prime factors of N of a multi-prime variant, then our result (Theorem 3) can be used to complete the attack, i.e., full factoring of N.

1.3   Related Works

The use of lattice reduction to find small roots of low-degree polynomial equations is discovered by Coppersmith[6] in 1996. At Eurocrypt 1996, Coppersmith showed that using LLL, we can find small roots of unimodular equations and bivariate polynomial equations over integer in polynomial time. However, the Coppersmith's method is generally difficult to put into practice. Howgrave-Graham[10] constructed a simplification of Coppersmith's method of solving unimodular equations such that it is easier to understand and more practical. This simplification has been broadly used in some applications[3]. Later, at Eurocrypt 2004, Coron[7] provided a simplification of Coppersmith's method of solving bivariate polynomial equations, ala Howgrave-Graham. Coron's algorithm also runs in polynomial time.

Note that our main theorem (Theorem 1) can be seen as extension of Coron [7], i.e., from a method of solving bivariate polynomial equations into that of k-variate polynomial equations. Also, although our result has a drawback, i.e., our algorithm runs in heuristic polynomial time instead of strict polynomial time as in the case of Coron's, our result still inherits the following features from Coron's: (1)the simple approach of Howgrave-Graham and (2)a polynomial-time proven running time until a certain step.

Organization of the paper. The next section is devoted to a brief introduction into lattice and LLL-reduced based related definitions. In section 3 we prove the upper bound of the norm of vectors in LLL-reduced basis (Lemma 2), and the upper bound of the factors of a k-variate polynomial (Lemma 3). Then, in the section 4 we will prove the Theorem 1. Finally, in section 5 we show an application of Theorem 1 to factor a composite integer N with k different prime factors with the same bit length, given several bits of high order bits of each prime factor. We present some discussion about our result in section 6. We close this paper by a brief conclusion and some directions for further research.

2   Preliminaries

Unless otherwise noted, any vector in this paper is represented as a single column matrix. A *k-variate integer polynomial* p *of independent degree* δ is denoted by: $p(x_1, \ldots, x_k) = \sum_{(i_1,\ldots,i_k)\in[0,\delta]^k} p_{i_1,\ldots,i_k} \prod_{j=1}^{k} x_j^{i_j}$ where $p_{i_1,\ldots,i_k} \in \mathbb{Z}$. We define $||p||^2 := \sum_{(i_1,\ldots,i_k)} |p_{i_1,\ldots,i_k}|^2$ and $||p||_\infty := \max_{(i_1,\ldots,i_k)} |p_{i_1,\ldots,i_k}|$. For a k-variate polynomial $p(x_1,\ldots,x_k)$, we will refer $x_1^{i_1} \ldots x_k^{i_k}$ as $(i_1,\ldots,i_k)$-*th term*, $(i_1,\ldots,i_k)$ as *power sequence* of $x_1^{i_1} \ldots x_k^{i_k}$, and $p_{i_1,\ldots,i_k}$ as *coefficient* of $(i_1,\ldots,i_k)$-th term.

2.1   The LLL-Reduced Basis and LLL Algorithm

Definition 1 (Lattice). *Let* $\mathbf{b}_1,\ldots,\mathbf{b}_\omega \in \mathbb{Z}^m$ *be linearly independent vectors with* $\omega \leqq n$. *The lattice spanned by them is defined as follows:* $\mathcal{L}(\mathbf{b}_1,\ldots,\mathbf{b}_\omega) := \left\{ \sum_{i=1}^{\omega} c_i \mathbf{b}_i \middle| c_i \in \mathbb{Z} \right\}$. *Equivalently, we can define a* $m \times \omega$ *matrix* B *whose columns are* $\mathbf{b}_1,\ldots,\mathbf{b}_\omega$, *and the lattice generated by* B *as* $\mathcal{L}(B) := \{Bc | \mathbf{c} \in \mathbb{Z}^\omega\}$. *Clearly,* $\mathcal{L}(B) = \mathcal{L}(\mathbf{b}_1,\ldots,\mathbf{b}_\omega)$ *holds.*

We refer to such a set of vectors b_i's or such a matrix B as a *basis* of the lattice $\mathcal{L}(B)$. In this case, the lattice $\mathcal{L}(B)$ is said to have *rank* m and *dimension* ω. If $m = \omega$, then $\mathcal{L}(B)$ is a *full-rank lattice*. Note that all bases of a lattice have the same rank and dimension. Let B' be any basis of a lattice $\mathcal{L}'$. The determinant of lattice $\mathcal{L}'$, denoted $\det(\mathcal{L}')$, is defined as $\det(\mathcal{L}') := \sqrt{\det(B'^T B')}$.

Definition 2 (LLL-reduced). *Let* $\mathcal{L}$ *be a lattice spanned by matrix* $B = (\mathbf{b}_1,\ldots,\mathbf{b}_\omega) \in \mathbb{Z}^{m\times\omega}$. *Let* $B^* = (\mathbf{b}_1^* \cdots \mathbf{b}_\omega^*)$ *denote a result of Gram-Schmidt orthogonalization process on* B *and* $\mu_{i,j} := \langle \mathbf{b}_i, \mathbf{b}_j^* \rangle / \langle \mathbf{b}_j^*, \mathbf{b}_j^* \rangle$. *The basis* B *is called a LLL-reduced basis if the following holds: (1)* $|\mu_{i,j}| \leqq 1/2$ *for* $j,i \in [1,\omega]$ *where* $j < i$, *and (2)* $(3/4)||\mathbf{b}_{i-1}^*||^2 \leqq ||\mathbf{b}_i^* + \mu_{i,i-1}\mathbf{b}_{i-1}^*||^2$ *for* $i \in (1,\omega]$. *Note that* $\mathbf{b}_1 = \mathbf{b}_1^*$ *and* $\mathbf{b}_i = \mathbf{b}_i^* + \sum_{j=1}^{i-1} \mu_{i,j}\mathbf{b}_j^*$ *for* $i \in (1,\omega]$ *hold.*

Theorem 4 (LLL[11]). *Let* $\mathcal{L}$ *be a lattice spanned by* $V = (\mathbf{v}_1 \cdots \mathbf{v}_\omega) \in \mathbb{Z}^{m\times\omega}$. *The LLL algorithm, given* V, *finds in polynomial time a LLL-reduced basis* $B = (\mathbf{b}_1,\ldots,\mathbf{b}_\omega)$.

2.2   Howgrave-Graham

The following lemma is introduced by Howgrave-Graham[10] to simplify the Coppersmith method of solving polynomial equations[6].

Lemma 1 (Howgrave-Graham). *Let* $p(x_1, \ldots, x_k) \in \mathbb{Z}[x_1, \ldots x_k]$ *have at most* ω *monomials. If* $p(x_{0_1}, \ldots x_{0_k}) \equiv 0 \pmod{n}$ *for some positive integer* n *where* $|x_{0_i}| \leq X_i$ *for any* $i \in [1, k]$ *and* $||p(x_1 X_1, \ldots, x_k X_k)||_\infty < n/\sqrt{\omega}$ *holds, then* $p(x_{0_1}, \ldots x_{0_k}) = 0$ *holds over* $\mathbb{Z}$.

3   Some Additional Tools

In this section we prove two lemmas (Lemma 2 and Lemma 3) which play important roles at proving Theorem 1. Lemma 2 gives the upper bound of norm of each vector in an LLL-reduced basis. Lemma 2 is especially useful when we need to get the upper bound of norm of the n-th shortest vector in an LLL reduced basis. Lemma 3 gives the norm's upper bound of the coefficients of the factors of a polynomial. We use Lemma 3 to prove that a polynomial is not a factor of another polynomial. As seen in the next section, this enables us to guarantee that Assumption 2 is sufficient to prove that the algorithm featured in Theorem 1 runs within polynomial time.

3.1   Upper Bound of LLL-Reduced Basis

Lemma 2. *Let* $\mathcal{L}$ *be a lattice spanned by* $B = (\mathbf{b}_1, \ldots, \mathbf{b}_\omega) \in \mathbb{Z}^{m \times \omega}$. *If* B *is LLL-reduced basis, then the following holds:*

$$||\mathbf{b}_i|| \leq 2^{\frac{\omega + i - 2}{4}} \det(\mathcal{L})^{\frac{1}{\omega - i + 1}} \tag{2}$$

for $i \in [1, \omega]$.

Proof. Let $B^* = (\mathbf{b}_1^* \cdots \mathbf{b}_\omega^*)$ denote a result of Gram-Schmidt orthogonalization process on B and $\mu_{i,j} := \langle \mathbf{b}_i, \mathbf{b}_j^* \rangle / \langle \mathbf{b}_j^*, \mathbf{b}_j^* \rangle$. Also let $b_i := ||\mathbf{b}_i||^2$ and $b_i^* := ||\mathbf{b}_i^*||^2$. From the definition of LLL-reduced basis (Definition 2), it is easy to see that $b_i^* \geq (3/4 - \mu_{i,i-1}^2) b_{i-1}^* \geq b_{i-1}^*/2$ holds since $|\mu_{i,i-1}| \leq 1/2$. Then, by induction, we can get $b_i^* \geq (1/2)^{i-j} b_j^*$ for any $i, j \in [1, \omega]$ where $j \leq i$.

Combining this with the fact that $\mathbf{b}_1 = \mathbf{b}_1^*$ and $\mathbf{b}_i = \mathbf{b}_i^* + \sum_{j=1}^{i-1} \mu_{i,j} \mathbf{b}_j^*$ for $i \in (1, \omega]$, we are able to derive the following relation: $b_i \leq b_i^* + (1/2)^2 \sum_{j=1}^{i-1} b_j^* \leq b_i^*(1 + (1/2)^2 \cdot \sum_{j=1}^{i-1} 2^{i-j}) = b_i^*(2^{-1} + 2^{i-2})$. Hence, for $i, j \in [1, \omega]$ where $1 \leq j \leq i$, we get $b_j \leq (2^{-1} + 2^{j-2}) \cdot 2^{i-j} \cdot b_i^* \leq (2^{i-j-1} + 2^{i-2}) \cdot b_i^* \leq 2^{i-1} b_i^*$. Consequently, it shows that $b_j^{(\omega - j + 1)} \leq 2^{\sum_{i=j}^{\omega}(i-1)} \prod_{i=j}^{\omega} b_i^*$. Finally, since $\sum_{i=j}^{\omega}(i-1) = (\omega - j + 1)(\omega + j - 2)/2$ and $\prod_{i=1}^{\omega} b_i^* = \det(\mathcal{L})^2$, we get $b_j \leq 2^{\frac{\omega + j - 2}{2}} \det(\mathcal{L})^{\frac{2}{\omega - j + 1}}$. ∎

Remark 1. We give a tighter bound than the one shown by Blömer and May[2]. In [2], it is stated that $||\mathbf{b}_i|| \leqq 2^{\frac{\omega+i-2+2(i-1)(i-2)/(\omega-i+1)}{4}} \det(\mathcal{L})^{\frac{1}{\omega-i+1}}$.

3.2 Upper Bound of the Factors of Polynomial

Lemma 3. *Let p and h be two k-variate integer polynomials of independent degree δ. Let $p_{0^k} \neq 0$ hold and h be divisible by a non-zero integer r such that $\gcd(r, p_{0^k}) = 1$. If h is a multiple of p in $\mathbb{Z}[x_1, \ldots, x_k]$, then : $||h|| \geq 2^{-(\delta+1)^k}$. $|r| \cdot ||p||_\infty$ holds.*

Proof. Our proof is an extension of the proof given by Coron[7]. Since from the assumption h is a multiple of p, then there exists a polynomial $g \in \mathbb{Z}[x_1, \ldots, x_k]$, such that $p(x_1, \ldots, x_k) \cdot g(x_1, \ldots, x_k) = h(x_1, \ldots, x_k)$.

As the first step, we prove that $r \cdot p$ divides h by showing that r divides $g(x_1, \ldots, x_k)$. Assume that r does not divide $g(x_1, \ldots, x_k)$. Thus, we can have a non-empty set $G := \{(i_{1_1}, \ldots, i_{1_k}), \ldots, (i_{n_1}, \ldots, i_{n_k})\}$, which contains all of k-integer sequences $(j_1, \ldots, j_k)$ where $g_{j_1, \ldots, j_k}$ is not divisible by r. Let G be lexicographically ordered. Notice that for $(i_{1_1}, \ldots, i_{1_k})$ we have $b_{i_{1_1}, \ldots, i_{1_k}} = g_{i_{1_1}, \ldots, i_{1_k}} \cdot a_{0^k} + \sum_{j_1, \ldots, j_k} g_{(i_{1_1}-j_1), \ldots, (i_{1_k}-j_k)} \cdot a_{j_1, \ldots, j_k}$. Also note that any $(i_{1_1}-j_1, \ldots, i_{1_k}-j_k)$ for any $(j_1, \ldots j_k) \notin 0^k$ can not be a member of G, because $(i_{1_1}, \ldots, i_{1_k})$ is the "least" member in G lexicographically. Thus, for any $(j_1, \ldots j_k) \notin 0^k$, $g_{(i_{1_1}-j_1), \ldots, (i_{1_k}-j_k)}$ must be divisible by r. Hence, we get $b_{i_{1_1}, \ldots, i_{1_k}} \equiv g_{i_{1_1}, \ldots, i_{1_k}} \cdot a_{0^k} \pmod{r}$. However, this a contradiction, because: (1) $b_{i_{1_1}, \ldots, i_{1_k}} \equiv 0 \pmod{r}$ holds, (2) $g_{i_{1_1}, \ldots, i_{1_k}}$ is not zero value, and (3) a_{0^k} has inverse in modulus r. Therefor, r must divide $g(x_1, \ldots, x_k)$.

The next step is based on a well-known equation which is called *Mignotte's bound*: let $f(x)$ and $g(x)$ be two non-zero polynomials over $\mathbb{Z}$, such that $\deg f \leqq \delta$, and f divides g in $\mathbb{Z}[x]$; then: $||g|| \geq 2^{-\delta}||f||_\infty$ holds.

Now let the followings: $f(x) := r \cdot p(x, x^{\delta+1}, x^{(\delta+1)^2}, \ldots, x^{(\delta+1)^{k-1}})$ and $g(x) := h(x, x^{\delta+1}, x^{(\delta+1)^2}, \ldots, x^{(\delta+1)^{k-1}})$. Note that f and $r \cdot p$ have the same list of non-zero coefficients, and so do g and h. This gives $||f||_\infty = |r| \cdot ||p||_\infty$ and $||g|| = ||h||$. Also, $\deg f, \deg g \leqq (\delta + 1)^k - 1$. Since $r \cdot p$ divides h, we can apply Mignotte's bound on $r \cdot p$ and h. ■

4 Proof of Theorem 1

Recall that p denotes a k-variate integer polynomial of independent degree δ for $k \geq 2$, $X_1, \ldots, X_k$ denote the upper-bounds of $x_{0_1}, \ldots x_{0_k}$ such that $p(x_{0_1}, \ldots x_{0_k}) = 0$, and $W := ||p(x_i X_i, \ldots, x_k X_k)||_\infty$.

Our proof procedure follows the line of the proof of Theorem 4 in [7]. First, we construct an algorithm $\mathbf{Solve}_{k,l}^{\text{LLL}}(p, X_1, \ldots X_k)$. Then we prove that the algorithm finds $x_{0_1}, \ldots, x_{0_k}$ such that $p(x_{0_1}, \ldots x_{0_k}) = 0$ within polynomial time if the equation (1) is satisfied, assuming that Assumption 2 holds.

We take here $l \geq 1$. We will show later how to determine the parameter l by using the value of ϵ. For simplicity, we only discuss here the case where $p_{0^k} \neq 0$ and $\gcd(p_{0^k}, \prod_{i=1}^k X_i) = 1$. For the general case, see Appendix A in [7].

Algorithm 1 $\mathbf{Solve}_{k,l}^{\mathsf{LLL}}(p, X_1, \ldots, X_k)$

(1) Define $\omega \stackrel{\text{def}}{=} (\delta + l + 1)^k$.
   Find u such that $\sqrt{\omega} 2^{-\omega} W \leq u < 2W$ and $\gcd(p_{0^k}, u) = 1$ hold.
(2) Define $n \stackrel{\text{def}}{=} u \cdot (\prod_{i=1}^{k} X_i)^l$, and set $q(x_1, \ldots, x_k) \leftarrow p_{0^k}^{-1} \cdot p(x_1, \ldots, x_k) \bmod n$.
(3) For all $(i_1, \ldots, i_k) \in [0, l]^k$, do
$$q_{[i_1 \cdots i_k]}(x_1, \ldots, x_k) \leftarrow \prod_{j=1}^{k} x_j^{i_j} X_j^{l-i_j} q(x_1, \ldots, x_k).$$
   For all $(i_1, \ldots, i_k) \in [0, \delta + l]^k \backslash [0, l]^k$, do
$$q_{[i_1 \cdots i_k]}(x_1, \ldots, x_k) \leftarrow n \cdot \prod_{j=1}^{k} x_j^{i_j}.$$
   Define $\tilde{q}_{[i_1 \cdots i_k]}(x_1, \ldots, x_k) \stackrel{\text{def}}{=} q_{[i_1 \cdots i_k]}(x_1 X_1, \ldots, x_k X_k)$ for all $(i_1, \ldots, i_k) \in [0, \delta + l]^k$,

(4) Generate a lattice basis $B = (\mathbf{b}_1 \cdots \mathbf{b}_\omega)$ where each $\mathbf{b}_i$ represents the coefficients of $\tilde{q}_{[i_{j_1} \cdots i_{j_k}]}(x_1, \ldots, x_k)$ of some $i_{j_1} \cdots i_{j_k}$. Put the order of $\tilde{q}_{[i_{j_1} \cdots i_{j_k}]}(x_1, \ldots, x_k)$ and the coefficients of its terms such that B forms a triangular matrix.
(5) Input B into LLL algorithm and obtain a LLL-reduced basis $B' = (\mathbf{b}'_1 \cdots \mathbf{b}'_\omega)$.
   Take the first (shortest) $(k-1)$ vectors of B' and construct a k-variate polynomial $h_i(x_1, \ldots, x_k)$ from each $\mathbf{b}'_1, \ldots, \mathbf{b}'_{k-1}$
(6) For $i = 1$ to $k - 1$ do
   Compute $h_{1,i}(x_2, \ldots, x_k) = Res_{x_1}(p, h_i)$
   For $t = 2$ to $k - 1$, do
     For $j = 1$ to $+k - t$,do
       Compute $h_{t,j}(x_{t+1}, \ldots, x_k) = Res_{x_t}(h_{t-1,j}, h_{t-1,j+1})$
   Solve $h_{k-1,1}(x_k)$ using standard root finding algorithms. And use the solution, x_{0_k}, to find $x_{0_1}, \ldots, x_{0_{k-1}}$ such that $p(x_{0_1}, \ldots, x_{0_{k-1}}, x_{0_k}) = 0$.

Two important notes about algorithm $\mathbf{Solve}_{k,l}^{\mathsf{LLL}}(p, X_1, \ldots X_k)$ above:

Note 1. $q(x_1, \ldots, x_k)$ is always in the form of $1 + \sum q_{t_1, \ldots, t_k} \prod_{j=1}^{k} x_j^{t_j}$. Therefore, for $(i_1, \ldots, i_k) \in [0, l]^k$, the form of $\tilde{q}_{[i_1, \ldots, i_k]}(x_1, \ldots, x_k)$ is $\left(\prod_{j=1}^{k} X_j\right)^l \prod_{j=1}^{k} x_j^{i_j} + \left(\prod_{j=1}^{k} X_j\right)^l \sum q_{t_1, \ldots, t_k} \prod_{j=1}^{k} x_j^{i_j + t_j}$.

Note 2. Any $\tilde{q}_{[i_1, \ldots, i_k]}(x_1, \ldots, x_k)$ where $(i_1, \ldots, i_k) \in [0, \delta + l]^k \backslash [0, l]^k$ is in the form of $n \cdot \prod_{j=1}^{k} (x_j X_j)^{i_j}$.

First, we will show a construction example of the triangular matrix B. Then, by assuming that *all resultant computations in step (6) are not vanished*, we derive the sufficient condition for algorithm $\mathbf{Solve}_{k,l}^{\mathsf{LLL}}(p, X_1, \ldots X_k)$ to work. We show that Eq. (1) satisfies this sufficient condition. At the end, we will prove that our previous assumption ,i.e., "all resultant computations in step (6) are not vanished" can be weakened into Assumption 2 by proving that any resultant computation involving p are not vanished, provided Eq. (1) holds.

We will show here one method of constructing B as follows. First we divide the set of all sequence $\{(i_1, \ldots, i_k) \in [0, \delta + l]^k\}$ into two sets: $S_I := \{(i_1, \ldots, i_k) \in [0, l]^k\}$ and $S_{II} := \{(i_1, \ldots, i_k) \in [0, \delta + l]^k \backslash [0, l]^k\}$. Note that $S_I \cap S_{II} = \emptyset$. Then we arrange S_I and S_{II} in lexicographical order ($x_1 < x_2 < \cdots < x_k$) respectively, and finally combine together S_I and S_{II} by putting S_I before S_{II}. We call this arrangement S_{I+II}. Finally, put $\{\tilde{q}_{[i_1,\cdots,i_k]}\}$ and its coefficients of all terms in the same arrangement according to S_{I+II} as follows: $\{\tilde{q}_{[i_1,\cdots,i_k]}\}$ is arranged in horizontal direction from left to right depending on the index, and the coefficient of terms is arranged in vertical direction start from top to bottom depending on the power sequence of each term. For detail illustration, see Figure 1. The detail proof that we can always get a triangular matrix with this arrangement (S_{I+II}) is in Appendix A.

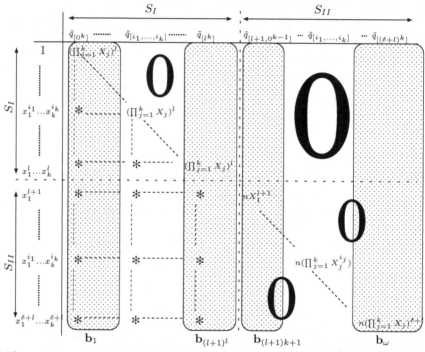

The entries marked with "*" and "$\cdots$" represent possible non-zero off-diagonal entries we may ignore.

Fig. 1. Matrix B

Now, we will derive the sufficient conditions in order to make algorithm $\mathbf{Solve}^{\mathsf{LLL}}_{k,l}(p, X_1, \ldots X_k)$ work, by first assuming that all resultant computations in step (6) are not vanished.

In step (2) of $\mathbf{Solve}^{\mathsf{LLL}}_{k,l}(p, X_1, \ldots X_k)$, one may set $u := W + ((1 - W) \bmod |p_{0^k}|)$. Clearly this u satisfies $\sqrt{\omega}2^{-\omega}W \leqq u < 2W$. Also, since $1 = u + \lfloor(1 - W)/|p_{0^k}|\rfloor \cdot |p_{0^k}|$ holds, $\gcd(u, |p_{0^k}|) = 1$ holds.

By the construction of $h_i(x_1, \ldots, x_k)$, we know that $p(x_1, \ldots, x_k)$ and $h_i(x_1, \ldots, x_k)$ have the same solution over $\mathbb{Z}_n$. However, there is no explicit guarantee that $p(x_1, \ldots, x_k)$ and $h_i(x_1, \ldots, x_k)$ have the same solution over $\mathbb{Z}$. Thus, we need to assure that any solution of $h_i(x_1, \ldots, x_k) \equiv 0 \pmod{n}$ is also a solution of $h_i(x_1, \ldots, x_k) = 0$ over $\mathbb{Z}$. Here we apply Howgrave-Graham lemma to all h_i. Howgrave-Graham requires us that all h_i satisfy $||h_i||_\infty < n/\sqrt{\omega}$. Since $n/\sqrt{\omega} \geq 2^{-\omega}(\prod_{j=1}^{k} X_j)^l W$ holds and $||h_i||_\infty \leq ||h_i||$ for any i, condition $||h_i|| < 2^{-\omega}(\prod_{j=1}^{k} X_j)^l W$ is sufficient to guarantee that we can use h_i to solve the equation over $\mathbb{Z}$ (by Howgrave-Graham). Note that the upper bound of $||h_i||$ is $||h_{k-1}||$, which is upper-bounded by Lemma 2 as follows: $||h_{k-1}|| \leq 2^{\frac{\omega+k-3}{4}} \det(B)^{\frac{1}{\omega-k+2}}$.

Combining all conditions: $\forall i : ||h_i|| < 2^{-\omega}(\prod_{j=1}^{k} X_j)^l W$ and $||h_{k-1}|| \leq 2^{\frac{\omega+k-3}{4}} \det(B)^{\frac{1}{\omega-k+2}}$, we get our final sufficient condition as follows.

$$2^{\frac{\omega+k-3}{4}} \det(B)^{\frac{1}{\omega-k+2}} < 2^{-\omega}(\prod_{j=1}^{k} X_j)^l W. \tag{3}$$

Now we will calculate $\det(B)$. $\left\{\tilde{q}_{[i_1,\ldots,i_k]}(x_1,\ldots,x_k) \middle| (i_1,\ldots,i_k) \in [0,l]^k\right\}$ gives contribution to $\det(B)$ as $\prod_{(i_1,\ldots,i_k)\in[0,l]^k}(\prod_{j=1}^{k} X_j)^l = (\prod_{j=1}^{k} X_j)^{(l+1)^k l}$, while $\left\{\tilde{q}_{[i_1,\ldots,i_k]}(x_1,\ldots,x_k) \middle| (i_1,\ldots,i_k) \in [0,\delta+l]^k\backslash[0,l]^k\right\}$ gives:

$$\prod_{\substack{(i_1,\ldots,i_k)\in \\ [0,\delta+l]^k\backslash[0,l]^k}} n\left((\prod_{j=1}^{k} X_j^{i_j})\right) = n^{(\delta+l+1)^k-(l+1)^k} \times$$

$$\left((\prod_{j=1}^{k} X_j)^{(\delta+l+1)^{k-1}\sum_{t=0}^{\delta+l} t - (l+1)^{k-1}\sum_{t=0}^{k} t}\right)$$

$$= n^{(\delta+l+1)^k-(l+1)^k}\left((\prod_{j=1}^{k} X_j)^{(\delta+l+1)^k(\delta+l)/2 - (l+1)^k l/2}\right).$$

Combine all the equations, we obtain:

$$\det(B) = n^{(\delta+l+1)^k-(l+1)^k}\left((\prod_{j=1}^{k} X_j)^{\frac{(\delta+l+1)^k(\delta+l)+(l+1)^k l}{2}}\right).$$

Since $n < 2W \cdot (\prod_{j=1}^{k} X_j)^l$, combining with $\det(B)$, we transform the final condition (3) into:

$$2^{\frac{\omega+k-3}{4}}\left((2W \cdot (\prod_{j=1}^{k} X_j)^l)^{(\delta+l+1)^k-(l+1)^k}(\prod_{j=1}^{k} X_j)^{\frac{(\delta+l+1)^k(\delta+l)+(l+1)^k l}{2}}\right)^{\frac{1}{\omega-k+2}}$$

$$< 2^{-\omega}(\prod_{j=1}^{k} X_j)^l W. \tag{4}$$

Further transformation of (4) results in the following. For detail, please refer Appendix B.

$$\left(\prod_{j=1}^{k} X_j\right)^{(k+1)\delta(\delta+l+1)^k+2l(k-2)}$$

$$< 2^{-2\{\frac{5}{4}\omega^2-(k-\frac{11}{4})\omega-\frac{(k-2)(k-3)}{4}-(l+1)^k\}}W^{2\{(l+1)^k-(k-2)\}}. \tag{5}$$

Thus, the sufficient condition for algorithm $\mathbf{Solve}_{k,l}^{\mathsf{LLL}}(k,p,X_1,\ldots X_k)$ to work is:

$$\left(\prod_{j=1}^{k} X_j\right) < 2^{-\beta}W^\alpha, \tag{6}$$

where

$$\alpha = \frac{2}{\delta(k+1)(\delta+l+1)^k} \left\{ \frac{(l+1)^k - (k-2)}{1 + 2l(k-2)\{\delta(k+1)(\delta+l+1)^k\}^{-1}} \right\}$$

$$\beta = \frac{2}{\delta(k+1)(\delta+l+1)^k} \times$$

$$\left\{ \frac{\frac{5}{4}(\delta+l+1)^{2k} - (k - \frac{11}{4})(\delta+l+1)^k - \frac{(k-2)(k-3)}{4} - (l+1)^k}{1 + 2l(k-2)\{\delta(k+1)(\delta+l+1)^k\}^{-1}} \right\}.$$

Using $\delta \geq 1$, $l \geq 1$ and $k \geq 2$, we can calculate the lower-bound of α and the upper-bound of β as follows:

$$\alpha \geq \frac{2}{\delta(k+1)} - \frac{2}{l+2}, \tag{7}$$

$$\beta < \delta^{k-1}2^{k+1}(l+2)^k + \frac{8}{k+1}. \tag{8}$$

Letting $\epsilon := \frac{2}{l+2}$ and taking from (6), (7), and (8), we obtain the following sufficient condition for $\prod_{j=1}^k X_j$ in order to make algorithm $\mathbf{Solve}^{\text{LLL}}(k,p,X_1,\ldots X_k)$ works:

$$\prod_{j=1}^k X_j < 2^{-\frac{2^{2k+1}\delta^{k-1}}{\epsilon^k} - \frac{8}{k+1}} W^{\frac{2}{\delta(k+1)} - \epsilon}. \tag{9}$$

Now, we will prove the following claim in order to weaken our assumption. Namely, from the assumption that all resultant computations of two polynomials in step (6) are not vanished, into Assumption 2, i.e., all resultant computations of two polynomials in step (6) *where neither is p* are not vanished.

Claim. For any $1 \leq i \leq k-1$, $h_{1,i}(x_2,\ldots,x_k) = Res_{x_1}(p,h_i)$ in step (6) does not vanished to zero.

Proof. We make use of Lemma 3 to guarantee that all $h_{1,i}(x_2,\ldots,x_k) = Res_{x_1}(p,h_i)$ are not vanished.

First, we will show that B has all its elements to be divisible by $(\prod_{j=1}^k X_j)^l$. From Note 1, it is easy to see that for $(i_1,\ldots,i_k) \in [0,l]^k$ any $\tilde{q}_{[i_1,\ldots,i_k]}(x_1,\ldots,x_k)$ is always divisible by $(\prod_{j=1}^k X_j)^l$. Also, since any $\tilde{q}_{[i_1,\ldots,i_k]}(x_1,\ldots,x_k)$ where $(i_1,\ldots,i_k) \in [0,\delta+l]^k \backslash [0,l]^k$ is in the form of $n \cdot \prod_{j=1}^k (x_j X_j)^{i_j}$, from Note 2, we can see that this is always divisible by $(\prod_{j=1}^k X_j)^l$. Thus, all coefficients of $h_i(x_1,\ldots,x_k)$ are also divisible by $(\prod_{j=1}^k X_j)^l$.

Now we will apply Lemma 3. Let $r = (\prod_{j=1}^k X_j)^l$ and $||p||_\infty = W$. Note that $\gcd(r,p_{0^k}) = 1$ holds since $p_{0^k} = 1$. Also, since $||h_i|| \leq ||h_{k-1}|| \leq 2^{\frac{w+k-3}{4}} \det(B)^{\frac{1}{w-k+2}}$ holds for any $1 \leq i \leq k-1$ (Lemma 2) and Eq. (9) implies Eq. (3), then h_i for any $1 \leq i \leq k-1$ satisfies $||h_i|| < 2^{-w}(\prod_{j=1}^k X_j)^l W = 2^{-(\delta+l+1)^k}(\prod_{j=1}^k X_j)^l W$. Thus, we can guarantee that h_i is not a multiple of p. And since p is irreducible, p and h_i do not have common factor. Therefore, $Res_{x_1}(p,h_i)$ is not vanished. $\square$

Therefore, for a fixed $k \geq 2$, when a k-variate integer polynomial $p(x_1, \ldots, x_k)$ satisfies (9), the algorithm $\mathbf{Solve}_{k,l}^{\mathrm{LLL}}(p, X_1, \ldots X_k)$ can find the root of $p(x_1, \ldots, x_k)$ within polynomial time in $(\log W, \delta, 1/\epsilon)$. To get the weaker condition (1), according to [7], we can exhaustively search total high order $\frac{2^{2k+1}\delta^{k-1}}{\epsilon^k} + \frac{8}{k+1}$ bits of $x_{1_0}, x_{2_0}, \ldots, x_{k_0}$ so that the condition (9) is satisfied, then apply the algorithm $\mathbf{Solve}_{k,l}^{\mathrm{LLL}}(p, X_1, \ldots, X_k)$ with new bounds $X_1, \ldots, X_k$. For a fixed $\epsilon > 0$, the running time is polynomial in $(\log W, 2^{\delta^{k-1}})$. This terminates the proof of Theorem 1. ∎

5   Factoring Square-Free Composite Integer with Balanced Prime Factors

In this section we apply Theorem 1 to do factoring of a square-free composite integer N which has k different prime factors with the same bit length, given several most of significant bits of each prime factor. Here we use $\ell(x)$ to denote the bit-length of an integer x.

Theorem 5. *Let N be a composite integer with k different prime factors, denoted by $p_1, \ldots, p_k$. Assume that all p_i have the same bit length. If for each prime p_i, we are given at least $\left(\frac{1}{k+2} + \frac{\epsilon}{k(k-1)}\right)\ell(N)$ most significant bits of p_i for some $\epsilon > 0$, then we have an algorithm to factor N with running time polynomial in $(\log N)$.*

Proof. Let N's prime factors be $p_1, \ldots, p_k$. Note that since $\ell(p_1) = \ell(p_2) = \ldots = \ell(p_k) = \ell_p$ bits, we can put $\ell(N) = k \cdot \ell_p$. Assume that for each prime p_i, we are given the $\gamma \cdot \ell_p$ most significant bits of p_i $(0 < \gamma < 1)$, denoted by $\tilde{p}_i/2^{(1-\gamma)\ell_p}$. Note that for each p_i, $\ell(p_i) = \ell(\tilde{p}_i)$ and $p_i - \tilde{p}_i < 2^{(1-\gamma)\ell_p}$ hold. In order to find the rest of unknown bits, we will solve the polynomial:

$$p(x_1, \ldots, x_k) = (\tilde{p}_1 + x_1)(\tilde{p}_2 + x_2)\cdots(\tilde{p}_k + x_k) - N, \qquad (10)$$

where $p_i = \tilde{p}_i + x_i$ holds for each p_i, using algorithm $\mathbf{Solve}_{k,l}^{\mathrm{LLL}}(p, X_1, \ldots X_k)$ of Theorem 1. Also, we will derive the lower bound of γ to satisfy the sufficient condition for applying Theorem 1.

Since $x_i = p_i - \tilde{p}_i < 2^{(1-\gamma)\ell_p} = 2^{\ell_p}2^{-\gamma\ell_p} \leq 2\tilde{p}_i N^{-\gamma/k}$ holds, we can set the upper bound of x_i, denoted by X_i, as follows:

$$X_i = 2\tilde{p}_i N^{-\gamma/k}. \qquad (11)$$

And we also obtain:

$$\begin{aligned} W &= \|p(x_1 X_1, x_2 X_2, \cdots, x_k X_k)\|_\infty \\ &= \|(X_1 x_1 + \tilde{p}_1)(X_2 x_2 + \tilde{p}_2)\cdots(X_k x_k + \tilde{p}_k) - N\|_\infty \\ &\geq \tilde{p}_1 \tilde{p}_2 \cdots \tilde{p}_{k-1} X_k. \end{aligned} \qquad (12)$$

Since $\ell(p_i) = \ell(\tilde{p}_i)$, we can assume that $p_i \geq \tilde{p}_i \geq p_i/2$ holds. Thus:

$$\tilde{p}_1 \tilde{p}_2 \cdots \tilde{p}_{k-1} X_k \geq \left(\frac{p_1}{2}\right)\left(\frac{p_2}{2}\right)\cdots\left(\frac{p_{k-1}}{2}\right) \times \left(\frac{p_k}{2}\right)2N^{-\gamma/k} = \frac{N^{1-(\gamma/k)}}{2^{k-1}}. \qquad (13)$$

Hence, we can conclude that:

$$W \gtrsim \frac{N^{1-(\gamma/k)}}{2^{k-1}} \qquad (14)$$

holds. We also know that the following holds:

$$X_1 X_2 \cdots X_k = 2^k N^{-k(\gamma/k)} \tilde{p}_1 \cdots \tilde{p}_k \lesssim 2^k N^{1-\gamma}. \qquad (15)$$

Now we want to derive the sufficient condition such that $X_1 X_2 \cdots X_k < W^{\frac{2}{k+1}-\epsilon}$, so that we can apply Theorem 1 (note that the maximum degree of each variable x_i in (10) is 1).

Defining $\tilde{\alpha} := \frac{2}{k+1} - \epsilon$ for some $\epsilon > 0$, if we set $N^{1-\gamma} < (N^{1-(\gamma/k)})^{\tilde{\alpha}}$, we have:

$$1 - \gamma < (1 - \frac{\gamma}{k})\tilde{\alpha} \Leftrightarrow \gamma > \frac{1-\tilde{\alpha}}{1-\frac{\tilde{\alpha}}{k}} \Leftrightarrow \gamma > \frac{k - \frac{2k}{k+1} + \epsilon}{k - \frac{2}{(k+1)} + \epsilon}$$

$$\Leftarrow \gamma > \frac{k(k+1) - 2k}{k(k+1) - 2} + \frac{\epsilon}{k - \frac{2}{(k+1)}}$$

$$\Leftarrow \gamma > \frac{k}{k+2} + \frac{\epsilon}{k-1}. \qquad (16)$$

Setting γ as above, combining with (14) and (15), we get:

$$2^{-k} \prod_{j=1}^{k} X_j < 2^{(k-1)\tilde{\alpha}} W^{\tilde{\alpha}} \Leftrightarrow \prod_{j=1}^{k} X_j < 2^{(k-1)\tilde{\alpha}+k} W^{\tilde{\alpha}}$$

$$\Leftarrow \prod_{j=1}^{k} X_j < 2^{(k-1)(1+\tilde{\alpha})} W^{\tilde{\alpha}}. \qquad (17)$$

In order to achieve the sufficient condition for $\prod_{j=1}^{k} X_j$ as shown in (1) of Theorem 1, it is sufficient to have additional exhaustive search of $\lceil 1 + \alpha \rceil$ most significant bits for each p_i. Since $\alpha < 1$, it is easy to see that this additional exhaustive search is at most 2 bits for each p_i. Finally, by Theorem 1, we can conclude that the total running time to factor N, given $\left(\frac{k}{k+2} + \frac{\epsilon}{k-1}\right)\ell_p = \left(\frac{1}{k+2} + \frac{\epsilon}{k(k-1)}\right)\ell(N)$ most significant bits of each prime factor p_i, is a polynomial in $(\log N)$ for a fixed $\epsilon > 0$. This terminates the proof of Theorem 5. ∎

Remark 2. Note that the upper-bound of $\prod_{j=1}^{k} X_j$ in (17) is larger than the upper bound of $\prod_{j=1}^{k} X_j$ in (9). This means that we also need to perform exhaustive search of about $(1/\epsilon)^k$ bits in order to solve p in (10).

6   Discussion

The total bits that has to be given in order to get our algorithm in Theorem 5 work is about $\frac{k}{k+2}\ell(N)$ bits. Thus, when k is getting larger, we need more given bits in order to get our algorithm work. Thus, when k gets larger, the success possibility of attacking the cryptosystem using our algorithm gets smaller.

Therefore, from the point of view of our algorithm, we can say that it is more secure to have a public modulus N with large number of different prime factors than that with small number of different prime factors. For instance, in the case of N with 2048 bits, when N only has two prime factors $(k = 2)$ our algorithm only needs 1024 bits to run, but when N has four prime factors $(k = 4)$ our algorithm needs 1365 bits to run. Thus, from the point of view of our algorithm, it is more secure to have $N = pqrs$ (4 different prime factors) than to have $N = pq$ (2 different prime factors). Of course one must remind that we can not generalize this argument since if N has too much prime factors, the prime factors will be mostly quite small, and thus it might be easy enough for ECM[12] to factor N.

7   Conclusion and Further Research

We have presented a factoring algorithm of a composite integer N which has k prime factors with the same bit-length, given several high-order bits of each prime factor. Also, we have presented a general method of solving k-variate polynomial equations. This method can be seen as a generalization of Coron [7]. In this paper we only deal with the case when each variable of the polynomial has independent maximum degree. It might be interesting to investigate about the other cases, e.g., the case when the total summation of the degree of all variables is given.

Acknowledgment

This research was supported in part by the Grants-in-Aid No. 16500009 for Scientific Research, JSPS.

References

1. Aciiçmez, O., Schindler, W., and Çetin Kaya Koç. Improving Brumley and Boneh timing attack on unprotected SSL implementations. In *ACM Conference on Computer and Communications Security* (2005), pp. 139–146.
2. Blömer, J., and May, A. New Partial Key Exposure Attacks on RSA. *In Advances in Cryptology (Crypto 2003), Lecture Notes in Computer Science Volume 2729, pages 27-43, Springer Verlag* (2003).
3. Boneh, D., and Durfee, G. Cryptanalysis of RSA with Private Key d Less than $N^{0.292}$. In *Eurocrypt* (1999), pp. 1–11.
4. Boneh, D., and Shacham, H. Fast Variants of RSA. CryptoBytes Volume 5 No. 1, Winter/Spring 2002.
5. Brumley, D., and Boneh, D. Remote timing attacks are practical. *Computer Networks 48*, 5 (2005), 701–716.
6. Coppersmith, D. Finding a Small Root of a Bivariate Integer Equation; Factoring with High Bits Known. In *Eurocrypt* (1996), pp. 178–189.
7. Coron, J.S. Finding Small Roots of Bivariate Integer Polynomial Equations Revisited. In *Eurocrypt* (2004), pp. 492–505.

8. Fouque, P.A., Poupard, G., and Stern, J. Sharing Decryption in the Context of Voting or Lotteries. In *Financial Cryptography* (2000), pp. 90–104.
9. Hinek, M.J., Low, M.K., and Teske, E. On Some Attacks on Multi-prime RSA. In *Selected Areas in Cryptography* (2002), pp. 385–404.
10. Howgrave-Graham, N. Finding Small Roots of Univariate Modular Equations Revisited. In *IMA Int. Conf.* (1997), pp. 131–142.
11. Lenstra, A.K., Lenstra, H.W., and Lovász, L. Factoring polynomials with rational coefficients. *Mathematische Annalen 261* (1982), 515–534.
12. Lenstra, H.W. Factoring integers with elliptic curves. *Annals of Mathematics 126* (1987), 649–673.
13. Poupard, G., and Stern, J. Fair Encryption of RSA Keys. In *Eurocrypt* (2000), pp. 172–189.
14. RSA Laboratories. PKCS #1 v2.1: RSA Cryptography Standard. http://www.rsasecurity.com/rsalabs/, June 2001.
15. Schindler, W. A Timing Attack against RSA with the Chinese Remainder Theorem. In *CHES* (2000), pp. 109–124.
16. Takagi, T. Fast RSA-Type Cryptosystem Modulo $p^k q$. In *Crypto* (1998), pp. 318–326.

A   Detail Proof of Construction of Triangular Matrix B in Figure 1

Here we prove that we can get a triangular matrix by arranging $\{\tilde{q}_{[i_1,\cdots,i_k]}\}$ and their terms according to S_{I+II}. We compare sequence $(i_1,\ldots,i_k)$ using lexicographical order $(x_1 < x_2 < \cdots < x_k)$. Let $\mathbf{b}_j$ be the column corresponding to $\tilde{q}_{[i_1,\ldots,i_k]}$ where $(i_1,\ldots,i_k) \in S_I$. Notice that from the form of $\tilde{q}_{[i_1,\ldots,i_k]}$ in this range (shown by Note 1), the least power sequence in $\tilde{q}_{[i_1,\ldots,i_k]}(x_1,\ldots,x_k)$ (lexicographically) is $(i_1,\ldots,i_k)$, which is exactly the same as the index of $\tilde{q}_{[i_1,\ldots,i_k]}(x_1,\ldots,x_k)$. Thus, it is clear that all coefficients of terms whose power sequences are less than $(i_1,\ldots,i_k)$ are zero in $\tilde{q}_{[i_1,\ldots,i_k]}(x_1,\ldots,x_k)$. Since the vertical direction is also arranged lexicographically, the upper part of $\mathbf{b}_j$ consists of zero values, from power sequence $(0,\ldots,0)$ until $(i_1^-,\ldots,i_k^-)$, where $(i_1^-,\ldots,i_k^-)$ is one less than $(i_1,\ldots,i_k)$ lexicographically. Now let $\mathbf{b}_{j+1}$ be the column corresponding to the immediate next column on the right side of $\mathbf{b}_j$. Let $\mathbf{b}_{j+1}$ corresponds to $\tilde{q}_{[i_1',\ldots,i_k']}(x_1,\ldots,x_k)$ where $(i_1',\ldots,i_k') \in S_I$. Note that $(i_1',\ldots,i_k')$ is lexicographically greater than $(i_1,\ldots,i_k)$. Thus, the coefficient of the term whose power sequence is $(i_1,\ldots,i_k)$ is zero in $\tilde{q}_{[i_1',\ldots,i_k']}(x_1,\ldots,x_k)$. This means that the number of zeroes in the upper part of $\mathbf{b}_{j+1}$ is at least one more than $\mathbf{b}_j$. Since $\mathbf{b}_{j+1}$ is the immediate next column on the right of $\mathbf{b}_j$, $(i_1',\ldots,i_k')$ is exactly the next of $(i_1,\ldots,i_k)$ in above lexicographical ordering. Thus, in vertical direction, the power sequence $(i_1',\ldots,i_k')$ is exactly one below $(i_1,\ldots,i_k)$. Hence, the number of zeroes in the upper part of $\mathbf{b}_{j+1}$ is exactly one more than $\mathbf{b}_j$. Therefore, the columns which corresponds to $\{\tilde{q}_{[i_1,\ldots,i_k]}(x_1,\ldots,x_k)\}$ where $(i_1,\ldots,i_k) \in S_I$, will make a stair-case shape such that the number of zeroes in the upper part of the column increases one by one from the left to the right.

Next, for any column $\mathbf{b}_{j'}$ where $\mathbf{b}_{j'}$ corresponds to $\tilde{q}_{[i_1,\ldots,i_k]}(x_1,\ldots,x_k)$, $(i_1,\ldots,i_k) \in S_{II}$, there is only one non-zero coefficient, that is the coefficient of term whose power sequence is $(i_1,\ldots,i_k)$, with the value of $n \cdot \prod_{j=1}^{k} X_j^{i_j}$. Since both horizontal and vertical directions are in the same arrangement of S_{I+II} and $S_I \cap S_{II} = \emptyset$, it is obvious that any two columns $\mathbf{b}_{j'}$ and $\mathbf{b}_{j'+1}$ will make a stair-case shape, such that the location of non-zero coefficient in $\mathbf{b}_{j'+1}$ is one step lower than $\mathbf{b}_{j'}$.

Combining the observation of both cases S_I and S_{II}, we conclude that for any two columns $\mathbf{b}_j$, $\mathbf{b}_{j+1}$ where $\mathbf{b}_{j+1}$ is on the right side of $\mathbf{b}_j$, the location of first non-zero coefficient of $\mathbf{b}_{j+1}$ is one step lower than $\mathbf{b}_j$. This terminates the proof that B is a triangular matrix. ∎

B Derivation of (17)

First we start from (4).

$$2^{\frac{\omega+k-3}{4}}\left((2W \cdot (\prod_{j=1}^{k} X_j)^l)^{(\delta+l+1)^k-(l+1)^k}(\prod_{j=1}^{k} X_j)^{\frac{(\delta+l+1)^k(\delta+l)+(l+1)^k l}{2}}\right)^{\frac{1}{\omega-k+2}}$$

$$< 2^{-\omega}(\prod_{j=1}^{k} X_j)^l W$$

$$\Leftrightarrow 2^{\frac{\omega+k-3}{4}(\omega-k+2)+(\delta+l+1)^k-(l+1)^k+\omega(\omega-k+2)}\left((\prod_{j=1}^{k} X_j)^{\frac{(\delta+l+1)^k(\delta+l)+(l+1)^k l}{2}}\right) \times$$

$$\left((\prod_{j=1}^{k} X_j)^{(\delta+l+1)^k-(l+1)^k}\right) < (\prod_{j=1}^{k} X_j)^{l(\omega-k+2)} W^{(\omega-k+2)+(l+1)^k-(\delta+l+1)^k}$$

$$\Leftrightarrow 2^{(\omega-k+2)\frac{5\omega+k-3}{4}+\omega-(l+1)^k}\left(\prod_{j=1}^{k} X_j\right)^{\frac{(\delta+l+1)^k(\delta+l)-(l+1)^k l}{2}+l(k-2)} < W^{(l+1)^k-(k-2)}$$

$$\Leftrightarrow 2^{\frac{5}{4}\omega^2-(k-\frac{11}{4})\omega-\frac{(k-2)(k-3)}{4}-(l+1)^k}\left(\prod_{j=1}^{k} X_j\right)^{\frac{(\delta+l+1)^k(\delta+l)-(l+1)^k l}{2}+l(k-2)} < W^{(l+1)^k-(k-2)}$$

Since $(\delta+l+1)^k(\delta+l) - l(l+1)^k < (k+1)\delta(\delta+l+1)^k$ holds for $k \geq 2$, $l \geq 1$, $\delta \geq 1$, it is sufficient to have:

$$\left(\prod_{j=1}^{k} X_j\right)^{(k+1)\delta(\delta+l+1)^k+2l(k-2)}$$

$$< 2^{-2\{\frac{5}{4}\omega^2-(k-\frac{11}{4})\omega-\frac{(k-2)(k-3)}{4}-(l+1)^k\}} W^{2\{(l+1)^k-(k-2)\}}. \quad \blacksquare$$

C Derivation of (7) and (8)

Note that $\ell \geq 1$, $\delta \geq 1$, and $k \geq 2$.

$$\alpha = \frac{2}{\delta(k+1)(\delta+l+1)^k} \left\{ \frac{(l+1)^k - (k-2)}{1 + 2l(k-2)\{\delta(k+1)(\delta+l+1)^k\}^{-1}} \right\}$$

$$\geqq \frac{2}{\delta(k+1)(\delta+l+1)^k} \{(l+1)^k - (k-2)\} \left(1 - \frac{2l(k-2)}{\delta(k+1)(\delta+l+1)^k}\right)$$

$$\geqq \frac{2}{(k+1)} \left\{ \frac{1}{\delta} - \sum_{j=0}^{k-1} \frac{(l+1)^j}{(\delta+l+1)^{j+1}} - \frac{k-2}{(\delta+l+1)^k} \right\} \left(1 - \frac{2l}{(\delta+l+1)^k}\right)$$

$$> \frac{2}{(k+1)} \left\{ \frac{1}{\delta} - \frac{k}{\delta+l+1} - \frac{k-2}{(\delta+l+1)^k} \right\} \left(1 - \frac{2}{(\delta+l+1)^{k-1}}\right)$$

$$> \frac{2}{(k+1)} \left\{ \frac{1}{\delta} - \frac{k+2}{\delta+l+1} \right\} \geqq \frac{2}{(k+1)} \left\{ \frac{1}{\delta} - \frac{2(k+1)}{\delta+l+1} \right\}$$

$$\geqq \frac{2}{\delta(k+1)} - \frac{2}{l+2}$$

$$\beta = \frac{2}{\delta(k+1)(\delta+l+1)^k} \times$$

$$\left\{ \frac{\frac{5}{4}(\delta+l+1)^{2k} - (k-\frac{11}{4})(\delta+l+1)^k - \frac{(k-2)(k-3)}{4} - (l+1)^k}{1 + 2l(k-2)\{\delta(k+1)(\delta+l+1)^k\}^{-1}} \right\}$$

$$\leqq \frac{2}{\delta(k+1)} \left\{ \frac{5}{4}(\delta+l+1)^k - (k-\frac{11}{4}) - \frac{(k-2)(k-3)}{4(\delta+l+1)^k} - \left(\frac{l+1}{\delta+l+1}\right)^k \right\} \times$$

$$\left(1 - \frac{2l(k-2)}{\delta(k+1)(\delta+l+1)^k} + \frac{4l^2(k-2)^2}{\delta^2(k+1)^2(\delta+l+1)^{2k}}\right)$$

$$\leqq \frac{2}{\delta(k+1)} \left\{ \frac{5}{4}(\delta+l+1)^k - (k-\frac{11}{4}) + \frac{1}{16 \times (l+2)^k} - \frac{2^k}{(l+2)^k} \right\} \times$$

$$\left(1 + \left(\frac{2l}{(l+2)^k}\right)^2\right)$$

$$\leqq \frac{2}{\delta(k+1)} \left\{ \frac{5}{4}(\delta+l+1)^k - (k-\frac{11}{4}) + \frac{3}{64 \times 3^k} \right\} \left(1 + \frac{4}{3^{2k-2}}\right)$$

$$\leqq \frac{2}{\delta(k+1)} \cdot \frac{13}{9} \left\{ \frac{5}{4}(\delta+l+1)^k + \frac{11}{4} + \frac{1}{64 \times 9} \right\}$$

$$\leqq \frac{4}{\delta}(1 + \frac{3}{k+1})(\delta+l+1)^k + \frac{1}{(k+1)16 \cdot 9}$$

$$\leqq \frac{4}{3\delta}(\delta+l+1)^k + \frac{8}{k+1} < \delta^{k-1}2^{k+1}(l+2)^k + \frac{8}{k+1}$$

Scalar Multiplication on Koblitz Curves Using Double Bases

Roberto Avanzi[1,*] and Francesco Sica[2,**]

[1] Institute for Cryptology and IT-Security
The Horst Görtz institute (HGI) of IT-Security
Ruhr-Universität Bochum
Universitätsstraße 150, D-44780 Bochum, Germany
roberto.avanzi@ruhr-uni-bochum.de
[2] Mount Allison University – AceCrypt
Department of Mathematics and Computer Science
67 York Street, Sackville, NB, E4L 1E6, Canada
fsica@mta.ca
http://www.acecrypt.com

Abstract. The paper is an examination of double-base decompositions of integers n, namely expansions loosely of the form $n = \sum_{i,j} \pm A^i B^j$ for some base $\{A, B\}$. This was examined in previous works [5,6], in the case when A, B lie in $\mathbb{N}$.

We show here how to extend the results of [5] to Koblitz curves over binary fields. Namely, we obtain a sublinear scalar algorithm to compute, given a generic positive integer n and an elliptic curve point P, the point nP in time $O\left(\frac{\log n}{\log \log n}\right)$ elliptic curve operations with essentially no storage, thus making the method asymptotically faster than any know scalar multiplication algorithm on Koblitz curves. In view of combinatorial results, this is the best type of estimate with two bases, apart from the value of the constant in the O notation.

Keywords: Scalar Multiplication, Elliptic Curves, Koblitz Curves, Double Base Number Systems, Sublinear Algorithms.

1   Introduction

In cryptographic protocols whose security relies on the hardness of the discrete logarithm problem on elliptic curves [14,12], the computationally most expensive part is the scalar multiplication nP, where P is a point on the curve and n is an integer, called the scalar. In order to speed up this computation, special types of curves where large multiples of P could be computed quickly have been proposed already very early in the history of elliptic curve cryptography:

---

* Partially supported by the European Commission through the IST Programme under Contract IST-2002-507932 ECRYPT.
** This work was partially supported by a NSERC Discovery Grant.

Notable examples are Koblitz curves [13] and more general curves with efficiently computable endomorphisms [9].

In this paper we will consider Koblitz curves. The coefficients of their defining equations lie in $\mathbb{F}_2$, but the curve are considered over the field extension $\mathbb{F}_{2^\mathbf{p}}/\mathbb{F}_2$. The Frobenius automorphism τ of the field extension, that sends each field elements to its square, induces an endomorphism of the group of points $E_a(\mathbb{F}_{2^\mathbf{p}})$ of the curve, called the Frobenius endomorphism and also denoted by τ. The evaluation of τP takes time $O(1)$ using normal bases or $O(\mathbf{p})$ using polynomial bases. The map τ is used to devise efficient scalar multiplication algorithms, see Section 3. But, those algorithms that compute nP without relying on (a variable amount of) precomputations require[1] $\Omega(\log n)$ groups operations (such as doublings or additions). We call such algorithms *linear*.

We study the use of τ in double base number systems, which were introduced in elliptic curve cryptography in [8]. We show how to find a decomposition

$$n = \sum_{i=1}^{\ell}(-1)^{e_i}\tau^{s_i}3^{t_i}$$

with s_i, t_i nonnegative integers and $e_i \in \{0, 1\}$. The length ℓ of this expansion is $O(\log n/\log\log n)$. Starting from here, we design a scalar multiplication algorithm with complexity $O(\log n/\log\log n)$ group operations, similarly to [5]. Our algorithm does not make use of tables of precomputations (nor of tables of accumulation registers as Yao's method [21]). Such an algorithm is said to be *sublinear*. The number of group operations over the bit size $\mathbf{p}$ of the field goes to zero as $\mathbf{p}$ increases.

This is a first instance of a sublinear scalar multiplication algorithm with very little precomputations (which depend only on $\mathbf{p}$, not the curve or the point P) or storage requirements ($O(\log\mathbf{p})$ bits).

Our algorithm has similar asymptotic complexity as windowed methods with optimal parameters, but whereas the latter require storage for $O(\log n/\log\log n)$ points, we need only one additional variable.

The techniques presented here, in contrast to, for example [20,19], are not based on computations with residue classes in number rings. Instead we use analytic methods to find successive approximations of the input by numbers of the form $\tau^s 3^t$ as in [5]. Approximation approaches are not entirely new: apart from the already cited paper [5], also [15] recodes an integer with respect to a single base by finding "close" elements that are representable with respect to the base - but the use of a single base does not require analytic methods.

The "analytic" approach to double base number systems is presented here for the first time with a complex base, and we hope to foster further research in this hitherto little explored direction.

The paper is structured as follows. In Section 2 we recall the definitions and make some assumptions that we shall need later. In Section 3 we recall the fundamentals and some recent developments in scalar multiplication on Koblitz

---

[1] We use the notation $\Omega(x)$ to mean $> cx$ for some positive c.

curves. Our new expansion and its analysis are the subject of Section 4. Finally, conclusions and a discussion of possible future research directions (Section 5) round off our presentation.

2 Preliminaries

2.1 Koblitz Curves

A Koblitz curve E_a is an elliptic curve defined over $\mathbb{F}_{2^\mathbf{P}}$, with Weierstrass equation

$$E_a \quad : \quad y^2 + xy = x^3 + ax^2 + 1 \ . \tag{1}$$

Here $a = 0$ or 1, and $\mathbf{p}$ is prime number. The parameters a and $\mathbf{p}$ are chosen in such a way that the order of the rational point group $E_a(\mathbb{F}_{2^\mathbf{P}})$ is a (large) prime r times a cofactor equal to 2 or 4. A point $P \in E_a(\mathbb{F}_{2^\mathbf{P}})$ is then randomly chosen with order equal to r. By Hasse's theorem, which states that $|\#E_a(\mathbb{F}_{2^\mathbf{P}}) - 2^\mathbf{P} - 1| < 2^{\frac{\mathbf{P}}{2}+1}$, $r = \operatorname{ord} P$ is then very close to $2^{\mathbf{P}-1}$ or $2^{\mathbf{P}-2}$, depending on the cofactor.

Being the coefficients of E_a in $\mathbb{F}_2$, the Frobenius map $\tau(x, y) = (x^2, y^2)$ is an endomorphism of $E_a(\mathbb{F}_{2^\mathbf{P}})$. Now, squaring is a linear operation in even characteristic and takes time $O(\mathbf{p})$. Hence, τ on the curve is also linear and takes time $O(\mathbf{p})$. If normal bases are used to represent the field extension $\mathbb{F}_{2^\mathbf{P}}/\mathbb{F}_2$, then computing τ on the curve is even faster, as it amounts to two bit-rotations, which are essentially free operations.

For any P on the curve it is easily verified that $\tau^2 P + 2P = (-1)^{1-a}\tau P$. Hence, τ can be identified with a complex number of norm 2 satisfying the quadratic equation $\tau^2 - (-1)^{1-a}\tau + 2 = 0$. Explicitly,

$$\tau = \frac{(-1)^{1-a} + \sqrt{-7}}{2} \ .$$

In what follows it does not matter which "determination" of the square root we use, hence we fix $\operatorname{Im}\sqrt{-7} > 0$.

2.2 Continued Fractions

Continued fractions are a way to find very good rational approximations p_s/q_s (in terms of the maximum of the absolute values of p_s and q_s) to arbitrary real numbers, by an algorithmic process which generalizes the computation of the greatest common divisor (gcd) of two integers.

We list the properties of p_s/q_s, called the s-th convergent to α, relevant to this paper. There exists a sequence of positive integers $(a_s)_{s\geq 1}$ with

$$p_s = a_s p_{s-1} + p_{s-2} \quad \text{and} \quad q_s = a_s q_{s-1} + q_{s-2} \quad \text{for all } s \geq 1 \ .$$

Therefore $q_s \geq q_{s-1} + q_{s-2}$ and similarly for p_s. These two sequences have at least a Fibonacci-like (exponential) growth. If $\alpha \notin \mathbb{Q}$, we have the following inequalities for all $s \geq 1$

$$0 < \alpha - \frac{p_{2s}}{q_{2s}} < \frac{1}{q_{2s}^2} \quad \text{and} \quad -\frac{1}{q_{2s-1}^2} < \alpha - \frac{p_{2s-1}}{q_{2s-1}} < 0 \ .$$

In particular, note that $\lim_{s \to \infty} p_s/q_s = \alpha$.

2.3   Measure of Irrationality

We begin with a famous result (usually proved with the "pigeon-hole" or box principle).

Theorem 1 (Dirichlet-Legendre, cfr. [10, Ch. 11]). *Let $Q > 1$ and $\alpha \in \mathbb{R}$. There exist integers $0 < q < Q$ and $p \in \mathbb{Z}$ such that*

$$|q\alpha - p| < \frac{1}{Q} \ .$$

The irrationality measure $\mu(\alpha)$ of $\alpha \in \mathbb{R} - \mathbb{Q}$ is defined as

$$\mu(\alpha) = \sup \left\{ x \in \mathbb{R} \colon \exists \infty \, (p, q) \in \mathbb{Z}^2 \text{ with } \left| \alpha - \frac{p}{q} \right| \leq \frac{1}{q^x} \right\} \ .$$

Notice that the convergents $\dfrac{p}{q}$ of the continued fraction expansion of α satisfy

$$\left| \alpha - \frac{p}{q} \right| \leq \frac{1}{q^2} \ ,$$

hence $\mu(\alpha) \geq 2$. It is known that the set of reals with irrationality measure greater than 2 has Lebesgue measure zero. Therefore, given α, we should conjecture that $\mu(\alpha) = 2$.

In the rest of the paper, we will then assume that the irrational numbers $\log_2 3$ and θ/π (see below for definition) have measure 2.

2.4   Double Bases

Following [6] we call a $\{A, B\}$-integer a number which can be written as $A^s B^t$ for some nonnegative integers s, t ($\pm A^s B^t$ if signed numbers are allowed). We extend the definition to algebraic integers, more precisely, integers in $\mathbb{Z}[\tau]$. We will also allow $A, B \in \mathbb{Z}[\tau]$. We define a $\{A, B\}$-integer expansion of n as a decomposition of n into a sum of (possibly signed) $\{A, B\}$-integers.

3   Scalar Multiplication on Koblitz Curves

In this section we are chiefly concerned with scalar multiplication techniques that do not make use of point precomputations or storage for on-the-fly computed tables of point multiples. Variants of these methods than can take advantage of such devices exist and in many cases have been extensively treated in the literature.

3.1 The τ-NAF

All facts used here are only stated. Proofs can be found in [20,19].

Let E_a denote the Koblitz curve defined over $\mathbb{F}_{2^p}$ by (1), and P a point in the subgroup of large prime order r of $E(\mathbb{F}_{2^p})$. Let τ denote the Frobenius endomorphism. We can view $\tau(P)$ as multiplication of the point P by τ, and thus the whole ring $\mathbb{Z}[\tau]$ operates on the subgroup $\langle P \rangle$ of $E(\mathbb{F}_{2^p})$ generated by P. In fact there exists an integer λ such that $\tau(P) = \lambda P$, and thus τ operates on $\langle P \rangle$ like multiplication by λ. (This λ satisfies $\lambda^2 + (-1)^a \lambda + 2 \equiv 0 \bmod r$.)

The τ-adic non-adjacent form (τ-NAF) of an integer $z \in \mathbb{Z}[\tau]$ is an expression $z = \sum_i z_i \tau^i$ where $z_i \in \{0, \pm 1\}$ and satisfying the non-adjacency property $z_j z_{j+1} = 0$ (similarly to the classical NAF [17]). The expected density (i.e. the ratio of non-zero bits w.r.t. to the total number of bits) of a τ-NAF is $1/3$. Each $z \in \mathbb{Z}[\tau]$ admits a unique τ-NAF.

The length of the τ-NAF of a randomly chosen scalar n is $\approx 2\mathbf{p}$, whereas the bit length of n is $\approx \mathbf{p}$. For any point $P \in E_a(\mathbb{F}_{2^p}) \setminus E_a(\mathbb{F}_2)$, it is $\tau^{\mathbf{p}} P = P$ and $\tau P \neq P$. Hence, $\mathbb{Z}[\tau]$ being an Euclidean ring, we can take the remainder ζ of $n \bmod (\tau^{\mathbf{p}} - 1)/(\tau - 1)$ and use it in place of n. This ζ will have smaller norm than that of $(\tau^{\mathbf{p}} - 1)/(\tau - 1)$, and thus its length will be at most $\mathbf{p}$. Its τ-NAF is called the reduced τ-NAF of n. We shall write $\sum_i z_i \tau^i$ for the τ-NAF of ζ in what follows.

Just as the classic double-and-add scalar multiplication algorithm is simply a Horner scheme for evaluating nP as $\sum_{i=0}^{\ell} n_i 2^i P$, it is possible to evaluate $zP = \sum_i z_i \tau^i(P)$ by a Horner scheme. The resulting method is called a τ-and-add algorithm. Let n be a randomly chosen integer in $[1 .. \#E(\mathbb{F}_{2^p})]$. Its binary expansion and its reduced τ-NAF have roughly the same expected length and density. This, and the fact that Frobenius evaluations are much faster than doublings, explain why the τ-and-add algorithm is much faster than a double-and-add scheme on Koblitz curves.

3.2 Inserting a Halving and Viewing It as a Second Base

Point halving [11,18] is the inverse operation to point doubling and applies to *all* elliptic curves over binary fields, not only to Koblitz curves. Its evaluation is 2 to 3 times faster than that of a doubling and it is possible to rewrite the scalar multiplication algorithm using halving instead of doubling. The resulting method is very fast, but on Koblitz curves it is slower than the τ-and-add method, because a halving is slower than a Frobenius evaluation.

In [1] a single point halving is inserted in the τ-and-add method. This allowed to reduce the amount of group operations by 14% with respect to the τ-NAF. A refinement in [4] brought the speed-up to 25%. The basic idea in both approaches is to express nP as $\sum_i e_i \tau^i(P) + \sum_i f_i \tau^i(Q)$ with $Q = \frac{1}{2}P$ and a smaller joint Hamming weight of the e_i, f_i's.

This approach can be viewed as a simple case of double base expansion, where the second base (the first being τ) is $\frac{1}{2}$, but it only appears with exponent at most 1. (For a different approach also based in part on the ideas of [1], see also [16], where "windows of width 5" are used essentially without storing precomputations. This cannot be formulated as a double base method.)

4   Scalar Multiplication with $\{\tau, 3\}$-Expansions

The aim of this section is to produce an efficient decomposition of a scalar n as a signed sum of $\{\tau, 3\}$-integers

$$n = \sum_{i=1}^{\ell} (-1)^{e_i} \tau^{s_i} 3^{t_i}$$

with s_i, t_i nonnegative integers, $(s_i, t_i) \neq (s_j, t_j)$ for $i \neq j$ and $e_i \in \{0, 1\}$. Here $\ell = O(\log n / \log \log n)$.

As a first simplification we replace n its reduced τ-NAF, ζ, as defined in Section 3. This allows to cut by half the representation of n, since $nP = \zeta P$ on the curve.

Using a lexicographic order on powers of 3 and τ one can rewrite such a $\{\tau, 3\}$ expansion as

$$\zeta = \sum_{i=1}^{\mathtt{J}} 3^{t_i} \sum_{j=1}^{\partial_i} (-1)^{e_{i,j}} \tau^{s_{i,j}} \tag{2}$$

where $e_{i,j} \in \{0, 1\}$,

$$\sum_{i=1}^{\mathtt{J}} \sum_{j=1}^{\partial_i} 1 = \ell , \quad t_i > t_{i+1} \quad \text{and} \quad s_{i,j} > s_{i,j+1} .$$

4.1   Preliminaries

Let p_s/q_s be the s-th convergent to $\log_3 2/2$ (this is a slight departure from [5]). Let $m = \mathbf{p}^{2/5}$. Fix s as the first odd index such that $p_s > m$. Then $0 < \frac{2}{m^{1+\epsilon} \log 2} < p_s \frac{2 \log 3}{\log 2} - q_s < \frac{1}{m} < \frac{2}{m \log 2}$. This shows the following lemma.

Lemma 1. *Using the above notations we have, as* $\mathbf{p} \to \infty$

$$\exp\left(\frac{1}{m^{1+\epsilon}}\right) < \frac{3^{p_s}}{2^{\frac{q_s}{2}}} < \exp\left(\frac{1}{m}\right) . \tag{3}$$

The authors of [5] then use this lemma to prove the following "reduction" theorem (which we cite after fixing the value of m).

Theorem 2 (Thm. 1 of [5]). *Let n be a large integer. There exists a $\{2, 3\}$-integer N satisfying*

$$|n - N| < \frac{n}{\log^{\frac{1}{3}} n} .$$

Repeated use of this theorem leads to an effective construction of a $\{2, 3\}$-integer decomposition of n as in the following.

Theorem 3 (Thm. 2 of [5]). *Every sufficiently large rational nonnegative integer* n *can be written as a sum* $n = \sum_{i=1}^{k} 2^{s_i} 3^{t_i}$ *where* $s_i, t_i \in \mathbb{N} \cup \{0\}$ *satisfy* $(s_i, t_i) \neq (s_j, t_j)$ *for* $i \neq j$ *and* $k \leq 3 \frac{\log n}{\log \log n} + o\left(\frac{\log n}{\log \log n}\right)$. *Furthermore, one can ensure that* $\max_i s_i \leq \log^{2/3+\epsilon} n$.

This last theorem allows to build a sublinear scalar multiplication algorithm (Algorithm 2 in [5]). That work forms the blueprint of our sublinear scalar multiplication algorithm for Koblitz curves.

4.2   The Expansion

We now generalize the algorithms of [5] to ordinary Koblitz curves defined over $\mathbb{F}_{2^p}$. The main difference is that we have to view τ as a complex number, which requires controlling the argument of the numbers thus involved if we want to find "close" $\{\tau, 3\}$-numbers.

Let τ be the Frobenius endomorphism of E_a. Put $\theta = \arg(\tau)$. We first prove the following easy result.

Lemma 2. $\dfrac{\theta}{\pi} \notin \mathbb{Q}$.

Proof. We want to show that $\tau^a \notin \mathbb{R}$ for any $a \in \mathbb{Z}$. Let there otherwise be some such a. Let $M = \tau^a$. Taking complex conjugates, we get $M = \bar{\tau}^a = \tau^a$, which is impossible, since $\mathbb{Z}[\tau]$ is a unique factorization domain (it is Euclidean) and τ and $\bar{\tau}$ are two non-associated irreducibles.   □

Theorem 4. *Let* $\zeta \in \mathbb{Z}[\tau]$ *be large. There exists a* $\{\tau, 3\}$-*number* N *satisfying either*

$$|\zeta - N| \leq \frac{|\zeta|}{\log^{\frac{2}{25}} |\zeta|} \qquad or \qquad |\zeta + N| \leq \frac{|\zeta|}{\log^{\frac{2}{25}} |\zeta|} .$$

Proof. In view of (3) we have[2]

$$\left| \frac{3^{p_s}}{\tau^{q_s}} \right| \asymp e^{\frac{1}{m}} .$$

We then take the largest power 2^ν less than or equal to $|\zeta|$. Define t as the largest integer such that

$$\left| \frac{3^{p_s}}{\tau^{q_s}} \right|^t \leq \frac{|\zeta|}{2^\nu} \tag{4}$$

and

$$q_s t < 2\nu . \tag{5}$$

---

[2] We write something is $\asymp f(m)$ for some function f to mean that it lies between $f(m^{1-\epsilon})$ and $f(m^{1+\epsilon})$. Similarly with $\mathfrak{m}$ instead of m. This will avoid notation cluttering, while giving enough indications for a complete technical proof.

Note that our choice of m will guarantee that (5) is automatically fulfilled, as in [5]. Then $\widetilde{N} := \tau^{2\nu - tq_s} 3^{tp_s}$ satisfies

$$1 \le \left| \frac{\zeta}{\widetilde{N}} \right| \le e^{\frac{1}{m}} .$$

Unlike in the supersingular case we cannot conclude that $|\zeta - \widetilde{N}|$ is small, because we need to adjust the argument of $\widetilde{N}$. We will rely on the following result.

Lemma 3. *Let* ξ_1, ξ_2 *be two nonzero complex numbers and* $m \ge 3$ *such that* $1 \le |\xi_1 / \xi_2| \le e^{\frac{1}{m}}$ *and* $\cos \arg(\xi_1 / \xi_2) \ge e^{-\frac{1}{m}}$. *Then*

$$|\xi_1 - \xi_2| \le \frac{2|\xi_2|}{\sqrt{m}} .$$

Proof. See Appendix A.

We now find an integer $u \ge 0$ such that there exists an integer v with

$$|u q_s \theta - 2v\pi| < \frac{1}{\sqrt{m}} .$$

We can do this by looking at the continued fraction expansion of $q_s \theta / 2\pi$ which is irrational by Lemma 2. The previous inequality becomes

$$\left| u \frac{q_s \theta}{2\pi} - v \right| < \frac{1}{2\pi \sqrt{m}} .$$

By the Dirichlet-Legendre theorem, v/u can be chosen as the convergent to $q_s \theta / 2\pi$ with $u < 2\pi \sqrt{m}$ closest to this bound. By our assumption on irrationality measures, actually

$$u \asymp \sqrt{m} \qquad \text{and} \qquad |u q_s \theta - 2v\pi| \asymp \frac{1}{\sqrt{m}} . \tag{6}$$

This u can actually be precomputed, as it will depend only on the size 2^P of the finite field (see below), not even on the curve. Define then

$$-k = \left\lfloor \frac{\arg_\pi \zeta - (2\nu - tq_s)\theta}{u q_s \theta - 2v\pi} \right\rceil \le 0 ,$$

(here $\lfloor x \rceil$ denotes the closest integer to x) where $-\pi < \arg_\pi \zeta - (2\nu - tq_s)\theta < \pi$ is defined modulo π to make k non-negative. Then $k = O(\sqrt{m})$ and

$$|k u q_s \theta + \arg_\pi \zeta - (2\nu - tq_s)\theta - 2kv\pi| < \frac{1}{\sqrt{m}} .$$

Define now

$$N = \widetilde{N} \left(\frac{3^{p_s}}{\tau^{q_s}} \right)^{ku} = \tau^{2\nu - (t+ku)q_s} 3^{(t+ku)p_s} . \tag{7}$$

Note that, if $\mathfrak{m}$ is small enough, N is a $\{\tau, 3\}$-integer. Also, either $|\arg(N/\zeta)| < \frac{1}{\sqrt{\mathfrak{m}}}$ or $|\arg(-N/\zeta)| < \frac{1}{\sqrt{\mathfrak{m}}}$ and thus we get $|\cos\arg(N/\zeta)| > e^{-\frac{1}{\mathfrak{m}}}$. Also,

$$1 \leq \left|\frac{N}{\zeta}\right| \leq \left|\frac{N}{\widetilde{N}}\right| \leq e^{\frac{O(\mathfrak{m})}{\mathfrak{m}}} = e^{O\left(\frac{\mathfrak{m}}{\mathfrak{m}}\right)} \ .$$

Thus choosing $\mathfrak{m} \leq m^{1/2-\epsilon}$, we can apply Lemma 3 to $\xi_1 = N$ or $\xi_1 = -N$ and $\xi_2 = \zeta$ to conclude that

$$|\zeta - N| \leq \frac{2|\zeta|}{m^{1/4-\epsilon}} \quad \text{or} \quad |\zeta + N| \leq \frac{2|\zeta|}{m^{1/4-\epsilon}}$$

thus proving Theorem 4 once we fix $\mathfrak{m} = m^{2/5} = \mathbf{p}^{4/25}$. $\qquad\square$

Repeated applications of this theorem will yield the next result, whose proof follows, mutatis mutandis, that of Theorem 3.

Theorem 5. *Every $\zeta \in \mathbb{Z}[\tau]$ with $\zeta\bar{\zeta} < \#E_a(\mathbb{F}_{2^\mathbf{p}})$ can be written as a sum*

$$\zeta = \sum_{i=1}^{\ell} (-1)^{e_i} \tau^{s_i} 3^{t_i}$$

with s_i, t_i nonnegative integers, $(s_i, t_i) \neq (s_j, t_j)$ for $i \neq j$ and $e_i \in \{0,1\}$. Furthermore the length of the expansion is

$$\ell \leq 12.5 \frac{\mathbf{p}}{\log_2 \mathbf{p}} + o\left(\frac{\mathbf{p}}{\log \mathbf{p}}\right)$$

and one can insure that $\max_i t_i \leq \mathbf{p}^{4/5}$.

In view of the fact that, by Hasse's theorem,

$$\mathbf{p} = \log_2 \#E_a(\mathbb{F}_{2^\mathbf{p}}) + O\left(\#E_a(\mathbb{F}_{2^\mathbf{p}})^{-1/2}\right) = \log_2 n + O(1)$$

on average for n, this is the analogue of Theorem 3 in our context. It is this constructive theorem which is responsible for the sublinear running time of a scalar multiplication algorithm similar to Algorithm 2 in [5]. See also the next subsection for details.

Remark 1. We should note that Theorem 5 also holds for unsigned expansions, with the same constant.

4.3 Practical Estimates

Algorithms 1 and 2 describe respectively the initial precomputation and the scalar recoding. We will draw some remarks concerning their application. Algorithm 3, that perform the actual scalar multiplication, is reported in Appendix D, as it is essentially identical to Algorithm 2 from [5].

Algorithm 1. Precomputations (depending on **p**)

Input: An integer **p**, the bit size of the ground field $\mathbb{F}_{2^\mathbf{P}}$.

Output: Three integers P_CONV, Q_CONV and U_CONV, and two floating point numbers MODULUS_RATIO and ANGLE_RATIO.

---

1. $m \leftarrow \mathbf{p}^{2/5}$
2. $s \leftarrow \min\{2j + 1 : p_{2j+1} > m\}$
3. P_CONV $\leftarrow p_s$
4. Q_CONV $\leftarrow q_s$
5. MODULUS_RATIO $\leftarrow 2p_s \log_2 3 - q_s$
6. MODULUS_RATIO $\leftarrow 1/\text{MODULUS\_RATIO}$
7. ANGLE_RATIO $\leftarrow uq_s\theta - 2v\pi$, as per (6)
8. ANGLE_RATIO $\leftarrow 1/\text{ANGLE\_RATIO}$
9. U_CONV $\leftarrow u$

Algorithm 2. Binumber Scalar Decomposition

Input: An integer $\zeta \in \mathbb{Z}[\tau]$ with $2^{\mathbf{P}/4} < |\zeta| < 2^{\mathbf{P}}$, and constants P_CONV, Q_CONV, U_CONV and MODULUS_RATIO, ANGLE_RATIO.

Output: A set $\mathcal{S} = \{(e_1, s_1, t_1), \ldots, (e_\ell, s_\ell, t_\ell)\}$ with s_i, t_i nonnegative integers, $e_i \in \{0, 1\}$ and $\ell = O(\log n / \log \log n)$ such that $\zeta = \sum_{i=1}^{\ell} (-1)^{e_i} \tau^{s_i} 3^{t_i}$

---

1. $\mathcal{S} = \emptyset$
2. **while** $|\zeta| > 2^{\mathbf{P}^{4/5}}$ **do**
3. $\quad \epsilon \leftarrow 0$
4. $\quad \nu \leftarrow \lfloor \log_2 |\zeta| \rfloor$
5. $\quad t \leftarrow \lfloor 2\text{MODULUS\_RATIO}(\log_2 |\zeta| - \nu) \rfloor$
6. $\quad k_0 \leftarrow \text{Arg}(\zeta / \tau^{2\nu - t\text{Q\_CONV}})$
7. $\quad$ **if** $k_0\text{ANGLE\_RATIO} > 0$ **then**
8. $\quad\quad k_0 \leftarrow k_0 - \text{sign}(\text{ANGLE\_RATIO})\pi$
9. $\quad\quad \epsilon \leftarrow 1$
10. $\quad k \leftarrow -\lfloor k_0\text{ANGLE\_RATIO} \rceil$
11. $\quad c \leftarrow t + k\text{U\_CONV}$
12. $\quad N \leftarrow \tau^{2\nu - c\text{Q\_CONV}} 3^{c\text{P\_CONV}}$
13. $\quad \mathcal{S} \leftarrow \mathcal{S} \cup \{(\epsilon, 2\nu - c\text{Q\_CONV}, c\text{P\_CONV})\}$
14. $\quad \zeta \leftarrow \zeta - (-1)^\epsilon N$
15. Find the τ-NAF of ζ, appending exponents and signs to $\mathcal{S}$.
16. **return** $\mathcal{S}$

Algorithm 2 differs somewhat from its counterpart, Algorithm 1 in [5], in that we are always using the same MODULUS_RATIO, until we reach a sufficiently low stage, and then give up and use a τ-NAF. In fact, while $|\zeta| > 2^{\mathbf{P}^{4/5}}$, on applying Theorem 4 we keep dividing moduli by a quantity at least $\log_2^{\frac{2}{25}} 2^{\mathbf{P}^{4/5}} = \mathbf{p}^{\frac{8}{125}}$. Therefore we need less than

$$\frac{\mathbf{p}}{\log_2\left(\mathbf{p}^{\frac{8}{125}}\right)} = 15.625\frac{\mathbf{p}}{\log_2 \mathbf{p}}$$

iterations to get down to $2^{\mathbf{p}^{4/5}}$. Below this threshold, a τ-NAF will have length $\leq$ $\mathbf{p}^{4/5}$. Altogether, this gives a new bound for ℓ similar to the bound of Theorem 5 with 15.625 replacing 12.5.

A remark is now due about the precision of the floating point quantities MODULUS_RATIO and ANGLE_RATIO. Note that even in the case of the curve K-571, where $\mathbf{p} = 571$, we have $m = \mathbf{p}^{2/5} \approx 12.664...$ and ANGLE_RATIO must be precise enough to enable us to resolve differences smaller than $\frac{1}{2\sqrt{m}}$, which for K-571 are around 0.3, in order to correctly compute k. in Algorithm 2. Therefore, single precision floating point values are more than sufficient for our purposes, and in fact we can even use low-precision fixed point numbers implemented via "short" (i.e. 16 or 32-bit) long integers. This also shows that the runtime of Algorithm 2 is negligible compared to a scalar multiplication algorithm like Algorithm 3.

Practical simulations have shown that a greedy-type algorithm gives considerably shorter expansions han the τ-NAF also for cryptographically relevant curves, such as NIST curves K-163, K-233 etc. Heuristically, if we use a greedy algorithm for the τ-NAF expansion then at each step the intermediate variable is halved, whereas in our new algorithm we divide it by a power of $\log|\zeta|$ (a small one, true, but for small values of $\mathbf{p}$ we use ad-hoc look-up which should at least find the smallest power of τ close to ζ). The constant in the bound of the expansion in practice seems much smaller than 15 (in [5], this is about 1, using only unsigned expansions, which means that the corresponding algorithm is 60% faster than the τ-and-add).

5   Conclusions and Perspectives

We have analyzed double-base expansions to the extent that we could generalize scalar multiplication algorithms to use some bases $\{A, B\}$ where one of A, B is complex. Our results are so far mainly of an asymptotic nature and rely on the assumption that certain irrational numbers are "generically" approximable by rationals. By combinatorial reasons, these algorithms cannot achieve a better asymptotic running time than $\mathbf{p}/\log \mathbf{p}$ elliptic curve additions (cf. Appendices B and C).

In the aftermath of this article, two works are now about to be published on the same topic. In [7], accepted at CHES 2006, the authors present practical measurements on FPGA and show that indeed one achieves a 50% speedup already on the smallest Koblitz curve K-163 (although they don't use our decomposition). Also [2], accepted at Asiacrypt 2006, provides a unified description of double base recoding which fills in a proof of the theoretical assumptions made in the previously cited work. The recoding is right-to-left, that is in ascending ordering of the powers of the fast endomorphism (in this case τ), as opposed to greedy-type algorithms which are naturally left-to-right. Moreover, the authors achieve the asymptotic lower bound described in Appendix C.

The paper [3], to appear in the proceedings of SAC 2006, is devoted to digit sets for τ-adic expansions of integers. In particular, a classification of the digit sets that allow a successful expansion of all inputs is given, and a scalar multiplication method of the same complexity as the one in [2] is proposed, but given in terms of single bases and suitably chosen digit sets.

References

1. R. Avanzi, M. Ciet, and F. Sica. Faster Scalar Multiplication on Koblitz Curves combining Point Halving with the Frobenius Endomorphism. In *Proceedings of PKC 2004*, volume 2947 of *Lecture Notes in Computer Science*, pages 28–40. Springer, 2004.
2. R. Avanzi, V. S. Dimitrov, C. Doche, and F. Sica. Extending Scalar Multiplication using Double Bases. In *Proceedings of Asiacrypt 2006*, Lecture Notes in Computer Science. Springer, 2006.
3. R. Avanzi, C. Heuberger, and H. Prodinger. On Redundant τ-adic Expansions and Non-Adjacent Digit Sets. In *Proceedings of SAC 2006 (Workshop on Selected Areas in Cryptography)*, Lecture Notes in Computer Science. Springer.
4. R. Avanzi, C. Heuberger, and H. Prodinger. Minimality of the Hamming Weight of the τ-NAF for Koblitz Curves and Improved Combination with Point Halving. In *Proceedings of SAC 2005*, volume 3897 of *Lecture Notes in Computer Science*, pages 332–344. Springer, 2006.
5. M. Ciet and F. Sica. An Analysis of Double Base Number Systems and a Sublinear Scalar Multiplication Algorithm. In E. Dawson and S. Vaudenay, editors, *Progress in Cryptology - Proceedings of Mycrypt 2005*, volume 3715 of *Lecture Notes in Computer Science*, pages 171–182. Springer, 2005.
6. V. S. Dimitrov, L. Imbert, and P. K. Mishra. Efficient and secure elliptic curve point multiplication using double-base chains. In *Advances in Cryptology - ASIACRYPT 2005*, volume 3788 of *Lecture Notes in Computer Science*, pages 59–78. Springer, 2005.
7. V. S. Dimitrov, K. Jarvinen, M. J. Jacobson Jr, W. F. Chan, and Z. Huang. FPGA Implementation of Point Multiplication on Koblitz Curves Using Kleinian Integers. In *Proceedings of CHES 2006*, Lecture Notes in Computer Science. Springer, 2006.
8. V. S. Dimitrov, G. A. Jullien, and W. C. Miller. An algorithm for modular exponentiation. *Information Processing Letters*, 66(3):155–159, 1998.
9. R. P. Gallant, J. L. Lambert, and S. A. Vanstone. Faster Point Multiplication on Elliptic Curves with Efficient Endomorphisms. In J. Kilian, editor, *Advances in Cryptology - Proceedings of CRYPTO 2001*, volume 2139 of *Lecture Notes in Computer Science*, pages 190–200. Springer, 2001.
10. G. H. Hardy and E. M. Wright. *An Introduction to the Theory of Numbers*. Oxford University Press, fifth edition, 1979.

11. E. W. Knudsen. Elliptic Scalar Multiplication Using Point Halving. In K.-Y.Lam, E. Okamoto, and C. Xing, editors, *Advances in Cryptography - Proceedings of ASIACRYPT 1999*, volume 1716 of *Lecture Notes in Computer Science*, pages 135–149. Springer, 1999.
12. N. Koblitz. Elliptic Curve Cryptosystems. *Mathematics of Computation*, 48(177):203–209, 1987.
13. N. Koblitz. CM-curves with good cryptographic properties. In Joan Feigenbaum, editor, *Advances in Cryptology - Proceedings of CRYPTO 1991*, volume 576 of *Lecture Notes in Computer Science*, pages 279–287, Berlin, 1991. Springer.
14. V.S. Miller. Use of Elliptic Curves in Cryptography. In H.C. Williams, editor, *Advances in Cryptology - Proceedings of CRYPTO 1985*, volume 218 of *Lecture Notes in Computer Science*, pages 417–426. Springer, 1986.
15. J.A. Muir and D.R. Stinson. New Minimal Weight Representations for Left-to-Right Window Methods. In *Topics in cryptology – CT-RSA 2005*, volume 3376 of *Lecture Notes in Comput. Sci.*, pages 366–383. Springer-Verlag, Berlin, 2005.
16. K. Okeya, T. Takagi, and C. Vuillaume. Short Memory Scalar Multiplication on Koblitz Curves. In *Proceedings of CHES 2005*, volume 3659 of *Lecture Notes in Computer Science*, pages 91–105. Springer, 2005.
17. G.W. Reitwiesner. Binary arithmetic. *Advances in Computers*, 1:231–308, 1960.
18. R. Schroeppel. Elliptic curves: Twice as fast!, 2000. Presentation at the Crypto 2000 Rump Session.
19. J. A. Solinas. An Improved Algorithm for Arithmetic on a Family of Elliptic Curves. In Burton S. Kaliski Jr., editor, *Advances in Cryptology - Proceedings of CRYPTO 1997*, volume 1294 of *Lecture Notes in Computer Science*, pages 357–371. Springer, 1997.
20. J. A. Solinas. Efficient arithmetic on Koblitz curves. *Designs, Codes and Cryptography*, 19:195–249, 2000.
21. A.C. Yao. On the evaluation of powers. *SIAM Journal on Computing*, 5:100–103, 1976.

Disclaimer: *The information in this document reflects only the authors' views, is provided as is and no guarantee or warranty is given that the information is fit for any particular purpose. The user thereof uses the information at its sole risk and liability.*

A   Proof of Lemma 3

After rescaling, we may suppose that $\xi_2 = e^{-1/m}$ and $\xi_1 = \xi$ has modulus $e^{-1/m} \le |\xi| \le 1$, with $\cos \arg(\xi) \ge e^{-1/m}$. This means that ξ is in the grayed out sector in the figure. Let $0 < \psi < \pi/2$ be the angle such that $\cos \psi = e^{-1/m}$, as shown in the figure. Then the maximum of the distance between ξ and $e^{-1/m}$ is the dotted length. Analytically,

$$|\xi_1 - \xi_2| = \left| \xi - e^{-\frac{1}{m}} \right| \le \sin \psi = \sqrt{1 - e^{-2/m}} \le \frac{\sqrt{2}}{\sqrt{m}} = \frac{\sqrt{2}\, e^{1/m}}{\sqrt{m}} |\xi_2| \ .$$

Since $\sqrt{2}\, e^{1/m} < 2$ for $m \ge 3$ this concludes the proof.

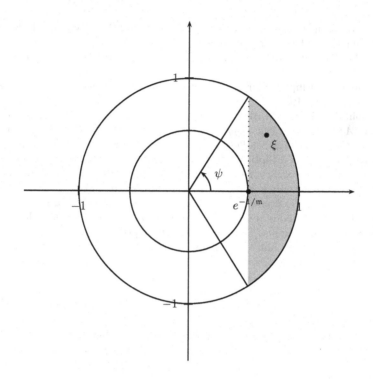

B Why Double-Base Algorithms Are Sublinear Exclusively on Curves with Fast Endomorphisms

We prove here that the maximum of the exponents in any $\{2,3\}$-integer expansion of n must be of order $\log n$. As a corollary we have that no such expansion can give rise to a sublinear scalar multiplication algorithm on a generic elliptic curve, where we can only hope to improve the scalar multiplication timings by a bounded factor.

Theorem 6. *Let*

$$n = \sum_{i=1}^{k} 2^{s_i} 3^{t_i}, \quad s_i, t_i \in \mathbb{N} \cup \{0\}$$

with $(s_i, t_i) \neq (s_j, t_j)$ for $i \neq j$. Then, as n goes to infinity,

$$\max_i (s_i, t_i) \geq \log_6 n + O(\log \log n) \ .$$

Proof. Let $s = \max_i (s_i, t_i)$. We have

$$n = \sum_{i=1}^{k} 2^{s_i} 3^{t_i} \leq k6^s \ .$$

Since $k \leq \log^2 n / (\log 2 \log 3)$ we must have from $\frac{\log^2 n}{\log 2 \log 3} 6^s \geq n$ that

$$s \geq \frac{\log n}{\log 6} - \frac{2 \log \log n}{\log 2 \log 3 \log 6} .$$

$\square$

Corollary 1. *A double base expansion of a generic scalar n cannot be converted into a sublinear scalar multiplication algorithm on a generic elliptic curve.*

Proof. Indeed, there are at least $\Omega(\log n)$ powers of 2 or 3, and on a generic elliptic curve these two operations are costly, hence Algorithm 2 in [5] in computing nP on the elliptic curve will have to perform $\Omega(\log n)$ elliptic curve operations. $\square$

Remark 2. The same goes of course for the $\{\tau, 3\}$-number algorithm described in this paper.

C   Limitations of Greedy-Type Algorithms

In this appendix, we take all logs to the base 2. We want to show the following theorem.

Theorem 7. *If we use a greedy algorithm to find a $\{\tau, 3\}$ expansion, then we must have*

$$\ell \geq \frac{\mathbf{p}}{\log \mathbf{p}} + o\left(\frac{\mathbf{p}}{\log \mathbf{p}}\right)$$

Remark 3. In particular, this shows that we cannot achieve a constant better than 1 in Theorem 5, at least with our method.

Proof. Since we are using a greedy algorithm to find all our $\{\tau, 3\}$-integers, we are restricting our pool of $\{\tau, 3\}$-integers to $\tau^s 3^t$ with $s \leq 2\mathbf{p}$ and $t \leq \mathbf{p}$, hence at most $2\mathbf{p}^2$ numbers. We know that the number of integers of norm less than $2^{\mathbf{p}}$ which can be represented by at most ℓ $\{\tau, 3\}$-numbers is upper bounded by

$$\sum_{i=1}^{\ell} 2^i \binom{2\mathbf{p}^2}{i} \leq \ell\, 2^{\ell} \binom{2\mathbf{p}^2}{\ell} = \frac{\Gamma(2\mathbf{p}^2 + 1)}{\Gamma(\ell + 1)\Gamma(2\mathbf{p}^2 - \ell)}\, \ell\, 2^{\ell} .$$

This is due to the fact that for any weight i, we can represent an integer by choosing i $\{\tau, 3\}$-integers among at most $2\mathbf{p}^2$ and each of them can have a positive or negative sign. The inequality follows from the ascertained fact that $\ell < \mathbf{p}$. Using Stirling's formula for $\Gamma(z)$, we arrive at the following asymptotic formula

$$\left(1 + \frac{\ell}{2\mathbf{p}^2 - \ell}\right)^{2\mathbf{p}^2 - \ell} \cdot 2^{\ell} \cdot \mathbf{p}^{2\ell} \cdot \frac{\ell^{3/2}}{\ell^{\ell}} \leq \frac{(2e)^{\ell}\, \mathbf{p}^{2\ell}\, \ell^{3/2}}{\ell^{\ell}} .$$

With $\ell = c\dfrac{\mathbf{p}}{\log \mathbf{p}}$ we transform the previous expression into

$$(2e)^{(c - c \log c)\frac{\mathbf{p}}{\log \mathbf{p}}} \cdot 2^{c\mathbf{p}} \cdot 2^{c\frac{\mathbf{p} \log \log \mathbf{p}}{\log \mathbf{p}}} \left(\frac{c\mathbf{p}}{\log \mathbf{p}}\right)^{3/2} < 2^{\mathbf{p} - 3}$$

when $\mathbf{p} \to \infty$, as soon as $c < 1$. This contradicts the fact that we must find a representation of all the integers of norm less than $2^{\mathbf{p}}$, which are at least $2^{\mathbf{p}-2}$ (at least all remainders ζ of all possible $n \bmod (\tau^{\mathbf{p}} - 1)/(\tau - 1)$). $\qquad\square$

Remark 4. The same theorem also holds for *any* unsigned $\{2,3\}$ expansion as in [5,6], or for those signed $\{2,3\}$ expansion obtained with a greedy algorithm, since the only ingredient we need to make this cardinality-type argument work is an upper bound on the double exponents (a, b) in $2^a 3^b$, which we automatically have in theses cases. It is not clear that the same holds in general (actually, it seems more plausible to have $o(\mathbf{p}/\log \mathbf{p})$ in signed expansions).

D   Sublinear Scalar Multiplication Algorithm

Algorithm 3. Sublinear Multiplication

Input: A point P on the Koblitz curve E_a and a sequence of triplets of exponents $(e_{i,j}, s_{i,j}, t_i)$ as in (2).
Output: The point Q on the elliptic curve such that $Q = \zeta P$.

---

1. $Q \leftarrow \mathcal{O}$
2. **for** $i = 1$ **to** $\mathfrak{I} - 1$ **do**
3. $\qquad R \leftarrow (-1)^{e_{i,1}} P$
4. $\qquad$ **for** $j = 1$ **to** $\mathfrak{J}_i$ **do**
5. $\qquad\qquad R \leftarrow \tau^{s_{i,j} - s_{i,j+1}} R + (-1)^{e_{i,j+1}} P$
6. $\qquad Q \leftarrow Q + R$
7. $\qquad Q \leftarrow 3^{t_i - t_{i+1}} Q$
8. $R \leftarrow (-1)^{e_{\mathfrak{I},1}} P$
9. **for** $j = 1$ **to** $\mathfrak{J}_{\mathfrak{I}}$ **do**
10. $\qquad R \leftarrow \tau^{s_{\mathfrak{I},j} - s_{\mathfrak{I},j+1}} R + (-1)^{e_{\mathfrak{I},j+1}} P$
11. $Q \leftarrow Q + R$
12. **return** Q

Compressed Jacobian Coordinates for OEF

Fumitaka Hoshino, Tetsutaro Kobayashi, and Kazumaro Aoki

NTT Information Sharing Platform Laboratories, NTT Corporation
{fhoshino,kotetsu,maro}@isl.ntt.co.jp

Abstract. This paper presents a new coordinate system for elliptic curves that accelerates the elliptic curve addition and doubling over an optimal extension field (OEF). Many coordinate systems for elliptic curves have been proposed to accelerate elliptic curve cryptosystems. This paper is a natural extension of these papers and the new coordinates are much faster when the elliptic curve is defined over an OEF. This paper also shows that the total computational cost is reduced by 28% when the elliptic curve is defined over $\mathbb{F}_{q^m}$, $q = 2^{61} - 1$ for $m = 5$ and the speed of a scalar multiplication on an elliptic curve becomes 41.9 μsec per operation on a 2.82-GHz Athlon 64 FX PC.

Keywords: elliptic curve cryptosystem, coordinate system, OEF, fast software implementation.

1 Introduction

Elliptic curve schemes such as EC-DSA and EC-ElGamal have been the focus of much attention since they provide smaller key sizes and faster operations compared to RSA. In particular, the use of an optimal extension field (OEF) [1] for software implementation has determined that an elliptic curve cryptosystem is faster than a public key cryptosystem based on modular exponentiations [1]. Moreover, elliptic curves over extension fields become much important because they are used in pairing based cryptosystems.

Points on an elliptic curve can be represented using different coordinate systems. Many coordinate systems have been studied to find a way to accelerate the computation of elliptic curve addition (ECADD) and doubling (ECDBL) calculations because the computational costs of ECADD and ECDBL are mainly determined by the coordinate systems. The most popular coordinates are affine coordinates, Jacobian coordinates, and its variation as shown in [2] and Appendix B. The best coordinate for fast software implementation depends on the computational environment. If an inversion is slower than 7.6 multiplications, Jacobian coordinates are faster than affine coordinates, for example. Previous implementations [1,3] used the Jacobian coordinates because multiplication algorithms are more efficient than inversion algorithms in many environments for an OEF.

Recently, there has been an increased availability of 64-bit word size processors. In OEF for these processors, inversion is slower than in the 32-bit environment; however, *pseudo-inversion* can be computed at a low cost. The basic

P.Q. Nguyen (Ed.): VIETCRYPT 2006, LNCS 4341, pp. 147–156, 2006.

concept of pseudo-inversion over $\mathbb{F}_{q^m}$ was first proposed in [4] to compute efficiently a pairing on elliptic curves. In order to apply the pseudo-inversion to ECADD and ECDBL, we propose new coordinates suitable for OEFs. These new coordinates, *compressed Jacobian coordinates*, represent an intermediate concept of the affine and Jacobian coordinates and make the elliptic curve addition faster than that for either type of coordinates.

The compressed Jacobian coordinates are a natural extension of conventional coordinates because the coordinates over $\mathbb{F}_q$ become equivalent to Jacobian coordinates and the coordinates over $\mathbb{F}_{2^m}$, where q is prime, become equivalent to affine coordinates.

We also show the efficiency of the proposed algorithm based on theoretical estimation and actual implementation using the assembly language on a PC. In our implementation, the speed of random scalar multiplications over an elliptic curve, which is almost equal to the speed of EC-DSA signature generation, is 41.9 μsec on a 2.82-GHz Athlon 64 FX PC.

This paper is organized as follows. Section 2 defines the concepts used in the paper. Section 3 describes the proposed algorithm. Section 4 presents an efficiency comparison of the proposed algorithm and the conventional one. Section 5 concludes this paper.

2   Preliminaries

2.1   OEF

Let an OEF [1] be finite field $\mathbb{F}_{q^m}$ that satisfies the following:

- q is a prime less than but close to the maximum integer word of the processor,
- $q = 2^n \pm c$ where $\log_2 c \leq n/2$ and
- an irreducible binomial $f(x) = x^m - \omega$ exists.

We consider the following polynomial based representation of element $A \in \mathbb{F}_{q^m}$,

$$A = a_{m-1}\alpha^{m-1} + \cdots + a_1\alpha + a_0$$

where $a_i \in \mathbb{F}_q$ and $\alpha \in \mathbb{F}_{q^m}$ is a primitive root of $f(x)$. Since we choose q to be less than the maximum integer word of the processor, we can represent A using m registers.

2.2   Computational Cost

Here, we define the following notation.

M is the computational cost of multiplication in $\mathbb{F}_{q^m}$
S is the computational cost of squaring in $\mathbb{F}_{q^m}$
I is the computational cost of inversion in $\mathbb{F}_{q^m}$
P is the computational cost of pseudo-inversion in $\mathbb{F}_{q^m}$
A is the computational cost of addition/subtraction in $\mathbb{F}_{q^m}$
v is the computational cost of multiplication $\mathbb{F}_q \times \mathbb{F}_{q^m} \to \mathbb{F}_{q^m}$

m is the computational cost of multiplication in $\mathbb{F}_q$
s is the computational cost of squaring in $\mathbb{F}_q$
i is the computational cost of inversion in $\mathbb{F}_q$
a is the computational cost of addition in $\mathbb{F}_q$
f is the computational cost of the Frobenius map in $\mathbb{F}_{q^m}/\mathbb{F}_q$

2.3   Representation of Extension Field Element

This section discusses the representations of extension field elements and the cost of arithmetics using the representations.

Consider an algorithm that consists of a series of arithmetic operations, which include inversion(s), in $\mathbb{F}_{q^m}$. For most implementations, inversion is a costly operation. So, rational representation $\dfrac{N}{D}$ $(N, D \in \mathbb{F}_{q^m}^{\times})$ can be used. Using the representation, division is free,

$$\left(\frac{N}{D}\right)^{-1} = \frac{D}{N}$$

On the other hand, multiplication and addition require additional multiplications,

$$\frac{N_1}{D_1}\frac{N_2}{D_2} = \frac{N_1 N_2}{D_1 D_2}$$
$$\frac{N_1}{D_1} + \frac{N_2}{D_2} = \frac{N_1 D_2 + N_2 D_1}{D_1 D_2}$$

That is, rational representation does not require an inversion in the algorithm, except for the final inversion(s). If the inversion is a very costly operation, the rational representation is an effective way to reduce the complexity of the algorithm. In some $\mathbb{F}_{q^m}$, when adopting a rational representation in the context of elliptic curve arithmetics, we observed that "complete inversion" is not always necessary. We can use pseudo-inversion, which is easy to compute compared to complete inversion, in some finite fields, $\mathbb{F}_{q^m}$. The definition of the pseudo-inversion algorithm is given hereafter.

Definition 1 (Pseudo-Inversion Algorithm)

Input: $X \in \mathbb{F}_{q^m}$
Output: $(\iota(X), \mathbf{N}(X)) \in \mathbb{F}_{q^m} \times \mathbb{F}_q^{\times}$ such that $\iota(X)X = \mathbf{N}(X)$

Moreover, we call $\iota(X)$ a pseudo-inversion of X and $\mathbf{N}(X)$ a co-pseudo-inversion of X. When the pseudo-inversion algorithm is light, the *compressed* rational representation, $\dfrac{N}{d}$ $(N \in \mathbb{F}_{q^m}, d \in \mathbb{F}_q^{\times})$, is superior to the common and the rational representation. The arithmetics for the compressed rational representation are

$$\left(\frac{N}{d}\right)^{-1} = \frac{\iota(N)d}{\mathbf{N}(N)}$$

$$\frac{N_1}{d_1}\frac{N_2}{d_2} = \frac{N_1 N_2}{d_1 d_2}$$

$$\frac{N_1}{d_1} + \frac{N_2}{d_2} = \frac{N_1 d_2 + N_2 d_1}{d_1 d_2}$$

When $m > 1$, the cost of multiplication between $\mathbb{F}_{q^m}$ and $\mathbb{F}_q$ is less than the cost of multiplication in $\mathbb{F}_{q^m}$. Utilizing the pseudo-inversion algorithm can reduce the cost of an algorithm that consists of a series of arithmetics. This paper applies this idea to elliptic curve scalar multiplication.

Two examples of pseudo-inversion algorithm implementations are given in Sections 2.4 and 2.5.

2.4   Pseudo-inversion by Norm

To compute the pseudo-inversion, we can use exponentiation by q^i.

Given $X \in \mathbb{F}_{q^m}$, we call

$$z = \prod_{i=0}^{m-1} X^{q^i}$$

the norm. The norm, z, is known to be $z \in \mathbb{F}_q$. By this definition,

$$z = X \left(\prod_{i=1}^{m-1} X^{q^i} \right)$$

and

$$Y = \prod_{i=1}^{m-1} X^{q^i}.$$

Output Y can be considered as the pseudo-inversion and z is the co-pseudo-inversion. We can reduce the computational cost of the above algorithm by using the Ito-Tsujii algorithm [5].

For example when $m = 5$,

$$Y \leftarrow X \cdot X^q$$
$$Y \leftarrow Y \cdot Y^{q^2}$$
$$Y \leftarrow Y^q$$
$$z \leftarrow X \cdot Y$$

The computational cost of the q^i-th power in $\mathbb{F}_{q^m}$ can be estimated as

$$\mathtt{f} \approx \mathtt{v} \cdot (m-1)/m = 0.8\mathtt{v},$$

if we use the polynomial basis using a binomial as an irreducible polynomial.

In addition, the computational cost of the last multiplication, $z \leftarrow X \cdot Y$, can be estimated as $\mathtt{v}$ because $z \in \mathbb{F}_q$. Therefore, the total computational cost is estimated as

$$\mathtt{P} = 2\mathtt{M} + 3\mathtt{f} + 1\mathtt{v} \approx 2\mathtt{M} + 3.4\mathtt{v}.$$

2.5   Pseudo-inversion by Euclidean Method

Let $\deg(X)$ be the degree of polynomial X, $\texttt{lead}(X)$ be the coefficient of polynomial X's leading term, and $\texttt{swap}(X,Y)$ be the procedure that exchanges the values of X and Y. The following algorithm computes a pseudo-inversion.

```
A = f(x) ; // irreducible polynomial.
B = g(x) ; // input polynomial (where deg(A) > deg(B)) ≠ 0

C = 0 ;
D = 1 ;

while(deg(B) > 0){
   b = lead(B) ;           // coefficient of B's leading term
   while(deg(A) >= deg(B)){
      a = lead(A) ;        // coefficient of A's leading term
      n = deg(A) - deg(B) ; // difference between degrees A and B

      A *= b ; C *= b ;
      A -= a*B*xⁿ ;
      C -= a*D*xⁿ ;
   } ;
   swap(A,B) ;
   swap(C,D) ;
} ;

// output D as pseudo-inversion
// output B as co-pseudo-inversion
```

Roughly speaking, the computational cost of this algorithm is $4mv$.

3   Compressed Jacobian Coordinates

In the Jacobian coordinate system, a point on an elliptic curve over an extension field is represented as an element of $\mathbb{F}_{q^m}^3$. We can decrease the redundancy of the Jacobian coordinates by using pseudo-inversion. In our new coordinates, a point is represented as an element of $\mathbb{F}_{q^m}^2 \times \mathbb{F}_q$. We call this the compressed Jacobian coordinate system. The compressed Jacobian coordinates can be used for both Koblitz curves[1] and random curves over OEF. Addition formulas in the compressed Jacobian coordinates are given hereafter.

Compressed Jacobian coordinates
$(X, Y, z) \in \mathbb{F}_{q^m}^2 \times \mathbb{F}_q,$ where $Y^2 = X^3 + az^4X + bz^6$

---

[1] Usually, Koblitz curve means elliptic curves over $\mathbb{F}_{2^m}$. In this paper, we use the word "Koblitz curve" as a general elliptic curve over $\mathbb{F}_{p^m}$ that the cofactor is in $\mathbb{F}_p$.

Addition formulas in compressed Jacobian coordinates
Input: $(X_1, Y_1, z_1), (X_2, Y_2, z_2)$, Output: (X_3, Y_3, z_3).

$$z_3' \leftarrow z_1 z_2$$
$$(X_1', Y_1') \leftarrow (X_1 z_2^2, Y_1 z_2^3)$$
$$(X_2', Y_2') \leftarrow (X_2 z_1^2, Y_2 z_1^3)$$
$$\Lambda_n \leftarrow \iota(X_2' - X_1') \cdot (Y_2' - Y_1')$$
$$\lambda_d \leftarrow \mathbf{N}(X_2' - X_1')$$
$$z_3 \leftarrow \lambda_d z_3'$$
$$(X_1'', Y_1'') \leftarrow (X_1' \lambda_d^2, Y_1' \lambda_d^3)$$
$$(X_2'', Y_2'') \leftarrow (X_2' \lambda_d^2, Y_2' \lambda_d^3)$$
$$X_3 \leftarrow \Lambda_n^2 - X_1'' - X_2''$$
$$Y_3 \leftarrow \Lambda_n(X_1'' - X_3) - Y_1''$$

We don't compute Y_2'' in real implementation because Y_2'' is not used to compute X_3, Y_3 and z_3. We estimate the computational cost of ECADD without computing Y_2'' hereafter.

Doubling formulas in compressed Jacobian coordinates
Input: (X_1, Y_1, z_1), Output: (X_3, Y_3, z_3).

$$\Lambda_n \leftarrow \iota(2Y_1) \cdot (3X_1^2 + az_1^4)$$
$$\lambda_d \leftarrow \mathbf{N}(2Y_1)$$
$$z_3 \leftarrow \lambda_d z_1$$
$$(X_1'', Y_1'') \leftarrow (X_1 \lambda_d^2, Y_1 \lambda_d^3)$$
$$X_3 \leftarrow \Lambda_n^2 - 2X_1''$$
$$Y_3 \leftarrow \Lambda_n(X_1'' - X_3) - Y_1''$$

4   Evaluation and Implementation

In this section, we evaluate the performance of addition and doubling in the compressed Jacobian coordinates. For simplicity, we neglect the cost of addition and subtraction in $\mathbb{F}_{q^m}$, and multiplication in $\mathbb{F}_q$. In addition, if $m = 5$, we can estimate that

$$\mathrm{v} \approx 0.2\mathrm{M}.$$
$$\mathrm{P} \approx 2\mathrm{M} + 3.4\mathrm{v} \approx 2.68\mathrm{M}.$$

See Section 2.4 for details.

If $\mathrm{I} > 3.68\mathrm{M}$ and $\mathrm{S} = 0.8\mathrm{M}$, the compressed Jacobian coordinate system is the fastest when $m = 5$.

Implementation Result
We chose the parameters in Table 3 and the environment as given in Table 4. We chose the cyclic window method [3] as the scalar multiplication algorithm and the results are given in Table 7. In Table 8, we refer to the previously

Table 1. Computational Cost of ECADD

Coordinate system	Cost	$m = 5$
Compressed Jacobian	2M + 1S + 1P + 7v	$\approx 6.08M + 1S$
Compressed Jacobian ($z_2 = 1$)	2M + 1S + 1P + 5v	$\approx 5.68M + 1S$
Affine	2M + 1S + 1I	
Jacobian	12M + 4S	
Jacobian ($Z_2 = 1$)	8M + 3S	
Modified Jacobian	13M + 6S	
Modified Jacobian ($Z_2 = 1$)	9M + 5S	
Modified Jacobian ($Z_2 = 1$) ($a = -3$)	8M + 5S	

Table 2. Computational Cost of ECDBL

Coordinate system	Cost	$m = 5$
Compressed Jacobian	2M + 2S + 1P + 3v	$\approx 5.28M + 2S$
Compressed Jacobian ($a \in \mathbb{F}_q$)	2M + 2S + 1P + 2v	$\approx 5.08M + 2S$
Affine	2M + 2S + 1I	
Jacobian	4M + 6S	
Jacobian ($a = -3$)	4M + 4S	
Modified Jacobian	4M + 4S	

Table 3. Parameters

q	$2^{61} - 1$ (prime)
$\mathbb{F}_{q^m}$	$\mathbb{F}_q[\alpha]/(\alpha^5 - 3)$ (polynomial basis)
Elliptic curve	$Y^2 = X^3 - 3X + 2023176626027320614$
Trace	$t = 2713676959$
Generator	$X = \begin{aligned}&1225397330577448427\ \alpha^4 + 110313758532384199\ \alpha^3 \\ &+ 1639881413522258503\ \alpha^2 + 547643109538786165\ \alpha \\ &+ 221493176281168480 9\end{aligned}$ $Y = \begin{aligned}&74533231004088031\ \alpha^4 + 1705584686783011420\ \alpha^3 \\ &+ 2159424991416008329\ \alpha^2 + 509248187364731537\ \alpha \\ &+ 570065311020511817\end{aligned}$
Order	2826955306972373196333094892835328944445537312030068865701569742858979617 1
	(244-bit prime)

Table 4. Environment

CPU	AMD Athlon 64 FX-57
Clock Frequency	2.82 GHz
OS	FreeBSD 5.4R
Compiler	gcc version 4.0.3 20051229 (prerelease)

reported fastest implementation results. We cannot find any previous work that uses similar parameter we used. We decided to show the references:

1. The security parameter is similar to ours. (122 bit)
2. The field construction is similar to ours. (OEF)

Table 5. Implementation Results of ECADD

Coordinate system	Cycles	μsec
Compressed Jacobian	939	0.33
Compressed Jacobian ($z_2 = 1$)	889	0.32
Affine	2216	0.79
Jacobian	1996	0.71
Jacobian ($Z_2 = 1$)	1481	0.53
Modified Jacobian ($a = -3$)	2213	0.79
Modified Jacobian ($a = -3, Z_2 = 1$)	1702	0.60

Table 6. Implementation Results of ECDBL

Coordinate system ($a = -3$)	Cycles	μsec
Compressed Jacobian	942	0.33
Affine	2337	0.83
Jacobian	1086	0.39
Modified Jacobian	1031	0.37

Table 7. Implementation Results of EC Scalar Multiplication

Coordinate system ($a = -3$)	kcycles	μsec
Compressed Jacobian	118	41.9
Affine	276	97.9
Jacobian	165	58.6
Modified Jacobian	174	61.9

Table 8. Comparison to Known Results

Curve		Security	kcycles	μsec	CPU	Clock
$\mathbb{F}_{q^m}$ Koblitz (This paper)	$q = 2^{61} - 1, m = 5$	122 bits	118	41.9	Athlon 64 FX	2.82 GHz
$\mathbb{F}_q$ Montgomery form [6]	$q = 2^{255} - 19$	126 bits	625	–	Athlon	–
$\mathbb{F}_{2^m}$ Koblitz [7]	$m = 233$	116 bits	897	2243	Pentium II	400 MHz
$\mathbb{F}_{q^m}$ Koblitz [3]	$q = 2^{29} - 3, m = 7$	87 bits	127	254	21264	500 MHz
$\mathbb{F}_{q^m}$ Koblitz, Hessian form [8]	$q = 2^{29} - 3, m = 7$	87 bits	213	66.6	Pentium 4	3.2 GHz

Our current results are faster than these both in terms of speed in CPU cycles and μsec. These numbers were obtained using various parameters and/or platforms. Please take note of the parameters when comparing these numbers.

5   Conclusion

In this paper, we proposed a new coordinate system, which we call the compressed Jacobian coordinates, to accelerate elliptic curve cryptosystems over extension fields. The main idea of the proposed coordinates is to utilize pseudo-inversion, and the coordinates are a natural extension of traditional affine and Jacobian coordinates. We estimated the computational cost of ECADD and ECDBL for the coordinates. We also implemented scalar multiplication as software on the x86-64 platform and confirmed that 244-bit OEF scalar multiplication can be computed in 41.9 μsec on an Athlon 64 FX-57 (2.82 GHz) using the compressed Jacobian coordinates. To the best knowledge of the authors, this is the fastest software implementation of elliptic curve scalar multiplications for the security level of 122 bits or higher. This means that we can exchange more than 10,000 ECDH keys per second on an inexpensive personal computer (if the communication cost is negligible).

The idea of the compressed Jacobian coordinates can be easily applied to the other coordinates, e.g., projective coordinates and Montgomery coordinates, and the other algebraic curves, e.g., the hyper-elliptic curves and C_{ab} curves.

References

1. Bailey, D.V., Paar, C.: Optimal extension fields for fast arithmetic in public-key algorithms. In Krawczyk, H., ed.: Advances in Cryptology — CRYPTO'98. Volume 1462 of Lecture Notes in Computer Science. Springer-Verlag, Berlin, Heidelberg, New York (1998) 472–485
2. Cohen, H., Miyaji, A., Ono, T.: Efficient elliptic curve exponentiation using mixed coordinates. In Ohta, K., Pei, D., eds.: Advances in Cryptology — ASIACRYPT'98. Volume 1514 of Lecture Notes in Computer Science. Springer-Verlag, Berlin, Heidelberg, New York (1998) 51–65
3. K.Aoki, F.Hoshino, T.: A cyclic window algorithm for ECC defined over extension fields. In: ICICS 2001. Volume 2229 of Lecture Notes in Computer Science., Springer-Verlag (2001) 62–73
4. Kobayashi, T., Aoki, K., Imai, H.: Efficient algorithms for Tate pairing. IEICE Transactions Fundamentals of Electronics, Communications and Computer Sciences (Japan) **E89-A** (2006) 134–143
5. Itoh, T., Tsujii, S.: A fast algorithm for computing multiplicative inverses in GF(2^m) using normal bases. In: Information and Computation. Volume 78. (1988) 171–177
6. Bernstein, D.J.: Curve25519: new Diffie-Hellman speed records. Lecture Notes in Computer Science **3958** (2006) 207–228
7. Hankerson, D., Hernandez, J.L., Menezes, A.: Software implementation of elliptic curve cryptography over binary fields. Lecture Notes in Computer Science **1965** (2001) 1–24
8. Kumagai, M.: Efficient implementation of Hessian-form elliptic curve cryptosystem with SIMD instructions. In: The 2005 Symposium on Cryptography and Information Security (SCIS2005), Maiko, Kobe, Japan, The Institute of Electronics, Information and Communication Engineers (2005) 1651–1656 (in Japanese).

A   Appendix: Frobenius Map

In this section, we recall the Frobenius map. Let $E/\mathbb{F}_q$ denote a non-supersingular elliptic curve defined over finite field $\mathbb{F}_q$ where q is a prime or any power of a prime. $P = (X, Y)$ is an $\mathbb{F}_{q^m}$-rational point of elliptic curve E defined over $\mathbb{F}_q$. The Frobenius map, ϕ, is defined as

$$\phi : (X, Y) \rightarrow (X^q, Y^q).$$

The Frobenius map is an endomorphism over $E(\mathbb{F}_{q^m})$. It satisfies the equation

$$\phi^2 - t\phi + q = 0, \qquad -2\sqrt{q} \le t \le 2\sqrt{q} \tag{1}$$

It takes a negligible amount of time to compute the Frobenius map in an endomorphism ring where t is the trace of $E/\mathbb{F}_q$ provided that element in $\mathbb{F}_{q^m}$ is represented using the polynomial basis of $\mathbb{F}_{q^m}$ over $\mathbb{F}_q$ using a binomial as a definition polynomial.

B   Appendix: Coordinates

Let

$$E : Y^2 = X^3 + aX + b \qquad (a, b \in \mathbb{F}_{q^m},\ 4a^3 + 27b^2 \ne 0)$$

be the equation of elliptic curve E over $\mathbb{F}_{q^m}$.

For Jacobian coordinates, with $X = X_J/Z_J^2$ and $Y = Y_J/Z_J^3$, a point on elliptic curve P is represented as $P = (X_J, Y_J, Z_J) = (X, Y)$. In order to accelerate ECADD, the Chudnovsky Jacobian coordinates [2] represent a Jacobian point as the quintuple $(X_J, Y_J, Z_J, Z_J^2, Z_J^3)$. On the other hand, in order to accelerate ECDBL, the modified Jacobian coordinates [2] represent a Jacobian point as the quadruple (X_J, Y_J, Z_J, aZ_J^4).

The number of operations required to compute ECDBL and ECADD is shown in Table 9.

Table 9. Number of Operations for Each Coordinate Systems

Coordinates	ECDBL	ECADD	ECADD with $Z = 1$
Affine	2 M+ 2 S+ I	2 M+ 1 S+ I	—
Chudnovsky Jacobian	5 M+ 6 S	12 M+ 4 S	8 M+ 3 S
Modified Jacobian	4 M+ 4 S	13 M+ 6 S	9 M+ 5 S

On the Definition of Anonymity
for Ring Signatures

Miyako Ohkubo[1] and Masayuki Abe[2]

[1] Information-Technology Promotion Agency, Japan
2-28-8, Honkomagome, Bunkyo-ku, Tokyo 113-6591 Japan
[2] NTT Laboratories
1-1 Hikari-no-oka, Yokosuka-shi, Kanagawa-ken, 239-0847 Japan

Abstract. This paper studies the relations among several definitions of anonymity in the same attack environment. It is shown that one intuitive and two technical definitions we consider are asymptotically equivalent, and the indistinguishability-based technical definition is the strongest, i.e., the most secure when achieved, when the exact reduction cost is taken into account. We then extend our result to the threshold case where a subset of members cooperate to create a signature. The threshold setting makes the notion of anonymity more complex and yields a greater variety of definitions. We explore several notions and observe certain relation does not seem hold unlike the simple single-signer case. Nevertheless, we see that an indistinguishability-based definition is the most favorable in the threshold case. We also present a notion of anonymity with linkability and a simple generic construction.

1 Introduction

BACKGROUND. Ring signatures feature unforgeability and anonymity in such a sense that anyone can be sure that the signature is created by one of the group members but cannot identify which member the real signer is. It can be seen as a variant of group signatures [11] without the group manager who registers the signing keys of legitimate group members. Without the group manager, a ring signature scheme allows ad-hoc members to be bound to a signature without any authorization of the members or preliminary setup.

The notion of ring signatures are introduced by Rivest, Shamir, and Tauman in [18] with the first instantiation based on the RSA assumption. [1] extends the result so that variety of public-keys can be involved in a signature. [10,2] addresses the threshold case where a group of the real signers cooperate to create a ring signature. [10] is the first threshold scheme with a group of logarithmic size, and [2] achieves polynomial size groups. [22] presents an ID-based ring signature scheme. A threshold variant is presented in [12] and more efficiency and generality is pursued in [13]. While all these constructions analyzes the security in the random oracle model [7] ([18] also relies on an ideal cipher), [9] presents a generic scheme without random oracles. We refer to [19] for more survey of ring signatures.

P.Q. Nguyen (Ed.): VIETCRYPT 2006, LNCS 4341, pp. 157–174, 2006.

The security (unforgeability and anonymity) of a ring signature scheme is defined by the combination of an attack environment and an attack goal as well as those for other cryptographic primitives. [9] studies the notion of security mainly from the viewpoint of the attack environment, and shows the strongest case where all the private keys are exposed to the adversary during the anonymity game. While existential unforgeability is widely accepted as the strongest security goal for unforgeability, anonymity is defined in different ways in the literature. For instance, in [18,1,13], the goal of an adversary is to identify the real signer among the group of members bound to the challenge signature, while in [9] adopts the idea of indistinguishability [15] like in the case of group signatures formally addressed in recent papers, e.g., [6,16,8]. The situation becomes more involved for threshold case where many variations are available.

Somewhat contrary to anonymity, *linkability* is useful in a scenario of ring signatures. Suppose that a whistle-blower publishes a ring signature for a message such as "We are doing insider trading" that surprises other involved members. Then someone in the involved members may issues "I'm just kidding" to disclaim the previous message. Therefore, it is important to allow anyone to verify the link between the signatures to convince him/herself that they are issued by the same signer and anyone else (even the involved members) cannot create the linked signatures. Unfortunately, it is known that strong sense of anonymity implies unlinkability [6]. However, extending the syntax of ring signature schemes changes the situation. In [17,3], the notion of linkable anonymous signatures are introduced. Their security, however, guarantees so limited sense of linkability that it cannot accommodate with anonymity under chosen real-signer attacks. [21] addresses the threshold case, but their security notion and instantiation allow any insider member involved in a signature can later create linked signatures. Hence it cannot be available for the scenario given above. Another result is in [20] but their security definition has ambiguity and unnecessarily complicated. For instance, they do not consider the case where insiders attempt to create linkable signatures. Furthermore, their scheme does not allow the real signer to issue unlinkable signatures with the same members involved. (This is their design goal, though. It may have other application.) Therefore, to our best knowledge, sufficient instantiation and rigorous treatment of the notion of anonymity with linkability is not known so far.

Our Contribution. We focus on the attack goals in the definitions of anonymity and show their relations in the same attack environment. To see the relation between intuitive and technical definitions, we consider the following attack goals, which will be defined formally in Sec 2. Given a challenge ring signature of an arbitrary group, the attacker:

1. identifies the real signer (WHO),
2. decides which of two specific signers is the real signer (IND),
3. decides if it is created by a specific signer or not (ROR).

The third notion (ROR:real-or-random) has not been used in the context of ring signatures but used to define secrecy in, for instance, symmetric encryption [4] and key encapsulation [14]. We show that the most intuitive notion (WHO) and other two technical definitions (IND, ROR) are asymptotically equivalent. Namely, if a ring signature scheme is anonymous in the sense of WHO, it is also anonymous in the sense of IND or ROR. Our detailed proof shows that IND is the strongest, i.e. the most secure when achieved, when the exact reduction cost is concerned. In [9], it is briefly pointed out that WHO is reducible to IND by a hybrid argument, which usually yield polynomial cost and hence places these two notions in a distance. We provide a direct reduction between these notions and shows that they are actually close each other.

We then extend the argument to the threshold case, where it is not just a simple replacement of the single real signer to a group of real signers. We consider variants of WHO used in the literature and show that a natural extension of IND is reducible to these notions. It means that an indistinguishability-based definition is the most favorable even in the threshold case. We do not know whether the reverse relation is also true or not. We present a further extended variant of WHO that is equivalent to the indistinguishability-based definition.

Finally, we address the issue of linkability. We first extend the syntax of ring signature schemes and show a suitable definition for link-unforgeability and anonymity with linkability. Then we present a simple and generic construction from a (non-linkable) ring signature scheme and ordinary signature scheme.

2   Definitions

Definition 1 (Ring Signature Scheme) [1]. *A 1-out-of-n ring signature scheme, $\Sigma^{1,n}$, is a triple of polynomial-time algorithms, $\Sigma^{1,n} = (\mathcal{G}^{1,n}, \mathcal{S}^{1,n}, \mathcal{V}^{1,n})$:*

$\mathcal{G}^{1,n} : (1^\lambda) \rightarrow (sk, vk)$. *A probabilistic algorithm that takes security parameter λ and outputs private signing key sk and public verification key vk.*

$\mathcal{S}^{1,n} : (sk, m, L) \rightarrow \sigma$. *A (probabilistic) algorithm that takes message m, and a list, say L, of verification-keys including the one that corresponds to sk, outputs signature σ.*

$\mathcal{V}^{1,n} : (L, m, \sigma) \rightarrow 1/0$. *An algorithm that takes message m and signature σ, and outputs either 1 or 0 meaning **accept** and **reject**, respectively. We require that $\mathcal{V}^{1,n}(L, m, \mathcal{S}^{1,n}(sk, m, L)) = 1$ for any message m, any (sk, vk) generated by $\mathcal{G}^{1,n}$, and any L that includes vk.*

It is assumed that each public-key is implicitly bound to the identity of its owner and one can see the identity by seeing the public-key. In the following, we give double-meaning to L as a set of public-key and a set of corresponding identity. Hence we may allow description such as choosing an identity from L.

The signature generation and verification functions may be limited to work only if L contains legitimate or legitimately-looking public-keys which are indistinguishable from the legitimate ones.

To define anonymity, we define the attack resources as an oracle given to the adversary. It accepts the following queries.

key-gen(i). A key generation query. Given this query with a unique identity i, a new public and private key pair is generated for that identity. The public-key is appended to list $\mathcal{L}^*$ and the private-key is kept secret from the adversary. List $\mathcal{L}^*$ is initially empty and available to the adversary.

add-pk(i, vk). A new member addition query. It allows the adversary to append an arbitrary public-key vk bound to identity i to list $\mathcal{L}^\dagger$ which is empty at the beginning. It is required that both i and vk are unique in $\mathcal{L}^* \cup \mathcal{L}^\dagger$.

get-sign(i, L, m). A signature generation query. It allows the adversary to obtain a ring signature for arbitrary signer, group, and message. Oracle $\mathcal{R}$ returns $\sigma \leftarrow \mathcal{S}^{1,n}(sk_i, m, L)$ if $i \in L \cap \mathcal{L}^*$. It returns $\bot$, otherwise.

key-expose(i). A key exposure query. It allows the adversary to obtain the private key of an arbitrary public-key. If $i \in \mathcal{L}^*$, oracle $\mathcal{R}$ returns corresponding sk.

corrupt(i). A corruption query. It allows the adversary to obtain all the inputs to signer i, which includes the private-key and all random choices. If $i \in \mathcal{L}^*$, oracle $\mathcal{R}$ returns corresponding sk and the random tape of i.

Remark 1. Requiring unique i and vk for making add-pk means that the adversary is not allowed to register someone else's public-key as his own. However, it is just for a notational simplicity and does not lose generality. It can be handled by separately treating identities and the public-keys if desired.

Remark 2. Through add-pk, the adversary is allowed to add a bogus public-key whose secret-key may not exist and cannot generate signatures. This affects to the anonymity issue as discussed in the sequel.

We next define anonymity in a various form. First we show the most intuitive notion, which we denote WHO. The idea is that the adversary guesses who is the actual signer when it is chosen randomly from the signature group specified by the adversary.

Definition 2 (WHO-Anonimity). *A ring signature scheme is WHO-anonymous if for any polynomial-time adversary $\mathcal{A}$ playing the following game, there exists a negligible function ϵ in λ such that $\Pr[\tilde{i} = i^*] < 1/|L^*| + \epsilon$ where L^* is $L \cap \mathcal{L}^*$ for $\mathcal{L}^*$ observed in step 3.*

[Attack Game: WHO]

1. *Global : $\mathcal{L}^*$ and $\mathcal{L}^\dagger$ (initially empty).*
2. *$(L, m, \omega) \leftarrow \mathcal{A}^{\mathcal{R}}(1^\lambda)$.*
3. *$i^* \leftarrow L \cap \mathcal{L}^*$, $\sigma \leftarrow \mathcal{S}^{1,n}(sk_{i^*}, m, L)$.*
4. *$\tilde{i} \leftarrow \mathcal{A}^{\mathcal{R}}(\omega, \sigma)$.*

It is required that $L \cap \mathcal{L}^* \neq \emptyset$ in step 2, and $\tilde{i} \in L \cap \mathcal{L}^*$ in step 4.

Variable ω is an internal state of $\mathcal{A}$.

Remark 3. When an honest signer, i, issues a signature with list L, she might naturally expect that her identity is hidden among the identities in L. This intuition is true if $L \subseteq \mathcal{L}^*$ in the above definition. However, the adversary might add fake public-keys that are indistinguishable from the legitimate ones but never work for generating signatures. It may result in $L \cap \mathcal{L}^*$ only contains her identity and the adversary can identify the real signer with probability 1. This is a practical threat that is hard to avoid and not captured by the definition. A possible solution is to choose L from the set of public-keys that are once used to create signatures. Self-signature in a public-key certificate would be useful for this purpose in practice.

We add two more technical definitions, IND and ROR, as mentioned in Sec. 1.

Definition 3 (IND-Anonimity). *A ring signature scheme is IND-anonymous if for any polynomial-time adversary $\mathcal{A}$ playing the following game, there exists a negligible function ϵ in λ such that* $\Pr[\tilde{\chi} = \chi] < 1/2 + \epsilon$.

[Attack Game: IND]

1. *Global :* $\mathcal{L}^*$ *and* $\mathcal{L}^\dagger$ *(initially empty).*
2. $(i_0, i_1, L, m, \omega) \leftarrow \mathcal{A}^{\mathcal{R}}(1^\lambda)$.
3. $\chi \leftarrow \{0,1\}$, $\sigma \leftarrow \mathcal{S}^{1,n}(sk_{i_\chi}, m, L)$.
4. $\tilde{\chi} \leftarrow \mathcal{A}^{\mathcal{R}}(\omega, \sigma)$.

It is required that $\{i_0, i_1\} \subseteq L \cap \mathcal{L}^*$ *in step 2, and* $\tilde{\chi} \in \{0,1\}$ *in step 4.*

Definition 4 (ROR-Anonymity). *A ring signature scheme is ROR-anonymous if for any polynomial-time adversary $\mathcal{A}$ playing the following game, there exists a negligible function ϵ in λ such that* $\Pr[\tilde{\xi} = \xi] < 1/2 + \epsilon$.

[Attack Game: ROR]

1. *Global :* $\mathcal{L}^*$ *and* $\mathcal{L}^\dagger$ *(initially empty).*
2. $(i, L, m, \omega) \leftarrow \mathcal{A}^{\mathcal{R}}(1^\lambda)$.
3. $\xi \leftarrow \{0,1\}$, $i_1 \leftarrow i$, $i_0 \leftarrow L \cap \mathcal{L}^* \setminus \{i\}$, $i^* = i_\xi$, $\sigma \leftarrow \mathcal{S}^{1,n}(sk_{i^*}, m, L)$.
4. $\tilde{\xi} \leftarrow \mathcal{A}^{\mathcal{R}}(\omega, \sigma)$.

It is required that $i \in L \cap \mathcal{L}^*$ *and* $|L \cap \mathcal{L}^*| \geq 2$ *in step 2, and* $\tilde{\xi} \in \{0,1\}$ *in step 4.*

3   Relation Among the Definitions

3.1   Theorems

For security notion A and B, we denote by A $\Rightarrow$ B that if a scheme is secure in A it is also secure in B with respect to the same attack resources. By (ϵ, t)-adversary, we denote an adversary that stops within time t, and wins the attack game with probability at least ϵ. Let L^* denote the legitimate public-keys in L appeared in step 3 of each attack game. Namely, $L^* = L \cap \mathcal{L}^*$ for L and $\mathcal{L}^*$ observed at step 3 of each attack game.

Theorem 1. *IND⇒WHO. In particular, if there exists an $(\epsilon_{who}, t_{who})$-adversary in the sense of WHO, then there exists an $(\epsilon_{ind}, t_{ind})$-adversary in the sense of IND such that $\epsilon_{ind} = \frac{|L^*|}{2(|L^*|-1)} \epsilon_{who}$ and $t_{ind} \approx t_{who}$.*

It is interesting to see that $\epsilon_{ind} \geq \frac{1}{2}\epsilon_{who}$ even for large L^*. Therefore, Theorem 1 says that an IND-secure scheme is also WHO-secure regardless of the size of the group bound to the signatures.

Theorem 2. *WHO ⇒ IND. In particular, if there exists an $(\epsilon_{ind}, t_{ind})$-adversary in the sense of IND, then there exists an $(\epsilon_{who}, t_{who})$-adversary in the sense of WHO such that $\epsilon_{who} = \frac{2}{|L^*|} \epsilon_{ind}$ and $t_{who} \approx t_{ind}$.*

Theorem 3. *IND ⇒ ROR. In particular, if there exists an $(\epsilon_{ror}, t_{ror})$-adversary in the sense of ROR, then there exists an $(\epsilon_{ind}, t_{ind})$-adversary in the sense of IND such that $\epsilon_{ind} = \epsilon_{ror}$ and running time is $t_{ind} \approx t_{ror}$.*

Theorem 4. *ROR ⇒ IND. In particular, if there exists an $(\epsilon_{ind}, t_{ind})$-adversary in the sense of IND, then there exists an $(\epsilon_{ror}, t_{ror})$-adversary in the sense of ROR such that $\epsilon_{ror} = \frac{|L^*|}{2(|L^*|-1)} \epsilon_{ind}$ and running time is $t_{ror} \approx t_{ind}$.*

From Theorem 1 and 2, we see that IND is a stronger notion than WHO. The larger L^* gets, the less secure WHO-security becomes than IND-security (unless Theorem 2 finds better reduction). This relation is similar to the one between indistinguishability and one-way for cryptosystems with poly-size message space. Indeed, since the number of the signers is polynomially limited, a scheme that hides the identity of the real signer can be seen as a cryptosystem with poly-size message space. In Sec. 4, we will see that the situation changes in the threshold case where the number of the real-signer group is exponential.

Looking at Theorem 3 and 4, one can see that IND and ROR are close together but IND is slightly better. In conclusion, IND is the most preferable as a security notion to achieve.

3.2   Proofs

(*Proof of Theorem 1.*) We construct adversary $\mathcal{A}_{ind}$ for IND by using adversary $\mathcal{A}_{who}$ for WHO as follows. First of all, $\mathcal{A}_{ind}$ forwards all queries from $\mathcal{A}_{who}$ to $\mathcal{R}$ to the corresponding $\mathcal{R}$ in IND and returns the answers to $\mathcal{A}_{who}$. $\mathcal{A}_{ind}$ simulates the challenge oracle in WHO in the following way; Given (m, L) from $\mathcal{A}_{who}$, $\mathcal{A}_{ind}$ chooses i_0 and i_1 randomly from L^*. It then send query (i_0, i_1, L, m) to the challenge oracle of IND. Then return the obtained signature to $\mathcal{A}_{who}$. Eventually, $\mathcal{A}_{who}$ outputs $\tilde{i} \in L^*$. If $\tilde{i} \in \{i_0, i_1\}$, then $\mathcal{A}_{ind}$ outputs $\tilde{\chi} \in \{0,1\}$ such that $i_{\tilde{\chi}} = \tilde{i}$. Otherwise, $\mathcal{A}_{ind}$ outputs randomly chosen $\tilde{\chi} \leftarrow \{0,1\}$.

Since all the oracle queries are handled correctly, the attack environment for $\mathcal{A}_{who}$ is simulated perfectly. (Note that the distribution of the challenge signatures also perfect because the i_χ is chosen uniformly from L^* from the viewpoint of $\mathcal{A}_{who}$.) Next, there are two cases when $\mathcal{A}_{ind}$ wins; 1) $\mathcal{A}_{who}$ wins, and 2) $\mathcal{A}_{who}$

fails but $\tilde{i}$ is not in $\{i_0, i_1\}$ and the succeeding random guess is true. Let $\bar{\chi}$ be $1 - \chi$. A crucial observation is that when $\mathcal{A}_{who}$ fails, $i_{\bar{\chi}}$ is independent of the view of $\mathcal{A}_{who}$. Since $i_{\bar{\chi}}$ is selected uniformly from $L^* \setminus \{i_\chi\}$, the case $\tilde{i} = i_{\bar{\chi}}$ happens only with probability $\frac{1}{|L^*|-1}$. Now, the success probability of $\mathcal{A}_{ind}$ is the following.

$$Pr[\mathcal{A}_{ind} \text{ wins}] = Pr[\mathcal{A}_{who} \text{ wins}] + \frac{1}{2} \cdot Pr[\mathcal{A}_{who} \text{ fails} \wedge \tilde{i} \neq i_{\bar{\chi}}]$$

$$= (\frac{1}{|L^*|} + \epsilon_{who}) + \frac{1}{2} \cdot \{1 - (\frac{1}{|L^*|} + \epsilon_{who})\} \cdot \frac{|L^*| - 2}{|L^*| - 1}$$

$$= \frac{1}{2} + \frac{|L^*|}{2(|L^*| - 1)} \cdot \epsilon_{who}$$

Hence we have a bound $\epsilon_{ind} = \frac{|L^*|}{2(|L^*|-1)} \cdot \epsilon_{who}$. It is obvious that $t_{ind} \approx t_{who}$ from the construction of $\mathcal{A}_{ind}$.   □

(*Proof of Theorem 2.*) We construct adversary $\mathcal{A}_{who}$ for WHO by using adversary $\mathcal{A}_{ind}$ for IND as follows. First of all, $\mathcal{A}_{who}$ forwards all queries from $\mathcal{A}_{ind}$ to $\mathcal{R}$ to the corresponding $\mathcal{R}$ in WHO and returns the answers to $\mathcal{A}_{ind}$. $\mathcal{A}_{who}$ simulates the challenge oracle in IND in the following way; Given (i_0, i_1, m, L) from $\mathcal{A}_{ind}$, $\mathcal{A}_{who}$, then send query (m, L) to the challenge oracle of WHO. Then return the obtained signature to $\mathcal{A}_{ind}$. If $\mathcal{A}_{ind}$ stops and outputs $\tilde{\chi} \in \{0, 1\}$, $\mathcal{A}_{who}$ outputs $\tilde{i} = i_{\tilde{\chi}}$.

Now, observe that the view of $\mathcal{A}_{ind}$ is simulated perfectly if challenge oracle $\mathcal{C}$ coincidently chooses i^* to i_0 or i_1. And it is clear by construction that $\mathcal{A}_{who}$ wins only in this case. Therefore,

$$Pr[\mathcal{A}_{who} \text{ wins}] = Pr[\mathcal{A}_{ind} \text{ wins} \wedge i^* \leftarrow \{i_0, i_1\}]$$

$$= (\frac{1}{2} + \epsilon_{ind}) \cdot \frac{2}{|L^*|}$$

$$= \frac{1}{|L^*|} + \frac{2}{|L^*|} \epsilon_{ind}.$$

Thus we have $\epsilon_{who} = \frac{2}{|L^*|} \epsilon_{ind}$. The running time is clearly almost the same.   □

(*Proof of Theorem 3.*) We construct adversary $\mathcal{A}_{ind}$ for IND by using adversary $\mathcal{A}_{ror}$ for ROR as follows. First of all, $\mathcal{A}_{ind}$ forwards all queries from $\mathcal{A}_{ror}$ to $\mathcal{R}$ to the corresponding $\mathcal{R}$ in IND and returns the answers to $\mathcal{A}_{ror}$. $\mathcal{A}_{ind}$ simulates the challenge oracle in ROR in the following way; Given (i, m, L) from $\mathcal{A}_{ror}$, adversary $\mathcal{A}_{ind}$ sets $i_1 = i$ and choose i_0 randomly from $|L^*| \setminus i$. It then sends query (i_0, i_1, L, m) to the challenge oracle of IND and returns the obtained signature to $\mathcal{A}_{ror}$. Eventually, if $\mathcal{A}_{ror}$ outputs $\tilde{\xi} \in \{0, 1\}$ then $\mathcal{A}_{ind}$ outputs $\tilde{\chi} = \tilde{\xi}$.

Observe that the challenge oracle of IND essentially chooses the target signer from i or randomly selected one because (i_0, i_1) is set to (i, random) by $\mathcal{A}_{ind}$. Hence the view of $\mathcal{A}_{ror}$ is perfectly simulated and it is clear that $\mathcal{A}_{ind}$ wins whenever $\mathcal{A}_{ror}$ wins. Thus we have $\epsilon_{ror} = \epsilon_{ind}$. The running time is clearly almost the same.   □

(*Proof of Theorem 4.*) We construct adversary $\mathcal{A}_{ror}$ for ROR by using adversary $\mathcal{A}_{ind}$ for IND as follows. First of all, $\mathcal{A}_{ror}$ forwards all queries from $\mathcal{A}_{ind}$ to $\mathcal{R}$ to the corresponding $\mathcal{R}$ in ROR and returns the answers to $\mathcal{A}_{ind}$. $\mathcal{A}_{ror}$ simulates the challenge oracle in IND in the following way; Given (i_0, i_1, m, L) from $\mathcal{A}_{ind}$, adversary $\mathcal{A}_{ror}$ flips coin $b \leftarrow \{0, 1\}$ and sets $i = i_b$. It then sends query (i, L, m) to the challenge oracle of ROR and returns the obtained signature to $\mathcal{A}_{ind}$. If $\mathcal{A}_{ind}$ stops and outputs $\tilde{\chi} \in \{0, 1\}$, adversary $\mathcal{A}_{ror}$ outputs $\tilde{\xi} = 1$ if $\tilde{\chi} = b$. It outputs $\tilde{\xi} = 0$, otherwise.

Observe that the view of $\mathcal{A}_{ind}$ is perfectly simulated in two cases; 1) the challenge oracle of ROR gets $\xi = 1$, i.e., it sets the real signer $i^* = i_b$, or 2) the challenge oracle of ROR gets $\xi = 0$ and i_{1-b} is coincidentally selected as a real signer, $i^* = i_{1-b}$. For these cases, $\mathcal{A}_{ror}$ wins whenever $\mathcal{A}_{ind}$ wins. $\mathcal{A}_{ror}$ could also win even when $\mathcal{A}_{ind}$ does not get a correct challenge signature. It is the case when the output $\tilde{\chi}$ from $\mathcal{A}_{ind}$ happens to $\tilde{\chi} \neq b$. This happens with probability $1/2$ because, in this case, randomly chosen b is independent from the view of $\mathcal{A}_{ind}$. From these observation, the success probability of $\mathcal{A}_{ror}$ is as follows.

$$
\begin{aligned}
\Pr[\mathcal{A}_{ror} \text{ wins}] &= \Pr[\xi = 1] \cdot \Pr[\mathcal{A}_{ind} \text{ wins}] + \Pr[\xi = 0] \cdot \Pr[i^* = i_{1-b}] \cdot \Pr[\mathcal{A}_{ind} \text{ wins}] \\
&\quad + \Pr[\xi = 0] \cdot \Pr[i^* \neq i_{1-b}] \cdot \Pr[\tilde{\chi} \neq b] \\
&= \frac{1}{2}(\frac{1}{2} + \epsilon_{ind}) + \frac{1}{2} \cdot \frac{1}{|L^*| - 1} \cdot (\frac{1}{2} + \epsilon_{ind}) + \frac{1}{2} \cdot \frac{|L^*| - 2}{|L^*| - 1} \cdot \frac{1}{2} \\
&= \frac{1}{2} + \frac{|L^*|}{2(|L^*| - 1)} \epsilon_{ind}
\end{aligned}
$$

Hence we have $\epsilon_{ror} = \frac{|L^*|}{2(|L^*|-1)} \epsilon_{ind}$. The running time is clearly almost the same from the construction.   □

4   Extension to Threshold Case

4.1   Definitions

Syntactical definition of threshold ring signature can be obtained by modifying the signing function to take a set of private keys, say $\{sk_i\}$, of the real signer group[1]. Then, the signing function is actually considered as a set of k or more interactive machines each of which has corresponding private key and cooperatively compute a signature. The verification function is also modified to take threshold k as an input.

We first consider a straightforward extension of WHO to the threshold case. The idea is that the adversary has to guess all the members of the real signing group. We denote it by WHO^k_{group}. This notion is used in [2].

---

[1] By $\{X_i\}$, we denote $\{X_1, X_2, \cdots, \}$. This convention will be used in the rest of this paper.

Definition 5 (WHO$^k_{\text{group}}$-Anonimity). *A ring signature scheme is* WHO$^k_{\text{group}}$-*anonymous if for any polynomial-time adversary $\mathcal{A}$ playing the following game, there exists a negligible function ϵ in λ such that* $\Pr[\{\tilde{i}\} = \{i^*\}] < 1/\binom{|L^*|}{k} + \epsilon$.

[Attack Game: WHO$^k_{\text{group}}$]

1. *Global : $\mathcal{L}^*$ and $\mathcal{L}^\dagger$ (initially empty).*
2. $(k, L, m, \omega) \leftarrow \mathcal{A}^{\mathcal{R}}(1^\lambda)$.
3. $\{i^*\} \leftarrow \{L \cap \mathcal{L}^*\}^{(k)}$, $\sigma \leftarrow \mathcal{S}^{k,n}(\{sk_i\}_{i\in\{i^*\}}, m, L)$.
4. $\{\tilde{i}\} \leftarrow \mathcal{A}^{\mathcal{R}}(\omega, \sigma)$.

It is required that $|L \cap \mathcal{L}^*| \geq k$ in step 2, and $\{\tilde{i}\} \subseteq L \cap \mathcal{L}^*$ and $|\{\tilde{i}\}| = k$ in step 4.

Here, $\{L \cap \mathcal{L}^*\}^{(k)}$ denotes all subsets of $L \cap \mathcal{L}^*$ of size k. The same applies to the definitions hereafter.

The above WHO$^k_{\text{group}}$ concerns the anonymity of the real signer group but does not seem to capture the anonymity for *each* member of the real signer group. The following one used in, e.g. [21], cares for this case. The idea is that the adversary wins even if it identifies only one of the real signer in the group. The Argument, Challenge, are the same as before. The wining condition and the security bound is relaxed as follows.

Definition 6 (WHO$^k_{\text{single}}$-Anonimity). *A ring signature scheme is* WHO$^k_{\text{single}}$-*anonymous if for any polynomial-time adversary $\mathcal{A}$ playing the following game, there exists a negligible function ϵ in λ such that* $\Pr[\tilde{i} \in \{i^*\}] < \frac{k}{|L^*|} + \epsilon$.

[Attack Game: WHO$^k_{\text{single}}$]

1. *Global : $\mathcal{L}^*$ and $\mathcal{L}^\dagger$ (initially empty).*
2. $(k, L, m, \omega) \leftarrow \mathcal{A}^{\mathcal{R}}(1^\lambda)$.
3. $\{i^*\} \leftarrow \{L \cap \mathcal{L}^*\}^{(k)}$, $\sigma \leftarrow \mathcal{S}^{k,n}(\{sk_i\}_{i\in\{i^*\}}, m, L)$.
4. $\tilde{i} \leftarrow \mathcal{A}^{\mathcal{R}}(\omega, \sigma)$.

It is required that $|L \cap \mathcal{L}^*| \geq k$ in step 2, and $\tilde{i}^* \in L \cap \mathcal{L}^*$ in step 4.

The notions IND and ROR are extended straightforwardly to the threshold case. Name them IND^k and ROR^k, respectively. We put them here just for notation and completeness.

Definition 7 (IND^k-Anonymity). *A ring signature scheme is IND-anonymous if for any polynomial-time adversary $\mathcal{A}$ playing the following game, there exists a negligible function ϵ in λ such that* $\Pr[\tilde{\chi} = \chi] < 1/2 + \epsilon$.

[Attack Game: IND^k]

1. *Global : $\mathcal{L}^*$ and $\mathcal{L}^\dagger$ (initially empty).*
2. $(\{i\}_0, \{i\}_1, L, m, \omega) \leftarrow \mathcal{A}^{\mathcal{R}}(1^\lambda)$.
3. $\chi \leftarrow \{0, 1\}$, $\sigma \leftarrow \mathcal{S}^{k,n}(\{sk_i\}_{i\in\{i\}_\chi}, m, L)$.
4. $\tilde{\chi} \leftarrow \mathcal{A}^{\mathcal{R}}(\omega, \sigma)$.

It is required that both $\{i\}_0$ and $\{i\}_1$ are subsets of $L \cap \mathcal{L}^*$ and $|\{i\}_0| = |\{i\}_1|$ in step 2, and $\tilde{\chi} \in \{0,1\}$ in step 4.

Definition 8 (ROR^k-Anonimity). *A ring signature scheme is ROR^k-anonymous if for any polynomial-time adversary $\mathcal{A}$ playing the following game, there exists a negligible function ϵ in λ such that $\Pr[\tilde{\chi} = \chi] < 1/2 + \epsilon$.*

[Attack Game: ROR^k]

1. *Global : $\mathcal{L}^*$ and $\mathcal{L}^\dagger$ (initially empty).*
2. *$(\{i\}, L, m, \omega) \leftarrow \mathcal{A}^{\mathcal{R}}(1^\lambda)$.*
3. *$\xi \leftarrow \{0,1\}$, $\{i\}_1 = \{i\}$, $\{i\}_0 \leftarrow \{L \cap \mathcal{L}^*\}^{(|\{i\}|)} \setminus \{i\}$. $\{i^*\} = \{i\}_\xi$, $\sigma \leftarrow \mathcal{S}^{k,n}(\{sk_i\}_{i \in \{i^*\}}, m, L)$.*
4. *$\tilde{\xi} \leftarrow \mathcal{A}^{\mathcal{R}}(\omega, \sigma)$.*

It is required that $\{i\} \subsetneq L \cap \mathcal{L}^*$ in step 2, and $\tilde{\xi} \in \{0,1\}$ in step 4.

4.2   Relation Among the Definitions

Since WHO^k_{group}, IND^k, and ROR^k are straightforward extension of corresponding notions in the single real signer case, their relation preserves naturally. Precise reduction cost can be given by replacing $|L^*|$ in Theorem 2, 1, and 3 with $\binom{|L^*|}{k}$. However, the threshold version of Theorem 2 that will state reduction from WHO^k_{group} to IND^k will suffer the factor of $\binom{|L^*|}{k}^{-1}$, which is exponentially small when k and $|L^*|$ are both polynomial in the security parameter. Hence we are not sure whether $WHO^k_{group} \Rightarrow IND^k$ holds or not. This situation bears resemblance to the relation between IND and OW for cryptosystems with exponentially large message space. (Recall that WHO^k_{group} demands to identify all the real signers. Hence it can be translated to notion of OW with exp-size message space.) We list the theorems among WHO^k_{group}, IND^k, and ROR^k as follows. Proof can be derived from the non-threshold case and omitted.

Theorem 5. *$IND^k \Rightarrow WHO^k_{group}$. In particular, if there exists an $(\epsilon^k_{ind}, t^k_{ind})$-adversary in the sense of IND^k, then there exists an $(\epsilon^k_{who}, t^k_{who})$-adversary in the sense of WHO^k_{group} where $\epsilon^k_{ind} = \binom{|L^*|}{k}/2(\binom{|L^*|}{k} - 1)\epsilon^k_{who}$ and $t_{kwho} \approx t_{kind}$.*

Theorem 6. *$IND^k \Rightarrow ROR^k$. In particular, if there exists an $(\epsilon_{kind}, t_{kind})$-adversary in the sense of IND, then there exists an $(\epsilon_{kror}, t_{kror})$-adversary in the sense of ROR where $\epsilon_{kind} = \epsilon_{kror}$ and running time is $t_{kind} \approx t_{kror}$.*

Theorem 7. *$ROR^k \Rightarrow IND^k$. In particular, if there exists an $(\epsilon_{kror}, t_{kror})$-adversary in the sense of ROR^k, then there exists an $(\epsilon_{kind}, t_{kind})$-adversary in the sense of IND where $\epsilon_{kror} = \binom{|L^*|}{k}/2(\binom{|L^*|}{k} - 1)\epsilon_{kind}$ and running time is $t_{kror} \approx t_{kind}$.*

An interesting case would be the relation between WHO^k_{single} and IND^k. We prove the following theorem holds.

Theorem 8. $IND^k \Rightarrow WHO^k_{single}$. *In particular, if there exists an* $(\epsilon^k_{who}, t^k_{who})$*-adversary in the sense of* WHO^k_{single}*, then there exists an* $(\epsilon^k_{ind}, t^k_{ind})$*-adversary in the sense of* IND^k *where* $\epsilon^k_{ind} = \epsilon^k_{who}/2$ *and* $t^k_{ind} \approx t^k_{who}$.

One may notice that Theorem 8 with $k = 1$ gives slightly worse bound compared to Theorem 1. This is nothing but a technical reason that we construct the adversary in IND^k in such a way that it may coincidently choose the same candidates when it chooses two. Detailed proof is given in Appendix A.

From Theorem 5 to 8, we conclude that IND^k is the most preferable as a starting point for the notion of anonymity in the threshold case as well as the single-signer case.

4.3 Discussion

It remains to consider the reverse relation from WHO^k_{group} and/or WHO^k_{single} to IND^k. Unfortunately, we are not aware of efficient reduction for this case. Observe that WHO^k_{single} has a resemblance to a special case of semantic security for cryptosystems. Namely, the adversary of WHO^k_{single} need to guess one signer from a poly-size group of real signers while the adversary in the special case of semantic security game guesses one bit of the polynomial length plaintext. A particular problem in showing $WHO^k_{single} \Rightarrow IND^k$ is that the adversary in WHO^k_{single} is not allowed to chose convenient distribution from which the challenge oracle chooses the real signer. Therefore, we strengthen the notion by giving more choice to the adversary. Concretely, the adversary is allowed to specify an access structure, say Γ, from which the set of real signers is selected. Formal description is as follows.

Definition 9 (WHO^k_{spec}-Anonimity). *A ring signature scheme is* WHO^k_{spec}*-anonymous if for any polynomial-time adversary $\mathcal{A}$ playing the following game, there exists a negligible function ϵ in λ such that* $\Pr[\{\tilde{i}\} = \{i^*\}] < \frac{1}{|\Gamma|} + \epsilon$.

[Attack Game: WHO^k_{spec}]

1. *Global :* $\mathcal{L}^*$ *and* $\mathcal{L}^\dagger$ *(initially empty).*
2. $(\Gamma, k, L, m, \omega) \leftarrow \mathcal{A}^{\mathcal{R}}(1^\lambda)$.
3. $\{i^*\} \leftarrow \Gamma, \sigma \leftarrow \mathcal{S}^{k,n}(\{sk_i\}_{i \in \{i^*\}}, m, L)$.
4. $\{\tilde{i}\} \leftarrow \mathcal{A}^{\mathcal{R}}(\omega, \sigma)$.

It is required that $\Gamma = \{\{i_1, \cdots, i_k\} | i_j \subseteq L \cap \mathcal{L}^*\}$ for some fixed $k < |L \cap \mathcal{L}^*|$. Furthermore, for every $\{i_1, \cdots, i_k\}, \{i'_1, \cdots, i'_k\} \in \Gamma, \{i_1, \cdots, i_k\} \neq \{i'_1, \cdots, i'_k\}$.

With this relaxed definition, it is not hard to see that WHO^k_{spec} and IND^k can be reduced each other. We state the following theorems without proofs that is straightforward.

Theorem 9. $WHO^k_{spec} \Rightarrow IND^k$. *In particular, if there exists an* $(\epsilon_{kind}, t_{kind})$*-adversary in the sense of* IND^k*, then there exists an* $(\epsilon_{kwho}, t_{kwho})$*-adversary in the sense of* WHO^k_{spec} *where* $\epsilon_{kwho} = \epsilon_{kind}$ *and running time is* $t_{kwho} \approx t_{kind}$.

Theorem 10. $IND^k \Rightarrow WHO^k_{\mathrm{spec}}$. In particular, if there exists an $(\epsilon_{kwho}, t_{kwho})$-adversary in the sense of WHO^k_{spec}, then there exists an $(\epsilon_{kind}, t_{kind})$-adversary in the sense of IND^k where $\epsilon_{kind} = \frac{|\Gamma|}{2(|\Gamma|-1)} \epsilon_{kwho}$ and running time is $t_{kind} \approx t_{kwho}$.

Allowing the adversary to chose a set of sets of the target message space plays an important role in showing relations among notions for cryptosystems. For instance, showing the relation between indistinguishability and non-malleability essentially uses this feature in [5].

Our conclusion is that WHO^k_{spec} would be the strongest notion. IND^k would be the second one and close to WHO^k_{spec}. IND^k is easy to use in security proof, so IND^k would be the best to work with. Avoid other weak variants, WHO^k_{group} and WHO^k_{single}.

5   On Unlinkability and Anonymity

5.1   Definitions

A linkable ring signature scheme would be formalized as a ring signature scheme with an additional function for verifying the link between two signatures. But such a simple extension does not really accommodate with anonymity when the adversary can launch chosen-message attacks by making signature-request queries **get-sign**. One can easily see that it is either the case where anonymity is achieved and the link verification function does not exist, or, where the link verification is possible and no anonymity is achieved.

To get around the situation, we first of all extend the syntax of ring signature so that the signing function takes additional parameter used for linking signatures. In this paper, we consider stateful schemes where a signer must privately memorize a parameter, which we call *link-key*, to yield a link. A link-key is used to tag signatures. So that anyone can see two signatures are related to the same private link-key and thus the same signer created the signatures. For this paradigm to work, private link-keys must be taken from an appropriately large space. By LK we denote the space for the link-keys.

Definition 10 (Linkable Ring Signature Scheme). *A linkable ring signature scheme, Σ^{lnk}, is a set of polynomial-time algorithms, $(\mathcal{G}^{lnk}, \mathcal{S}^{lnk}, \mathcal{V}^{lnk}, \mathcal{C}^{lnk})$:*

$\mathcal{G}^{lnk} : (1^\lambda) \to (sk, vk)$. *A probabilistic algorithm that takes security parameter λ and outputs private signing key sk and public verification key vk.*

$\mathcal{S}^{lnk} : (sk, m, L, z) \to (\sigma, z')$. *A probabilistic algorithm that takes link-key z, message m, and list L of verification-keys that includes the one that corresponds to sk, outputs signature σ and (possibly updated) link-key z'.*

$\mathcal{V}^{lnk} : (L, m, \sigma) \to 1/0$. *An algorithm that takes message m and signature σ, and outputs either 1 or 0 meaning accept and reject, respectively. We require that $\mathcal{V}^{lnk}(L, m, \mathcal{S}^{lnk}(sk, m, L, z)) = 1$ for any link-key $z \in LK$, message m, any (sk, vk) generated by $\mathcal{G}^{lnk}$, and any L that includes vk.*

$\mathcal{C}^{lnk} : (\Sigma_0, \Sigma_1) \rightarrow 1/0$. *An algorithm that takes two signed messages, $\Sigma_i = (m_i, \sigma_i, L_i)$ for $i = 0, 1$, and outputs 1 if there exists a single link-key, say $z \in LK$, such that $\sigma_0 \in \mathcal{S}^{lnk}(sk_{i_0}, m_0, L_0, z)$ and $\sigma_1 \in \mathcal{S}^{lnk}(sk_{i_1}, m_1, L_1, z)$ for some $i_0 \in L_0$ and $i_1 \in L_1$. It outputs 0, otherwise.*

Intuition behind the syntax of signature generation function $\mathcal{S}^{lnk}$ is that it takes empty link-key, denoted by "null", when the signer first generate a signature. $\mathcal{S}^{lnk}$ then outputs a new private link-key for later use. The syntax allows $\mathcal{S}^{lnk}$ to update the link-key if needed.

On top of ordinary notion of existential unforgeability against chosen-message attacks, it is generally expected that linkable signatures can be created only by the same signer. For ring signatures, it is very important that it is also true even for a members in a ring bind to a signature. This notion will formally be captured by the following definition of *link unforgeability*. As well as the attack game for anonymity, adversary $\mathcal{A}$ has access to $\mathcal{R}$. Resource oracle $\mathcal{R}$ accepts the queries listed in Section 2 with following modifications;

- Signature Request Query get-sign is modified to take arbitrary link-key z. It however returns the generated signature only, i.e., updated or newly generated link-key is kept private unless otherwise requested.
- Linked Signature Request Query linked-sign is newly allowed. It takes a reference signature $\Sigma = (m, \sigma, L)$ and target (m, L, i). If Σ is once appeared in the history of get-sign, get the corresponding link-key z and call get-sign(m, L, i, z). If no such z is found, call get-sign(m, L, i, null). Then return the obtained signature.
- Link-Key Exposure Query link-key is newly allowed. It takes a signed message (m, σ, L) and return corresponding z if (m, σ, L) once appeared among the input/output of get-sign. Otherwise, it returns nothing.

By using these queries, the signer can obtain arbitrary linked signatures and have full information about the link-keys if desired. On the contrary, the challenge oracle does not take a link-key from the adversary. In fact, the goal of the adversary is to generate a new signature linked to the challenge signature. Formally, the link unforgeability is described as follows.

Definition 11 (Link Unforgeability). *A linkable ring signature scheme is link unforgeable if, for any polynomial-time adversary $\mathcal{A}$ playing the following game, there exists a negligible function ϵ in λ such that $\Pr[\mathcal{A} \text{ wins.}] < \epsilon$.*

[Link Forgery Attack]

1. *Global : $\mathcal{L}^*$ and $\mathcal{L}^\dagger$ (initially empty).*
2. *$(m_0, L_0, i^*, \omega) \leftarrow \mathcal{A}^{\mathcal{R}}(1^\lambda)$.*
3. *$(\sigma_0, z^*) \leftarrow \mathcal{S}^{lnk}(sk_{i^*}, m_0, L_0, \text{null})$.*
4. *$(m_1, L_1, \sigma_1) \leftarrow \mathcal{A}^{\mathcal{R}}(\omega, \sigma_0)$.*

It is required that $i^* \in L_0 \cap \mathcal{L}^*$ Let $\Sigma_0 = (m_0, \sigma_0, L_0)$ and $\Sigma_1 = (m_1, \sigma_1, L_1)$. The adversary is restricted not to make link-key query with regard to Σ_0. (It means that z^* is not available to the adversary. But the adversary can obtain signatures linked to Σ_0 through linked-sign queries.) The adversary wins if

- Σ_0 *is fresh, i.e., not observed in the input-output history of* get-sign *(including the ones internally called from* linked-sign*),*
- $1 \leftarrow \mathcal{V}^{lnk}(L_1, m_1, \sigma_1)$, *and*
- $1 \leftarrow \mathcal{C}^{lnk}(\Sigma_0, \Sigma_1)$.

Now we addresses the definition of anonymity with linkability. Intuition is that the real signer is indistinguishable even if the adversary is free to make any link and/or even to obtain the link-keys.

Definition 12 (Anonymity with Linkability). *A ring signature scheme is anonymous with linkability if, for any polynomial-time adversary $\mathcal{A}$ playing the following game, there exists a negligible function ϵ in λ such that $\Pr[\tilde{\chi} = \chi] < 1/2 + \epsilon$.*

[Attack Game: INDwithLINK]

1. Global : $\mathcal{L}^*$ and $\mathcal{L}^\dagger$ (initially empty).
2. $(i_0, i_1, m, L, \omega) \leftarrow \mathcal{A}^{\mathcal{R}}(1^\lambda)$.
3. $\chi \leftarrow \{0, 1\}$, $(\sigma, z) \leftarrow \mathcal{S}^{lnk}_{sk_{i_\chi}}(m, L, null)$.
4. $\tilde{\chi} \leftarrow \mathcal{A}^{\mathcal{R}}(\omega, \sigma)$.

It is required that $\{i_0, i_1\} \subseteq L \cap \mathcal{L}^*$ in step 2, and $\tilde{\chi} \in \{0, 1\}$ in step 4.

Note that there is no restriction on the access to the resource oracle. For instance, the adversary can obtain link-key z of the target signature σ through a link-key query. This implies that link-keys have to be independent of the identity of the signer.

5.2   Instantiation

We present a generic construction of linkable ring signature schemes from any ordinary ring signature scheme and an ordinary signature scheme. The idea is simple; To create a signature, first generate an ordinary ring signature and sign it by ordinary signature with a newly generated random public-key. Then, to create a linked signature, do the same but use the same random public-key. One can see the link by observing the same random public-key included in the signatures. Anonymity and unforgeability are preserved by the underlying ring signature and the independence of the random public-key. Link-unforgeability is achieved if the underlying ordinary signature scheme is existentially unforgeable against adaptive chosen message attacks. We show detailed description as follows.

Let $\Sigma = (\mathcal{G}, \mathcal{S}, \mathcal{V})$ and $\Sigma^{1,n} = (\mathcal{G}^{1,n}, \mathcal{S}^{1,n}, \mathcal{V}^{1,n})$ be an ordinary signature scheme and a ring signature scheme, respectively. We construct a linkable ring signature scheme $\Sigma^{lnk} = (\mathcal{G}^{lnk}, \mathcal{S}^{lnk}, \mathcal{V}^{lnk}, \mathcal{C}^{lnk})$ as follows.

Function: $\mathcal{G}^{\mathsf{lnk}}(1^\lambda)$

$(sk, vk) \leftarrow \mathcal{G}^{1,n}(1^\lambda)$
Output (sk, vk).

Function: $\mathcal{S}^{\mathsf{lnk}}_{sk_i}(m, L, z)$

$\sigma' \leftarrow \mathcal{S}^{1,n}_{sk_i}(m, L)$
If z is empty, $(sk, vk) \leftarrow \mathcal{G}(1^\lambda)$.
Otherwise, parse $z \rightarrow (sk, vk)$.
$\sigma'' \leftarrow \mathcal{S}_{sk}(\sigma' \| m \| L)$
Set $\sigma = (\sigma', \sigma'', vk)$, $z' = (sk, vk)$.
Output (σ, z').

Function: $\mathcal{V}^{\mathsf{lnk}}_L(m, \sigma)$

Parse $\sigma \rightarrow (\sigma', \sigma'', vk)$.
$v' \leftarrow \mathcal{V}^{1,n}_L(m, \sigma')$
$v'' \leftarrow \mathcal{V}_{vk}(m, \sigma'')$
Set $v = v' * v''$.
Output v.

Function: $\mathcal{C}^{\mathsf{lnk}}(\Sigma_0, \Sigma_1)$

Parse $\Sigma_0 \rightarrow (\sigma'_0, \sigma''_0, vk_0)$.
Parse $\Sigma_1 \rightarrow (\sigma'_1, \sigma''_1, vk_1)$.
Set $u = 1$ if $vk_0 = vk_1$.
Otherwise, $u = 0$.
Output u.

The security parameter given to the key generation function $\mathcal{G}$ in $\mathcal{S}^{\mathsf{lnk}}$ will be decided from the longest public-key in L (assuming that the size of public-key allows one to compute λ).

The following theorem holds. The proof is rather straightforward and hence omitted for this version of the paper.

Theorem 11. *If Σ is unforgeable and $\Sigma^{1,n}$ is unforgeable and anonymous, then the above signature scheme Σ^{lnk} is unforgeable and anonymous in the same sense, and it is link-unforgeable, too.*

The use of random public-key is also found in [20], though their scheme requires simulation-sound non-interactive zero-knowledge proofs without common reference string, which is not possible in the standard model. Furthermore, their scheme does not allow the signer to issue two or more unlinkable signatures for the same group because the scheme generates one random public-key for each group to be involved. So the signatures are always linkable for the same group.

Finally we remark that this generic construction requires the signer to securely store the random private link-key, which has high entropy. As mentioned above, it is impossible to achieve linkable anonymity with stateless schemes. It is an open question if one can achieve linkable anonymity with memorable low-entropy status.

References

1. M. Abe, M. Ohkubo, and K. Suzuki. 1-out-of-n signatures from variety of keys. In Y. Zheng, editor, *Advances in Cryptology — ASIACRYPT 2002*, volume 2501 of *Lecture Notes in Computer Science*, pages 415–432. Springer-Verlag, 2002.
2. M. Abe, M. Ohkubo, and K. Suzuki. Efficient threshold signer-ambiguous signatures from variety of keys. *IEICE Trans. Fundamentals*, E87-A(3):471–479, March 2004.

3. M. Au, S. Chow, W. Susilo, and P. Tsang. Short linkable ring signatures revisited. In A. Atzeni and A. Lioy, editors, *EuroPKI 2006*, volume 4043 of *Lecture Notes in Computer Science*, pages 101–115. Springer-Verlag, 2006.

4. M. Bellare, A. Desai, E. Jokipii, and P. Rogaway. A concrete security treatment of symmetric encryption. In *Proceedings of the 38th IEEE Annual Symposium on Foundations of Computer Science*, pages 394–403. IEEE Computer Society, October 1997. Full version available from http://www-cse.ucsd.edu/users/mihir/papers/sym-enc.html.

5. M. Bellare, A. Desai, D. Pointcheval, and P. Rogaway. Relations among notions of security for public-key encryption schemes. In H. Krawczyk, editor, *Advances in Cryptology — CRYPTO '98*, volume 1462 of *Lecture Notes in Computer Science*, pages 26–45. Springer-Verlag, 1998.

6. M. Bellare, D. Micciancio, and B. Warinschi. Foundations of group signatures: Formal definitions, simplified requirements and a construction based on general assumptions. In E. Biham, editor, *Advances in Cryptology - EUROCRPYT 2003*, volume 2656 of *Lecture Notes in Computer Science*, pages 614–629. Springer-Verlag, 2003.

7. M. Bellare and P. Rogaway. Random oracles are practical: a paradigm for designing efficient protocols. In *First ACM Conference on Computer and Communication Security*, pages 62–73. Association for Computing Machinery, 1993.

8. M. Bellare, H. Shi, and C. Zhang. Foundations of group signatures: The case of dynamic groups. In A. Menezes, editor, *Topics in Cryptology - CT-RSA 2005*, volume 3376 of *Lecture Notes in Computer Science*, pages 136–154. Springer-Verlag, 2005. Full version available at IACR e-print 2004/077.

9. A. Bender, J. Katz, and R. Morselli. Ring signatures: Stronger definitions, and constructions without random oracles. In S. Halevi and T. Rabin, editors, *Theory of Cryptography - TCC 2006*, volume 3876 of *Lecture Notes in Computer Science*, pages 60–79. Springer-Verlag, 2006.

10. E. Bresson, J. Stern, and M. Szydlo. Threshold ring signatures and applications to ad-hoc groups. In M. Yung, editor, *Advances in Cryptology — CRYPTO 2002*, volume 2442 of *Lecture Notes in Computer Science*, pages 465–480. Springer-Verlag, 2002.

11. D. Chaum and E. V. Heyst. Group signatures. In D. W. Davies, editor, *Advances in Cryptology — EUROCRYPT '91*, volume 547 of *Lecture Notes in Computer Science*, pages 257–265. Springer-Verlag, 1991.

12. S. S. M. Chow, S. Yiu, and L. C. K. Hui. Identity based threshold ring signature. In C. Park and S. Chee, editors, *Information Security and Cryptology - ICISC 2004*, volume 3506 of *Lecture Notes in Computer Science*, pages 218–232. Springer-Verlag, 2004.

13. S. S. M. Chow, S. Yiu, and L. C. K. Hui. Efficient identity based ring signature. In J. Ioannidis, A. Keromytis, and M. Yung, editors, *Applied Cryptography and Network Security - ACNS 2005*, volume 3531 of *Lecture Notes in Computer Science*, pages 499–512. Springer-Verlag, 2005. Also available at IACR e-print 2004/327.

14. R. Cramer and V. Shoup. Design and analysis of practical public-key encryption schemes secure against adaptive chosen ciphertext attack. *SIAM Journal on Computing*, 33(1):167–226, 2003.

15. S. Goldwasser and S. Micali. Probabilistic encryption. *Journal of Computer and System Sciences*, 28:270–299, 1984.

16. A. Kiayias and M. Yung. Group signatures: Provable security, efficient constructions and anonymity from trapdoor-holders. IACR e-print 2004/076, 2004.

17. J. K. Liu, V. K. Wei, and D. S. Wong. Linkable spontaneous anonymous group signature for ad hoc groups (extended abstract). In H. W. et al., editor, *Information Security and Privacy: 9th Australasian Conference, ACISP 2004*, volume 3108 of *Lecture Notes in Computer Science*, pages 325–335. Springer-Verlag, 2004.
18. R. Rivest, A. Shamir, and Y. Tauman. How to leak a secret. In C. Boyd, editor, *Advances in Cryptology – Asiacrypt 2001*, volume 2248 of *Lecture Notes in Computer Science*, pages 552–565. Springer-Verlag, 2001.
19. R. Rivest, A. Shamir, and Y. Tauman. How to leak a secret: Theory and applications of ring signatures. In O. Goldreich, A. Rosenberg, and A. Selman, editors, *Theoretical Computer Science – Essays in Memory of Shimon Even*, volume 3895 of *Lecture Notes in Computer Science*, pages 164–186. Springer-Verlag, 2006.
20. P. P. Tsang and V. K. Wei. Short linkable ring signatures for e-voting e-cash and attestation. In R. H. D. et al., editor, *IPSEC 2005*, volume 3439 of *Lecture Notes in Computer Science*, pages 48–60. Springer-Verlag, 2005.
21. P. P. Tsang, V. K. Wei, T. K. Chan, M. H. Au, J. K. Liu, and D. S. Wong. Separable linkable threshold ring signatures. In A. Canteaut and K. Viswanathan, editors, *Indocrypt 2004*, volume 3348 of *Lecture Notes in Computer Science*, pages 384–398. Springer-Verlag, 2004.
22. F. Zhang and K. Kim. ID-based blind signature and ring signature from pairings. In Y. Zheng, editor, *Advances in Cryptology – Asiacrypt 2002*, volume 2501 of *Lecture Notes in Computer Science*, pages 533–547. Springer-Verlag, 2002.

A   Proof of Theorem 8

We construct adversary $\mathcal{A}_{ind}^k$ for IND^k by using adversary $\mathcal{A}_{who}^k$ for WHO_{single}^k as follows. First of all, $\mathcal{A}_{ind}^k$ forwards all queries from $\mathcal{A}_{who}^k$ to $\mathcal{R}$ to the corresponding $\mathcal{R}$ in IND^k and returns the answers to $\mathcal{A}_{who}^k$. $\mathcal{A}_{ind}^k$ simulates the challenge oracle in WHO_{single}^k in the following way; Given argument (k, m, L) from $\mathcal{A}_{who}^k$, $\mathcal{A}_{ind}^k$ chooses $\{i\}_0$ and $\{i\}_1$ of size k randomly and independently from L^*. (Note that it might be the case that they are coincidently identical. Such a case will anyway be handled in our success probability assessment.) It then sends query $(\{i\}_0, \{i\}_1, L, m)$ to the challenge oracle of IND^k. Then return the obtained signature to $\mathcal{A}_{who}^k$. Eventually, $\mathcal{A}_{who}^k$ outputs $\tilde{i} \in L^*$. Then, $\mathcal{A}_{who}^k$ sets $\tilde{\chi}$ as follows;

- For $b = 0, 1$, if $\tilde{i}$ is in $\{i\}_b \setminus \{i\}_{1-b}$, set $\tilde{\chi} = b$. (Namely, if $\tilde{i}$ is an exclusive member in one of the specified signer groups, identify the group as the real signer group.)
- If $\tilde{i} \in \{i\}_0 \cap \{i\}_1$, then randomly select $\tilde{\chi} \leftarrow \{0, 1\}$. (That is, if $\tilde{i}$ is common in the specified signer groups, just make a random guess.)
- If $\tilde{i} \notin \{i\}_0 \cup \{i\}_1$, then randomly select $\tilde{\chi} \leftarrow \{0, 1\}$. (That is, if $\mathcal{A}_{who}^k$ obviously fails, just make a random guess.)

Note that the attack environment for $\mathcal{A}_{who}^k$ is simulated perfectly since the candidate signer group $\{i\}_0$ and $\{i\}_1$ are randomly and independently chosen. Next we consider when $\mathcal{A}_{ind}^k$ wins. From the way how $\mathcal{A}_{ind}^k$ sets $\tilde{\chi}$, there are three cases; 1) $\mathcal{A}_{who}^k$ wins and $\tilde{i}$ is an exclusive member, or 2) $\mathcal{A}_{who}^k$ wins but $\tilde{i}$ is a common member, but the random guess succeeds anyways, or 3) $\mathcal{A}_{who}^k$

fails and it is clearly known since $\tilde{i}$ does not belong to either $\{i\}_0$ or $\{i\}_1$, and the random guess succeeds eventually. A crucial observation is that unused $\{i\}_{1-\chi}$ is independent from the view of $\mathcal{A}_{who}^k$. Therefore, we can give the following probabilities;

$$\Pr[\tilde{i} \text{ is an exclusive member.} \,|\, \mathcal{A}_{who}^k \text{ wins.}] = \Pr[\tilde{i} \notin \{i\}_{1-\chi}] = \binom{|L^*|-1}{k} \Big/ \binom{|L^*|}{k}.$$

$$\Pr[\tilde{i} \text{ is a common member.} \,|\, \mathcal{A}_{who}^k \text{ wins.}] = \Pr[\tilde{i} \in \{i\}_{1-\chi}] = \binom{|L^*|-1}{k-1} \Big/ \binom{|L^*|}{k}.$$

$$\Pr[\tilde{i} \text{ is not in } \{i\}_{1-\chi}. \,|\, \mathcal{A}_{who}^k \text{ fails.}] = \binom{|L^*|-1}{k} \Big/ \binom{|L^*|}{k}.$$

From these observation, the success probability of $\mathcal{A}_{ind}^k$ is the following.

$$\begin{aligned}
\Pr[\mathcal{A}_{ind}^k \text{ wins.}] =& \Pr[\mathcal{A}_{who}^k \text{ wins.} \wedge \tilde{i} \text{ is an exclusive member.}] \\
&+ \frac{1}{2} \Pr[\mathcal{A}_{who}^k \text{ wins.} \wedge \tilde{i} \text{ is a common member.}] \\
&+ \frac{1}{2} \Pr[\mathcal{A}_{who}^k \text{ fails.} \wedge \tilde{i} \notin \{i\}_0 \cup \{i\}_1] \\
=& \left(\frac{k}{|L^*|} + \epsilon_{who}^k\right) \cdot \binom{|L^*|-1}{k} \Big/ \binom{|L^*|}{k} \\
&+ \frac{1}{2} \cdot \left(\frac{k}{|L^*|} + \epsilon_{who}^k\right) \cdot \binom{|L^*|-1}{k-1} \Big/ \binom{|L^*|}{k} \\
&+ \frac{1}{2} \cdot \left(1 - \left(\frac{k}{|L^*|} + \epsilon_{who}^k\right)\right) \cdot \binom{|L^*|-1}{k} \Big/ \binom{|L^*|}{k} \\
=& \frac{1}{2} + \frac{\epsilon_{who}^k}{2}
\end{aligned}$$

Hence we have a bound $\epsilon_{ind}^k = \epsilon_{who}^k/2$. It is obvious that $t_{ind}^k \approx t_{who}^k$ from the construction. $\qquad\square$

Escrowed Linkability of Ring Signatures and Its Applications

Sherman S.M. Chow[1], Willy Susilo[2], and Tsz Hon Yuen[3]

[1] Department of Computer Science
Courant Institute of Mathematical Sciences
New York University, NY 10012, USA
schow@cs.nyu.edu
[2] Center for Information Security Research
School of Information Technology and Computer Science
University of Wollongong, Australia
wsusilo@uow.edu.au
[3] Department of Information Engineering
The Chinese University of Hong Kong
thyuen4@ie.cuhk.edu.hk

Abstract. Ring signatures allow a user to sign anonymously on behalf of a group of spontaneously conscripted members. Two ring signatures are linked if they are issued by the same signer. We introduce the notion of *Escrowed Linkability* of ring signatures, such that only a *Linking Authority* can link two ring signatures; otherwise two ring signatures remain unlinkable to anyone. We give an efficient instantiation, and discuss the applications of escrowed linkability, like spontaneous traceable signature and anonymous verifiably encrypted signature. Moreover, we propose the *first* short identity-based *linkable* ring signatures from bilinear pairings. All proposals are provably secure under the random oracle model.

Keywords: ring signature, identity-based, linkability, pairings.

1 Introduction

Ring signature schemes allow a user to sign anonymously on behalf of a group of spontaneously conscripted members. They offer simple group formation procedures that can be executed by any user individually. In contrast to group signatures [11], where each of the group members is required to join the group before giving a signature; the group formation of ring signature can happen in an ad-hoc manner and does not need the help of a group manager. Traditional ring signatures offer unconditional anonymity. It is impossible to identify the signer beyond the very fact that the signer belongs to the group in question.

Consider the situation where one of the parliamentarians wants to leak a secret to the public. He wants to remain anonymous. On the other hand, he wants the public to be convinced that the secret is actually leaked from the parliament, to make sure the secret is reliable. A ring signature scheme is a cryptographic primitive specially designed for this purpose.

P.Q. Nguyen (Ed.): VIETCRYPT 2006, LNCS 4341, pp. 175–192, 2006.

The idea of ring signature scheme seems to be a "malicious" one that protects whistleblower's privacy while leaking secrets. However, its evolution may be beyond the imagination of those proposed it. Apart from whistle blowing [22], ring signature schemes can be used for ad-hoc groups anonymous authentication [10] or identification [18], non-interactive deniable ring authentication [24], and many other applications that do not want complicated group formation stage but require anonymity. Ring signature schemes can be used to derive other signature schemes with privacy concerns as well, such as concurrent signature [12,17] and multi-designated verifiers signature [13]. Short ring signatures were proposed by Dodis *et al.* [18]. The term "short" means that the sub-linear dependency between the signature size and the number of diversion group members.

Ring signatures may also be used in fighting phishing attacks [1]. Digital identity thefts are prevalent nowadays, it often take place via email due to the widespread use of email. The attacker masquerades as a trustworthy person of an established organization, and sends a deceptive email to the victims. The email often looks like an official electronic notification, which tricks the victims to reveal their sensitive information, such as passwords and credit card details. This social engineering technique is known as email phishing, carding or spoofing attacks. Digitally signed emails could mitigate these attacks. However, traditional email system is repudiable and does not assume a widespread adoption of a public key infrastructure (PKI). Identity-based (ID-based) ring signature is best suited for the above application. Moreover, in the scenario where all signers have a published identity, such as a MAC address or an IP address (which is a typical scenario in a mobile ad-hoc system), ID-based ring signatures are more applicable. All users can enjoy the benefits brought by the cryptographic applications made possible by ring signatures, without having a PKI setup.

Linkable ring signature schemes allow anyone to identify the *linkage* between *two* ring signatures if they are issued by the *same* signer. This special property prohibits all group members from signing more than once, otherwise, his/her identity will be linked. Short ring signature scheme with linkability is studied in [2] and [25], which can be applied in e-voting, e-cash and attestation. We briefly describe the case for e-voting here. Three obvious security requirements of an e-voting system are anonymity, verifiability and double voting detection. If the voters use ring signatures to cast votes, their anonymity is preserved by the anonymity property of ring signatures. The vote can be verified by the verification of ring signatures. Linkability comes into play in double voting detection as it makes two votes cast by the same voter linked. Note that using signature as a vote supports write-in votes in a more straightforward way.

Our Contributions

We present a number of contributions to the ring signature paradigm. Firstly, to avoid the public linkability that restricts everyone from issuing more than one ring signature without being linked, we introduce the notion of *Escrowed Linkability* – a ring signature will remain anonymous until a *Linking Authority* performs its action to "link" the signatures, and the signatures will become linkable. A signer can use the ring to sign more than once normally, but it ensures that there is no abuse in the ring signature used. In the case where such minimum level of revocability is needed, the escrowed linkability is very useful.

Spontaneity is a key property of ring signatures; with identity-based cryptography, the signer can involve members by just knowing their identities. We instantiate *Escrowed Linkability* of ring signature in identity-based setting; the signer can involve members who do not have public keys in traditional PKI.

Our scheme supports identity escrow, and other options like threshold extension naturally. We also suggest applications of escrowed linkability, including *spontaneous traceable signature* and *anonymous verifiably encrypted signature*.

Finally, we propose the *first* short identity-based *linkable* ring signatures from bilinear pairings. The signature size is *independent* of the size of diversion group.

2   Related Works

Ring signature schemes were first formalized by Rivest *et al.* in [22]. Bresson *et al.* [10] extended [22] in threshold setting – any group of t entities spontaneously conscript arbitrarily $n - t$ entities to generate a publicly verifiable t-out-of-n signature. Liu *et al.* [20] proposed the notion of a *linkable* ring signature. They also provided a trivial threshold version by concatenating a threshold number of linkable ring signatures. Tsang *et al.* [26] provided a better solution to linkable threshold ring signature. The linkability is based on each signing member, so two ring signatures are linked if they are signed with the help of the same signer. Dodis *et al.* [18] firstly proposed the short ring signature. Tsang and Wei [25] incorporated linkability into Dodis *et al.*'s work, while Wu *et al.* [29] provided an improvement and an extension of the notion to blind ring signature. A comprehensive survey of ring signatures can be found in [23].

There are many pairing-based ring signature schemes like [8,15]. Utilizing pairings, ID-based ring signature and threshold ring signature were introduced in [30] and in [14] respectively. Chow *et al.*'s survey [16] summarized the study of ID-based ring signature. Recently, Bender *et al.* [5] proposed a ring signature scheme without random oracles based on general assumptions, and efficient constructions for two users. Chow *et al.* [15] proposed an efficient construction for n users.

2.1   Different Levels of Anonymity

We make a comparison on the levels of anonymity provided by different notions. Ring signature provides the strongest sense of anonymity. Linkable ring signature scheme [20] allows anyone to link signatures from the same signer. Such a linkage cannot revoke the signer's anonymity. Tracing-by-linking group signature schemes [27] revoke the anonymity when a signer double signs. In group signature schemes, there exists an open authority that can revoke the anonymity at anytime, even the signer has only given one group signature. Our "Escrow Linkability" introduces a minimum level to revoke the anonymity in case of dispute, by employing the idea of a linking authority. This idea can be seen as originated from the open authority in the group signature paradigm. Another key difference between group signatures and "Escrow Linkability" is that a group signature is related to a group of users registered with the same group manager; while in the latter case, the formation of a group is as dynamic as ring signatures.

2.2   Ring Signatures Using ID-Based Keys or Certified Keys

Ring signatures differ from group signatures in that no group manager is required to handle the joining and leaving of group members. For ring signature schemes under a traditional public key infrastructure (PKI), it is assumed that all the diversion members have registered for a certificate. In an ID-based cryptosystem, the public key is derived from any string that can act as an identifier of the user. Each user is already implicitly associated with a public key, hence ID-based ring signature effectively removes the above assumption [16]. All diversion group members can be totally unaware of being conscripted into the group.

Notice the subtle difference of "ID-based key" between a normal ring signature scheme and one with escrowed linkability and/or escrowed identity. Apart from the master private key held by the trusted authority, the only kind of private key in a normal ring signature scheme is the user's signing key. For schemes supporting escrow, the revocation can only be done by the knowledge of some private keys. We describe a scheme as "truly-ID-based" if these private keys are also ID-based, i.e. a trusted authority generates the corresponding private key when given an identity as the public key. A similar idea is previously considered in [28], which studied group signature schemes with ID-based group manager, ID-based group members and ID-based open authority.

3   Preliminaries

3.1   Bilinear Pairing

Let $\mathbb{G}_1$ and $\mathbb{G}_2$ be two (multiplicative) cyclic groups of prime order p. Let g_1 be a generator of $\mathbb{G}_1$ and g_2 be a generator of $\mathbb{G}_2$. We also let ψ be an efficiently computable isomorphism from $\mathbb{G}_2$ to $\mathbb{G}_1$, with $\psi(g_2) = g_1$, and $\hat{e}$ be a bilinear map such that $\hat{e} : \mathbb{G}_1 \times \mathbb{G}_2 \to \mathbb{G}_T$ with the following properties:

1. *Bilinearity:* For all $u \in \mathbb{G}_1$, $v \in \mathbb{G}_2$ and $a, b \in \mathbb{Z}$, $\hat{e}(u^a, v^b) = \hat{e}(u, v)^{ab}$.
2. *Non-degeneracy:* $\hat{e}(g_1, g_2) \neq 1$.
3. *Computability:* It is efficient to compute $\hat{e}(u, v)$ for all $u \in \mathbb{G}_1, v \in \mathbb{G}_2$.

3.2   Diffie-Hellman Problems

A few problems are assumed to be intractable for the security our constructions. The following q-SDH problem is proven secure in the generic group model in [6]. We introduce a new decisional problem which is the variant of the q-SDH problem in $(\mathbb{G}_1, \mathbb{G}_2, \mathbb{G}_T)$, and review the decisional BDH problem.

Definition 1 (q-SDH). *The q-Strong Diffie-Hellman Problem in $(\mathbb{G}_1, \mathbb{G}_2)$ is defined as follows: Given a $(q+2)$-tuple $(g_1, g_2, g_2^x, g_2^{x^2}, \cdots, g_2^{x^q}) \in \mathbb{G}_1 \times \mathbb{G}_2^{q+1}$, output a pair (A, c) such that $A^{(x+c)} = g_1 \in \mathbb{G}_1$ where $c \in \mathbb{Z}_p^*$. We say that the (q, τ, ϵ)-SDH assumption holds in $(\mathbb{G}_1, \mathbb{G}_2)$ if no τ-time algorithm has advantage at least ϵ in solving the q-SDH problem.*

Definition 2 (q-DSDH). *The q-Decisional Strong Diffie-Hellman Problem in $(\mathbb{G}_1, \mathbb{G}_2, \mathbb{G}_T)$ is defined as follows: Given a $(q+4)$-tuple follows the form of $(g_1, g_2, g_2^x, g_2^{x^2}, \cdots, g_2^{x^q}, R, \gamma) \in \mathbb{G}_1 \times \mathbb{G}_2^{q+1} \times \mathbb{G}_T \times \mathbb{Z}_p$, decide if $R = \hat{e}(g_1, g_2)^{1/(\gamma+x)}$. We say that the (q, τ, ϵ)-DSDH assumption holds in $(\mathbb{G}_1, \mathbb{G}_2, \mathbb{G}_T)$ if no τ-time algorithm has advantage at least ϵ in solving the q-DSDH problem.*

Definition 3 (DBDH). *The Decisional Bilinear Diffie-Hellman Problem in $(\mathbb{G}_1, \mathbb{G}_2, \mathbb{G}_T)$ is defined as follow: Given a sextuple $(g_1, g_2, g_2^\alpha, g_2^\beta, g_2^\gamma, R) \in \mathbb{G}_1 \times \mathbb{G}_2^4 \times \mathbb{G}_T$, decide if $R = \hat{e}(g_1, g_2)^{\alpha\beta\gamma}$. We say that the (τ, ϵ)-DBDH assumption holds in $(\mathbb{G}_1, \mathbb{G}_2, \mathbb{G}_T)$ if no τ-time algorithm has advantage at least ϵ in solving the DBDH problem.*

4   Security Definition

4.1   Ring Signature

Definition 4 (Ring Signature Scheme). *A ring signature scheme is a quadruple (*Init, UKg, Sign, Vfy*) where:*

- *param $\leftarrow$ Init(1^k) is a probabilistic polynomial time (PPT) algorithm that takes as input a security parameter and produces public parameters param.*
- *$(pk, sk) \leftarrow$ UKg(param) is a PPT algorithm that takes as input param and produces the user public key pk and private key sk.*
- *$\sigma \leftarrow$ Sign(param, X, sk, M) is a PPT algorithm that accepts as inputs param, a set of public keys X including the one that corresponds to the private key sk and a message $M \in \{0, 1\}^*$ and produces a signature σ.*
- *$1/0 \leftarrow$ Vfy(param, X, M, σ) is a PPT algorithm that accepts as inputs param, a set of public keys X, a message $M \in \{0, 1\}^*$ and a signature σ and returns 1 or 0 for* accept *or* reject, *respectively.*

We say that a ring signature scheme is *secure* if it satisfies **Correctness**, **Anonymity** and **Unforgeability**. Our security definition is similar to the strongest security level of [5].

Definition 5 ((Verification) Correctness). Vfy(param, X, M, σ) = 1 *with probability 1 for arbitrary* param, X, sk, M *such that $\sigma \leftarrow$ Sign(param, X, sk, M).*

We have the following oracles for an adversary to query:

- *Random Oracle $\mathcal{RO}$*: simulate the random oracle normally.
- *Corruption Oracle $\mathcal{CO}$*: (param, pk) $\rightarrow$ sk. Upon input $pk \in X$, the set of public keys given by simulator, output the private key sk.

- *Signing Oracle* $\mathcal{SO}$: $(\text{param}, \hat{X}, pk, M) \rightarrow \sigma$. Upon input any public keys set $\hat{X}$, a designated signer $pk \in \hat{X}$ and a message M, output a valid signature $\sigma \leftarrow \text{Sign}(\text{param}, \hat{X}, sk, M)$, where sk is the private key of pk.

Definition 6 (Anonymity). *Experiment* Anon *is defined as:*

1. *A simulator $\mathcal{S}$ invokes* Init *and* UKg, *gives* param *and a set of public keys X to an adversary $\mathcal{A}$.*
2. *$\mathcal{A}$ queries $\mathcal{RO}, \mathcal{CO}, \mathcal{SO}$ in arbitrary interleaf.*
3. *$\mathcal{A}$ selects a set of public keys X', two users $pk_0, pk_1 \in X' \cap X$, and a message M and gives them to $\mathcal{S}$. Then $\mathcal{S}$ randomly chooses $b \in \{0,1\}$ and returns the challenge signature $\sigma \leftarrow \mathcal{SO}(\text{param}, X', sk_b, M)$.*
4. *$\mathcal{A}$ queries $\mathcal{RO}, \mathcal{CO}, \mathcal{SO}$ in an arbitrary sequence.*
5. *$\mathcal{A}$ delivers an estimate $\hat{b} \in \{0,1\}$ of b.*

$\mathcal{A}$ wins the Experiment Anon *if $\hat{b} = b$. $\mathcal{A}$'s advantage is its probability of winning Experiment* Anon *minus half. A ring signature scheme is anonymous if no PPT adversary has a non-negligible advantage in Experiment* Anon.

Definition 7 (Unforgeability). *Experiment* Unf *is defined as:*

1. *A simulator $\mathcal{S}$ invokes* Init *and* UKg, *gives* param *and a set of public keys X to an adversary $\mathcal{A}$.*
2. *$\mathcal{A}$ queries $\mathcal{RO}, \mathcal{CO}, \mathcal{SO}$ in an arbitrary sequence.*
3. *$\mathcal{A}$ delivers (σ, M, X') with $X' \subseteq X$.*

$\mathcal{A}$ wins the Experiment Unf *if* $\text{Vfy}(\text{param}, X', M, \sigma) = 1$, *the public keys in X' have never been queried to $\mathcal{CO}$ and σ is not the output from $\mathcal{SO}(\text{param}, X', pk, M)$ for all $pk \in X'$. A ring signature scheme is unforgeable if no PPT adversary has a non-negligible probability of winning in Experiment* Unf.

4.2 Linkable Ring Signature

Definition 8 (Linkable Ring Signature Scheme). *A linkable ring signature scheme is a quintuple (*Init, UKg, Sign, Vfy, Link*) where*

- Init, UKg, Sign, Vfy *are the same as in the ring signature scheme, except that an event identity e is included as part of the message.*
- $1/0/\bot \leftarrow \text{Link}(\text{param}, \sigma_0, \sigma_1, e)$ *is a PPT algorithm which takes as inputs* param, *two signatures σ_0, σ_1 for the same event identity e, returns 1, 0 or $\bot$ for* linked, unlinked *or* invalid, *respectively.*

The definition for correctness includes the linking correctness: suppose $\sigma_i \leftarrow \text{Sign}(\text{param}, X_i, sk, M_i, e)$ where $i = 0, 1$ for arbitrary param, X_0, X_1, sk, M_0, M_1 e, then $\text{Link}(\text{param}, \sigma_0, \sigma_1, e) = 1$ with probability 1.

The definition for the Experiment Anon is the same as above, except that $\mathcal{A}$ cannot query id_0, id_1 to the $\mathcal{CO}$, and to the $\mathcal{SO}$ as the designated signer with the event identity e^* in the challenge ciphertext.

The definition for the Experiment Unf is the same as above.

Definition 9 (Linkability). *Experiment* Link *is defined as:*

1. *A simulator S invokes* Init *and* UKg, *gives* param *and a set of public keys X to an adversary $\mathcal{A}$.*
2. *$\mathcal{A}$ queries $\mathcal{RO}, \mathcal{CO}, \mathcal{SO}$ in an arbitrary sequence.*
3. *$\mathcal{A}$ delivers (σ_0, M_0, X_0, e), (σ_1, M_1, X_1, e), with $X_0, X_1 \subseteq X$.*

$\mathcal{A}$ wins the Experiment Link *if* Vfy$(\text{param}, X_b, M_b, \sigma_b) = 1$ *for $b = 0, 1$,* Link$(\text{param}, \sigma_0, \sigma_1) = 0$, *not both message-signature pairs from $\mathcal{A}$ are the $\mathcal{SO}$'s output, and*

- *if one message-signature pair from $\mathcal{A}$ is the $\mathcal{SO}$ output for a signer id in X_b, then no identity in X_{1-b} should have been queried to $\mathcal{CO}$.*
- *otherwise at most one public key in $X_0 \cup X_1$ has been queried to $\mathcal{CO}$.*

A ring signature scheme is linkable if no PPT adversary has a non-negligible probability of winning in Experiment Link.

We say that a linkable ring signature scheme is *secure* if it satisfies **Correctness, Anonymity, Unforgeability** and **Linkability**.

4.3   Ring Signature with Escrowed Linkability

Definition 10 (Ring Signature Scheme with Escrowed Linkability). *A ring signature scheme with escrowed linkability is a sextuple (*Init, UKg, LAKg, Sign, Vfy, Link*) where*

- Init, UKg *are the same as ring signature scheme.*
- *$(pk', sk') \leftarrow$* LAKg(param) *is a PPT algorithm which takes as input* param, *produces the Linking Authority (LA) public key pk' and private key sk'.*
- Sign, Vfy *are similar to that in the linkable ring signature scheme, with extra input LA public key pk'.*
- Link *are similar to that in the linkable ring signature scheme, with extra input LA private key sk'.*

The definition for anonymity is the same as that of the ring signature scheme. The definitions for correctness, unforgeability and linkability are basically the same as those for the linkable ring signature scheme, with some extra oracles. $\mathcal{A}$ can query an extra "Linking Oracle" $\mathcal{LO}$ which takes as inputs two signatures, outputs whether or not the two signatures are linked. We also allow the adversary to query a "LA Corruption Oracle" $\mathcal{LCO}$, such that the oracle returns the private key of LA with particular identity. Then $\mathcal{A}$ can also query $\mathcal{LO}$ in Experiment Anon, except that the challenge LA identity has never been queried to $\mathcal{LCO}$.

We say that a ring signature scheme with escrowed linkability is *secure* if it satisfies **Correctness, Anonymity, Unforgeability** and **Linkability**.

In some applications, the linking authority may be required to convince another one about the linkability of two signatures. In this case, we need an extra algorithm Judge that takes $(\text{param}, \sigma_0, \sigma_1, sk')$ as input. Judge will produce a signature from the proof of knowledge of sk' in the linkability tag of σ_0 and σ_1. For simplicity of the paper, we omit the protocol here.

4.4   Identity-Based Version

To add identity-based key for users and authority for the above three schemes, we incorporate the following addition and modification.

- Addition: TAKg: $(x, y) \leftarrow$ TAKg(param) is a PPT algorithm which takes as input param, outputs the TA public key y and private key x.
- Modification: $sk \leftarrow$ UKg(param, x, id) is a PPT algorithm which takes as input param, TA private key x and user identity id, outputs the user private key sk. (LAKg is also modified as above for escrowed linkability).

The definitions for the Experiments Anon, Unf, Link are the same as those in the original scheme, except in the first step $\mathcal{S}$ invokes Init and TAKg, and gives param, y to $\mathcal{A}$. Notice that the corruption oracle $\mathcal{CO}$ returns all user's private keys derived from the TA's private key x, but not x itself.

5   Identity-Based Linkable Ring Signature

In this section, we construct a secure ID-based linkable ring signature scheme. We use the pairing accumulator in [21] to accumulate the public keys into the ring and produce the witness proving that the signer's public key is included in the accumulator. We use the concept of "event identity" [26] to link signatures in the same event. For example, in a voting scenario we can use "Vote #2: 01/01/2005" as the event identity; so that voting in a different day or other events held in the same day by a same signer cannot be linked.

Let $(\mathbb{G}_1, \mathbb{G}_2)$ be bilinear groups where $|\mathbb{G}_1| = |\mathbb{G}_2| = p$ for some prime p. For simplicity, we suppose id $\in \mathbb{Z}_p$ below, which refers to the identity. In the security proof, we define id $= \mathcal{H}(identity)$ where $\mathcal{H} : \{0,1\}^* \to \mathbb{Z}_p$ is a cryptographic hash function. The signing algorithm will be described using the notation of "a signature based on a proof of knowledge" (SPK) with concrete instantiation.

Init. Select a pairing $\hat{e} : \mathbb{G}_1 \times \mathbb{G}_2 \to \mathbb{G}_T$. Let g_1 be a generator of $\mathbb{G}_1$, g_2 be a generator of $\mathbb{G}_2$ and $\psi(g_2) = g_1$. Randomly pick $s, u \in \mathbb{Z}_p^*$ and compute $g_2^s, g_2^{s^2}, \ldots, g_2^{s^q}$, where q is the maximum number of members in a ring signature. The auxiliary information s can be safely deleted. Randomly pick $g_3, g_4 \in \mathbb{G}_1$. Set hash function $H : \mathbb{G}_1^3 \times \mathbb{G}_2 \times \mathbb{G}_T \times \mathbb{G}_1^2 \times \mathbb{G}_T^3 \times \{0,1\}^* \to \mathbb{Z}_p$ and $H_0 : (\{0,1\}^*)^2 \to \mathbb{G}_2$. The public parameters param are $(\hat{e}, \psi, g_1, g_2, g_2^s, \ldots, g_2^{s^q}, g_3, g_4, u, H, H_0)$. For efficiency reasons, $\hat{e}(g_1, g_2)$ and $\hat{e}(g_1, g_2^s)$ can be included in the public parameters.

TAKg. The TA picks $x \in_R \mathbb{Z}_p^*$ as the master key, the public key is $y = g_2^x$.

UKg. On input an identity id, the TA computes the private key $S_{\text{id}} = g_1^{1/(\text{id}+x)}$. The user can verify the private key by checking if $\hat{e}(S_{\text{id}}, g_2^{\text{id}} y) = \hat{e}(g_1, g_2)$.

Sign. The user with identity id_1 and private key S_{id_1} who wants to sign a ring signature for message M with users $\text{id}_2, \ldots, \text{id}_n$ firstly computes $W =$

$g_1^{u(\mathsf{id}_2+s)\cdots(\mathsf{id}_n+s)}$ and $V = g_1^{u(\mathsf{id}_1+s)(\mathsf{id}_2+s)\cdots(\mathsf{id}_n+s)}$. ($W$ and V can be computed efficiently like the pairing accumulator in [21]). Let $h = H_0(\mathsf{param}, e)$ where e is the event identity, he/she then computes the signature:

$$SPK\{\ (\,\mathsf{id}_1, S_{\mathsf{id}_1}, W\,) : \hat{e}(V, g_2) = \hat{e}(W, g_2^{\mathsf{id}_1+s})$$
$$\wedge\ \hat{e}(S_{\mathsf{id}_1}, g_2^{\mathsf{id}_1} y) = \hat{e}(g_1, g_2)\ \wedge\ \tilde{y} = \hat{e}(S_{\mathsf{id}_1}, h)\}(M)$$

Detailed scheme of the SPK:

1. Randomly generate $t_1, t_2, t_3 \in \mathbb{Z}_p^*$ and compute:

$$T_1 = S_{\mathsf{id}_1} g_1^{t_1}, \qquad T_2 = W g_1^{t_2}, \qquad T_3 = g_3^{t_1} g_4^{t_2} g_1^{t_3}, \qquad \tilde{y} = \hat{e}(S_{\mathsf{id}_1}, h)$$

2. Randomly generate $r_1, r_2, \ldots, r_7 \in \mathbb{Z}_p^*$ and compute:

$$R_1 = g_3^{r_2} g_4^{r_4} g_1^{r_6}, \qquad R_2 = g_3^{r_3} g_4^{r_5} g_1^{r_7} T_3^{-r_1}, \qquad R_5 = \hat{e}(g_1, h)^{-r_2},$$
$$R_3 = e(T_1, g_2)^{r_1} e(g_1, g_2)^{-r_3} e(g_1, y)^{-r_2},$$
$$R_4 = \hat{e}(T_2, g_2)^{r_1} \hat{e}(g_1, g_2)^{-r_5} \hat{e}(g_1, g_2^s)^{-r_4}$$

3. Compute $c = H(T_1, T_2, T_3, h, \tilde{y}, R_1, \ldots, R_5, M)$
4. Compute $s_1 = r_1 + c\,\mathsf{id}_1, s_2 = r_2 + ct_1, s_3 = r_3 + ct_1\mathsf{id}_1, s_4 = r_4 + ct_2, s_5 = r_5 + ct_2\mathsf{id}_1, s_6 = r_6 + ct_3, s_7 = r_7 + ct_3\mathsf{id}_1$.
5. Output the signature $\sigma = (T_1, T_2, T_3, e, \tilde{y}, c, s_1, \ldots, s_7)$ and the group public key V (or the set of identities $\mathsf{id}_1, \mathsf{id}_2, \ldots, \mathsf{id}_n$).

Vfy. Given a signature $\sigma = (T_1, T_2, T_3, e, \tilde{y}, c, s_1, \ldots, s_7)$, the group public key V (or the identity set $\{\mathsf{id}_1, \ldots, \mathsf{id}_n\}$) and a message M, the verification is done by:

1. Compute:

$$h = H_0(\mathsf{param}, e), \qquad R_1 = g_3^{s_2} g_4^{s_4} g_1^{s_6} T_3^{-c}, \qquad R_2 = g_3^{s_3} g_4^{s_5} g_1^{s_7} T_3^{-s_1}$$
$$R_3 = e(T_1, g_2)^{s_1} e(g_1, g_2)^{-s_3} e(g_1, y)^{-s_2} \big(e(T_1, y)/e(g_1, g_2)\big)^c$$
$$R_4 = \hat{e}(T_2, g_2)^{s_1} \hat{e}(g_1, g_2)^{-s_5} \hat{e}(g_1, g_2^s)^{-s_4} \big(\hat{e}(T_2, g_2^s)/\hat{e}(V, g_2)\big)^c$$
$$R_5 = \hat{e}(g_1, h)^{-s_2} \big(\hat{e}(T_1, h)/\tilde{y}\big)^c$$

2. Accept the message iff $c = H(T_1, T_2, T_3, h, \tilde{y}, R_1, \ldots, R_5, M)$.

Link. On input σ_1, σ_2, output $\perp$ if one or both of the signatures do not pass Vfy. Output 1 if their corresponding values of $\tilde{y}, e$ are the same, 0 otherwise.

Theorem 1. *Our proposed scheme is anonymous if the q-DSDH assumption holds in $(\mathbb{G}_1, \mathbb{G}_2, \mathbb{G}_T)$ in the random oracle model.*

Theorem 2. *Our proposed scheme is unforgeable if the q-SDH assumption holds in $(\mathbb{G}_1, \mathbb{G}_2)$ in the random oracle model.*

Theorem 3. *Our proposed scheme is linkable if the q-SDH assumption holds in $(\mathbb{G}_1, \mathbb{G}_2)$ in the random oracle model.*

Correctness of our scheme is easy to show. Proofs are in the appendix.

6   Escrowed Linkability, Identity Escrow, and Extensions

6.1   Escrowed Linkability

For "escrowed linkability", only a linking authority has the power to link signatures from the same signer. To illustrate this idea, we give similar instantiation for ID-based ring signature. We remark that our technique can be integrated with non-ID-based ring signature scheme from pairing [21], too.

Init, UKg. Same as above.

TAKg. The TA randomly picks $x, x' \in_R \mathbb{Z}_p^*$ as the master key, the corresponding public key is $(y = g_2^x, y' = g_2^{x'})$.

LAKg. On input the Linking Authority (LA) with identity ℓa, the TA computes the private key $S_{\ell a} = H_1(\ell a)^{x'} \in \mathbb{G}_1$. Notice that the linking authority is also equipped with an ID-based key, making our scheme a truly-ID-based solution.

Sign. The user with identity id_1 with private key S_{id_1} who wants to sign a ring signature for message M with users $\mathsf{id}_2, \ldots, \mathsf{id}_n$ firstly computes $W = g_1^{u(\mathsf{id}_2+s)\cdots(\mathsf{id}_n+s)}$, $V = g_1^{u(\mathsf{id}_1+s)(\mathsf{id}_2+s)\cdots(\mathsf{id}_n+s)}$, and $h = H_0(\mathsf{param}, e)$ where e is the event identity. Then he/she computes the signature by:

$$SPK\{\ (\mathsf{id}_1, S_{\mathsf{id}_1}, W, d) : \hat{e}(V, g_2) = \hat{e}(W, g_2^{\mathsf{id}_1+s}) \ \wedge \ \hat{e}(S_{\mathsf{id}_1}, g_2^{\mathsf{id}_1} y) = \hat{e}(g_1, g_2)$$
$$\wedge \ \tilde{y} = \hat{e}(S_{\mathsf{id}_1}, h)\hat{e}(H_1(\ell a), y')^d \ \wedge \ U = g_2^d\}(M)$$

Detailed scheme of the SPK:

1. Randomly generate $t_1, t_2, t_3, d \in \mathbb{Z}_p^*$ and compute:

$$T_1 = S_{\mathsf{id}_1} g_1^{t_1}, \quad T_2 = W g_1^{t_2}, \quad T_3 = g_3^{t_1} g_4^{t_2} g_1^{t_3},$$
$$\tilde{y} = \hat{e}(S_{\mathsf{id}_1}, h)\hat{e}(H_1(\ell a), y')^d, \quad U = g_2^d$$

2. Randomly generate $r_1, r_2, \ldots, r_8 \in \mathbb{Z}_p^*$ and compute:

$$R_1 = g_3^{r_2} g_4^{r_4} g_1^{r_6}, \quad R_2 = g_3^{r_3} g_4^{r_5} g_1^{r_7} T_3^{-r_1},$$
$$R_3 = e(T_1, g_2)^{r_1} e(g_1, g_2)^{-r_3} e(g_1, y)^{-r_2}, \quad R_5 = \hat{e}(H_1(\ell a), y')^{r_8} \hat{e}(g_1, h)^{-r_2}$$
$$R_4 = \hat{e}(T_2, g_2)^{r_1} \hat{e}(g_1, g_2)^{-r_5} \hat{e}(g_1, g_2^s)^{-r_4}, \quad R_6 = g_2^{r_8}$$

3. Compute $c = H(T_1, T_2, T_3, h, \tilde{y}, U, R_1, \ldots, R_6, M)$
4. Compute $s_1 = r_1 + c\mathsf{id}_1, s_2 = r_2 + ct_1, s_3 = r_3 + ct_1\mathsf{id}_1, s_4 = r_4 + ct_2, s_5 = r_5 + ct_2\mathsf{id}_1, s_6 = r_6 + ct_3, s_7 = r_7 + ct_3\mathsf{id}_1, s_8 = r_8 + cd$.
5. Output the signature $\sigma = (T_1, T_2, T_3, e, \tilde{y}, c, s_1, \ldots, s_8, U)$ and the group public key V or the set of identity $\{\mathsf{id}_1, \mathsf{id}_2, \ldots, \mathsf{id}_n\}$.

Vfy. Given a signature $\sigma = (T_1, T_2, T_3, e, \tilde{y}, c, s_1, \ldots, s_8, U)$ and the group public key V and a message M, the verification can be done by:

1. Compute $h = H_0(\mathsf{param}, e)$ and:

$$R_1 = g_3^{s_2} g_4^{s_4} g_1^{s_6} T_3^{-c}, \qquad R_2 = g_3^{s_3} g_4^{s_5} g_1^{s_7} T_3^{-s_1},$$

$$R_3 = e(T_1, g_2)^{s_1} e(g_1, g_2)^{-s_3} e(g_1, y)^{-s_2} \left(e(T_1, y)/e(g_1, g_2)\right)^c$$

$$R_4 = \hat{e}(T_2, g_2)^{s_1} \hat{e}(g_1, g_2)^{-s_5} \hat{e}(g_1, g_2^s)^{-s_4} (\hat{e}(T_2, g_2^s)/\hat{e}(V, g_2))^c,$$

$$R_5 = \hat{e}(H_1(\ell a), y')^{s_8} \hat{e}(g_1, h)^{-s_2} (\hat{e}(T_1, h)/\tilde{y})^c, \qquad R_6 = g_2^{s_8} U^{-c}$$

2. Accept the message if $c = H(T_1, T_2, T_3, h, \tilde{y}, U, R_1, \ldots, R_6, M)$.

Link. On input signatures σ_b for $b = 0, 1$, output $\perp$ if they do not pass Vfy. Else compute $y_b = \tilde{y}/\hat{e}(S_{\ell a}, U_b)$. Output 1 if $y_0 = y_1$ with the same e, 0 otherwise. Note that the correctness of $\hat{e}(S_{\ell a}, U_b)$ can be easily proven the linking authority can thus convince any other parties about the linkage between the signatures.

Theorem 4. *Our proposed extension is anonymous if the DBDH assumption holds in* $(\mathbb{G}_1, \mathbb{G}_2)$ *in the random oracle model.*

Theorem 5. *Our proposed extension is unforgeable if the q-SDH assumption holds in* $(\mathbb{G}_1, \mathbb{G}_2)$ *in the random oracle model.*

Theorem 6. *Our proposed extension is linkable if the q-SDH assumption holds in* $(\mathbb{G}_1, \mathbb{G}_2)$ *in the random oracle model.*

Correctness of our scheme is easy to show. Proofs are in the appendix.

6.2   Ring Signatures with Identity Escrow

Identity escrow means a certain party can revoke the anonymity of the signer of a ring signature. Since our SPK includes the proof of signer's identity, it can be done easily by attaching a verifiably encrypted ciphertext, similar to the way we escrow the linkability. Our scheme is "truly-ID-based", which offers flexibility in the sense that any user of the system (i.e. with an ID-based key) can be designated as the revocation manager for identity escrow; in contrast with other solutions like [3], where a single revocation manager is assumed.

On the other hand, our scheme can naturally support a single system-wide revocation manager. The linkability tag in our construction is always a deterministic function of the private key of the signer (such that it can be used to link). There is a trusted authority to generate user private keys in ID-based paradigm. Identity escrow can simply be achieved by asking the trusted authority to compare the linkability tags of the given signature with all possible linkability tags, generated according to the list of purported diversion group members.

6.3   Extensions

In our instantiation, the linking authority's ID-based keys come from the Boneh-Franklin paradigm [7]. It is easy to integrate extensions of identity-based encryption [7] to our scheme, such as the threshold decryption [4].

7   Applications of Escrowed Linkability

7.1   Spontaneous Traceable Signatures

The notion of traceable signatures was introduced in [19], as an *added feature* to the group signature schemes. This notion allows tracing of all signatures, produced by group signatures, by a single misbehaving party *without opening* the signatures and revealing identities of any other user in the system. In contrast to group signatures, which requires the opening of signatures of *all* users.

The concept of escrowed linkability in our scheme allows us to obtain *spontaneous* traceable signatures. The main difference between this concept and the traceable signatures due to Kiayias *et al.* is that the traceable signatures require all users to join the group prior to producing the traceable signatures. In contrast to traceable signatures, in our *spontaneous traceable signatures*, the users can be conscripted in an ad-hoc manner, and the resulting signatures satisfy the requirements of traceable signatures, namely the signatures can be *traced* whenever required, by a designated party namely the Linking Authority.

7.2   Anonymous Verifiably Encrypted Signatures

Verifiably encrypted signature (VES) [8] enables a signer to give a signature on a message M, where the signature is encrypted using a third party's public key. The recipient cannot do the decryption, but can make sure that a third party can decrypt the VES and recover the original signature on M. This class of signature schemes provides a solution for fair exchange of signature as follows. Alice creates a VES encrypting the signature on messages M_A and sends it to Bob. After verification, Bob returns a VES of M_B that Alice wants. Alice performs the verification as Bob does. If the verification passes, Alice sends Bob the original signature, and expects Bob will return the unencrypted version of the signature she wants. In case any signer does not reveal the original signature finally, the recipient can seek help from the third party to do the decryption.

Notice that even only a single designated party can decrypt the VES and get back the signature, anyone can use the VES's verification algorithm to check whether the VES encrypts a signature by a certain signer on a certain message.

Our schemes can be applied to give a variant of anonymous verifiably encrypted signature, for fair exchange of signature in an anonymous way (or "how to *exchange* a secret", under the original motivation in [22]).

In this variant, the signature is not encrypted, but the linkability tag. From the property of ring signature, no one can be convinced whether the sender or the recipient is the real signer of a particular message given two ring signatures. So even the signature is not encrypted, it is not convincing as the case of VES.

The exchange protocol is as follows. Alice gives a ring signature on message M_A, with both Alice and Bob in the diversion group, using e as the event identity. Bob, after verification, returns a ring signature on message M_B using the same event identity e. In case of dispute, the designated party can reveal the linkability and the real signer of any signature, such that both signatures are binding to the real signer, providing non-repudiation property as normal signatures.

8   Conclusion

We close the open problem in [16] asking for a linkable identity-based ring signature scheme from bilinear pairings. Our proposed scheme produces signature of small size, which is independent of the diversion group size. Furthermore, we introduce the idea of *escrowed linkability* and *linking authority* that provides the minimum level of anonymity-revocability in the literature. We also show how to incorporate identity escrow into our scheme. All escrow can be decrypted by identity-based key. Supported by a number of applications, we believe that our new consideration of anonymity is an important contribution to the literature.

References

1. Ben Adida, Susan Hohenberger, and Ronald L. Rivest. Separable Identity-Based Ring Signatures: Theoretical Foundations for Fighting Phishing Attacks. In *DIMACS Workshop on Theft in E-Commerce: Content, Identity, and Service*, 2005.
2. Man Ho Au, Sherman S. M. Chow, Willy Susilo, and Patrick P. Tsang. Short Linkable Ring Signatures Revisited. In *EuroPKI 2006, Proceedings*, volume 4043 of *Lecture Notes in Computer Science*, pages 110–115. Springer, 2006.
3. Man Ho Au, Joseph K. Liu, Patrick P. Tsang, and Duncan S. Wong. A Suite of ID-Based Threshold Ring Signature Schemes with Different Levels of Anonymity. Cryptology ePrint Archive, Report 2005/326, 2005.
4. Joonsang Baek and Yuliang Zheng. Identity-based Threshold Decryption. In *PKC 2004*, volume 2947 of *Lecture Notes in Computer Science*, pages 262–276. Springer.
5. Adam Bender, Jonathan Katz, and Ruggero Morselli. Ring Signatures: Stronger Definitions, and Constructions without Random Oracles. In *TCC 2006, Proceedings*, volume 3876 of *Lecture Notes in Computer Science*, pages 60–79. Springer.
6. Dan Boneh and Xavier Boyen. Short Signatures Without Random Oracles. In *EUROCRYPT 2004, Proceedings*, volume 3027 of *Lecture Notes in Computer Science*, pages 56–73. Springer, 2004.
7. Dan Boneh and Matt Franklin. Identity-Based Encryption from the Weil Pairing. *SIAM J. Comput*, 32(3):586–615, 2003.
8. Dan Boneh, Craig Gentry, Ben Lynn, and Hovav Shacham. Aggregate and Verifiably Encrypted Signatures from Bilinear Maps. In *EUROCRYPT 2003, Proceedings*, volume 2656 of *Lecture Notes in Computer Science*, pages 416–432. Springer.
9. Stefan Brands. Untraceable Off-line Cash in Wallets with Observers (Extended Abstract). In *CRYPTO '93, Proceedings*, volume 773 of *Lecture Notes in Computer Science*, pages 302–318. Springer, 1993.
10. Emmanuel Bresson, Jacques Stern, and Michael Szydlo. Threshold Ring Signatures and Applications to Ad-hoc Groups. In *CRYPTO 2002, Proceedings*, volume 2442 of *Lecture Notes in Computer Science*, pages 465–480. Springer, 2002.
11. David Chaum and Eugène van Heyst. Group Signatures. In *EUROCRYPT '91, Proceedings*, volume 547 of *Lecture Notes in Computer Science*, pages 257–265.
12. Liqun Chen, Caroline Kudla, and Kenneth G. Paterson. Concurrent Signatures. In *EUROCRYPT 2004, Proceedings*, volume 3027 of *Lecture Notes in Computer Science*, pages 287–305. Springer, 2004.

13. Sherman S.M. Chow. Identity-based Strong Multi-Designated Verifiers Signatures. In *EuroPKI '06*, volume 4043 of *Lecture Notes in Computer Science*, pages 257–9.
14. Sherman S.M. Chow, Lucas C.K. Hui, and S.M. Yiu. Identity Based Threshold Ring Signature. In *ICISC 2004*, volume 3506 of *Lecture Notes in Computer Science*, pages 218–232. Springer-Verlag, 2004.
15. Sherman S.M. Chow, Joseph K. Liu, Victor K. Wei, and Tsz Hon Yuen. Ring Signatures without Random Oracles. In *ASIACCS 2006*, pages 297–302.
16. Sherman S.M. Chow, Richard W.C. Lui, Lucas C.K. Hui, and S.M. Yiu. Identity Based Ring Signature: Why, How and What Next. In *EuroPKI 2005, Proceedings*, volume 3545 of *Lecture Notes in Computer Science*, pages 144–161. Springer, 2005.
17. Sherman S.M. Chow and Willy Susilo. Generic Construction of (Identity-based) Perfect Concurrent Signatures. In *ICICS 2005*, volume 3783 of *Lecture Notes in Computer Science*, pages 194–206. Springer, 2005.
18. Yevgeniy Dodis, Aggelos Kiayias, Antonio Nicolosi, and Victor Shoup. Anonymous Identification in Ad Hoc Groups. In *EUROCRYPT 2004, Proceedings*, volume 3027 of *Lecture Notes in Computer Science*, pages 609–626. Springer, 2004.
19. Aggelos Kiayias, Yiannis Tsiounis, and Moti Yung. Traceable Signatures. In *Eurocrypt '04*, volume 3027 of *Lecture Notes in Computer Science*, pages 571–589.
20. Joseph K. Liu, Victor K. Wei, and Duncan S. Wong. Linkable Spontaneous Anonymous Group Signature for Ad Hoc Groups. In *ACISP 2004, Proceedings*, volume 3108 of *Lecture Notes in Computer Science*, pages 325–335. Springer, 2004.
21. Lan Nguyen. Accumulators from Bilinear Pairings and Applications. In *CT-RSA 2005*, volume 3376 of *Lecture Notes in Computer Science*, pages 275–292.
22. Ronald L. Rivest, Adi Shamir, and Yael Tauman. How to Leak a Secret. In *ASIACRYPT 2001, Proceedings*, volume 2248 of *Lecture Notes in Computer Science*, pages 552–565. Springer, 2001.
23. Ronald L. Rivest, Adi Shamir, and Yael Tauman. How to Leak a Secret: Theory and Applications of Ring Signatures. In *Theoretical Computer Science, Essays in Memory of Shimon Even*, pages 164–186. Springer, 2006.
24. Willy Susilo and Yi Mu. Non-Interactive Deniable Ring Authentication. In *ICISC 2003*, volume 2971 of *Lecture Notes in Computer Science*, pages 386–401.
25. Patrick P. Tsang and Victor K. Wei. Short Linkable Ring Signatures for E-Voting, E-Cash and Attestation. In *ISPEC 2005, Proceedings*, volume 3439 of *Lecture Notes in Computer Science*, pages 48–60. Springer, 2005.
26. Patrick P. Tsang, Victor K. Wei, Tony K. Chan, Man Ho Au, Joseph K. Liu, and Duncan S. Wong. Separable Linkable Threshold Ring Signatures. In *INDOCRYPT 2004, Proceedings*, Lecture Notes in Computer Science, pages 384–398. Springer.
27. Victor K. Wei. Tracing-by-Linking Group Signatures. In *ISC 2005, Proceedings*, volume 3650 of *Lecture Notes in Computer Science*, pages 149–163. Springer, 2005.
28. Victor K. Wei, Tsz Hon Yuen, and Fangguo Zhang. Group Signature Where Group Manager, Members and Open Authority Are Identity-Based. In *ACISP 2005, Proceedings*, volume 3574 of *Lecture Notes in Computer Science*, pages 468–480.
29. Qianhong Wu, Fangguo Zhang, Willy Susilo, and Yi Mu. An Efficient Static Blind Ring Signature Scheme. In *ICISC 2005, Revised Selected Papers*, volume 3935 of *Lecture Notes in Computer Science*, pages 410–423. Springer, 2005.
30. Fangguo Zhang and Kwangjo Kim. ID-Based Blind Signature and Ring Signature from Pairings. In *ASIACRYPT 2002, Proceedings*, volume 2501 of *Lecture Notes in Computer Science*, pages 533–547. Springer, 2002.

A Security Proofs

Lemma 1. *The detailed SPK protocol in Section 5 is an honest verifier zero knowledge proof of knowledge, provided that the discrete logarithm assumption holds in $\mathbb{G}_1$ in the random oracle model.*

Proof. The completeness is straightforward. For soundness, suppose the protocol accepts two signatures $(c, s_1, \ldots, s_7)$ and $(c', s'_1, \ldots, s'_7)$ for the same commitment $(T_1, T_2, T_3, R_1, \ldots, R_5)$. Let $\delta s_i = (s_i - s'_i)/(c - c')$ for $i = 1, \ldots, 7$. We have:

$$T_3 = g_3^{\delta s_2} g_4^{\delta s_4} g_1^{\delta s_6}$$
$$T_3^{\delta s_1} = g_3^{\delta s_3} g_4^{\delta s_5} g_1^{\delta s_7}$$
$$\hat{e}(g_1, g_2)/\hat{e}(T_1, g_2^x) = \hat{e}(T_1, g_2)^{\delta s_1} \hat{e}(g_1, g_2)^{-\delta s_3} \hat{e}(g_1, g_2^x)^{-\delta s_2}$$
$$\hat{e}(V, g_2)/\hat{e}(T_2, g_2^s) = \hat{e}(T_2, g_2)^{\delta s_1} \hat{e}(g_1, g_2)^{-\delta s_5} \hat{e}(g_1, g_2^s)^{-\delta s_4}$$
$$\tilde{y}/\hat{e}(T_1, h) = \hat{e}(g_1, h)^{-\delta s_2}$$

From the first two equations, we have $\delta s_3 = \delta s_1 \delta s_2$, $\delta s_5 = \delta s_1 \delta s_4$, $\delta s_7 = \delta s_1 \delta s_6$ by the discrete logarithm assumption in $\mathbb{G}_1$.

Let $\mathsf{id} = \delta s_1$, $S_{\mathsf{id}} = T_1 g_1^{-\delta s_2}$ and $W = T_2 g_1^{-\delta s_4}$. Then $\hat{e}(V, g_2) = \hat{e}(W, g_2^{\mathsf{id}+s})$, $\hat{e}(S_{\mathsf{id}}, g_2^{\mathsf{id}} y) = \hat{e}(g_1, g_2)$ and $\tilde{y} = \hat{e}(S_{\mathsf{id}}, h)$. Hence, the soundness is proved.

For zero-knowledge, the simulator randomly chooses $c, s_1, \ldots, s_7 \in \mathbb{Z}_p$, T_1, T_2, $T_3, \tilde{y} \in \mathbb{G}_1$. Then he computes $R_1, \ldots, R_5$ as in Vfy. He sets $c = H(T_1, T_2, T_3, h, \tilde{y}, R_1, \ldots, R_5, M)$. We can obviously see that it is zero-knowledge.   □

A.1 Proof of Theorem 1

Proof. By the above lemma, we can see that the SPK itself provides unconditional anonymity for $(T_1, T_2, T_3, R_1, \ldots, R_5)$. Only the linkability tag $(h, \tilde{y})$ may affect the anonymity of the scheme. The proof is as follows.

Now suppose $\mathcal{A}$ can break the anonymity of the proposed scheme. We construct an algorithm $\mathcal{B}$ that uses $\mathcal{A}$ to solve the q-DSDH problem. The simulator $\mathcal{B}$ is given the q-DSDH instance $(g_1, g_2, g_2^x, \ldots, g_2^{x^q}, R, \gamma)$. $\mathcal{B}$ randomly picks $\mathsf{id}_1, \ldots, \mathsf{id}_{q-1} \in \mathbb{Z}_p$ and sets $\mathsf{id}_* = \gamma$. Let $f(y)$ be the polynomial $f(z) = \prod_{i=1}^{q-1}(z + \mathsf{id}_i)$. Expand $f(z)$ and write $f(z) = \sum_{i=0}^{q-1} \alpha_i z^i$ where $\alpha_0, \ldots, \alpha_{q-1} \in \mathbb{Z}_p$ are the coefficients of the polynomial $f(z)$. Then $\mathcal{B}$ computes:

$$g'_2 = \prod_{i=0}^{q-1}(g_2^{x^i})^{\alpha_i} = g_2^{f(x)} \quad \text{and} \quad y = \prod_{i=1}^{q}(g_2^{x^i})^{\alpha_{i-1}} = g_2^{xf(x)} = (g'_2)^x$$

Let $g'_1 = \psi(g'_2)$. We assume that $f(x) \neq 0$, otherwise $x = -\mathsf{id}_i$ for some i which means that $\mathcal{B}$ obtained the private key x for the q-DSDH problem. $\mathcal{B}$ randomly picks $u, s, \in \mathbb{Z}_p$ and $g_3, g_4 \in \mathbb{G}_1$. $\mathcal{B}$ gives $\mathcal{A}$ param $= (\hat{e}, \psi, g'_1, g'_2, g'^s_2, \ldots, g'^{s^q}_2, u$, $g_3, g_4, H, H_0)$, y, and the set of public keys $X = \{\mathsf{id}_1, \ldots, \mathsf{id}_{q-1}, \mathsf{id}_*\}$.

For the $\mathcal{RO}$ query, simulate as random oracles.

For the $\mathcal{CO}$ query for id_i, $\mathcal{B}$ computes $S = g_1'^{1/(x+\mathrm{id}_i)} = \psi(g_2'^{f(x)/(x+\mathrm{id}_i)})$. If the query is for id_*, $\mathcal{B}$ declares failure and exits. $\mathcal{B}$ can compute $g_2'^{f(x)/(x+\mathrm{id}_i)}$ by using $g_2, g_2^x, \ldots, g_2^{x^q}$. The private key S satisfies $\hat{e}(S, g_2'^{\mathrm{id}_i}y) = \hat{e}(g_1', g_2')$.

Replies to $\mathcal{SO}$ query are simulated as the zero-knowledge proof in Lemma 1.

At some point $\mathcal{A}$ gives a set of public keys X', two users $id_0, id_1 \in X \cap X'$, and a message M to $\mathcal{B}$. If both $id_0, id_1 \neq \mathrm{id}_*$, $\mathcal{B}$ declares failure and exits. Suppose $id_b = \mathrm{id}_*$ for $b = 0/1$. $\mathcal{B}$ randomly picks $\alpha \in \mathbb{Z}_p$ and patches $h = g_2^\alpha = H_0(\mathrm{param}, e)$. He has to compute $\tilde{y} = \hat{e}(g_1'^{1/(x+id_b)}, h) = \hat{e}(g_1, g_2)^{\alpha f(x)/(x+id_b)}$. As $f(x)$ is not divisible by $(x + id_b)$, we can write $\tilde{y} = \hat{e}(g_1, g_2^{\sum_{i=0}^{q-1} \beta_i x^i + \beta_{-1}/(x+id_b)})$ for some $\beta_i \in \mathbb{Z}_p$, $\beta_{-1} \neq 0$. Therefore $\tilde{y} = \hat{e}(g_1, g_2^{\sum_{i=0}^{q-1} \beta_i x^i}) \cdot R^{\beta_{-1}}$, and the signature can be computed using id_b as in the zero-knowledge proof in Lemma 1. $\mathcal{B}$ then returns the signature to $\mathcal{A}$ and $\mathcal{A}$ returns a guess $\hat{b}$. If $\hat{b} = b$, $\mathcal{B}$ returns true for the q-DSDH problem. Otherwise, he returns false. For this simulation to succeed, it suffices that $\mathcal{A}$ never asks $\mathcal{CO}$ query on id_*, which holds with probability $1/q$; and for the challenge users, not both $id_0, id_1 \neq \mathrm{id}_*$, which holds with probability at least $2/q$. If $\mathcal{A}$ has probability ϵ of breaking the anonymity, $\mathcal{B}$ has probability at least $2\epsilon/q^2$ of breaking the q-DSDH assumption.   □

A.2   Proof of Theorem 2

Proof. The security proof is similar as above. The oracle simulations are similar. Finally $\mathcal{A}$ returns a signature σ for message M and public keys set $X' \subseteq X$. $\mathcal{B}$ rewinds and extract $(\mathrm{id}, S_{\mathrm{id}})$ as in Lemma 1. If S_{id} is not the output from $\mathcal{CO}(\mathrm{id})$, then $\mathcal{B}$ returns the new key pairs $(\mathrm{id}, S_{\mathrm{id}})$ as the solution of the q-SDH problem. If S_{id} is the output from $\mathcal{CO}(\mathrm{id})$, it means $\mathrm{id} \notin X'$, which breaks the collision resistance property of the pairing accumulator, and hence the q-SDH assumption by theorem 2 of [21]. If we simulate each of the above two cases with probability $1/2$, and $\mathcal{A}$ has probability ϵ of breaking the unforgeability, $\mathcal{B}$ has probability ϵ of breaking the q-SDH assumption.   □

A.3   Proof of Theorem 3

Proof. The security proof is similar as above. The oracle simulations are similar. $\mathcal{A}$ returns signatures σ_0, σ_1, such that the value h is the same while they are not linked. $\mathcal{B}$ rewinds two signatures to obtain S_0, S_1 respectively. Therefore $\hat{e}(S_0, h) \neq \hat{e}(S_1, h)$ and hence $S_0 \neq S_1$. For $b = 0/1$, at least one valid key pairs (id_b, S_b) has never been queried to $\mathcal{CO}$ for some $\mathrm{id}_b \in X_b$. Then $\mathcal{B}$ returns the new key pairs (id_b, S_b) as the solution of the q-SDH problem. If $\mathcal{A}$ has probability ϵ of breaking the linkability, $\mathcal{B}$ has probability ϵ of breaking the q-SDH assumption.   □

A.4   Proof of Theorem 4

Proof. Similar to Lemma 1, we can see that the SPK itself provides unconditional anonymity for $(T_1, T_2, T_3, R_1, \ldots, R_6)$. Only the linkability tag $(h, \tilde{y}, U)$

may affect the anonymity of the scheme. Suppose $\mathcal{A}$ can break the anonymity of the proposed extension. We construct an algorithm $\mathcal{B}$ that uses $\mathcal{A}$ to solve the DBDH problem for $(g_1, g_2, g_2^\alpha, g_2^\beta, g_2^\gamma, R)$.

$\mathcal{B}$ randomly picks $s, x \in \mathbb{Z}_p$ and computes param and y as in Init and TAKg. $\mathcal{B}$ sets $y' = g_2^\alpha$. $\mathcal{B}$ randomly picks $id_1, \ldots, id_{q_C} \in \mathbb{Z}_p$, where q_C is the number of query to the $\mathcal{CO}$. $\mathcal{B}$ gives $\mathcal{A}$ param, y, y' and the set of public keys $X = \{id_1, \ldots, id_{q_C}\}$. $\mathcal{B}$ randomly picks $\mu \in \{1, \ldots, q_R\}$, where q_R is the maximum number of query to the random oracle H_1.

For the $\mathcal{RO}$ query to H_1, $\mathcal{B}$ randomly picks $\lambda \in \mathbb{Z}_p$ and returns g_1^λ, except the μ-th query returns $\psi(g_2^\beta)$. Record the (id, λ) on tape $\mathcal{L}_1$. Other hash queries are simulated as random oracles.

For the $\mathcal{CO}$ query, $\mathcal{B}$ computes the private key using x.

For the $\mathcal{TCO}$ query for id, $\mathcal{B}$ computes $H_1(id)$. If (id, λ) is in $\mathcal{L}_1$, he returns $\psi(g_2^\alpha)^\lambda$. If $(id, \psi(g_2^\beta))$ is in $\mathcal{L}_1$, $\mathcal{B}$ declares failure and exits.

Replies to $\mathcal{SO}$ query are simulated as the zero-knowledge proof in Lemma 1.

For the $\mathcal{LO}$ query for LA id, if $(id, \psi(g_2^\beta))$ is in $\mathcal{L}_1$, $\mathcal{B}$ declares failure and exits. Otherwise, $\mathcal{B}$ extracts the private key as in $\mathcal{TCO}$ and runs as in Link.

At some point $\mathcal{A}$ selects a set of public keys X', two users $id_0, id_1 \in X' \cap X$, a LA id_ℓ and a message M and gives them to $\mathcal{B}$. If $(id_\ell, \psi(g_2^\beta)) \notin \mathcal{L}_1$, $\mathcal{B}$ declares failure and exits. Otherwise, he randomly picks $b \in \{0,1\}$ and extracts the private key of id_b as in $\mathcal{CO}$. Then $\mathcal{B}$ computes as in Sign with $U = g_2^\gamma$ and $\tilde{y} = \hat{e}(S_{id_b}, h) \cdot R$. Then $\mathcal{B}$ returns the signature to $\mathcal{A}$.

Finally, $\mathcal{A}$ returns a guess $\hat{b}$. If $\hat{b} = b$, $\mathcal{B}$ returns true, otherwise false. We can easily see that if $\mathcal{A}$ has non-negligible probability ϵ of winning the game, $\mathcal{B}$ solves the DBDH problem if $\mathcal{A}$ never ask $\mathcal{TCO}$ for $\psi(g_2^\beta)$ with probability $1/q_R$; $\mathcal{A}$ never asks $\mathcal{LO}$ for $\psi(g_2^\beta)$ with probability $1/q_L$ (q_L is the maximum number of query to $\mathcal{LO}$); and the challenge LA is $\psi(g_2^\beta)$ with probability $1/q_R$. $\mathcal{B}$ can decide if $R = \hat{e}(g_1, g_2)^{\alpha\beta\gamma}$ with probability at least $\epsilon/q_L q_R^2$. $\square$

A.5   Proof of Theorem 5

Proof. Now suppose $\mathcal{A}$ can forge in the proposed extension. We construct an algorithm $\mathcal{B}$ that uses $\mathcal{A}$ to solve the q-SDH problem. The simulator $\mathcal{B}$ is given the q-SDH instance $(g_1, g_2, g_2^x, \ldots, g_2^{x^q})$.

$\mathcal{B}$ randomly picks $id_1, \ldots, id_{q-1} \in \mathbb{Z}_p$. Let $f(y)$ be the polynomial $f(z) = \prod_{i=1}^{q-1}(z + id_i)$. Expand $f(z)$ and write $f(z) = \sum_{i=0}^{q-1} \alpha_i z^i$ where $\alpha_0, \ldots, \alpha_{q-1} \in \mathbb{Z}_p$ are the coefficients of the polynomial $f(z)$. Then $\mathcal{B}$ computes:

$$g_2' = \prod_{i=0}^{q-1} (g_2^{x^i})^{\alpha_i} = g_2^{f(x)} \quad \text{and} \quad y = \prod_{i=1}^{q} (g_2^{x^i})^{\alpha_{i-1}} = g_2^{xf(x)} = (g_2')^x$$

Let $g_1' = \psi(g_2')$. We assume that $f(x) \neq 0$, otherwise $x = -id_i$ for some i which means that $\mathcal{B}$ obtained the private key x for the q-SDH problem. $\mathcal{B}$ randomly picks $u, s, x' \in \mathbb{Z}_p$ and $g_3, g_4 \in \mathbb{G}_1$. $\mathcal{B}$ gives $\mathcal{A}$ param $= (\hat{e}, \psi, g_1', g_2', g_2'^s, \ldots, g_2'^{s^q}, u, g_3, g_4, H, H_0, H_1), y, y' = g_2'^{x'}$, and the set of public keys $X = \{id_1, \ldots, id_{q-1}\}$.

For the $\mathcal{RO}$ query, simulate as random oracles.

For the $\mathcal{CO}$ query for id_i, $\mathcal{B}$ computes $S = g_1'^{1/(x+\mathsf{id}_i)} = \psi(g_2^{f(x)/(x+\mathsf{id}_i)})$ by using $g_2, g_2^x, \ldots, g_2^{x^q}$. The private key S satisfies $\hat{e}(S, g_2'^{\mathsf{id}_i}y) = \hat{e}(g_1', g_2')$.

Replies to $\mathcal{SO}$ query are simulated as the zero-knowledge proof in Lemma 1. The $\mathcal{TCO}$ and $\mathcal{LO}$ queries can be answered using the private key x'.

Finally $\mathcal{A}$ returns (σ, M, X). $\mathcal{B}$ rewinds and extracts $(\mathsf{id}, S_{\mathsf{id}})$ as in Lemma 1. If S_{id} is not the output from $\mathcal{CO}(\mathsf{id})$, then $\mathcal{B}$ returns the new key pairs $(\mathsf{id}, S_{\mathsf{id}})$ as the solution of the q-SDH problem. If S_{id} is the output from $\mathcal{CO}(\mathsf{id})$, it means $\mathsf{id} \notin X$, which breaks the collision resistance property of the pairing accumulator and hence the q-SDH assumption by theorem 2 of [21]. If we simulate each of the above two cases with probability $1/2$, and $\mathcal{A}$ has probability ϵ of breaking the unforgeability, $\mathcal{B}$ has probability ϵ of breaking the q-SDH assumption. □

A.6 Proof of Theorem 6

Proof. The security proof is similar as above. The oracle simulations are similar. Finally $\mathcal{A}$ returns signatures σ_0, σ_1, such that the value h is the same while they are not linked. $\mathcal{B}$ rewinds two signatures to obtain S_0, S_1 respectively. Therefore $\hat{e}(S_0, h) \neq \hat{e}(S_1, h)$ and hence $S_0 \neq S_1$. For $b = 0/1$, at least one valid key pairs (id_b, S_b) has never been queried to $\mathcal{CO}$ for some $\mathsf{id}_b \in X_b$. Then $\mathcal{B}$ returns the new key pairs (id_b, S_b) as the solution of the q-SDH problem. If $\mathcal{A}$ has probability ϵ of breaking the linkability, $\mathcal{B}$ has probability ϵ of breaking the q-SDH assumption. □

B   Au *et al.*'s Construction

Independent of our work, Au *et al.* [3] proposed an ID-based linkable threshold ring signature scheme with extensions supporting revocable-iff-linked and identity escrow. The former idea originates from an e-cash system [9], which is also similar to the tracing-by-linking concept in some group signature schemes [27].

Their scheme goes a step further than ours in the sense that a threshold number of signers (which is greater than one) is supported. However, the signature size is *not* short: the signature size is proportional to the number of signers. The choice of RSA-based construction instead of elliptic curve based contributes to a larger constant factor. These shortcomings make their scheme unsuitable for ring signature enabled applications involving a larger number of possible signers like e-cash. It is true that pairing computation is still rather expensive. All our schemes use a constant number of pairing operation in signing and verification.

Regarding identity escrow, their RSA-based design makes it difficult to verifiably encrypt the signer's identity using ID-based encryption; so the revocation manager (the party revoking the identity of a ring signature) is not using an ID-based key (i.e. not "fully ID-based"). Moreover, the discrete logarithm of an element of the system parameters, that is not supposed to be known to any body, is given. Consequently, a high level of trust is placed that the revocation manager will not abuse this knowledge to do any other things except revocation.

Dynamic Fully Anonymous
Short Group Signatures

Cécile Delerablée[1] and David Pointcheval[2]

[1] France Telecom Division R&D, Issy-les-Moulineaux, France
cecile.delerablee@orange-ftgroup.com
[2] CNRS-ENS, Paris, France
david.pointcheval@ens.fr

Abstract. Group signatures allow members to sign on behalf of a group. Recently, several schemes have been proposed, in order to provide more *efficient* and *shorter* group signatures. However, this should be performed achieving a strong security level. To this aim, a formal security model has been proposed by Bellare, Shi and Zang, including both dynamic groups and concurrent join. Unfortunately, very few schemes satisfy all the requirements, and namely the shortest ones needed to weaken the anonymity notion.

We present an extremely short dynamic group signature scheme, with concurrent join, provably secure in this model. It achieves stronger security notions than BBS, and namely the full anonymity, while still shorter. The proofs hold under the q-SDH and the XDH assumptions, in the random oracle model.

1 Introduction

Group signature schemes (thereafter denoted GSS) have been introduced by Chaum and van Heyst [12], in order to provide revocable anonymity to the signer, who is allowed to sign on behalf of a group. In such a scheme, an authority is able, in exceptional cases, to "open" any group signature, and thus recover the actual signer. Properties of group signature schemes make them very important cryptographic tools, with lots of applications (voting, bidding, *anonymous attestation*).

For many years, several GSS have been introduced, and namely the famous ACJT [1], which was the first provably secure coalition-resistant scheme, under the Strong RSA and DDH assumptions. More recently, Boneh, Boyen and Shacham (BBS) [6], and Camenisch and Lysyanskaya [11], proposed very efficient group signature schemes using bilinear maps. The former provides very short group signatures. Independently, Nguyen and Safavi-Naini (NS) [19] also proposed another group signature scheme using bilinear maps. Note that all these schemes were analyzed in the random oracle model [3].

Bellare, Micciancio and Warinschi (BMW) [2] gave formal definitions of the security properties of group signatures, and proposed the first scheme provably secure in the standard model (while totally unpractical). Independently, Kiayias

P.Q. Nguyen (Ed.): VIETCRYPT 2006, LNCS 4341, pp. 193–210, 2006.
© Springer-Verlag Berlin Heidelberg 2006

and Yung [16] (and later [17]), also defined a security model. Bellare, Shi and Zhang (BSZ) [4] extended the BMW model to the case of dynamic groups. Unforgeability and anonymity are indeed crucial security notions, but they should be guaranteed even if the adversary is allowed to play various attack games: adaptively open signatures, join any user of his choice (dynamic group [4]), possibly concurrently (concurrent join [17]).

However, in several schemes, this model has been "weakened", to obtain better efficiency, or to fit with the actually achieved security notions, as done in BBS with CPA-full-anonymity, a weaker version of anonymity where the adversary is not allowed to open signatures when trying to break the anonymity notion. Very recently, Boyen and Waters [8] proposed the first *efficient* GSS that is provably secure without random oracles, but with an important loss of efficiency. Indeed, the length of group signatures grows according to the number of users, and the group public key too.

1.1   Motivations and Related Work

Recently, several schemes have been proposed, in order to reduce the computational cost and the size of group signatures. In particular, BBS [6] is the most efficient one, and provides the shortest signatures so far. But they are still quite large if one compares to short classical signatures [7], and very short group signatures would be of great interest too.

Furthermore, the security level provided by BBS signatures does not fit in the security models proposed by Bellare et al. [2,4]. Namely, *anonymity* is no longer formally guaranteed as soon as one signature is open. However, such an opening process is expected to happen, hence the importance of anonymity as defined in [2]: it must be guaranteed, even if the adversary can see/ask for several openings. Moreover, *non-frameability*, as defined in BSZ is not guaranteed, because the group manager is able to sign on behalf of any group member. However, the authors suggest a possible way to fix this security problem, what we exploited, as explained below. In NS [19], the (full) anonymity is guaranteed, but the computational cost and the size of the group signatures are larger, compared to BBS. Furthermore, while NS claims to be in the BSZ security model, an adaptive access to the join oracle is not properly dealt in the security proofs, and namely for the traceability.

Adaptive, together with concurrent join is specifically considered by Kiayias and Yung [17]. It is indeed a very attractive property since it allows for several users to register at the same time, which could not be avoided (without a drastic efficiency reduction) in many applications (Internet-based for example) However, their scheme provides quite long signatures, with quite high computational cost.

A weakness in the BSZ model is the lack of revocation procedure. They gave some reasons for that, however, revocation of group members is usually a major issue in practice, one has to deal with for an actual scheme.

1.2 Contribution

In this paper, we deal with all the above problems together (therefore in the full BSZ security model, and we even address revocation in the full version [13]). We thus present a new GSS, which provides the strongest security level (under by now classical computational assumptions) in the random oracle model, with quite practical features: concurrent join, very efficient signing and verification procedures, and eXtremely Short (XS) signatures. The short size is also due to an original application of the Forking Lemma [21], which is of independent interest.

Our signature scheme, named XSGS (eXtremely Short Group Signatures), provides anonymity (which is a better security level than BBS [6], and most of the other schemes, except NS [19] and KY [17]), with still very short signatures: 1444 bits, that is almost 70% shorter than a NS-Signature. Furthermore, it is more efficient than NS [19], which provides the same security level (and even better than BBS for verification.) Concurrent join and revocation are possible, which make our schemes very attractive for dynamic groups.

2 Group Signature Schemes

According to the BSZ security model [4], group signature schemes involve distinct authorities, with various rights, and should satisfy several security notions. Even if in many GSS, there is a single authority, holding both *Issuing* and *Opening* capacities, it is preferable to separate those two capabilities. One reason is the fact that for security proofs, we can consider one of the authority corrupted, or partially corrupted, and the other not corrupted (we detail this point later).

2.1 Entities

A group signature scheme involves several entities: the **group manager** $\mathcal{GM}$, which can add new members to the group, by issuing new certificates (we could extend the model to include the revocation of certificates, but here we only consider the group manager as a certificate issuer. In the full version [13] we briefly deal with the revocation process); the **opening manager** $\mathcal{OM}$, which can revoke the anonymity of any group signature; **users** $\mathcal{U}$'s which are group members; and **outsiders**, which do not belong to the group, but just have access to the group public key.

2.2 PKI Environment

We assume that each user $\mathcal{U}_i$, before joining the group, obtains a personal secret key usk[i], associated to a personal certified public key upk[i] (in a PKI). The group manager will also have a certified pair of keys (gmpk, gmsk). This PKI environment is separated from the group environment, and thus the certification authority will be assumed fully trusted (the only one). Indeed, this PKI will provide the non-repudiation, but also the non-frameability property: even if the group authorities are corrupted, they cannot frame a group member. Such a PKI can be formalized by a *user-key generation* algorithm which generates a personal public and private key pair (upk[i], usk[i]) for a user $\mathcal{U}_i$.

2.3 Algorithms

In this section we recall the definitions regarding group signature algorithms, according to [4]:

- GKg– the key generation algorithm GKg generates, according to a security parameter, the group manager's secret key ik, the opening manager's secret key ok, and the group public key gpk.
- Join– running the *join* or issue algorithm Join($\mathcal{U}_i, \mathcal{GM}$), the group manager provides the member $\mathcal{U}_i$ with his secret key gsk[i]. The group manager makes an entry reg[i] in the registration table reg, with the entire transcript of the process (unless explicitly stated).
- GSig– the *group signing* algorithm GSig(gsk[i], m) generates a signature σ on a message m, in the name of the group, using the user's secret key gsk[i].
- GVf– the *group signature verification* algorithm GVf(gpk, m, σ): takes as input the group public key, a group signature σ on message m. It decides whether the signature has been generated by a member of the group (this is a deterministic algorithm).
- Open– the *opening* algorithm Open(ok, m, σ) revokes the anonymity of a signature, granted the opening manager's secret key. More precisely, with read-access to the registration table reg, the opening manager is able to recover the identity of the actual signer (this is a deterministic algorithm). The algorithm outputs an identity $\mathcal{U}_i$, and a proof τ of this claim (which will be used by the Judge algorithm).
- Judge– the *judge* algorithm takes as input the group public key gpk, the public key upk[i] of the user $\mathcal{U}_i$, a message m, a valid signature σ of m, and a proof τ. It is used to check that $\mathcal{U}_i$ produced the signature σ on the message m. This algorithm does not require any private information, and thus, the verification of the opening process is public.

2.4 Security Notions

The Oracles. In [4], the correctness and security definitions are formulated via experiments, which involve oracle access to the adversary. We briefly describe the oracles provided to adversaries in the security experiments.

- AddU(·) – *add user* oracle, which on input an identity $\mathcal{U}_i$ of a new user runs the Join algorithm. This user $\mathcal{U}_i$ is added to the list HU of the Honest Users;
- CrptU(·, ·) – *corrupt user* oracle, which on input an identity $\mathcal{U}_i$ of a new user and a string upk sets upk as the public key upk[i] of $\mathcal{U}_i$. This user is added to the list CU of the Corrupted Users;
- SndToI(·, ·) – *send to issuer (group manager)* oracle, which allows a corrupted user to run the Join algorithm with the issuer;

- SndToU$(\cdot, \cdot)$ – *send to user* oracle, which allows a corrupted group manager to run the Join algorithm with an honest user (which is important for the non-frameability);
- USK$(\cdot)$ – *user secret key* oracle, which converts an honest user (in HU) into a corrupted user (in CU) by leaking the private keys upk$[i]$ and gsk$[i]$;
- RReg$(\cdot)$ – *read registration table* oracle, which gives a read access to the registration table reg;
- WReg$(\cdot, \cdot)$ – *write registration table* oracle, which gives a write access to the registration table reg;
- GSig$(\cdot, \cdot)$ – *signing* oracle, which on input the identity i of an honest user and a message outputs the signature the user would produce;
- Ch$_b(\cdot, \cdot, \cdot)$ – *challenge* oracle, which on input the identities $\mathcal{U}_{i_0}$ and $\mathcal{U}_{i_1}$ of two honest users and a message m, outputs the signature the user $\mathcal{U}_{i_b}$ would produce on m. The message–signature pair generated by this oracle is appended to the list Gset (initially set to empty);
- Open$(\cdot, \cdot)$ – *opening* oracle, which on input a message–signature pair, not generated by the challenge oracle, and thus not in Gset, runs the Open algorithm to get the identity of the actual signer.

Security Notions. We review, in the full version [13], the formal experiments [4] which model the security notions of a dynamic group signature scheme $\mathcal{GSS}$:

Correctness. Signatures generated by a honest member should be accepted, and the open algorithm should correctly identify the signer (and the judge should accept the proof returned by the opening algorithm).

Anonymity. Given signatures produced by a user (among two of his choice – left-or-right) the adversary should not be able to have a significant advantage in guessing which users (the left or the right) provided the signatures. The adversary has a full and adaptive access to the Open oracle, except on the signatures produced by the left-or-right signing oracle.

Traceability. It must be impossible to produce a valid signature such that either the honest opener is unable to identify the signer, or the opener believes it has identified the origin but is unable to produce a correct proof of its claim.

Non-frameability. Even the authorities (group manager and opener) are not able to wrongly accuse someone for having signed a message. For this security level, we assume a colluding subset of users and both authorities to be corrupted.

3 Preliminaries

Since our schemes use classical assumptions and notations, let us introduce them, and review the most famous pairing-based group signature scheme, proposed by Boneh, Boyen, and Shacham [6].

3.1   Computational Assumptions

All the protocols below will apply in three isomorphic cyclic groups of prime order p: $\mathbb{G}_1$, $\mathbb{G}_2$ and $\mathbb{G}_T$. We furthermore assume that there exists an admissible bilinear map $e : \mathbb{G}_1 \times \mathbb{G}_2 \to \mathbb{G}_T$, which can be evaluated efficiently. We denote by ψ the isomorphism from $\mathbb{G}_2$ onto $\mathbb{G}_1$, that we assume to be one-way (easy to compute, but hard to invert).

The Decisional Diffie-Hellman Problem (DDH)

Definition 1. *Let us consider any group $\mathbb{G}$ of prime order p, the decisional Diffie-Hellman problem is defined as follows: given a random generator $G \in \mathbb{G}$, two random elements aG, bG in $\mathbb{G}$, and a candidate $X \in \mathbb{G}$, one has to decide whether $X = abG$ or not.*

We denote by $\mathsf{Adv}_{\mathbb{G}}^{\mathsf{ddh}}(\mathcal{A})$ the advantage of any adversary $\mathcal{A}$ in distinguishing the two distributions: (G, aG, bG, abG) and (G, aG, bG, cG). As usual, we also denote by $\mathsf{Adv}_{\mathbb{G}}^{\mathsf{ddh}}(t)$ the maximal advantage that any adversary can get within time t.

In our context, because of the efficient bilinear map $e : \mathbb{G}_1 \times \mathbb{G}_2 \to \mathbb{G}_T$, and the isomorphism $\psi : \mathbb{G}_2 \to \mathbb{G}_1$, the DDH problem is easy in $\mathbb{G}_2$: given a tuple $(G, aG, bG, cG) \in \mathbb{G}_2^4$, one simply checks whether $e(\psi(aG), bG) = e(\psi(G), cG)$.

The eXternal Diffie-Hellman Assumption (XDH). Note that we furthermore assumed this isomorphism to be one-way. This gives the chance for the following XDH assumption to be true. Such an assumption has been introduced by Camenisch, Hohenberger and Lysyanskaya in the full version of [9], and suggested in the full version of [6].

Definition 2. *Given three groups $\mathbb{G}_1$, $\mathbb{G}_2$ and $\mathbb{G}_T$, as well as a bilinear map $e : \mathbb{G}_1 \times \mathbb{G}_2 \to \mathbb{G}_T$, while the DDH problem is easy in $\mathbb{G}_2$, the XDH assumption states that the DDH problem is hard in $\mathbb{G}_1$.*

Note that the above assumption does not only imply the one-wayness of ψ, but also that there is not efficiently computable isomorphism from $\mathbb{G}_1$ onto $\mathbb{G}_2$. For supersingular curves, such an assumption is known to be false [15], however, it is conjectured to hold, using the Weil or Tate pairing on MNT curves (choosing curves with embedded degree > 1 and $\mathbb{G}_1$ to be the points defined over the ground field. In this case, one can use the Trace map to go from $\mathbb{G}_2$ to $\mathbb{G}_1$). This is reason why it has already been used in recent works [9], and the full version of [6].

The Strong Diffie-Hellman Assumption (SDH). A new assumption, similar to the Strong-RSA one, has been recently introduced by Boneh and Boyen [5]: the Strong Diffie-Hellman Assumption.

Definition 3. *Let us be given two isomorphic groups $\mathbb{G}_1$ and $\mathbb{G}_2$ (together with the isomorphism $\psi : \mathbb{G}_2 \to \mathbb{G}_1$.) The q-Strong Diffie-Hellman problem consists, on input a $(q + 2)$-tuple $(G_1, G_2, \gamma G_2, \gamma^2 G_2,\ldots,\gamma^q G_2)$, for a random element*

$\gamma \in \mathbb{Z}_p$ and a random generator G_2 of $\mathbb{G}_2$, and $G_1 = \psi(G_2)$, in outputting a pair $\left(x, \frac{1}{\gamma+x}G_1\right)$, with $x \in \mathbb{Z}_p^{\star}$.

We denote by $\mathsf{Succ}^{\mathsf{sdh}}_{(\mathbb{G}_1,\mathbb{G}_2)}(q, \mathcal{A})$ the success of any adversary $\mathcal{A}$ in outputting such a solution on a random input instance. We also denote by $\mathsf{Succ}^{\mathsf{sdh}}_{(\mathbb{G}_1,\mathbb{G}_2)}(q, t)$ the maximal success that any adversary can get within time t.

Definition 4. *The q-SDH assumption states that this problem is intractable for a given q.*

3.2 Common Parameters

One chooses a random generator G_2 in $\mathbb{G}_2$, and we denote by G_1 its transformation by ψ: therefore, $G_1 = \psi(G_2)$ is a generator of $\mathbb{G}_1$. We also need additional, and independent generators G, H and K in $\mathbb{G}_1$, whose relative discrete logarithms as well as discrete logarithms in basis G_1 are unknown (unless something else is made more precise). We will denote by $W = \gamma G_2$ the public key of the group (with all the above public informations: the groups and the generators). The value $\gamma \in \mathbb{Z}_p$ is kept secret by the group manager. It will be used to issue membership certificates.

3.3 BBS: Short Group Signatures

The idea of the BBS group signature [6] consists in providing a signature of knowledge of a solution to the SDH problem: (A, x) such that $(x + \gamma)A = G_1$. The latter is generated granted the help of the group manager who knows γ. However, in order to allow the anonymity revocation (the opening operation by the group manager), the proof must not be totally zero-knowledge but partially only: the group manager should be able to recover A.

Therefore, in order to sign a message m, the user first encrypts A with the encryption key of the group manager; he then provides a zero-knowledge proof that the plaintext actually contains an A for which he knows the corresponding x. The security analysis didnot follow the above BSZ model [4], because of some restrictions:

- the unforgeability of the certificates directly comes from the q-SDH assumption. However, the proposed format of the membership certificate does not make any value private to the group manager. Therefore, he can sign on behalf of any user: the non-frameability cannot be guaranteed.
- with a semantically secure encryption scheme, anonymity is guaranteed. However, since one works in groups subject to efficiently computable bilinear maps, the DDH problem may not be hard. Therefore, they prefer to use a new encryption scheme (linear encryption) instead of the classical El-Gamal encryption (the ciphertext is larger: 3 group elements, instead of 2). Furthermore, it is semantically secure against chosen-*plaintext* attacks only: the semantic security (and even the one-wayness) can be broken if the adversary has access to the decryption oracle: the above definition of anonymity

does not hold if the adversary has access to the Open-oracle. This is the reason why they defined a weaker notion of anonymity, the so-called CPA-full-anonymity, or weak anonymity.

On the other hand, the main goal was a *short* signature, which indeed consists of three elements of $\mathbb{G}_1$ (the encryption) and six elements of $\mathbb{Z}_p$ (the proof of knowledge). Hence, the size is just 1533 bits.

3.4   Improvements

In order to improve the security (anonymity and non-frameability), it seems natural that we have to enhance the scheme, and thus to degrade the size:

- make the encryption scheme IND-CCA2 [22], by adding a proof to ensure security against chosen-ciphertext attacks;
- involve an extra parameter in the membership certificate, known to the user only.

Actually, it is possible to make these security improvements without loosing anything from the efficiency point of view (and even improving it too), making the XDH assumption.

4   XS Group Signatures

Note that for simplicity, we will use "certificate" to designate $(A_i, \mathsf{gsk}[i])$ in general, and just A_i at some specific time. In our scheme, we exploit and study two suggestions from [6], also used in [19], together with new tricks:

- First, we make the assumption that the DDH holds in the group $\mathbb{G}_1$, which is true under the XDH assumption. This will then allow a compact IND-CCA2 ElGamal-based encryption scheme;
- Then, we enhance the membership certificate with an additional secret y, known to the user only: (A, x, y), with $A \in \mathbb{G}_1$, $x, y \in \mathbb{Z}_p$, such that $(x + \gamma)A = G_1 + yH$. Applying $e(\cdot, G_2)$ on both sides, one gets that a triple (A, x, y) is a valid certificate if and only it satisfies the relation:

$$e(A, G_2)^x \cdot e(A, W) \cdot e(H, G_2)^{-y} = e(G_1, G_2).$$

- Finally, we revisit the forking lemma [21] in order to even shorten the signatures.

4.1   Concurrent Join Protocol and Revocation

In order to guarantee the non-frameability, one needs a specific Join procedure which provides a group member with a certificate such that the group manager does not know the private key. During the Join protocol, a future group member interacts with the group manager, in order to obtain a valid group certificate (A, x, y), with a private y. This Join protocol is presented on figure 1, where

$\mathcal{U}$ (upk, usk)		$\mathcal{GM}$ (γ, gmsk)
$y \xleftarrow{R} \mathbb{Z}_p,\ C \leftarrow yH$		
$U \leftarrow \begin{pmatrix} c = \text{Ext-Commit}(y), \\ \text{NIZKPEqDL}(c, C, H) \end{pmatrix}$	$\xrightarrow{\ C,U\ }$	Verifies $C \in \mathbb{G}_1$, checks U
		$x \xleftarrow{R} \mathbb{Z}_p,\ A \leftarrow (\frac{1}{\gamma + x})(G_1 + C)$
		$B \leftarrow e(G_1 + C, G_2)/e(A, W)$
		$D \leftarrow e(A, G_2)$
	$\xleftarrow{\ A,V\ }$	$V \leftarrow \text{NIZKPoKDL}(B, D)$
$B \leftarrow e(G_1 + C, G_2)/e(A, W)$		
$D \leftarrow e(A, G_2)$		
Verifies $A \in \mathbb{G}_1$, checks V		
$S \leftarrow Sign_{\text{usk}}(A)$	$\xrightarrow{\ S\ }$	Checks S w.r.t. upk and A
Checks that $(x + \gamma)A \overset{?}{=} G_1 + yH$	$\xleftarrow{\ x\ }$	adds (upk, A, x, S)
i.e. $e(A, G_2)^x \cdot e(A, W) \cdot e(H, G_2)^{-y} = e(G_1, G_2)$		

Fig. 1. Join Protocol

- Ext-Commit is an extractable commitment, that is a commitment which is perfectly binding, and computationally hiding, and a trapdoor allows to open it. Actually, the trapdoor will not be known to anybody, except to our simulator in the security proofs of the traceability and non-frameability. A good example, well-suited to our situation, is the Paillier's encryption scheme [20]: as any encryption scheme, injectivity implies the unconditional binding property, while the computational hiding relies on the semantic security, the high-residuosity assumption. The decryption key allows the extraction;

- NIZKPEqDL(c, C, H) denotes a zero-knowledge proof of equality of the discrete logarithm of C in basis H with the committed value in c, non-interactive in the random oracle model. We won't detail such a proof, but it can be efficiently done with the Paillier's encryption scheme, since it is an equality of discrete logarithms (in different groups). Note that such a proof of membership together with an extractable commitment becomes a proof of knowledge: the user necessarily built C knowing y.

- NIZKPoKDL(B, D) denotes a zero-knowledge proof of knowledge of the discrete logarithm of B in basis D, non-interactive in the random oracle model.

Let us explain the steps in this protocol: This protocol is concurrently secure since all the proofs are non-interactive (NIZKPEqDL, NIZKPoKDL, and the signature), and everything is defined in the first move (the 2 first flows), while the second move (the 2 last flows) involves a signature before revealing the certificate to the user. It ensures the non-frameability property. Indeed, the signature $Sign_{\text{usk}}(A)$ ensures that $\mathcal{U}$ owns the certificate A, in a non-repudiable way. But such a signature is provided by $\mathcal{U}$ only after having checked V: $\mathcal{GM}$ actually knows x, and thus used the C chosen by $\mathcal{U}$. Therefore, he cannot know the associated y.

The *revocation*, which allows the group manager to remove a member from the group, works almost exactly as in [6] (inspired by [10]). We describe it in more details in the full version [13].

4.2   XSGS: An eXtremely Short Group Signature Scheme

Since we make the XDH assumption, it is reasonable to apply a classical ElGamal encryption, to hide the certificate, in a revocable way. In order to reach the (full) anonymity property, we enhance the encryption scheme with the IND-CCA2 security, using the Naor-Yung methodology [18], but in the random-oracle model [14]. In order not to increase too much the size of the signature, the above H (involved in the certificate) will be used as one of the opening manager's public keys. Actually, the secret key of the opening manager is the pair (ξ_1, ξ_2) such that $H = \xi_1 K$ and $G = \xi_2 K$.

Parameters. We thus have:

- group public key: gpk $= (\mathbb{G}_1, \mathbb{G}_2, \mathbb{G}_T, e, \psi; G_1, K, H = \xi_1 K, G = \xi_2 K;$ $G_2, W = \gamma G_2)$;
- group manager's secret key: ik $= \gamma$, which helps to generate the certificate triples (A, x, y), with $A \in \mathbb{G}_1$, $x, y \in \mathbb{Z}_p$, such that $(x + \gamma)A = G_1 + yH$;
- opening manager's secret key: ok $= (\xi_1, \xi_2)$, which will help to decrypt ElGamal ciphertexts;
- an extractable commitment scheme. In the case of the Paillier's encryption scheme [20], one has to choose an RSA modulus n, and an element g of maximal order in $\mathbb{Z}_{n^2}^\star$, without knowing/keeping the factorization.

Double ElGamal Encryption. The signer who owns a certificate (A, x, y), randomly chooses $\alpha, \beta \in \mathbb{Z}_p$ and computes: $T_1 = \alpha K \quad T_2 = A + \alpha H \quad T_3 = \beta K \quad T_4 = A + \beta G$.

First, in order to make the encryption scheme resistant to chosen-ciphertext attacks, one has to prove that (T_1, T_2) and (T_3, T_4), which are two ciphertexts with independent keys and independent random coins, encrypt the same plaintext: there exist α and β such that

$$T_1 = \alpha K \qquad T_3 = \beta K \qquad T_2 - T_4 = \alpha H - \beta G.$$

Secondly, as before, (T_1, T_2) is the encryption of a valid certificate (A, x, y) if and only if there exists an α such that (with $z = x\alpha + y$)

$$T_1 = \alpha K \quad \text{and} \quad e(T_2, G_2)^x \cdot e(H, W)^{-\alpha} \cdot e(H, G_2)^{-z} = e(G_1, G_2)/e(T_2, W).$$

Signature. The signer has thus to prove the knowledge of (α, β, x, z) which satisfies the 4 above relations. Such a proof of knowledge clearly shows that, both there exist convenient α and β values, and the prover knows a certificate. It can be performed with classical techniques, and the non-interactive version uses the Fiat-Shamir paradigm, in the random-oracle model: in order to sign m, $\mathcal{U}$ randomly chooses 4 elements r_α, r_β, r_x and r_z in $\mathbb{Z}_p$ and computes

- $R_1 = r_\alpha K$ $R_2 = e(T_2, G_2)^{r_x} \cdot e(H, W)^{-r_\alpha} \cdot e(H, G_2)^{-r_z}$
  $R_3 = r_\beta K$ $R_4 = r_\alpha H - r_\beta G.$
- $c = \mathcal{H}(m, T_1, T_2, T_3, T_4, R_1, R_2, R_3, R_4)$, where $\mathcal{H}$ outputs k-bit long elements;
- $s_\alpha = r_\alpha + c\alpha \bmod p$ $s_\beta = r_\beta + c\beta \bmod p$
  $s_x = r_x + cx \bmod p$ $s_z = r_z + cz \bmod p.$

A signature therefore consists of the tuple $(T_1, T_2, T_3, T_4, c, s_\alpha, s_\beta, s_x, s_z)$, and the verifier finally checks whether the following relations are satisfied or not:

$$s_\alpha K = R_1 + cT_1 \qquad s_\beta K = R_3 + cT_3 \qquad s_\alpha H - s_\beta G = R_4 + c(T_2 - T_4)$$
$$e(T_2, G_2)^{s_x} \cdot e(H, W)^{-s_\alpha} \cdot e(H, G_2)^{-s_z} = R_2 \cdot (e(G_1, G_2)/e(T_2, W))^c$$

Open. To open a signature, $\mathcal{OM}$ uses the decryption key ok to recover A (and provides a publicly-verifiable proof τ that he did it well —which is a simple proof of equality of discrete logarithms in $\mathbb{G}_1$)—, and then the actual signer, using his read-access to the registration table reg (to prove, in τ, that the designated user $\mathcal{U}_i$ has not be framed, $\mathcal{OM}$ uses $S = Sign_{usk[i]}(A))$.

4.3   Properties

Such a signature contains 4 elements from $\mathbb{G}_1$ (over 171 bits) and 4 scalars (modulo p) of 170 bits. The challenge can just be on 80 bits: the signature can thus be encoded on 1444 bits (less than 181 bytes). From the effiency point of view:

- for the signature, one can compute

$$R_2 = e(A, G_2)^{r_x} \cdot e(H, W)^{-r_\alpha} \cdot e(H, G_2)^{\alpha r_x - r_z}.$$

Since all the pairing values can be precomputed, the signature globally requires 7 multi-exponentiations in $\mathbb{G}_1$ and 1 multi-exponentiation in $\mathbb{G}_T$.
- to verify a signature one has to compute

$$R_2 = e(T_2, s_x G_2 + cW) \cdot e(H, W)^{-s_\alpha} \cdot e(H, G_2)^{-s_z} \cdot e(G_1, G_2)^{-c}.$$

Most of the pairing values can be precomputed: the verification requires 3 multi-exponentiations in $\mathbb{G}_1$, 1 multi-exponentiation in $\mathbb{G}_2$, 1 pairing computation and 1 multi-exponentiation in $\mathbb{G}_T$.

4.4   Without the XDH Assumption

One should note that the XDH assumption helps to get the very short signature, but is not crucial for our construction: if the XDH assumption does not hold, one can use a double variant of the Linear Encryption (the Linear Encryption has been introduced in [6], and is secure assuming the Decision Linear Diffie-Hellman Assumption). Thus we can obtain group signatures of 2126 bits (6 elements from $\mathbb{G}_1$, 6 scalars (modulo p) of 170 bits, and the challenge).

4.5   Security Analysis of XSGS

In order to prove the **correctness** of the group signature scheme, we first need to show that the interactive proof of knowledge is complete, then, the correctness of the Open algorithm immediately leads to the result. Actually, in order to prove the **traceability**, we furthermore need to show that the interactive proof of knowledge is an honest-verifier zero-knowledge and sound proof of knowledge. Thereafter, a simple application of the forking lemma 7 leads to the expected result. This means that we first need the following lemma, which proof can be found in the full version [13].

Lemma 5 (Honest-Verifier Zero-Knowledge Proof of Knowledge). *The interactive proof is a honest-verifier zero-knowledge proof of knowledge.*

From the correctness of the proof of knowledge and the use of a correct encryption scheme, one gets the correctness of the group signature scheme:

Theorem 6 (Correctness). *The group signature scheme XSGS is correct.*

If one gets a closer look at the proof of the forking lemma [21], with a random oracle $\mathcal{H}$ which outputs k-bit elements, one can claim the following lemma:

Lemma 7 (Forking Lemma). *Let $\mathcal{A}$ be a probabilistic polynomial time Turing machine whose input only consists of public data and which can ask q_H queries to the random oracle, with $q_H > 0$. We assume that, within the time bound T, $\mathcal{A}$ produces a valid signature $(m, \sigma_1, h, \sigma_2)$, with probability $\varepsilon \geq 1/2^k + \eta$ for some $\eta > 240q_H/2^k$. Then, within time $T' \leq 9q_H T/\varepsilon$, and with probability $\varepsilon' \geq \frac{1}{6}$, a replay of this machine outputs two valid signatures $(m, \sigma_1, h, \sigma_2)$ and $(m, \sigma_1, h', \sigma_2')$ such that $h \neq h'$.*

Proof. First, with probability greater than η, $\mathcal{A}$ outputs a signature $(m, \sigma_1, h, \sigma_2)$ that is valid, such that h has been obtained as an $\mathcal{H}$ answer on (m, σ_1). Therefore, if we run the attacker $2/\eta$ times with different random tapes, we get a success with probability greater than $1 - e^{-2} \geq \frac{6}{7}$, such that the query $\mathcal{H}(m, \sigma_1)$ has been asked, and answered by h: the crucial query.

By applying the Splitting-Lemma [21], we know that with probability of $1/4$, for each replay, we have a new success with probability greater than $\eta/4q_H$: we thus replay the attack $8q_H/\eta$ times with a new random oracle (but the same answers until the crucial query). With probability greater than $\frac{6}{7}$, we get another success. The challenge is different from the previous one with probability $8q_H/\eta 2^k$.

Finally, after less than $2(1 + 4q_H)/\eta$ replays of the attack, with probability greater $1/5 - 8q_H/\eta 2^k$, which is greater than $1/6$ as soon as $\eta \geq 240q_H/2^k$, we get two valid signatures $(m, \sigma_1, h, \sigma_2)$ and $(m', \sigma_1', h', \sigma_2')$ with $h' \neq h$. □

The following lemma shows that the signature is unforgeable without the knowledge of a certificate, even if the hash function outputs 80-bit values:

Lemma 8 (Unforgeability). *It is computationally impossible to produce a valid signature, without the knowledge of a membership certificate, even under chosen-message attacks, in the random oracle model: if there exists an adversary $\mathcal{A}$ able to build a valid signature within time t, with probability $\varepsilon \leq 1/2^k + \eta + q_S(q_H + q_S)/p^4$, for some $\eta > 240 q_H/2^k$, after q_H queries to the random oracle $\mathcal{H}$ and q_S queries to the signing oracle, then one can build a membership certificate in expecting time $\mathcal{O}(q_H t/\eta)$.*

Proof. We remind the signature consists of $((m, T_1, T_2, T_3, T_4), c, (s_\alpha, s_\beta, s_x, s_z))$, which is not exactly the framework used in the above Forking Lemma. Anyway, with a few extra computation (but no new hash-query), one can make such a signature of the more classical form $((m, T_1, T_2, T_3, T_4), (R_1, R_2, R_3, R_4), c, (s_\alpha, s_\beta, s_x, s_z))$, where $c = \mathcal{H}(m, T_1, T_2, T_3, T_4, R_1, R_2, R_3, R_4)$.

Furthermore, one can efficiently simulate the signing algorithm, in the name of any user (with a statistically negligible probability of failure when setting a random oracle value: less than $(q_H + q_S)/p^4$ for each signature simulation.) If the adversary succeeds with probability $\varepsilon \leq 1/2^k + \eta + q_S(q_H + q_S)/p^4$, then including the signature simulation, we build a *no-message* adversary that makes a forgery with probability greater than $1/2^k + \eta$. According to the above forking lemma, one can extract two related signatures, with the same hash-query but different challenges

$$(m, T_1, T_2, T_3, T_4), \quad (R_1, R_2, R_3, R_4), \quad c, (s_\alpha, s_\beta, s_x, s_z) \quad c', (s'_\alpha, s'_\beta, s'_x, s'_z)$$

in expected time $\mathcal{O}(q_H t/\eta)$. Thereafter, simply applying the same technique as the one used to prove the soundness, one gets a valid certificate (A, x, y). □

Theorem 9 (Traceability). *The group signature scheme* XSGS *is traceable.*

Proof. Suppose there is an adversary $\mathcal{A}$ that wins with probability ε the traceability game against our scheme. We describe an algorithm $\mathcal{B}$ that can break the SDH problem, with the help of the adversary $\mathcal{A}$. Let $\{(A_i, x_i, y_i)\}_{i=1}^q$, with $q = q_1 + q_2$ be the set of certificates generated during the whole attack. More precisely, $\{(A_i, x_i, y_i)\}_{i=1}^{q_1}$ is any set of certificates corresponding to honest users, who can be (all) corrupted by the adversary during the attack (using USK oracle). Furthermore, the adversary is allowed to add new (and possibly corrupted) members, which certificates are denoted $\{(A_i, x_i, y_i)\}_{i=q_1+1}^{q_2}$.

If the adversary can generate a signature which opens to a new $A^\star$ (not associated to an existing user), using the "unforgeability" technique (see lemma 8), one can find a new certificate $(A^\star, x^\star, y^\star)$ in reasonable expected time. ¿From the success of the adversary in the attack game, we know that $A^\star$ does not belong to $\{A_i\}_{i=1}^q$. Let $\mathcal{B}$ be given a q-SDH instance $(G, G', \Theta G', \dots, \Theta^q G')$. It

- randomly chooses $\alpha \xleftarrow{R} \mathbb{Z}_p$ and $x_i \xleftarrow{R} \mathbb{Z}_p$, for $i = 1, \dots, q$, such that the x_i's are pairwise distinct
- randomly chooses $y_i \xleftarrow{R} \mathbb{Z}_p$, for $i, \dots, q_1$, and $k \xleftarrow{R} \{1, \dots, q_1\}$
  (lets us formally define $\gamma \leftarrow \Theta - x_k$, which is unknown)

- computes from the challenge q-SDH instance (since all the formula involve polynomials in Θ of degree at most q times G' or G, and $G = \psi(G')$),

$$G_2 \leftarrow \alpha \left[\prod_{i=1}^{q}(\Theta + x_i - x_k)\right] G' - y_k \left[\prod_{\substack{i=1 \\ i \neq k}}^{q}(\Theta + x_i - x_k)\right] G'; \quad G_1 \leftarrow \psi(G_2)$$

$$H \leftarrow \left[\prod_{\substack{i=1 \\ i \neq k}}^{q}(\Theta + x_i - x_k)\right] G; \qquad\qquad\qquad\qquad W \leftarrow \gamma G_2$$

- randomly generates ok and compute the corresponding encryption keys
- generates the extractable commitment, but *knowing* the trapdoor,
- simulates the first set of users $\{(A_i, x_i, y_i)\}_{i=1}^{q_1}$, computing $A_i = \frac{1}{x_i + \gamma}(G_1 + y_i H)$ according to i:

  • if $i = k$, $A_k = \frac{1}{x_k + \gamma}(\alpha\Theta H) \leftarrow \alpha \left[\prod_{i=1, i \neq k}^{q}(\Theta + x_i - x_k)\right] G$

  • if $i \neq k$, since

  $$y_i = (y_i - y_k) + y_k \text{ and } (y_i - y_k)H = (y_i - y_k)\left[\prod_{j=1, i \neq k}^{q}(\Theta + x_i - x_k)\right] G,$$

  $$A_i \leftarrow (y_i - y_k)\left[\prod_{j=1, j \neq i, k}^{q}(\Theta + x_j - x_k)\right] G + \alpha \left[\prod_{j=1, j \neq i}^{q}(\Theta + x_j - x_k)\right] G.$$

- simulates oracles that $\mathcal{A}$ needs to access (AddU, Open, GSig...). To simulate the Join protocol (engaged with $\mathcal{A}$ via the SndToI-oracle to add a (corrupted) user $\mathcal{U}_i$ with $i \in \{q_1 + 1, \ldots, q_2\}$), $\mathcal{B}$ uses the trapdoor to extract the value y_i committed by the adversary at the beginning of the join, and computes A_i as described before: $i \neq k$, and all the x_i have been chosen ahead)

Finally, the certificate satisfies

$$A^\star = \frac{1}{x^\star + \gamma}(G_1 + y^\star H) = \frac{1}{x^\star + \Theta - x_k}(\alpha\Theta + y^\star - y_k)\left[\prod_{i=1, i \neq k}^{q}(\Theta + x_i - x_k)\right] G.$$

Since $A^\star \notin \{A_i\}_{i=1}^{q}$, and namely $A^\star \neq A_k$, $y^\star - y_k \neq \alpha(x^\star - x_k)$. When we extract $A^\star \notin \{A_i\}_{i=1}^{q}$ (in reasonable expected time), two cases may happen:

1. $x^\star \in \{x_1, \ldots, x_q\}$ with probability greater than $1/2$. Since no information leaks about k, $x^\star = x_j$ with $j \neq k$ with probability greater than $(q_1 - 1)/2q$, and then

   $$\frac{1}{y_j - y^\star}(y_j A^\star - y^\star A_j) = \frac{1}{x_j + \Theta - x_k} G,$$

   and $\mathcal{B}$ has obtained $(\frac{1}{x+\Theta}G, x)$, solution to the q-SDH problem (with $x = x_j - x_k$).

2. $x^\star \notin \{x_1, \ldots, x_q\}$ with probability greater than $1/2$. By an Euclidean division, one can express $A^\star = (C/(\Theta + x^\star - x_k) + P(\Theta)) G$, with

$$C = (\alpha(x_k - x^\star) + y^\star - y_k) \left[\prod_{i=1, i \neq k}^{q} (x_i - x^\star) \right] \neq 0$$

and P a polynomial of degree $q-1$. And thus, $\mathcal{B}$ can compute C and $P(\Theta)G$ from the initial instance, and therefore $(\frac{1}{x+\Theta}G, x)$, a solution to the q-SDH problem (with $x = x^\star - x_k$).

$\square$

The **anonymity** property (not only CPA-full-anonymity [6]) is achieved granted the Double ElGamal encryption scheme, which is IND-CCA [14].

Theorem 10 (Anonymity). *Under the XDH assumption, the group signature scheme* XSGS *is anonymous: if there exists an adversary $\mathcal{A}$ able to break the anonymity game, with advantage ε, and within time t (in the random oracle model), after q_H queries to the random oracle $\mathcal{H}$ and q_S queries to the challenge oracle, then one can break the DDH problem in $\mathbb{G}_1$ with advantage $\varepsilon/4 - (q_H + q_S)/p^4$, within time $t' \leq t + 4q_S T_{\mathsf{pairing}} + (2 + 8q_S)T_{\mathsf{exp}}$, where T_{pairing} is the time of a pairing computation, and T_{exp} is the time of a (multi)-exponentiation.*

Proof. We are given a quadruple $(K, T, U = uK, V = vT)$ in $\mathbb{G}_1$ such that either $u = v$ (DDH quadruple) or v is random (random quadruple). ¿From such a tuple, using the classical random self-reducibility, one can derive many independent tuples: $(K, T, U_i = u_i U + v_i K, V_i = u_i V + v_i K)$. We will choose the group manager's secret key γ, and compute $W = \gamma G_2$. Then, we flip a coin d, and define, either $H = \xi_1 K$ and $G = T$ (if $d = 0$), or $H = T$ and $G = \xi_2 K$ (if $d = 1$), to set the group public key as $(\mathbb{G}_1, \mathbb{G}_2, \mathbb{G}_T, e, \psi; G_1, K, H, G; G_2, W = \gamma G_2)$. Actually, we only know half of the opening manager's key. We will show it is enough to simulate the decryption process (while still allowing to extract something from the adversary). All the other queries can be perfectly answered since using γ, we can build certificates (Join-queries), and thus answer GSig-queries too. Hash-queries can be simulated as usual with a new random value in $\{0, 1\}^k$ for any new query. For the challenge queries, the simulator $\mathcal{B}$ chooses two independent random bits b and d', and will try to simulate the signature from A_b: the i-th request is answered, given a message m, and two certificates (A_0, x_0, y_0) and (A_1, x_1, y_1)

- the encryption: according to d, by choosing an additional random bit d':
  - if $d = 0$, $T_1 \leftarrow \alpha K$, $T_2 \leftarrow A_{d'} + \alpha H$, $T_3 \leftarrow U_i$ and $T_4 \leftarrow A_b + V_i$, for a random α;
  - if $d = 1$, $T_1 \leftarrow U_i$, $T_2 \leftarrow A_b + V_i$, $T_3 \leftarrow \beta K$ and $T_4 \leftarrow A_{d'} + \beta G$, for a random β.
- the proof of validity can be simulated (see the simulator for the zero-knowledge property in the full version [13]). The latter may fail only with a negligible probability when setting the random oracle value, but less than $(q_H + q_S)/p^4$.

In case of failure, $\mathcal{B}$ exits, otherwise, $(T_1, T_2, T_3, T_4, c, s_\alpha, s_\beta, s_x, s_z)$ is the signature of m given back to $\mathcal{A}$. Eventually, the latter returns its guess b' for b. Our algorithm $\mathcal{B}$ answers $\beta = (b' = b)$ as its guess about the tuple (K, T, U, V).

If this is a DDH tuple, and $d' = b$, the 2 ElGamal encryptions always encrypt A_b, this is a valid signature (the advantage of $\mathcal{A}$ in guessing b is ε.) However, if $d' \neq b$, both certificates are encrypted, $\mathcal{A}$ has thus no advantage in guessing b (we will indeed show below that decryption queries do not reveal any information about d.) If this is a random tuple, the signature is independent of b, thus the adversary's advantage is 0. As a consequence, our algorithm $\mathcal{B}$ has an advantage $\varepsilon/4$ in distinguishing DDH quadruples (in a group subject to bilinear maps, hence breaking XDH.)

Now, since we know half of the opening manager's key, as soon as a signature is valid (with identical plaintexts), the decryption of half of the ciphertext is enough. The soundness of the proof of validity (see in the full version [13]) showed that incorrect proofs are very unlikely. □

The final property is the **non-frameability**, which is important for honest users: it guarantees that neither the group manager or the opening manager can cheat, and frame him.

Theorem 11 (Non-Frameability). *The group signature scheme* XSGS *is non-frameable.*

Proof. First, it is clear, from the Open protocol is publicly verifiable, that a wrong open procedure is statistically negligible (even for a powerful adversary).

Second, suppose there is an adversary $\mathcal{A}$ that breaks the non-frameability of our scheme. We describe an algorithm $\mathcal{B}$ that can break the DL problem. Let $\{(A_i, x_i, y_i)\}_{i=1}^{q+q'}$ be the set of certificates generated during the attack. γ and all the (A_i, x_i) are given to the adversary (so the adversary has access to all $y_i H$), but only $\{y_i\}_{i=1}^q$, for the insider colluders. If the adversary can generate a signature which opens to a $A^\star \in \{A_i\}_{i=q+1}^{q+q'}$ (outside the collusion), using the "unforgeability" technique (replay attack and soundness), one can find the whole certificate $(A^\star, x^\star, y^\star)$, and two cases may happen:

Case 1: $A^\star \in \{A_i\}_{i=q+1}^{q+q'}$ and $(A^\star, x^\star) \notin \{(A_i, x_i)\}_{i=q+1}^{q+q'}$ more than half of the time. We show that given a discrete logarithm instance (G, G') in $\mathbb{G}_2$, $\mathcal{B}$ can compute $\Theta = \log_G G'$. In this case, $\mathcal{B}$ computes the group public key: $(\mathbb{G}_1, \mathbb{G}_2, \mathbb{G}_T, e, \psi; G_1 = \psi(G), K = \xi H, H = \psi(G'); G_2 = G, W = \gamma G)$ with $(\gamma, \xi) \xleftarrow{R} \mathbb{Z}_p^{*2}$, randomly chosen by $\mathcal{B}$. It can furthermore simulate any kind of join procedure using γ. We then have $H = \Theta G_1$. Since $A^\star = A_j \in \{A_i\}_{i=q+1}^{q+q'}$ (with $x^\star \neq x_j$, and thus $y^\star \neq y_j$), we have

$$A^\star = \frac{1}{x^\star + \gamma}(G_1 + y^\star H) = v\frac{1}{x^\star + \gamma}(G_1 + y^\star \Theta G_1)$$

$$= \frac{1}{x_j + \gamma}(G_1 + y_j H) = \frac{1}{x_j + \gamma}(G_1 + y_j \Theta G_1).$$

Therefore, $(x_j + \gamma)(1 + y^\star\Theta) = (x^\star + \gamma)(1 + y_j\Theta)$ and $y^\star \neq y_j$, which easily leads to Θ.

Case 2: $(A^\star, x^\star) \in \{(A_i, x_i)\}_{i=q+1}^{q+q'}$ more than half of the time. We show that given a discrete logarithm instance (G, G') in $\mathbb{G}_1$, $\mathcal{B}$ can compute $\Theta = \log_G G'$. In this case, $\mathcal{B}$ computes the group public key: $(\mathbb{G}_1, \mathbb{G}_2, \mathbb{G}_T, e, \psi; G_2 \xleftarrow{R} \mathbb{G}_1, K = \xi H, H = G; G_1 = \psi(G_2), W = \gamma G)$ with $(\gamma, \xi) \xleftarrow{R} \mathbb{Z}_p^{\star 2}$, randomly chosen by $\mathcal{B}$. It can simulate any join procedure as above, with γ, but also chooses a random $j \in \{q+1, ..., q+q'\}$, for which honest user it makes $A_j = \frac{1}{x_j + \gamma}(G_1 + G')$. Since $(A^\star, x^\star) \in \{(A_i, x_i)\}_{i=q+1}^{q+q'}$, we have $(A^\star, x^\star) = (A_j, x_j)$ whith probability $1/q'$, and thus:

$$A^\star = \frac{1}{x^\star + \gamma}(G_1 + y^\star G) = A_j = \frac{1}{x_j + \gamma}(G_1 + G')$$
$$= \frac{1}{x_j + \gamma}(G_1 + \Theta G) = \frac{1}{x^\star + \gamma}(G_1 + \Theta G)$$

which easily leads to Θ ($\Theta = y^\star$.) $\square$

Acknowledgements

The authors would like to thank the anonymous referees for their helpful comments. This work has been done thanks to the French RNRT Crypto++ contract.

References

1. G. Ateniese, J. Camenisch, M. Joye, and G. Tsudik. A Practical and Provably Secure Coalition-Resistant Group Signature Scheme. In *Crypto '00*, LNCS 1880, pages 255–270. Springer-Verlag, Berlin, 2000.
2. M. Bellare, D. Micciancio, and B. Warinschi. Foundations of Group Signatures: Formal Definitions, Simplified Requirements, and a Construction Based on General Assumptions. In *Eurocrypt '03*, LNCS 2656, pages 614–629. Springer-Verlag, Berlin, 2003.
3. M. Bellare and P. Rogaway. Random Oracles Are Practical: a Paradigm for Designing Efficient Protocols. In *Proc. of the 1st CCS*, pages 62–73. ACM Press, New York, 1993.
4. M. Bellare, H. Shi, and C. Zang. Foundations of Group Signatures: The Case of Dynamic Groups. In *CT – RSA '05*, LNCS 3376, pages 136–153. Springer-Verlag, Berlin, 2005.
5. D. Boneh and X. Boyen. Short Signatures without Random Oracles. In *Eurocrypt '04*, LNCS 3027, pages 56–73. Springer-Verlag, Berlin, 2004.
6. D. Boneh, X. Boyen, and H.Shacham. Short Group Signatures. In *Crypto '04*, LNCS 3152, pages 41–55. Springer-Verlag, Berlin, 2004.
7. D. Boneh, B. Lynn, and H. Shacham. Short Signatures from the Weil Pairing. In *Asiacrypt '01*, LNCS 2248, pages 514–532. Springer-Verlag, Berlin, 2001.
8. D. Boneh and B. Waters. Compact group signatures without random oracles. In *Eurocrypt '06*, LNCS 4004, pages 427–444. Springer-Verlag, Berlin, 2006.

9. J. Camenisch, S. Hohenberger, and A. Lysyanskaya. Compact E-cash. In *Eurocrypt '05*, LNCS 3494, pages 302–321. Springer-Verlag, Berlin, 2005.

10. J. Camenisch and A. Lysyanskaya. Dynamic Accumulators and Application to Efficient Revocation of Anonymous Credentials. In *Crypto '02*, LNCS 2442, pages 61–67. Springer-Verlag, Berlin, 2002.

11. J. Camenisch and A. Lysyanskaya. Signature Schemes and Anonymous Credentials from Bilinear Maps. In *Crypto '04*, LNCS 3152, pages 52–72. Springer-Verlag, Berlin, 2004.

12. D. Chaum and E. van Heyst. Group Signatures. In *Eurocrypt '91*, LNCS 547, pages 257–265. Springer-Verlag, Berlin, 1992.

13. C. Delerablée and D. Pointcheval. Dynamic Fully Anonymous Short Group Signatures. In P. Q. Nguyen, editor, *Vietcrypt '06*, LNCS, Hanoi, Vietnam, 2006. Springer-Verlag, Berlin. Full version available from http://www.di.ens.fr/users/pointche/.

14. P. A. Fouque and D. Pointcheval. Threshold Cryptosystems Secure against Chosen-Ciphertext Attacks. In *Asiacrypt '01*, LNCS 2248. Springer-Verlag, Berlin, 2001.

15. S. Galbraith and V. Rotger. Easy Decision-Diffie-Hellman Groups. *LMS Journal of Computation and Mathematics*, 7:201–218, 2004.

16. A. Kiayias and M. Yung. Extracting Group Signatures from Traitor Tracing Schemes. In *Eurocrypt '03*, LNCS 2656, pages 630–648. Springer-Verlag, Berlin, 2003.

17. A. Kiayias and M. Yung. Group Signatures with Efficient Concurrent Join. In *Eurocrypt '05*, LNCS 3494, pages 198–214. Springer-Verlag, Berlin, 2005.

18. M. Naor and M. Yung. Universal One-Way Hash Functions and Their Cryptographic Applications. In *Proc. of the 21st STOC*, pages 33–43. ACM Press, New York, 1989.

19. L. Nguyen and R. Safavi-Naini. Efficient and Provably Secure Trapdoor-free Group Signature Schemes from Bilinear Pairings. In *Asiacrypt '04*, LNCS 3329, pages 372–386. Springer-Verlag, Berlin, 2004.

20. P. Paillier. Public-Key Cryptosystems Based on Discrete Logarithms Residues. In *Eurocrypt '99*, LNCS 1592, pages 223–238. Springer-Verlag, Berlin, 1999.

21. D. Pointcheval and J. Stern. Security Arguments for Digital Signatures and Blind Signatures. *Journal of Cryptology*, 13(3):361–396, 2000.

22. C. Rackoff and D. R. Simon. Non-Interactive Zero-Knowledge Proof of Knowledge and Chosen Ciphertext Attack. In *Crypto '91*, LNCS 576, pages 433–444. Springer-Verlag, Berlin, 1992.

Formalizing Human Ignorance
Collision-Resistant Hashing Without the Keys

Phillip Rogaway[1,2]

[1] Dept. of Computer Science, University of California, Davis, California 95616, USA
[2] Dept. of Computer Science, Chiang Mai University, Chiang Mai 50200, Thailand

Abstract. There is a rarely mentioned foundational problem involving collision-resistant hash-functions: common constructions are keyless, but formal definitions are keyed. The discrepancy stems from the fact that a function $H\colon \{0,1\}^* \to \{0,1\}^n$ *always* admits an efficient collision-finding algorithm, it's just that us human beings might be unable to write the program down. We explain a simple way to sidestep this difficulty that avoids having to key our hash functions. The idea is to state theorems in a way that prescribes an explicitly-given reduction, normally a black-box one. We illustrate this approach using well-known examples involving digital signatures, pseudorandom functions, and the Merkle-Damgård construction.

1 Introduction

FOUNDATIONS-OF-HASHING DILEMMA. In cryptographic practice, a collision-resistant hash-function (an object like SHA-1) maps arbitrary-length strings to fixed-length ones; it's an algorithm $H\colon \{0,1\}^* \to \{0,1\}^n$ for some fixed n. But in cryptographic theory, a collision-resistant hash-function is always *keyed*; now $H\colon \mathcal{K} \times \{0,1\}^* \to \{0,1\}^n$ where each $K \in \mathcal{K}$ names a function $H_K(\cdot) = H(K, \cdot)$. In this case H can be thought of as a *collection* or *family* of hash functions $H = \{H_K\colon K \in \mathcal{K}\}$, each key (or *index*) $K \in \mathcal{K}$, naming one.[1]

Why should theoretical treatments be keyed when practical constructions are not? The traditional answer is that a rigorous treatment of collision resistance for unkeyed hash-functions just doesn't work. At issue is the fact that for *any* function $H\colon \{0,1\}^* \to \{0,1\}^n$ there is *always* a simple and compact algorithm that outputs a collision: the algorithm that has one "hardwired in." That is, by the pigeonhole principle there must be distinct strings X and X' of length at most n such that $H(X) = H(X')$, and so there's a short and fast program that outputs such an X, X'. The difficulty, of course, is that us human beings might not *know* any such pair X, X', so no one can actually write the program down.

Because of the above, what is meant when someone says that a hash function $H\colon \{0,1\}^* \to \{0,1\}^n$ is collision resistant cannot be that there is no efficient

---

[1] (a) We call K a key, but it is *not* secret; one chooses K from $\mathcal{K}$ and then makes it public. (b) Writing $H\colon \mathcal{K} \times \{0,1\}^* \to \{0,1\}^n$ assumes a concrete-security formalization; early formalizations were instead asymptotic. We'll discuss both. (c) Alternative terms for *collision-resistant* are *collision-free* and *collision-intractable*.

P.Q. Nguyen (Ed.): VIETCRYPT 2006, LNCS 4341, pp. 211–228, 2006.

adversary that outputs a collision in H. What is meant is that there is no efficient algorithm *known to man* that outputs a collision in H. But such a statement would seem to be unformalizable—outside the realm of mathematics. One can't hope to construct a meaningful theory based on what Xiaoyun Wang [28,29] does or doesn't know. Regarding a hash function like SHA-1 as a random element from a family of hash functions has been the traditional way out of this quandary.

Let us call the problem we've been discussing the *foundations-of-hashing dilemma*. The question is how to state definitions and theorems dealing with collision-resistant hashing in a way that makes sense mathematically, yet accurately reflects cryptographic practice. The treatment should respect our understanding that what makes a hash function collision resistant is *humanity's* inability to find a collision, not the computational complexity of printing one.

OUR CONTRIBUTIONS. First, we bring the foundations-of-hashing dilemma out into the open. To the best of our knowledge, the problem has never received more than passing mention in any paper. Second, we resolve the dilemma. We claim that an answer has always been sitting right in front of us, that there's never been any real difficulty with providing a rigorous treatment of unkeyed collision-resistant hash-functions. Finally, we reformulate in a significantly new way three fundamental results dealing with collision-resistant hashing.

Suppose a protocol Π uses a collision-resistant hash-function H. Conventionally, a theorem would be given to capture the idea that the existence of an effective adversary A against Π implies the existence of an effective adversary C against H. But this won't work when we have an unkeyed $H\colon \{0,1\}^* \to \{0,1\}^n$ because such an adversary C will *always* exist. So, instead, the theorem statement will say that there is an explicitly given reduction: given an adversary A against Π there is a corresponding, explicitly-specified adversary C, as efficient as A, for finding collisions in H. So if someone knows how to break the higher-level protocol Π then they know how to find collisions in H; and if nobody can find collisions in H then nobody can break Π. In brief, our solution to the foundations-of-hashing dilemma is to recast results so as to assert the existence of an explicitly given reduction. We call this the *human-ignorance approach* (or, less colorfully, the *explicit-reduction approach*).

We illustrate the approach with three well-known examples. The first is the hash-then-sign paradigm, where a signature scheme is constructed by hashing a message and then applying an "inner" signature to the result. Our second example is the construction of an arbitrary-input-length PRF by hashing and then applying a fixed-input-length PRF. Our third example is the Merkle-Damgård construction, where a collision-resistant compression-function is turned into a collision-resistant hash-function. In all cases we will give a simple theorem that captures the security of the construction despite the use of an unkeyed formalization for the underlying hash function.

We provide a concrete-security treatment for all the above. Giving our hash functions a security parameter and then looking at things asymptotically would only distance us, we feel, from widely-deployed, real-world hash-functions. That said, we will also point out that unkeyed hash-functions work fine in the asymptotic setting for the case of *uniform* adversaries. One eliminates keys but not the security parameter, making it the length of the hash-function's output.

RELATED WORK. The rigorous treatment of collision-resistant hash-functions begins with Damgård [7]. A concrete-security treatment of these objects is given by Bellare, Rogaway, and Shrimpton [3,26]. Practical and widely-deployed cryptographic hash-functions were first developed by Rivest [25] and later constructions, like SHA-1 [21], have followed his approach. Bellare et al.'s [1, Theorem 4.2] is an early example of an explicitly constructive provable-security theorem-statement. Using a simulator to model what an adversary must know or be able to do is from Goldwasser, Micali, and Rackoff [14], while black-box reductions come from Goldreich, Krawczyk, and Oren [22,12,11]. Brown [5, see footnote 10] and Devanbu et al. [10] prove the security of a protocol that employs an unkeyed hash-function by constructively transforming a successful adversary against it into a successful collision-finding one. Using such a transformation to evidence a hash-function-based protocol's security goes back to Merkle [19,17]. The option of speaking about reductions as a way of not having to key a hash function is hinted at in footnote 5 of Halevi and Krawczyk [16]. In general, it is well understood that one can rephrase provable-security results as assertions about explicitly given reductions, and probably a few researchers have understood, at some level, that this can be used to make formal sense of unkeyed hash-functions. What we do in this paper is to raise these ideas beyond the level of footnotes, offhand comments, and undocumented folklore.

2   Keyed Hash-Functions

We first give a *conventional* definition, in the concrete-security setting, for a (keyed) collision-resistant hash-function. Beginning with the syntax, a *keyed hash-function* is a pair of algorithms $(\mathcal{K}, H)$, the first probabilistic and the second deterministic. Algorithm $\mathcal{K}$, the *key-generation algorithm*, takes no input and produces a string K, the *key*. As a special case, $\mathcal{K}$ uniformly samples from a finite set, the *key space*, also denoted $\mathcal{K}$. Algorithm H takes as input a string K, the *key*, and a string X, the *message*, and it outputs a string of some fixed length n, *the output length*, or the distinguished value $\perp$. We write $H_K(X)$ for $H(K, X)$. We assume there is a set $\mathcal{X}$, the *message space*, such that $H_K(X) = \perp$ iff $X \notin \mathcal{X}$. We assume that $\mathcal{X}$ contains some string of length greater than n and that $X \in \mathcal{X}$ implies every string of length $|X|$ is in $\mathcal{X}$. We write a hash function as $H \colon \mathcal{K} \times \mathcal{X} \to \{0, 1\}^n$ instead of saying "the keyed hash-function $(\mathcal{K}, H)$ with message space $\mathcal{X}$ and output length n." Hash functions and all other algorithms in this paper are given by code relative to some fixed and reasonable encoding.

We define hash functions as algorithms, not functions, to enable providing them as input to other algorithms and speaking of their computational complexity. But a hash function $H \colon \mathcal{K} \times \mathcal{X} \to \{0, 1\}^n$ induces a function H from $\mathcal{K} \times \mathcal{X}$ to $\{0, 1\}^n$, where $\mathcal{K}$ is now the support of the key-generation algorithm, and usually it is fine to regard the hash function as being this function.

To measure the collision-resistance of hash function $H\colon \mathcal{K} \times \mathcal{X} \to \{0,1\}^n$ let C (for *collision-finder*) be an adversary, meaning, in this case, an algorithm that takes in one string (the key) and outputs a pair of strings (the purported collision). We let the *advantage* of C in finding collisions in H be the real number

$$\mathbf{Adv}_H^{\mathrm{coll}}(C) = \Pr[K \xleftarrow{\$} \mathcal{K};\ (X, X') \xleftarrow{\$} C(K)\colon\ X \neq X' \text{ and } H_K(X) = H_K(X')]$$

that measures the chance that C finds a collision in $H_K = H(K, \cdot)$ if a random key K is provided to it. Above and henceforth we assume that an adversary will never output a string outside the message space $\mathcal{X}$ of the hash function it is attacking (that is, $H_K(X) = H_K(X') = \perp$ never counts as a collision).

As usual, an advantage of 1 means that C does a great job (it always finds a collision) while an advantage of 0 means that C does a terrible job (it never finds a collision). Since we are in the concrete-security setting we do not define any absolute (yes-or-no) notion for H being coll-secure; instead, we regard a hash function H as good only to the extent that reasonable adversaries C can obtain only small advantage $\mathbf{Adv}_H^{\mathrm{coll}}(C)$. In order to obtain a useful theory, "reasonable" and "small" need never be defined.

TRYING TO REGARD FUNCTIONS LIKE SHA-1 AS KEYED. How can a real-world hash-function like SHA-1 be seen as fitting into the framework above? One possibility is that the intended key is the initial chaining vector; the constant $K = 67452301\ \mathrm{EFCDAB89}\ 98\mathrm{BADCFE}\ 10325476\ \mathrm{C3D2E1F0}$ can be regarded as the key. In this case the key space is $\mathcal{K} = \{0,1\}^{160}$ and what NIST did in choosing SHA-1 was to randomly sample from this set. The problem with this viewpoint is that, first of all, NIST never indicated that they did any such thing. Indeed the constant K above does not "look" random (whatever that might mean), and it seems as though the specific constant should hardly matter: likely any method that would let one construct collisions in SHA-1 with respect to the actual K-value would work for other K-values, too.

A second way one might regard SHA-1 as keyed is to say that NIST, in designing SHA-1, considered some universe of hash functions $\{H_K\ :\ K \in \mathcal{K}\}$ and randomly selected this one hash function, SHA-1, from it. But, once again, NIST never indicated that they did any such thing; all we know is that they selected this *one* hash function. And it's not clear what $\mathcal{K}$ would even be in this case, or what H_K would be for "other" functions in the family.

Fundamentally, both explanations seem disingenuous. They make random sampling a crucial element to a definition when no random sampling ostensibly took place. They disregard the basic intuition about what SHA-1 is supposed to be: a fixed map that people shouldn't be able to find collisions in. And they distance the definition from the elegantly simple goal of the cryptanalyst: publish a collision for the (one) function specified by NIST.

3   Unkeyed Hash-Functions

An *unkeyed hash-function* is a deterministic algorithm H that takes as input a string X, the *message*, and outputs a string of some fixed length n, *the output* *length*, or the distinguished value $\perp$. The *message space* of H is the set $\mathcal{X} = \{X \in \{0,1\}^* : H(X) \neq \perp\}$. We assume that $\mathcal{X}$ contains some string of length greater than n and that $X \in \mathcal{X}$ implies every string of length $|X|$ is in $\mathcal{X}$. We will write a hash function as $H: \mathcal{X} \rightarrow \{0,1\}^n$, or simply H, instead of saying "the unkeyed hash-function H with message space $\mathcal{X}$ and output length n."

Let C be an adversary for attacking $H: \mathcal{X} \rightarrow \{0,1\}^n$, meaning an algorithm that, with no input, outputs a pair of strings X and X' in $\mathcal{X}$. We let the *advantage* of C in finding collisions in H be the real number

$$\mathbf{Adv}_H^{\text{col}}(C) = \Pr[(X, X') \xleftarrow{\$} C : X \neq X' \text{ and } H(X) = H(X')]$$

that measure the chance that C finds a collision. Note the spelling of superscript *col* verses the earlier *coll* (the number of l's is the number of arguments to H).

Following the discussion in the Introduction, we observe that for any unkeyed hash-function H there is an efficient algorithm C (it runs in cn time and takes cn bits to write down, for some small c) for which $\mathbf{Adv}_H^{\text{col}}(C) = 1$. We're not going to let that bother us.

4   Three Styles of Provable-Security Statements

PROVABLE-SECURITY FORMULATIONS. Let Π be a cryptographic protocol that employs a (keyed or unkeyed) hash function H. Imagine, for now, that H is the only cryptographic primitive that Π employs. To prove security of Π using a reduction-based approach and assuming the collision-resistance of H one would typically make a theorem statement that could be paraphrased like this:

existential form (C0): If there's an effective algorithm A for attacking protocol Π then there's an effective algorithm C for finding collisions in H.

When cryptographic reductions were first introduced [13], theorems were stated with this kind of existential-only guarantee. To this day, people almost always state their provable-security results in such a manner.

Formalizing statement C0 works fine in the keyed setting but not in the unkeyed one, because, there, the conclusion vacuously holds. But in the unkeyed setting we can switch to a theorem statement that could be paraphrased as:

code-constructive form (C1): If you know an effective algorithm A for attacking protocol Π then you know an effective algorithm C for finding collisions in H.

We are asserting the existence of a known "compiler" that turns A into C. Now your belief in the security of Π stems from the fact that if some human being can break Π then he can exhibit collisions in H. Statement C1 can be regarded as a constructive version of C0. Continuing on this trajectory, we could say that it's enough to have access to A's functionality, you don't actually need the code:

blackbox-constructive form (C2): If you possess effective means A for attacking protocol Π then you possess effective means C for finding collisions in H.

Here, "possessing effective means" might mean owning a tamper-resistant device, or being able to run some big executable program, or it might even mean having a brain in your head that does some task well. Possessing effective means does not imply knowing the internal structure of those means; I might not know what happens within the tamper-resistant device, the big program, or in my own brain. Statement C2 is stronger than C1 because knowledge of an algorithm implies access to its functionality, but having access to an algorithm's functionality does not imply knowing how it works.

The main observation in this paper is that, in the concrete-security setting, it's easy to give provable-security results involving unkeyed hash-functions as long as you state your results in the code-constructive (C1) or blackbox-constructive (C2) format. In the asymptotic setting, all three formats work fine as long as you stick to uniform adversaries.

In high-level expositions, provable-security results are often summarized in what would appear to be a code-constructive or blackbox-constructive manner; people say things like "our result shows that if someone could break this signature scheme then he could factor large composite numbers." But when we write out our theorem statements, it has been traditional to adopt the existential format. Usually the proof is constructive but the theorem statement is not.

FORMALIZING C1 AND C2. In the next section we'll formalize C1, in an example setting, by asking for an explicitly given algorithm C that, given the code for A (and also H and Π) provides us our collision finder C. We likewise formalize C2, in three example settings, by asking for an explicitly given algorithm C that, given black-box access to A (and also H and Π) is itself our collision finder.

When an algorithm C has black-box access to an algorithm F we write the latter as a subscriptor superscript, C_F or C^F. We do not allow for C to see or control the internal coins of F; when C runs F, the latter's coins are random and externally provided. We do not object to C resetting F, so long as fresh (secret) coins are issued to it each time that it is run.

RESOURCE ACCOUNTING. Let F be an algorithm (possibly stateful, probabilistic, and itself oracle-querying). The algorithm F might be provided as an oracle to some other algorithm. Let $t_F(\ell)$ be the maximum amount of time (in a conventional, non-blackbox model) to compute F on strings that total ℓ or fewer bits (but count the empty string ε as having length 1). We simplify to t_F for an overall maximum. Let ℓ_F be the maximum of the total number of bits read or written by F (over F's input, output, oracle queries, and their responses) (but regard ε as having length 1). Let q_F be the maximum number of queries made by F before it halts (but no less than 2, to simplify theorem statements). We assume that all algorithms halt after some bounded amount of time. When an algorithm A calls out to an oracle for F, we charge to A the time to compute F (even though the internal computation of F seems, to the caller, unit time).

As an example of the above, for a keyless hash-function $H\colon \{0,1\}^* \to \{0,1\}^n$ we have that $t_H(\ell)$ is the maximal amount of time to compute H on any sequence of inputs $X_1, \ldots, X_q$ comprising ℓ total bits (where $X_i = \varepsilon$ counts as 1-bit). As a second example, for an adversary A attacking a signature scheme, the number ℓ_A

includes the length of the public-key provided to A, the length of the signing queries that A asks, the length of the signatures A gets in response, and the length A's forgery attempt. Since we insisted that A is bounded-time, if it is provided an overly-long input or oracle response, it should only read (and is only charged for) a bounded-length prefix.

5   Hash-Then-Sign Signatures

The usual approach for digital signatures, going back to Rabin [23], is to sign a message by first hashing it and then calling an underlying signature scheme. The purpose of this *hash-then-sign* approach is two-fold. First, it extends the domain of the "inner" signature scheme from $\{0,1\}^n$ to $\{0,1\}^*$ (where the hash-function's output is n bits). Second, it may improve security by obscuring the algebraic structure of the inner signature scheme. We focus only on the first of these intents, establishing the folklore result that the hash-then-sign paradigm securely extends the domain of a signature scheme from $\{0,1\}^n$ to $\{0,1\}^*$. Our purpose is not only to prove this (admittedly simple) result, but also to illustrate the human-ignorance approach for dealing with collision-resistant hash-functions.

First we establish the notation, using concrete-security definitions. A *signature scheme* is a three-tuple of algorithms $\Pi = (Gen, Sign, Verify)$. Algorithm Gen is a probabilistic algorithm that, with no input, outputs a pair of strings (PK, SK). (One could, alternatively, assume that Gen takes input of a security parameter k.) Algorithm $Sign$ is a probabilistic algorithm that, on input (SK, X), outputs either a string $\sigma \xleftarrow{\$} Sign(SK, X)$ or the distinguished value $\perp$. We require the existence of a *message space* $\mathcal{X} \subseteq \{0,1\}^*$ such that, for any SK, we have that $\sigma \xleftarrow{\$} Sign(SK, X)$ is a string exactly when $X \in \mathcal{X}$. We insist that $\mathcal{X}$ contain all strings of a given length if it contains any string of that length. Algorithm $Verify$ is a deterministic algorithm that, on input (PK, X, σ), outputs a bit. We require that if $(PK, SK) \xleftarrow{\$} Gen$ and $X \in \mathcal{X}$ and $\sigma \xleftarrow{\$} Sign(SK, X)$ then $Verify(PK, X, \sigma) = 1$. We sometimes write $Sign_{SK}(X)$ and $Verify_{PK}(X, \sigma)$ instead of $Sign(SK, X)$ and $Verify(PK, X, \sigma)$.

Let B be an adversary and $\Pi = (Gen, Sign, Verify)$ a signature scheme. Define $\mathbf{Adv}_{\Pi}^{\text{sig}}(B) = \Pr[(PK, SK) \xleftarrow{\$} Gen : B^{Sign_{SK}(\cdot)}(PK) \text{ forges}]$ where B is said to *forge* if it outputs a pair (X, σ) such that $Verify_{PK}(X, \sigma) = 1$ and B never asked a query X during its attack.

We now define the hash-then-sign construction. Let $H: \{0,1\}^* \to \{0,1\}^n$ be an unkeyed hash-function and let $\Pi = (Gen, Sign, Verify)$ be a signature scheme with message space of at least $\{0,1\}^n$. Define from these primitives the signature scheme $\Pi^H = (Gen, Sign^H, Verify^H)$ by setting $Sign_{SK}^H(X) = Sign_{SK}(H(X))$ and $Verify_{PK}^H(X, \sigma) = Verify_{PK}(H(X), \sigma)$. The message space for Π^H is $\{0,1\}^*$.

We are now ready to state a first theorem that describes the security of the hash-then-sign paradigm.

Theorem 1 (hash-then-sign, unkeyed, concrete, C1-form). *There exist algorithms* $\mathcal{B}$ *and* $\mathcal{C}$, *explicitly given in the proof of this theorem, such that for any unkeyed hash-function* $H: \{0,1\}^* \to \{0,1\}^n$, *signature scheme* $\Pi = (Gen, Sign, Verify)$ *with message space at least* $\{0,1\}^n$, *and adversary* A, *adversaries* $B = \mathcal{B}(\langle A, H \rangle)$ *and* $C = \mathcal{C}(\langle A, H, \Pi \rangle)$ *satisfy*

$$\mathbf{Adv}_{\Pi}^{\mathrm{sig}}(B) + \mathbf{Adv}_{H}^{\mathrm{col}}(C) \geq \mathbf{Adv}_{\Pi^H}^{\mathrm{sig}}(A).$$

Adversary B *runs in time at most* $t_A + t_H(\ell_A) + t_{Sign}(nq_A) + c(\ell_A + nq_A)$ *and asks at most* q_A *queries entailing at most* $\ell_A + n$ *bits. Adversary* C *runs in time at most* $t_A + t_{Gen} + t_H(\ell_A) + t_{Sign}(nq_A + n) + c(\ell_A + nq_A) \lg(q_A)$. *Functions* $\mathcal{B}$ *and* $\mathcal{C}$ *run in time* c *times the length of their input. The value* c *is an absolute constant implicit in the proof of this theorem.* ◇

The theorem says that if you know the code for A, H, and Π then you know the code for B and C. You know that code because it's given by *reduction functions* $\mathcal{B}$ and $\mathcal{C}$. These reduction functions are explicitly specified in the proof of the theorem. Reduction function $\mathcal{B}$ takes in an encoding of A and H and outputs the code for adversary B. Reduction function $\mathcal{C}$ takes in an encoding of A, H, and $\Pi = (Gen, Sign, Verify)$ and outputs the code for adversary C.

One might argue that we don't really care that B is constructively given—we might have demanded only that it exist whenever A does. But it seems simpler and more natural to demand that both adversaries B and C be constructively given when we are demanding that one adversary be. Besides, it is nicer to conclude *you know a good algorithm to break* Π than to conclude *there exists a good algorithm to break* Π; it would, in fact, be an unsatisfying proof that actually gave rise to a nonconstructive attack on the inner signature scheme Π.

Theorem 1 does not capture statement C2 because access to the functionality of adversary A might be more limited than possessing its code. To capture the intent of statement C2, we can strengthen our theorem as follows:

Theorem 2 (hash-then-sign, unkeyed, concrete, C2-form). *There exist adversaries* B *and* C, *explicitly given in the proof of this theorem, such that for any unkeyed hash-function* $H: \{0,1\}^* \to \{0,1\}^n$, *signature scheme* $\Pi = (Gen, Sign, Verify)$ *with message space at least* $\{0,1\}^n$, *and adversary* A, *we have that*

$$\mathbf{Adv}_{\Pi}^{\mathrm{sig}}(B_{A,H}) + \mathbf{Adv}_{H}^{\mathrm{col}}(C_{A,H,\Pi}) \geq \mathbf{Adv}_{\Pi^H}^{\mathrm{sig}}(A). \tag{1}$$

Adversary B *runs in time at most* $t_A + t_H(\ell_A) + t_{Sign}(nq_A) + c(\ell_A + nq_A)$ *and asks at most* q_A *queries entailing at most* $\ell_A + n$ *bits. Adversary* C *runs in time at most* $t_A + t_{Gen} + t_H(\ell_A) + t_{Sign}(nq_A + n) + c(\ell_A + nq_A) \lg(q_A)$. *The value* c *is an absolute constant implicit in the proof of this theorem.* ◇

The theorem asserts the existence of an explicitly known forging adversary B (for attacking Π) and an explicitly known collision-finding adversary C (for attacking H), at least one of which must do well if the original adversary A does well (in attacking Π^H). Algorithm C employs A, as well as H, Gen, $Sign$, and $Verify$, in a black-box manner. (Writing $\Pi = (Gen, Sign, Verify)$ as a subscript to C means giving each component algorithm as an oracle.) We may not care that the dependency on H, Gen, $Sign$, and $Verify$ is black-box, for there is no question there about having access to the code, but it seems simpler to demand that all dependencies be black-box when we require one to be. As with Theorem 1, the final set of lines in Theorem 2 explain that the time and communications complexity of algorithms B and C is insignificantly more than that of A.

Proof (of Theorem 2 and then Theorem 1). In the following exposition, computations of A, H, Gen, $Sign$, and $Verify$ are done via oracle queries.

Construct collision-finding adversary $C_{A,H,\Pi}$ as follows. It calls Gen to determine output $(PK, SK) \xleftarrow{\$} Gen$. Then it calls adversary A on input PK. When A makes its ith query, X_i, a request to sign the string X_i, algorithm C calls H to compute $x_i = H(X_i)$, it calls $Sign$ on input x_i to compute $\sigma_i \xleftarrow{\$} Sign_{SK}(x_i)$, and it returns σ_i in answer to A's query. When A halts with output (X_*, σ_*) algorithm C invokes H to compute $x_* = H(X_*)$. If x_* is equal to x_i for some prior i, and $X_* \neq X_i$, then algorithm C outputs the collision (x_i, x_*) and halts. Otherwise, algorithm C fails; it outputs an arbitrary pair of strings. The reader can check that C has the claimed time complexity. The log-term accounts for using a binary search tree, say, to lookup if x_* is equal to some prior x_i.

Construct forging-adversary $B^{Sign}_{A,H}(PK)$ as follows. Algorithm B, which is provided a string PK, runs black-box adversary A on input of PK. When A makes its ith query, X_i, a request for a signature of X_i, algorithm B uses its oracle H to compute $x_i = H(X_i)$. It then uses its $Sign$-oracle to compute $\sigma_i \leftarrow Sign(x_i)$. It returns σ_i in answer to the adversary A. When A halts with output (X_*, σ_*) algorithm B uses its H-oracle to compute $x_* = H(X_*)$. Algorithm B halts with output (x_*, σ_*). The reader can check that B has the claimed time and communications complexity. (The t_{Sign} term is because of our convention to consistently charge algorithms for their oracle calls.)

We must show (1). Let a be the probability that A, in carrying out its attack in the experiment defining $\mathbf{Adv}^{\text{sig}}_{\Pi^H}(A)$, outputs a valid forgery (X_*, σ_*) where $H(X_*) = H(X_i)$ for some i. Let b be the probability that A, in carrying out its attack in the experiment defining $\mathbf{Adv}^{\text{sig}}_{\Pi^H}(A)$, outputs a valid forgery (X_*, σ_*) where $H(X_*) \neq H(X_i)$ for all i. Then $a + b = \mathbf{Adv}^{\text{sig}}_{\Pi^H}(A)$. We also have that $\mathbf{Adv}^{\text{col}}_H(C_{A,H,\Pi}) \geq a$ and $\mathbf{Adv}^{\text{sig}}_{\Pi}(B_{A,H}) \geq b$, establishing Theorem 2.

As for Theorem 1, the reduction functions $\mathcal{B}$ and $\mathcal{C}$ are what is spelled out in the definition of B and C, above, except that computation by code replaces oracle invocations. (One can now see why we have selected our earlier conventions about how to charge-out oracle calls: it is convenient that it has no impact on the running time if one imagines calling an oracle for H, say, verses running that code oneself.) It is a simple, linear-time algorithm that takes in A and H (which are code) and outputs B (which is also code), or that produces C from A, H and each component of Π. ∎

For the remainder of our examples we will use the stronger, black-box style of theorem statement corresponding to Statement C2 and Theorem 2.

6   Hash-Then-PRF

As a second example of using our framework we consider a symmetric-key analog of hash-then-sign, where now we aim to extend the domain of a pseudorandom function (PRF) from $\{0,1\}^n$ to $\{0,1\}^*$. The algorithm, which we consider to be folklore, is to hash the message X and then apply a PRF, setting $F_K^H(X) = F_K(H(X))$ where $H\colon \{0,1\}^* \to \{0,1\}^n$ is the hash function and $F\colon \mathcal{K} \times \{0,1\}^n \to \{0,1\}^m$ is the PRF. A special case of this construction is using a hash function $H\colon \{0,1\}^* \to \{0,1\}^n$ and an n-bit blockcipher to make an arbitrary-input-length message authentication code (MAC). A second special-case is using a hash function $H\colon \{0,1\}^* \to \{0,1\}^{2n}$ and the two-fold CBC MAC of an n-bit blockcipher to make an arbitrary-input-length MAC.

First the definitions, following works like [2]. An (m-bit output) pseudorandom function (PRF) is an algorithm $F\colon \mathcal{K} \times \mathcal{X} \to \{0,1\}^m$ where $\mathcal{K}$ and $\mathcal{X}$ are sets of strings. We assume that there is an algorithm associated to F, which we also call $\mathcal{K}$, that outputs a random element of $\mathcal{K}$. For $\mathcal{X}, \mathcal{Y} \subseteq \{0,1\}^*$ and $\mathcal{Y}$ finite, let $\mathrm{Func}(\mathcal{X}, \mathcal{Y})$ be the set of all functions from $\mathcal{X}$ to $\mathcal{Y}$. Endow this set with the uniform probability distribution for each input. For a PRF $F\colon \mathcal{K} \times \mathcal{X} \to \{0,1\}^m$ let $\mathbf{Adv}_F^{\mathrm{prf}}(B) = \Pr[K \xleftarrow{\$} \mathcal{K}\colon B^{F_K(\cdot)} \Rightarrow 1] - \Pr[f \xleftarrow{\$} \mathrm{Func}(\mathcal{X}, \{0,1\}^m)\colon B^{f(\cdot)} \Rightarrow 1]$. The following quantifies the security of the hash-then-PRF construction F^H.

Theorem 3 (hash-then-PRF, unkeyed, concrete, C2-form). *There exist adversaries B and C, explicitly given in the proof of this theorem, such that for any unkeyed hash-function $H\colon \{0,1\}^* \to \{0,1\}^n$, pseudorandom function $F\colon \mathcal{K} \times \{0,1\}^n \to \{0,1\}^m$, and adversary A,*

$$\mathbf{Adv}_F^{\mathrm{prf}}(B_{A,H}) + \mathbf{Adv}_H^{\mathrm{col}}(C_{A,H,F}) \geq \mathbf{Adv}_{F^H}^{\mathrm{prf}}(A). \tag{2}$$

Adversary B runs in time at most $t_A + t_H(\ell_A) + t_F(nq_A) + c(\ell_A + nq_A + mq_A)$ and asks at most q_A queries entailing at most ℓ_A bits. Adversary C runs in time at most $t_A + t_H(\ell_A) + c(\ell_A + nq_A + mq_A + t_{\mathcal{K}}) \lg(q_A)$. The value c is an absolute constant implicit in the proof of this theorem. ◇

Proof. Construct collision-finding algorithm $C_{A,H,F}$ as follows. The algorithm runs adversary A, which is given by an oracle. When A makes its ith oracle query, X_i, algorithm C uses its H oracle to compute $x_i = H(X_i)$ and then, if $x_i \neq x_j$ for all $j < i$, adversary C returns a random $y_i \xleftarrow{\$} \{0,1\}^m$ in response to A's query. If $x_i = x_j$ for some $j < i$, adversary C returns $y_i = y_j$. When A finally halts, outputting a bit a, algorithm C ignores a and looks to see if there were distinct queries X_i and X_j made by A such that $x_i = x_j$. If there is such a pair, algorithm C outputs an arbitrary such pair (X_i, X_j) and halts. Otherwise, algorithm C *fails* and outputs an arbitrary pair of strings. The time of C is at most that which is stated in the theorem. Note that C does not actually depend on F beyond employing the values n and m.

Construct distinguishing algorithm $B_{A,H}^f$ as follows. It begins by running algorithm A, which is given by an oracle. When A makes its ith query, X_i, algorithm B computes $x_i = H(X_i)$ and then asks its f oracle x_i, obtaining return value $y_i = f(x_i)$. Algorithm B returns y_i to A. When A finally halts, outputting a bit a, algorithm B halts without output a. The resources of B are as given by the theorem statement.

We have that $\mathbf{Adv}_{F^H}^{\text{prf}}(A) - \mathbf{Adv}_F^{\text{prf}}(B_{A,H}) = \Pr[A^{F_K^H} \Rightarrow 1] - \Pr[A^R \Rightarrow 1] - \Pr[B_{A,H}^{F_K} \Rightarrow 1] + \Pr[B_{A,H}^{\rho} \Rightarrow 1]$ where $\rho \xleftarrow{\$} \mathrm{Func}(n,m)$ and $R \xleftarrow{\$} \mathrm{Func}(\{0,1\}^*, m)$ and $K \xleftarrow{\$} \mathcal{K}$. Now, from our definition of B, the first and third addend are equal, $\Pr[A^{F_K^H} \Rightarrow 1] = \Pr[B_{A,H}^{F_K} \Rightarrow 1]$, and so $\mathbf{Adv}_{F^H}^{\text{prf}}(A) - \mathbf{Adv}_F^{\text{prf}}(B_{A,H}) = \Pr[B_{A,H}^{\rho} \Rightarrow 1] - \Pr[A^R \Rightarrow 1]$.

Let C be the event that, during B's attack, there are distinct queries X_i and X_j made by B such that $H(X_i) = H(X_j)$. Let $c = \Pr[\mathsf{C}]$ where the probability is taken over B's oracle being a random function $\rho \xleftarrow{\$} \mathrm{Func}(n,m)$. Observe that, from C's definition, $c = \mathbf{Adv}_H^{\text{col}}(C_{A,H,F})$. Now note that $\Pr[B_{A,H}^{\rho} \Rightarrow 1] - \Pr[A^R \Rightarrow 1] \leq c$ because in the second experiment a random m-bit value is returned for each new X_i and in the first experiment a random m-bit value is returned for each new X_i except when $x_i = H(X_i)$ is identical to a prior $x_j = H(X_j)$. This establishes Equation (2). ∎

A result similar to Theorem 3, but for MACs instead of PRFs, can easily be established. That is, if $H\colon \{0,1\}^* \to \{0,1\}^n$ is an unkeyed hash-function and $\mathrm{MAC}\colon \{0,1\}^n \to \{0,1\}^m$ is a good MAC [2] then MAC^H is a good MAC. Here, as before, MAC^H is defined by $\mathrm{MAC}_K^H(M) = \mathrm{MAC}_K(H(M))$. The weaker assumption (F is a good MAC instead of a good PRF) suffices to get the weaker conclusion (F^H is a good MAC).

7   Merkle-Damgård Without the Keys

We adapt the Merkle-Damgård paradigm [8,18] to the unkeyed hash-function setting. To get a message space of $\{0,1\}^*$ and keep things simple we adopt the length-annotation technique known as Merkle-Damgård *strengthening*.

First we define the mechanism. Let $H\colon \{0,1\}^{b+n} \to \{0,1\}^n$ be an unkeyed hash-function, called a *compression function*, and define from it the unkeyed hash-function $H^*\colon \{0,1\}^* \to \{0,1\}^n$ as follows. On input $X \in \{0,1\}^*$, algorithm H^* partitions $\mathsf{pad}(X) = X \parallel 0^p \parallel [|X|]_b$ into b-bit strings $X_1 \cdots X_m$ where $p \geq 0$ is the least nonnegative number such that $|X| + p$ is a multiple of b and where $[|X|]_b$ is $|X| \bmod 2^b$ encoded as a b-bit binary number. Then, letting $Y_0 = 0^n$, say, define $Y_i = H(X_i \parallel Y_{i-1})$ for each $i \in [1 .. m]$ and let $H^*(X)$ return Y_m. Note that $\mathbf{Adv}_H^{\text{col}}(C) = \Pr[(X, X') \xleftarrow{\$} C : X \neq X' \text{ and } H(X) = H(X')]$ where C must output $X, X' \in \{0,1\}^{b+n}$. We now show that if H is a collision-resistant compression-function then H^* is a collision-resistant hash-function.

Theorem 4 (Merkle-Damgård, unkeyed, concrete, C2-form). *Fix positive numbers b and n. There exists an adversary C, explicitly given in the proof of this theorem, such that for any unkeyed hash-function H: $\{0,1\}^{b+n} \to \{0,1\}^n$ and any adversary A that outputs a pair of strings each of length less than 2^b,*

$$\mathbf{Adv}_H^{\mathrm{col}}(C_{A,H}) \geq \mathbf{Adv}_{H^*}^{\mathrm{col}}(A) \ . \tag{3}$$

Adversary C runs in time at most $t_A + (\ell_A/b + 4)t_H + c(\ell_A + b + n)$. The value c is an absolute constant implicit in the proof of this theorem. ◇

Proof. Construct the collision-finding adversary $C_{A,H}$ as follows. It runs the adversary A, which requires no inputs and halts with and output X, X', each string having fewer than 2^b bits. Swap X and X', if necessary, so that X is at least as long as X'. Adversary C then computes $X_1 \cdots X_m = \mathsf{pad}(X)$ and $X_1' \cdots X_{m'}' = \mathsf{pad}(X')$ where each X_i and X_j' is b-bits long. Using its H-oracle, adversary C computes Y_i-values by way of $Y_0 = 0^n$ and, for each $i \in [1..m]$, $Y_i = H(X_i \| Y_{i-1})$. It similarly computes Y_j'-values, defining $Y_0' = 0^n$ and $Y_j' = H(X_j' \| Y_{j-1}')$ for each $j \in [1..m']$. Now if $X = X'$ or $Y_m \neq Y_{m'}'$ then adversary C fails, outputting an arbitrary pair of strings. Otherwise, adversary C computes the largest value $i \in [1..m]$ such that $Y_i = Y_{i-\Delta}'$ but $X_i \| Y_{i-1} \neq X_{i-\Delta}' \| Y_{i-1-\Delta}'$ where $\Delta = m - m'$. (We prove in a moment that such an i exists.) Adversary C outputs the pair of strings $(X_i \| Y_{i-1}, X_{i-\Delta}' \| Y_{i-1-\Delta}')$, which collide under H.

We must show that this value of i, above, is well defined. To do so, distinguish two cases in which the adversary might succeed in finding a collision. For the first case, $|X| \neq |X'|$. In this case the definition of pad (together with the requirement that $|X|, |X'| < 2^b$) ensures that $X_m \neq X_m'$ and so we will have $i = m$ as the index for a collision. In the second case, $|X| = |X'|$ and so, in particular, $m = m'$ and $\Delta = 0$. Because $X \neq X'$ there is a largest value $j \in [1..m]$ such that $X_j \neq X_j'$. It must be the case that $Y_j = Y_j'$ because the messages X and X', being identical on later blocks, would otherwise yield $Y_m = Y_m'$. But $X_j \neq X_j'$ and $Y_j = Y_j'$ and so $j = i$ satisfies the definition above.

We have shown that whenever A outputs a collision of H^*, adversary $C_{A,H}$ outputs a collision of H. The running time of $C_{A,H}$ is as claimed (the $+4$ accounts for 0-padding and length annotation in the scheme), so we are done. ∎

8   Asymptotic Treatment of Unkeyed Hash Functions

DEFINITION. The traditional treatment of cryptographic hash-functions [7] is asymptotic. In this section we show that as long as one is willing to ask for security only against *uniform* adversaries, we don't need the keys in the asymptotic formalization of collision-resistant hash-functions either.

An *asymptotic-and-unkeyed hash-function* is a deterministic, polynomial-time algorithm H that takes as input an integer n, the *output length*, encoded in unary, and a string X, the *message*. It outputs either a string of length n or the distinguished value $\perp$. When we say that H is polynomial-time we mean that it is polynomial-time in its first input. We write H_n for the induced function

$H(1^n, \cdot)$. Define the *message space* of H_n as $\mathcal{X}_n = \{X \in \{0,1\}^* : H_n(X) \neq \bot\}$ and that of H as the indexed family of sets $\langle X_n : n \in \mathbb{N} \rangle$. We assume $X \in \mathcal{X}_n$ implies every string of length $|X|$ is in $\mathcal{X}_n$, and we assume that $\mathcal{X}_n$ contains a string of length exceeding n.

Let C be an adversary for attacking asymptotic-and-unkeyed hash-function H, meaning that C is an algorithm (*not* a family of circuits; we are in the uniform setting) that, on input 1^n, outputs a pair of strings $X, X' \in \mathcal{X}_n$. We let the *advantage* of C in finding collisions in H be the function (of n) defined by

$$\mathbf{Adv}_H^{\mathrm{col}}(C, n) = \Pr[(X, X') \xleftarrow{\$} C(1^n) : X \neq X' \text{ and } H_n(X) = H_n(X')]$$

measuring, for each n, the probability that $C(1^n)$ finds a collision in H_n. We say that H is *collision-resistant* if for every polynomial-time adversary C, the function $\mathbf{Adv}_H^{\mathrm{col}}(C, n)$ is negligible. As usual, function $\epsilon(n)$ is *negligible* if for all $c > 0$ there exists an N such that $\epsilon(n) < n^{-c}$ for all $n \geq N$.

AN ASYMPTOTIC TREATMENT OF HASH-THEN-SIGN. With a definition in hand it is easy to give an asymptotic counterpart for hash-then-sign, say. The existential (C0-style) statement would say that if Π is a secure signature scheme with message space $\langle \{0,1\}^n : n \in \mathbb{N} \rangle$ and H is a collision-resistant asymptotic-and-unkeyed hash-function with message space $\langle X_n \rangle$ then Π^H, the hash-then-sign construction using H and Π, is a secure signature scheme with message space $\langle X_n \rangle$. Details follow, beginning with the requisite definitions.

Now in the asymptotic setting [15], a *signature scheme* is a three-tuple of algorithms $\Pi = (Gen, Sign, Verify)$. Algorithm Gen is a probabilistic polynomial-time (PPT) algorithm that, on input 1^n, outputs a pair of strings (PK, SK). Algorithm $Sign$ is a PPT algorithm that, on input (SK, X), outputs either a string $\sigma \xleftarrow{\$} Sign(SK, X)$ or the distinguished value $\bot$. For each $n \in \mathbb{N}$ we require the existence of a *message spaces* $\mathcal{X}_n \subseteq \{0,1\}^*$ such that, for any SK that may be output by $Gen(1^n)$, we have that $\sigma \xleftarrow{\$} Sign(SK, X)$ is a string exactly when $X \in \mathcal{X}_n$. We insist that $\mathcal{X}_n$ contains all strings of a given length if it contains any string of that length. Algorithm $Verify$ is a deterministic polynomial-time algorithm that, on input (PK, X, σ), outputs a bit. We require that if $(PK, SK) \xleftarrow{\$} Gen(1^n)$ and $X \in \mathcal{X}_n$ and $\sigma \xleftarrow{\$} Sign(SK, X)$ then $Verify(PK, X, \sigma) = 1$. We sometimes write $Sign_{SK}(X)$ and $Verify_{PK}(X, \sigma)$ instead of $Sign(SK, X)$ and $Verify(PK, X, \sigma)$. The message space of Π is the collection $\langle \mathcal{X}_n : n \in \mathbb{N} \rangle$. Throughout, an algorithm is *polynomial time* if it is polynomial time in the length of its first input. Now let B be an adversary for a signature scheme $\Pi = (Gen, Sign, Verify)$ as above. Then define $\mathbf{Adv}_\Pi^{\mathrm{sig}}(B, n)) = \Pr[(PK, SK) \xleftarrow{\$} Gen(1^n) : B^{Sign_{SK}(\cdot)}(PK) \text{ forges}]$ where B is said to *forge* if it outputs a pair (X, σ) such that $Verify_{PK}(X, \sigma) = 1$ and B never asked a query X during its attack. We say that Π is *secure* (in the sense of existential unforgeability under an adaptive chosen-message attack) if for any polynomial-time adversary B the function $\mathbf{Adv}_\Pi^{\mathrm{sig}}(B, n)$ is negligible.

Let $\Pi = (Gen, Sign, Verify)$ be a signature scheme (for the asymptotic setting) with message space $\langle \mathcal{M}_n \rangle$ where $\mathcal{M}_n \supseteq \{0,1\}^n$. In this case we say that the message space of Π is "at least" $\langle \{0,1\}^n \rangle$. Let H be an asymptotic-and-unkeyed hash-function with message space $\langle \mathcal{X}_n \rangle$. Then define the hash-then-sign construction $\Pi^H = (Gen, Sign^H, Verify^H)$ by setting $Sign^H_{SK}(X) = Sign_{SK}(H(X))$ and $Verify^H_{PK}(X, \sigma) = Verify_{PK}(H(X), \sigma)$. The message space for Π^H is the message space for H. The security of the construction is captured by the following theorem. We omit a proof because it only involves writing down the asymptotic counterpart to the proof of Theorem 2.

Theorem 5 (hash-then-sign, unkeyed, asymptotic, C0-form). *If Π is a secure signature scheme with message space at least $\langle \{0,1\}^n \rangle$ and H is a collision-resistant asymptotic-and-unkeyed hash-function having message space $\langle \mathcal{X}_n \rangle$ then Π^H is a secure signature scheme with message space $\langle \mathcal{X}_n \rangle$.* $\Diamond$

Comparing Theorem 5 with Theorem 1 or 2, note that in stepping back to the asymptotic setting we also reverted to the existential style of theorem statement. But these choices are independent; one can given explicitly constructive (C1- or C2-style) theorem statements for the asymptotic setting.

EXISTENCE AND CONSTRUCTIONS. We do not investigate the complexity assumption necessary to construct a collision-resistant asymptotic-and-unkeyed hash-function, but we do regard this as an interesting question. Natural constructions and cryptographic assumptions would seem to present themselves by adapting prior work like that in [7,27].

9   Discussion

Using unkeyed hash-functions is no more complex than using keyed ones. For ease of comparison, we recall Damgård's definition of a collection of collision-free hash-functions [7] in Appendix A, and we provide a keyed treatment of hash-then-sign, in the concrete-security setting, in Appendix B.

Some readers may instinctively feel that there is something fishy about the approach advocated in this paper. One possible source of uneasiness is that, under our concrete-security treatment, no actual definition was offered for when an unkeyed hash-function is collision-resistant. But concrete-security treatments of cryptographic goals *never* define an absolute notion for when a cryptographic object is secure. Similarly, it might seem fishy that, in the asymptotic setting, we restricted attention to uniform adversaries. We proffer that collision-resistance of an unkeyed output-length-parameterized hash-function makes intuitive sense, but only in the uniform setting. Regardless, we suspect that the greater part of any sense of unease stems from our community having internalized the belief that an unkeyed treatment of collision-resistance just cannot work. In Damgård's words, *Instead of considering just one hash function, we will consider families of them, in order to make a complexity theoretic treatment possible* [7]. This refrain has been repeated often enough to have become undisputed fact. But Damgård was thinking in terms of asymptotic complexity and nonuniform adversaries; when one moves away from this, and makes a modest shift in viewpoint about what our theorem statements should say, what was formerly impossible becomes not just possible, but easy.

Going further, one could make the argument that it is historical tradition that has made our hash functions keyed more than the specious argument from Section 1 about the infeasibility of formalizing what human beings do not know. When Damgård defined collision-resistance [7] we already had well-entrenched traditions favoring asymptotic notions, non-uniform security, number-theoretic constructions, assumptions like claw-free pairs, and existential-format (C0-style) theorem statements. These traditions point away from the human-ignorance approach. Besides, it was never Damgård's goal to demonstrate how to do provable-security cryptography with an unkeyed hash-function $H: \{0,1\}^* \to \{0,1\}^n$. While such hash functions were known (eg, [23,20,30]), they probably were not looked upon as suitable starting points for doing rigorous cryptographic work.

Protocols that use cryptographic hash-functions are often proved secure in the random-oracle (RO) model [4]. In such a case, when one replaces the RO-modeled hash-function H by some concrete function one would like to preserve the function's domain and range, $H: \mathcal{X} \to \mathcal{Y}$ for $\mathcal{X}, \mathcal{Y} \subseteq \{0,1\}^*$. So replacing a RO by a concrete hash-function *always* takes you away from the keyed-hash-function setting. Concretely, one can prove security for hash-then-sign in the RO model but one can't instantiate the RO with a keyed hash-function without changing the protocol first.

One could argue that mandating an explicitly-specified reduction is a sensible way to state provable-security results in general. After all, if a reduction actually were nonconstructive it would provide a less useful guarantee. That's because a constructive reduction says something meaningful about cryptographic practice *now*, independent of mathematical truth. A constructive statement along the lines of "if you know how to break this signature scheme then you know how to factor huge numbers" tells us that, right now, he who can do the one task can already do the other. If it takes 100 years until anyone can factor huge numbers then signature schemes that enjoy the constructive provable-security guarantee are guaranteed to protect against forgeries for all those intervening years.

The human-ignorance approach can be used for cryptographic goals beyond collision resistance. For example, one might assume of a blockcipher $E: \{0,1\}^k \times \{0,1\}^n \to \{0,1\}^n$ that nobody can find distinct (K,X) and (K',X') such that $E_K(X) \oplus X = E_{K'}(X') \oplus X'$. Unlike the ideal-cipher model, the assumption is meaningful for a concretely instantiated blockcipher.

The topic of this paper is largely about *language*: how, exactly, to express provable-security results. Some may interpret this to mean that the topic is insignificant, being *only* an issue of language. But language is key. In a case like this, language shapes our basic ideas, their development, and their utility.

In recent years, MD4-family hash functions (MD4, MD5, SHA-0, SHA-1, RIPEMD) have suffered an onslaught of successful attacks. This paper provides no guidance in how to recognize or build unkeyed hash-functions for which mankind will *not* find collisions. It only illustrates how, when you *do* have such a hash function in hand, you can formulate the security of a higher-level protocol that uses it, obtaining the usual benefits of provable-security cryptography.

Acknowledgments

Some ideas in this paper go back to long-ago discussions with Mihir Bellare. Well over a decade ago we talked about the significance of making theorem statements explicitly constructive, which we did, for example, in [1, Theorem 4.2]. Mihir also provided his typically astute comments on this paper's first draft. I also received good comments from Dan Brown and Tom Shrimpton. Jesse Walker, and others later on, asked me about the foundations-of-hashing dilemma (of course not in this language), motivating me to produce this writeup. Andy Okun pointed me to the work of Carlo Cipolla, who considers human ignorance from a rather different perspective [6]. This work was supported by NSF grant CCR-0208842 and a generous gift from Intel Corporation.

References

1. M. Bellare, R. Guérin, and P. Rogaway. XOR MACs: New methods for message authentication using finite pseudorandom functions. Full version of CRYPTO '95 paper. Available on-line from the author's web page.
2. M. Bellare, J. Kilian, and P. Rogaway. The security of the cipher block chaining message authentication code. *J. of Computer and System Sciences* (JCSS), vol. 61, no. 3, pp. 362–399, 2000.
3. M. Bellare and P. Rogaway. Collision-resistant hashing: towards making UOWHFs practical. *Advances in Cryptology – CRYPTO '97*, LNCS, Springer, 1997.
4. M. Bellare and P. Rogaway. Random oracles are practical: a paradigm for designing efficient protocols. *First ACM Conference on Computer and Communications Security* (CCS '93), ACM Press, pp. 62–73, 1993.
5. D. Brown. Generic groups, collision resistance, and ECDSA. *Designs, Codes and Cryptography*, vol. 35, no. 1, pp. 119–152, 2005.
6. C. Cipolla. *Le leggi fondamentali della stupidità* (The fundamental laws of human stupidity). In *Allegro ma non troppo con Le leggi fondamentali della stupidità*, Società editrice il Malino, Bologna, 1988.
7. I. Damgård. Collision free hash functions and public key signature schemes. *Advance in Cryptology – EUROCRYPT '87*, LNCS vol. 304, Springer, pp. 203–216, 1987.
8. I. Damgård. A design principle for hash functions. *Advances in Cryptology – CRYPTO '89*, LNCS vol. 435, Springer, 1990.
9. A. De Santis and M. Yung. On the design of provably secure cryptographic hash functions. *Advance in Cryptology – EUROCRYPT '90*, LNCS vol. 473, Springer, pp. 412–431, 1991.
10. P. Devanbu, M. Gertz, A. Kwong, C. Martel, G. Nuckolls, and S. Stubblebine. Flexible authentication of XML documents. *J. of Computer Security*, vol. 12, no. 6, pp. 841–864, 2004.
11. O. Goldreich and H. Krawczyk. On the composition of zero-knowledge proof systems. *SIAM Journal on Computing*, vol. 25, no. 1, pp. 169–192, Feb 1997.
12. O. Goldreich and Y. Oren. *Definitions and properties of zero-knowledge proof systems. J. of Cryptology*, vol. 7, no. 1, pp. 1–32, 1994.
13. S. Goldwasser and S. Micali. Probabilistic encryption. *J. Comput. Syst. Sci.*, vol. 28, no. 2, pp. 270–299, 1984. Earlier version in *STOC 82*.

14. S. Goldwasser, S. Micali, and C. Rackoff. The knowledge complexity of interactive proof systems. *SIAM Journal on Computing*, vol. 18, no. 1, pp. 186–208, 1989.
15. S. Goldwasser, S. Micali, and R. Rivest. A digital signature scheme secure against adaptive chosen-message attacks. *SIAM J. on Comp.*, vol. 17, pp. 281–308, 1988.
16. S. Halevi and H. Krawczyk. Strengthening digital signatures by randomized hashing. Manuscript dated 6 June 2006. Proceedings version in *Advances in Cryptology – CRYPTO 06*, LNCS, Springer, 2006.
17. R. Merkle. Method of providing digital signatures. US Patent #4,309,569, 1982.
18. R. Merkle. One way hash functions and DES. *Advances in Cryptology – CRYPTO 89*, LNCS vol. 435, Springer, pp. 428–446, 1990.
19. R. Merkle. Protocols for public key cryptosystems. *Proceedings of the 1980 IEEE Symposium on Security and Privacy*, IEEE Press, pp. 122–134, 1980.
20. S. Matyas, C. Meyer, and J. Oseas. Generating strong one-way functions with cryptographic algorithm. *IBM Tech. Disclosure Bulletin*, 27, pp. 5658–5659, 1985.
21. National Institute of Standards and Technology. FIPS PUB 180-2, Secure Hash Standard, Aug 1, 2002.
22. Y. Oren. On the cunning power of cheating verifiers: some observations about zero-knowledge proofs. *28th Annual Symposium on the Foundations of Computer Science* (FOCS 87), IEEE Press, pp. 462–471, 1987.
23. M. Rabin. Digital signatures. In *Foundations of secure computation*, R. DeMillo, D. Dobkin, A. Jones, and R. Lipton, editors, Academic Press, pp. 155–168, 1978.
24. O. Reingold, L. Trevisan, and S. Vadhan. Notions of reducibility between cryptographic primitives. *Theory of Cryptography Conference, TCC 2004*, LNCS vol. 2951, Springer, pp. 1–20, 2004.
25. R. Rivest. The MD4 message digest algorithm. *Advance in Cryptology – CRYPTO '90*, LNCS vol. 537, Springer, pp. 303–311, 1991.
26. P. Rogaway and T. Shrimpton. Cryptographic hash-function basics: definitions, implications, and separations for preimage resistance, second-preimage resistance, and collision resistance. *Fast Software Encryption* (FSE 2004), LNCS vol. 3017, Springer, pp. 371–388, 2004.
27. A. Russell. Necessary and sufficient conditions for collision-free hashing. *J. of Cryptology*, vol. 8, no. 2, pp. 87–99, 1995.
28. X. Wang, X. Lai, D. Feng, H. Chen, and X. Yu. Cryptanalysis of the hash functions MD4 and RIPEMD. *Advances in Cryptology – EUROCRYPT '05*, LNCS vol. 3494, Springer, pp. 1–18, 2005.
29. X. Wang, Y. Yin, and H. Yu. Finding Collisions in the Full SHA-1. *Advances in Cryptology – CRYPTO '05*, LNCS vol. 3621, Springer, pp. 17–36, 2005.
30. R. Winternitz. A secure one-way hash function built from DES. *Proceedings of the IEEE Symposium on Inf. Security and Privacy*, pp. 88–90, IEEE Press, 1984.

A   The Traditional Definition of Collision Resistance

In this section we recall, for comparison, the traditional definition for a collision-resistant hash function, as given by Damgård [7]. The notion is keyed (meaning that the hash functions have an *index*) and asymptotic. Our wording and low-level choices are basically from [27].

A *collection of collision-free hash-functions* is a set of maps $\{h_K : K \in \mathcal{I}\}$ where $\mathcal{I} \subseteq \{0,1\}^*$ and $h_K : \{0,1\}^{|K|+1} \to \{0,1\}^{|K|}$ and where:

1. There is an EPT algorithm $\mathcal{K}$ that, on input 1^n, outputs an n-bit string $K \overset{\$}{\leftarrow} \mathcal{K}(1^n)$ in $\mathcal{I}$.

2. There is an EPT algorithm H that, on input $K \in \mathcal{I}$ and $X \in \{0,1\}^{|K|+1}$, computes $H(K, X) = h_K(X)$.

3. For any EPT adversary A, $\epsilon(n) = \Pr[K \overset{\$}{\leftarrow} \mathcal{K}(1^n); (X, X') \overset{\$}{\leftarrow} A(K) : X \neq X'$ and $H_K(X) = H_K(X')]$ is negligible.

Above, EPT stands for *expected polynomial time*, and a function $\epsilon(n)$ is *negligible* if for every $c > 0$ there exists an N such that $\epsilon(n) < n^{-c}$ for all $n \geq N$. For simplicity, we assumed that the domain of each h_K is $\{0,1\}^{|K|+1}$. This can be relaxed in various ways.

B   Hash-Then-Sign with a Keyed Hash-Function

In this section we provide a concrete-security treatment of the hash-then-sign paradigm using a keyed hash-function instead of an unkeyed one. Our purpose is to facilitate easy comparison between the keyed and unkeyed form of a theorem.

First we must modify our formalization of the hash-then-sign construction to account for the differing syntax of a keyed and unkeyed hash function. Let $H: \mathcal{K} \times \{0,1\}^* \to \{0,1\}^n$ be a keyed hash-function. Let $\Pi = (Gen, Sign, Verify)$ be a signature scheme with message space of at least $\{0,1\}^n$. Define from these the signature scheme $\Pi^H = (Gen^H, Sign^H, Verify^H)$ by saying that Gen^H samples $K \overset{\$}{\leftarrow} \mathcal{K}$ and $(PK, SK) \overset{\$}{\leftarrow} Gen$ and then outputs $(\langle PK, K \rangle, \langle SK, K \rangle)$; define $Sign^H_{\langle SK, K \rangle}(M) = Sign_{SK}(H_K(M))$; and define $Verify^H_{\langle PK, K \rangle}(M, \sigma) = Verify_{PK}(H_K(M), \sigma)$. The message space for Π^H is $\{0,1\}^*$. We have reused the notation Π^H and $Sign^H$ and $Verify^H$ because the "type" of the hash function H makes unambiguous what construction is intended.

The proof of the following, little changed from Theorem 2, is omitted.

Theorem 6 (hash-then-sign, keyed, concrete, C0). *Let $H: \mathcal{K} \times \{0,1\}^* \to \{0,1\}^n$ be a keyed hash-function, let $\Pi = (Gen, Sign, Verify)$ be a signature scheme with message space at least $\{0,1\}^n$, and let A be an adversary. Then there exist adversaries B and C such that*

$$\mathbf{Adv}^{\mathrm{sig}}_{\Pi}(B) + \mathbf{Adv}^{\mathrm{col}}_{H}(C) \geq \mathbf{Adv}^{\mathrm{sig}}_{\Pi^H}(A).$$

Adversary B runs in time at most $t_A + t_{\mathcal{K}} + t_H(\ell_A) + t_{Sign}(nq_A) + c(\ell_A + nq_A)$ and asks at most q_A queries entailing at most $\ell_A + n$ bits. Adversary C runs in time at most $t_A + t_{Gen} + t_{\mathcal{K}} + t_H(\ell_A) + t_{Sign}(nq_A+n) + c(\ell_A + nq_A)\lg(q_A)$. The value c is an absolute constant implicit in the proof of this theorem. ◇

Discrete Logarithm Variants of VSH

Arjen K. Lenstra[1,2], Daniel Page[3], and Martijn Stam[1]

[1] EPFL / IC - LACAL, INJ3.33,
Station 14, CH-1015 Lausanne, Switzerland
martijn.stam@epfl.ch
[2] Bell Laboratories
[3] Dept. Computer Science, University of Bristol, Merchant Venturers Building,
Woodland Road, Bristol, BS8 1UB, United Kingdom
page@cs.bris.ac.uk

Abstract. Recent attacks on standardised hash functions such as SHA1 have reawakened interest in design strategies based on techniques common in provable security. In presenting the VSH hash function, a design based on RSA-like modular exponentiation, the authors introduce VSH-DL, a design based on exponentiation in DLP-based groups. In this article we explore a variant of VSH-DL that is based on cyclotomic subgroups of finite fields; we show that one can trade-off performance against bandwidth by using known techniques in such groups. Further, we investigate a variant of VSH-DL based on elliptic curves and extract a tighter reduction to the underlying DLP in comparison to the original VSH-DL proposal.

Keywords: Hash Functions, Cyclotomic Subgroup, Collision Resistance.

1 Introduction

Hash function design. Hash functions can be considered, together with block ciphers, to be the core primitives on which modern applied cryptography is based. The design of block ciphers is guided by a fairly mature and well understood background, see for example linear [14] and differential [3] cryptanalysis and the wide-trail design strategy of the AES [7]. In contrast, standardised hash functions such as SHA1 are constructed using somewhat ad-hoc techniques and they are essentially derived from the same family. This fact has, in part, contributed to a number of recent collision attacks against designs including SHA1 [20,21].

Ideally, a hash function with output length n is a parameterised, deterministic function $\mathcal{H} : \{0,1\}^* \rightarrow \{0,1\}^n$ that takes an arbitrary length bitstring and maps it to a bitstring of length n. A good hash function satisfies several properties, the three most important of which are stated informally below.

1st-Preimage resistance. Given a random image $x \in \{0,1\}^n$, it should take time $\approx 2^n$ to find $m \in \{0,1\}^*$ such that $\mathcal{H}(m) = x$.

2nd-Preimage resistance. Given a 'random' $m \in \{0,1\}^*$, it should take time $\approx 2^n$ to find $m' \in \{0,1\}^*$ such that $\mathcal{H}(m) = \mathcal{H}(m')$ and $m \neq m'$.

P.Q. Nguyen (Ed.): VIETCRYPT 2006, LNCS 4341, pp. 229–242, 2006.
© Springer-Verlag Berlin Heidelberg 2006

Collision resistance. It should take time $\approx 2^{n/2}$ to find $m, m' \in \{0,1\}^*$ such that $\mathcal{H}(m) = \mathcal{H}(m')$, yet $m \neq m'$.

Since generic black box attacks are known that find collisions in time $\approx 2^{n/2}$ or preimages in time $\approx 2^n$, the above requirements are very strong. In many scenarios it suffices to achieve a relaxed notion of collision resistance, in the sense that attackers who can invest only time 2^k cannot find collisions, where possibly the output length n is larger than $2k$. Thus, these hash functions might not strictly satisfy the standard security notion, even though collision resistance may provably be linked to a well studied hard problem, using the type of exact reduction also known from provable security. The tightness of the reduction and our belief or current understanding of the hardness of the underlying problem then lead to a parameter choice for which the resulting hash function has the desired collision resistance.

Hash functions based on modular exponentiation. One of the first provably secure collision resistant hash functions is based on exponentiation modulo an RSA modulus, that is $\mathcal{H}(m) = x^m \bmod N$ where m is the message, N is the RSA modulus and x is some predefined value in $\mathbb{Z}_N^*$. If m and m' form a collision such that $\mathcal{H}(m) = \mathcal{H}(m')$, then $x^{m-m'} = 1 \bmod N$ which implies that $(m - m')$ is a multiple of the order of x. This order will necessarily be a divisor of $\phi(N)$ and if certain conditions hold, knowing any (nonzero) multiple of the order of x suffices to factor N in deterministic polynomial time. Note that there is no restriction on the length of m which means that there is no need for Merkle-Damgård [15,8] type constructions.

This scheme was recently extended by Contini et al. [6] who essentially propose to use multi-exponentiation for the compression function instead of single exponentiation, thus obtaining an improvement in performance by processing more message bits at the same time. Let p_i be the i-th prime number, for $i = 1, \ldots, k$, where the product of the k primes should be smaller than the RSA modulus. A message m is then split up into l blocks M_i of equal length and the hash is computed as the multi-exponentiation $\mathcal{H}(m) = \prod_i p_i^{M_i} \bmod N$. An additional requirement is that the total bitlength of the message m is smaller than 2^k.

One of the disadvantages of VSH is the need for a secret RSA-modulus N. Someone who knows how to factor N can construct collisions easily. A side-effect of this trap-door against collision resistance is that the modified Cramer-Shoup signature scheme [6] based on VSH does not provide non-repudiation as one might expect (cf. 'Creating Collisions' [6, p. 171]). Another disadvantage is the relatively large output length, namely the size of the RSA modulus. This means that to provide 80-bit security, one needs to use a hash function outputting approximately 1024 bits, rather than the desired 160 bits needed to thwart generic birthday attacks. To address these problems, Contini et al. mentioned the possibility of building VSH-DL, a hash function based on multi-exponentiation in DLP-based groups allowing short representations, such as elliptic curves or cyclotomic subgroups (allowing trace or torus-based methods). This design extends the corpus of previous work on DLP based hash functions, see [1,2] for example.

Computation in finite field extensions. The possibility to use finite fields with extension degree higher than one for public key cryptography has been known since the birth of public-key cryptography. However, for a long time nobody paid much attention to the subject since it was unclear whether the higher extension degree would offer any significant advantage over prime fields. It was not until Lenstra and Verheul showed the potential of working in a smaller subgroup of a larger field using their trace-based method called XTR [13] that interest increased.

Since then, Stam and Lenstra [19] showed how to efficiently work in the cyclotomic subgroup of a degree six extension field (provided that the characteristic satisfies a mild congruency relation), and Rubin and Silverberg [16] showed how to compress and decompress elements in this same subgroup using the theory of algebraic tori. The method of Rubin and Silverberg, called CEILIDH, differs from XTR in that compression is injective allowing full and exact decompression (in XTR conjugates are mapped to the same element). The downside of CEILIDH is that it is only a compression and decompression mechanism: it does not support direct computation on the compressed elements. Efficient arithmetic is still possible though, for instance by the method developed by Stam and Lenstra or the more involved hybrid methods by Granger et al. [10].

Main contributions. Since methods known from the study of arithmetic and schemes using cyclotomic subgroups can provide computational efficiency and reduced bandwidth due to their compression properties, it is a natural question to ask to what extent they can be used to implement VSH-DL. To address this question, in this paper we investigate VSH-DL type schemes based on the cyclotomic subgroup of a sixth degree extension field and on elliptic curves.

Such schemes provide natural efficiency in terms of bandwidth which leads to a smaller hash-output without compromising security against collision attacks; through an experimental implementation we reason that this benefit is balanced against decreased performance compared to the original VSH-DL proposal. We do not make any claims about other security properties of our proposed hash functions, although it is easy to see that finding preimages is at least as hard as finding a collision. Thus it is possible to pick a (longer) output length of the hash function such that one also has the desired level of security against these two attacks. We believe that in many applications the level of security against collision attacks and preimage attacks can be set the same. Because our hash functions essentially depend on $n > 2k$, it is not recommended to truncate the output of the hash function (cf. [17]).

The paper is organised as follows. After our introduction of VSH in Section 2, we explore the possibility to base a hash function on a problem related to the discrete logarithm problem in the cyclotomic subgroup of a sixth degree extension field in Section 3, where we achieve a compression by a factor of three which represents a trade-off against decreased performance versus the original VSH-DL under a similar assumption. In Section 4 we discuss the possibility to use elliptic curves over prime fields. In that case the collision resistance of the hash function can be based directly on ECDLP. Finally we present some experimental results and analysis in Section 5 before concluding in Section 6.

2 VSH and VSH-DL

Contini et al. [6] define and analyse a hash function called Very Smooth Hash (VSH), which is a multi-exponentiation generalisation of the well known RSA-based hash.

They write down the multi-exponentiation as a square-and-repeated-multiply algorithm, where they consider the processing of k bits (Step 5 in Algorithm 1 below) as a compression primitive and view the full compression function as repeated application of this compression primitive, which allows the computation of the hash function in a streaming fashion. Recall that for $i \in \mathbb{Z}_{>0}$, we let p_i is the i-th prime number.

Algorithm 1. VSH compression function [6].

Let N be an RSA modulus, let the block length k be the largest k for which $\prod_{i=1}^{k} p_i < N$. The VSH compression function $\mathcal{H}_{VSH} : \{0,1\}^{<2^k} \to \mathbb{Z}_N^*$ is defined as follows for an ℓ-bit message m consisting of bits $m_1, m_2, \ldots, m_\ell$, where $\ell < 2^k$.

1.   [Initialise] Set $x_0 \leftarrow 1$, $\mathcal{L} \leftarrow \lceil \frac{\ell}{k} \rceil$ and $j \leftarrow 0$.
2.   [Padding] Set $m_i \leftarrow 0$ for $\ell < i \le \mathcal{L}k$.
3.   [Merkle-Damgård Strengthening] Let $\ell = \sum_{i=1}^{k} \ell_i 2^{i-1}$ with $\ell_i \in \{0,1\}$ be the binary representation of the message length ℓ. Set $m_{\mathcal{L}k+i} \leftarrow \ell_i$ for $0 < i \le k$.
4.   [Finished] If $j = \mathcal{L} + 1$ terminate with output $x_{\mathcal{L}+1}$.
5.   [Hash next block] Set $x_{j+1} \leftarrow x_j^2 \times \prod_{i=1}^{k} p_i^{m_{j \cdot k+i}} \pmod{N}$.
6.   [Increase j] Increase j by one. Go back to Step 4.

It is not too hard to see that if we define $M_i = \sum_{j=0}^{\mathcal{L}} 2^{\mathcal{L}-j} m_{jk+i}$, taking into account the padding and the strengthening, then the hash is computed as the multi-exponentiation $H(m) = \prod_i p_i^{M_i} \bmod N$. In particular this means that one might be able to achieve some speedups by using techniques known from the theory of addition chains.

Contini et al. mention precomputing products of primes: indeed, if k primes (or bases) are given and a small positive integer b divides k, we can partition the bases in k/b sets of b primes each and for each set precompute all 2^b products of the different primes in that set. During the actual hashing bits are processed in chunks of b bits so that only k/b multiplications will be needed to process k bits of message (this is essentially a simplified version of Pippenger's algorithm). Contini et al. observe that instead of using precomputed products of primes for chunks of bits, one can also use fresh primes instead. Although this leads to a different hash function, called Fast VSH, it is based on the same hardness assumption as standard VSH but, as the name suggests, considerably faster (also compared to VSH with precomputation). A full description can be found in Appendix A.

The collision resistance of VSH can be reduced to the VSSR problem.

Definition 2. *(VSSR: Very Smooth number nontrivial modular Square Root [6, Def. 3]) Let N be the product of two unknown primes of approximately the same size and let $k \le (\log N)^c$. VSSR is the following problem: Given N, find $x \in \mathbb{Z}_N^*$ such that $x^2 = \prod_{i=1}^{k} p_i^{e_i}$ and at least one of $e_1, \ldots, e_k$ is odd.*

Contini et al. note that, given the existing known factoring algorithms, it seems as hard to solve the VSSR problem as it is to factor N (though they base their analysis on a more conservative relation). They also define a discrete log analogue to VSSR leading to VSH-DL. An important advantage of VSH-DL over VSH is the lack of a trapdoor.

Definition 3. *(VSDL: Very Smooth number Discrete Log [6, Def. 4]) Let p, q be primes with $p = 2q + 1$ and let $k \leq (\log p)^c$. VSDL is the following problem: given p, find integers $e_1, e_2, \ldots, e_k$ such that $2^{e_1} \equiv \prod_{i=2}^{k} p_i^{e_i} \bmod p$ with $|e_i| < q$ for $i = 1, 2, \ldots, k$, and at least one of $e_1, e_2, \ldots, e_k$ is non-zero (where p_i is to be understood to be the i-th prime number).*

Algorithm 4. VSH-DL compression function.

Let p be an S-bit prime of form $2q + 1$ for prime q, let k be a fixed integer length, typically $k \approx S/\log S$. The VSH-DL compression function $\mathcal{H}_{DL} : \{0,1\}^{<(S-2)k} \rightarrow \mathbb{Z}_p^*$ is defined as follows for an ℓ-bit message m consisting of bits $m_1, m_2, \ldots, m_\ell$, with $\ell < (S-2)k$.

1. [Initialise] Set $x_0 \leftarrow 1$, $\mathcal{L} \leftarrow \lceil \frac{\ell}{k} \rceil$ and $j \leftarrow 0$.
2. [Padding] Set $m_i \leftarrow 0$ for $\ell < i \leq \mathcal{L}k$.
3. [Merkle-Damgård Strengthening] Let $\ell = \sum_{i=1}^{k} \ell_i 2^{i-1}$ with $\ell_i \in \{0, 1\}$ be the binary representation of the message length ℓ. Set $m_{\mathcal{L}k+i} \leftarrow \ell_i$ for $0 < i \leq k$.
4. [Finished] If $j = \mathcal{L} + 1$ terminate with output $x_{\mathcal{L}+1}$.
5. [Hash next block] Set $x_{j+1} \leftarrow x_j^2 \times \prod_{i=1}^{k} p_i^{m_{j \cdot k+i}} \pmod{p}$.
6. [Increase j] Increase j by one. Go back to Step 4.

3 A Cyclotomic Subgroup Variant of VSH-DL

We begin with a brief overview of the mathematics underlying CEILIDH and XTR. This overview is specifically tailored to our needs, for a more general introduction see [9] and the references contained therein.

Let p be a prime and let $\mathbb{F}_p$ denote a finite field of order p and $\mathbb{F}_{p^6}$ a sixth degree extension thereof. The multiplicative group $\mathbb{F}_{p^6}^*$ is cyclic of order $p^6 - 1$, which factors as $(p^2 - p + 1)(p^2 + p + 1)(p + 1)(p - 1)$. Let G be the unique subgroup of order $p^2 - p + 1$ in $\mathbb{F}_{p^6}^*$. We call G the cyclotomic subgroup of $\mathbb{F}_{p^6}^*$. Alternatively, it can be regarded as a specific algebraic torus of dimension 2 over $\mathbb{F}_p$. It is argued [12] that the computational complexity of the discrete logarithm problem in $\mathbb{F}_{p^6}^*$ resides in this subgroup G of order $p^2 - p + 1$, since the subgroups of order dividing $(p^2 + p + 1)(p + 1)(p - 1)$ can be efficiently embedded in proper subfields of $\mathbb{F}_{p^6}^*$, thus allowing to run a sub-exponential algorithm in the smaller field.

Using a standard representation in $\mathbb{F}_{p^6}$ consumes $\approx 6 \log p$ bits which seems wasteful given that there are only $\approx p^2$ elements in G. This problem can be solved using either XTR [13] or CEILIDH [16]. With XTR, the trace map

$$\mathrm{Tr} : \mathbb{F}_{p^6} \rightarrow \mathbb{F}_{p^2} : x \rightarrow x^{p^4} + x^{p^2} + x$$

is used to compress an element in G to an element in $\mathbb{F}_{p^2}$. This map is not injective; since conjugates over $\mathbb{F}_{p^2}$ map to the same value in $\mathbb{F}_{p^2}$ it is essentially 3-to-1. One of the significant advantages of XTR is that it is possible to work directly with compressed elements when performing an exponentiation. Unfortunately, this method does not generalise very well to multi-exponentiation on more than two bases which makes XTR unsuitable for direct use in a VSH-DL variant.

CEILIDH is an alternative to XTR that offers only compression; that is, one cannot compute directly with compressed elements. Formally, CEILIDH is a bijection between $G\backslash\{a\}$ and $(\mathbb{F}_p)^2\backslash V(f)$, where a is some particular element of G and $V(f)$ is a well-defined subset of $(\mathbb{F}_p)^2$ (the notation $V(f)$ stems from the fact that it is a variety defined by a single polynomial). It is straightforward to extend CEILIDH into an injection from G to $(\mathbb{F}_p)^2$. Clearly, given any collision-resistant hash function, applying an injection to the result is not going to reduce its collision resistance.

One of the advantages of VSH-DL as described by Contini et al. is its use of small elements p_i. When we work directly with elements in G, there do not seem to be elements that are as efficient to work with. Since each group element is represented by six base field elements and the order of the group is $p^2 - p + 1$ one can prescribe at most two base field elements to be small, the other four will follow and be of normal size. However, an alternative presents itself by not working with elements in G, but rather elements in the full field $\mathbb{F}_{p^6}$ and only map to G at the very end through powering by $(p^3 - 1)(p + 1)$, i.e. the cofactor of G in $\mathbb{F}_{p^6}^*$). Thus our hash function will be

$$\text{Ceilidh}((\prod_i h_i^{M_i})^{(p^3 - 1)(p+1)})$$

for suitably chosen $h_i \in \mathbb{F}_{p^6}^*$.

In the following we will motivate this approach by showing that, when the elements h_i are chosen uniformly at random and then somehow replaced by a smaller sibling in the same coset (under exponentiation with the cofactor of G), the hash function is collision-resistant if the discrete logarithm problem in (a subgroup of) G is hard. Thus, contrary to the original VSH-DL, we reduce from a standard discrete logarithm assumption. Our proof uses a slightly more general notation than as above. We note that the order of G itself need not be prime, although to provide collision resistance the size of G's largest DLP-hard subgroup of prime order will be relevant for determining the maximal allowable message length. Our reduction is similar to that of Bellare and Micciancio [2], but tighter because we reduce from the DLP in a prime order subgroup.

Definition 5. *(DLP) Let G_q be a finite cyclic group of known prime order q and with generator f. The discrete logarithm problem in G_q is to find, given y drawn uniformly at random from G_q, the unique value $0 \le x < q$ such that $y = f^x$.*

Definition 6. *(k-modified DLP) Let H be a finite cyclic group with generator h. Let G be a subgroup of H with generator $g = h^{|H|/|G|}$. Let $\psi : H \to H$ be a map such that $\psi(h_i)^{|H|/|G|} = (h_i)^{|H|/|G|}$ for all $h_i \in H$. The k-modified discrete logarithm problem for (H, G, ψ) is to find, given h_i drawn uniformly at random from H for $i = 1, \dots, k$, a nonzero solution $(e_1, \dots, e_k) \in [0, q)^k$ of*

$$(\prod_{i=1}^{k} \psi(h_i)^{e_i})^{|H|/|G|} = 1$$

Theorem 7. *Assume q divides $|G|$, but q^2 does not divide $|H|$ and that $k > 1$. An attacker that solves the k-modified discrete logarithm problem for (H, G, ψ) in time t with probability ϵ can be used to solve the discrete logarithm problem in $G_q \subseteq G$ in* *time $t + t'$ with probability $\epsilon - 1/q$, where t' is essentially the time to perform a k-fold double exponentiation in* H.

Proof: Given $y \in G_q$, we need to find x such that $f^x = y$ with the help of an attacker that solves the k-modified discrete logarithm problem. Let $h_1 = y^{b_i} h^{a_i}$ for $i = 1, \ldots, k$, where the a_i and b_i are drawn uniform at random from $[0, |H|)$. As a result the h_i are distributed uniformly as expected by the k-modified DLP attacker, so on input of these h_i the attacker will, with probability ϵ, return $(e_1, \ldots, e_k)$ such that

$$(\prod_{i=1}^{k} \psi(y^{b_i} h^{a_i})^{e_i})^{|H|/|G|} = 1$$

hence

$$g^{\sum_{i=1}^{k} a_i e_i} = y^{-(|H|/|G|) \sum_{i=1}^{k} b_i e_i} .$$

Note that $y = f^x$ for some (yet unkown) value of $0 \le x < q$ and that, w.l.o.g., $f = g^{|G|/q}$. Since g is a generator of $|G|$, this implies that

$$\sum_{i=1}^{k} a_i e_i = -x(|H|/q) \sum_{i=1}^{k} b_i e_i \bmod |G|$$

and because q divides $|G|$ also

$$\sum_{i=1}^{k} a_i e_i = -x(|H|/q) \sum_{i=1}^{k} b_i e_i \bmod q .$$

Now we can compute x if $(|H|/q) \sum_{i=1}^{k} b_i e_i \ne 0 \bmod q$. Since q^2 does not divide $|H|$, we know that $|H|/q$ is invertible modulo q, so we need to show that $\sum_{i=1}^{k} b_i e_i \ne 0 \bmod q$ with probability at most $1/q$.

Because, given any $h_i \in H$ and $0 \le b_i < q$, there is exactly one a_i such that $h_i = h^{a_i} y^{b_i}$, it follows that the adversary given the h_i has no Shannon information on b_i when announcing the e_i. Consequently, unless all e_i are congruent to 0 modulo q (which is not allowed by the restrictions on the e_i), $\sum_{i=1}^{k} b_i e_i$ is uniformly randomly distributed modulo q, so the probability of it being 0 mod q is $1/q$.

Q.E.D.

Algorithm 8. Modified VSH-DL compression function.

Let H be a finite cyclic group of known, factored order and generator h. Let G be a subgroup of H with generator $g = h^{|H|/|G|}$. Let q be a prime dividing $|G|$ but not $|H|/q$. Let $\psi : H \to H$ be a map such that $\psi(h_i)^{|H|/|G|} = (h_i)^{|H|/|G|}$ for all $h_i \in H$. Let Compress : G $\to R$ be an efficiently computable injection. Let k be a fixed integer length such that $(\lfloor \log_2 q \rfloor - 1)k < 2^k$. The modified VSH-DL compression function $\mathcal{H}_{MDL} : \{0,1\}^{\le (\lfloor \log_2 q \rfloor - 1)k} \to R$ is defined for an ℓ-bit message m consisting of bits $m_1, m_2, \ldots, m_\ell$ with $\ell \le (\lfloor \log_2 q \rfloor - 1)k$ as follows.

1.   [Initialise] Set $x_0 \leftarrow 1$, $\mathcal{L} \leftarrow \lceil \frac{\ell}{k} \rceil$ and $j \leftarrow 0$.
2.   [Padding] Set $m_i \leftarrow 0$ for $\ell < i \leq \mathcal{L}k$.
3.   [Merkle-Damgård Strengthening] Let $\ell = \sum_{i=1}^{k} \ell_i 2^{i-1}$ with $\ell_i \in \{0,1\}$ be the binary representation of the message length ℓ. Set $m_{\mathcal{L}k+i} \leftarrow \ell_i$ for $0 < i \leq k$.
4.   [Finished?] If $j = \mathcal{L} + 1$ terminate with output $\text{Compress}(x_{\mathcal{L}}^{|H|/|G|})$.
5.   [Hash next block] Set $x_{j+1} \leftarrow x_j^2 \times \prod_{i=1}^{k} \psi(h_i)^{m_{j \cdot k + i}}$.
6.   [Increase j] Increase j by one. Go back to Step 4.

For completeness, we note that the collision resistance of the hash function above can be related to k-modified DLP by observing that the (implicit) map turning messages into the relevant exponents for h_i is injective on its domain (this is the reason for the length restrictions on the message).

As mentioned before, we will instantiate Algorithm 8 with $H = \mathbb{F}_{p^6}^*$ and G a subgroup of order $p^2 - p + 1$ which has cofactor $(p^3 - 1)(p + 1)$. Moreover for Compress we will substitute Ceilidh. For efficient field arithmetic we restrict ourselves to $p \equiv 2 \bmod 9$. In this case p will generate $\mathbb{Z}_9^*$ and $\Phi_9(x) = x^6 + x^3 + 1$ is irreducible in $\mathbb{F}_p$. Hence if γ is a root of $\Phi_9(x)$ (i.e., a ninth root of unity), then $(\gamma, \gamma^2, \gamma^3, \gamma^4, \gamma^5, \gamma^6)$ is a basis for the extension field $\mathbb{F}_{p^6} = \mathbb{F}_p[\gamma]$. The arithmetic based on this extension also lies at the basis of the fast implementation [10] of XTR and CEILIDH.

Let $a = \sum_{j=0}^{5} a_j \gamma^{j+1} \in \mathbb{F}_{p^6}$. We are interested in finding a small representation $\psi(a)$ of a such that multiplication of an arbitrary field element by $\psi(a)$ will be relatively cheap. We do this implicitly. Instead of giving a straightforward definition of ψ we show how to sample efficiently and (almost) uniformly from $\psi(\mathbb{F}_{p^6}^*)$. This is done by ensuring that $\psi(\mathbb{F}_{p^6}^*)^{(p^3-1)(p+1)}$ gives rise to the (almost) uniform distribution over G.

We sample from $\psi(\mathbb{F}_{p^6}^*)$ as follows. Draw a_0 and a_5 uniformly and independently at random from $\mathbb{F}_p$. Let $\psi(a) = a_0 \gamma + \gamma^2 + a_5 \gamma^6$. That this works follows from the following observation:

Lemma 9. *The map $\psi' : \mathbb{F}_p^2 \to G$ defined by $\psi'(a_0, a_5) = (a_0 \gamma + \gamma^2 + a_5 \gamma^6)^{(p^3-1)(p+1)}$ has a range of at least $p^2/3$ elements.*

Proof: To prove the statement we will upper bound the number of collisions, that is, sets of distinct pairs (a_0, a_5) and (b_0, b_5) such that $\psi'(a_0, a_5) = \psi'(b_0, b_5)$. Equivalently, we are counting the sets a and b of prescribed form $a = a_0 \gamma + \gamma^2 + a_5 \gamma^6$ and $b = b_0 \gamma + \gamma^2 + b_5 \gamma^6$ such that $a^{(p^3-1)(p+1)} = b^{(p^3-1)(p+1)}$. The latter equation can be rewritten as $(a^{p^3} b)^{p+1} - (a b^{p^3})^{p+1} = 0$. This can be computed algebraically, giving rise to initially six equations (one for each coordinate), but that can be simplified to

$$(a_0 - b_0)(1 + a_0 b_0 + a_5 b_5) = 0$$
$$(a_5 - b_5 + 2(a_0 - b_0) + a_0 b_5 - a_5 b_0)(1 + a_0 b_0 + a_5 b_5) = 0$$

plus a third condition. Simultaneously satisfying the above two equations can only be done if either $(a_0, a_5) = (b_0, b_5)$ or if $(1 + a_0 b_0 + a_5 b_5) = 0$. Unless $(b_0, b_5) = (0, 0)$ the latter solution allows us substitution of either a_0 or a_5 in the third and final equation to be satisfied (and note that only $(0, 0)$ is in the preimage of $\psi'(0, 0)$). This third equation will then yield a quadratic equation in either a_0 or a_5 that is only degenerate (i.e. equal to zero) if both $b_0 = 0$ and $b_5 = 0$. Since a quadratic equation over a finite field has at most two solutions, we have shown that any preimage of ψ' has cardinality at most three, from which the claim follows.   *Q.E.D.*

One can improve performance considerably by picking small a_0 and a_5. The caveat is that the security is no longer directly related to a clean DLP assumption. Our choice will be to set $(a_5)_i$ equal to 1 and let $(a_0)_i$ depend on i, i.e. simply range through a number of a_0 values that are easy to multiply with. In particular, we use $(a_0)_i = i + 1$ for a given i.

For completeness we note that the preimage under ψ' of $1 \in G$ in this case is restricted to $a_0 = 0$ or $a_0 = 1$, which can be seen by using that $a^{(p^3-1)(p+1)} = 1$ is equivalent to $a^{(p^2+p+1)(p+1)} \in \mathbb{F}_p^*$. Moreover, if $a = a_0\gamma + \gamma^2 + \gamma^6$ and $b = b_0\gamma + \gamma^2 + \gamma^6$, then $a^{(p^3-1)(p+1)} = b^{(p^3-1)(p+1)}$ iff $a = b$ or a, b are in the preimage of 1.

Thus we are insured that as long as we pick the a_0 distinct and unequal to 0 or 1 our system is not obviously flawed and there is no reason to assume that the choice of our $\psi(h_i)$ is weak. The resulting hash function can be proven secure assuming that solving the following, admittedly tailor-made, problem is hard:

Definition 10. *(Small Element DLP) Let p be a prime congruent to 2 mod 9 such that $p^2 - p + 1$ has at least one big prime factor q. Let γ be a ninth-root of unity and let $h_i = (i + 1)\gamma + \gamma^2 + \gamma^6 \in \mathbb{F}_{p^6}^*$ for $i = 1, \ldots, k$. The small element discrete logarithm problem is to find, given p, a nonzero solution $(e_1, \ldots, e_k) \in [0, q)^k$ of*

$$\left(\prod_{i=1}^{k} \psi(h_i)^{e_i}\right)^{(p^3-1)(p+1)} = 1 .$$

4   An Elliptic Curve Variant of VSH-DL

Since their introduction to cryptography by Koblitz and Miller, elliptic curves have become ever more popular as a replacement for finite fields to base DLP-based schemes on. This is mainly due to the fact that there is no known algorithm to solve the DLP on a general elliptic curve faster than the generic Pollard-ρ method. This allows one to use curves with group sizes quadratic in the security one wishes to offer.

An immediate result of this is that to obtain 2^k bit security against collision-finding, one can actually use a hash function based on VSH-DL that outputs just over $2k$ bits (by using affine representation and standard point compression, where the Y-coordinate is replaced by a single bit to resolve any square root ambiguity). Moreover, the computations are relatively fast. In this article we concentrate on curves over prime fields, though similar results are expected to hold for curves over binary or ternary fields.

An elliptic curve over a prime field $\mathbb{F}_p$ with $p > 3$ can be represented using the short Weierstrass form

$$Y^2 = X^3 + a_4 X + a_6$$

where it is common to use $a_4 = -3$, for example in the NIST standard curves, in order to extract some performance benefits. The set of points $(X, Y) \in \mathbb{F}_p^2$ satisfying the equation above together with a point at infinity form an abelian group under the addition operation also known as the chord-tangent process. The point at infinity serves as group identity and the negation of a point (X, Y) is $(X, -Y)$.

Optimising elliptic curve arithmetic has been the focus of a large number of articles. Excellent overviews are given by Brown et al.[5] and Hankerson et al.[11]. One of the most efficient methods is the use of mixed coordinates, where the fixed multiplicands are kept in affine representation but where computations are done using the Jacobian representation. Point doubling with the Jacobian representation costs 4 field multiplications and 4 field squarings. Adding an affinely represented point to a point in Jacobian representation costs 8 field multiplications and 3 field squarings. The result again is in the Jacobian representation.

One could consider the use of small points P_i. However, it seems that only one of the coordinates of P_i can be picked small, since the other coordinate typically follows from the curve equation (indeed, if $a_4 = -3$ and a_6 has full size, it is impossible for both the X and the Y-coordinate of a point on the curve to be small). It is easy to see from [11, Algorithm 3.22] that both coordinates are used only once during the point addition, so picking small P_i's will reduce the cost of a point addition by at most one field multiplication.

A possible solution to this problem is the use of a Montgomery representation. Recently Brown [4] showed how to perform a multi-exponentiation efficiently in this setting.

5   Experimental Results

Strategy. The crucial operation in the implementation of VSH is

$$x_{j+1} \leftarrow x_j^2 \prod_{i=1}^{k} p_i^{m_{j \cdot k+1}} \pmod{n}.$$

This product can be computed in several ways by reshuffling the order in which the multiplications take place. One option is to start with x_j^2 and then multiply by the relevant p_i terms one by one, reducing each time. The option recommended by Contini et al. is to first compute the product

$$P = \prod_{i=1}^{k} p_i^{m_{j \cdot k+1}}$$

and then multiply by x_j^2. Due to the choice of parameters, the product P will be smaller than the modulus N when computed over the integers, so modular reductions are not necessary. VSH relies on this feature of the p_i, or small element values to enable the construction of high performance implementations. It also acts as the bottleneck for our compressed VSH-DL variants. Specifically, it is much harder to reason about how one would delay reductions in either the cyclotomic subgroup or elliptic curve cases: although one can attempt to construct some notion of small elements, it is difficult to imagine how performing computation with such elements will be as efficient as in the original VSH case.

Table 1. A comparison between the original VSH design and three variants of VSH-DL. Results are given in clock cycles per byte of input.

	slow	fast $b = 2$	fast $b = 4$	fast $b = 8$
SHA1-160	26.29 (or 16.89 with SIMD)			
VSH	632.36	622.11	370.37	277.60
VSH-DL-$\mathcal{A}$	715.71	676.04	382.68	274.75
VSH-DL-$\mathcal{B}$	5507.54	6244.37	3126.34	1338.18
VSH-DL-$\mathcal{C}$	16080.26	11777.89	7542.05	4105.31

Results. In order to evaluate the relative performance characteristics of our VSH-DL designs versus the original proposal by Contini et al. [6], we produced some experimental results using an implementation in C. Our platform for these experiments was a 2.8Ghz Intel Pentium 4; we used GCC 4.0.1 to compile our implementation which relied on NTL [18] for the underlying arithmetic. We produced an implementation of the original VSH scheme and three variants of the VSH-DL scheme as detailed below:

VSH. The original VSH scheme operates modulo an RSA number $N = pq$ for primes p and q. The parameters p and q were selected such that $\log_2(N) = 1024$.

VSH-DL-$\mathcal{A}$. The first variant represents the original VSH-DL scheme of Contini et al. by working in the group of integers modulo a prime p such that $p = 2q + 1$ for some prime q. The parameters p and q were selected such that $\log_2(p) = 1152$.

VSH-DL-$\mathcal{B}$. The second variant represents the cyclotomic subgroup based VSH-DL design as described in Section 3. It works in a group G which is a subgroup of $\mathbb{F}_{p^6}^*$. The parameter p was selected such that $\log_2(p) = 192$. In this implementation we were careful to use delayed reduction techniques to improve arithmetic in G, and to construct a dedicated multiplication function for multiplication by small elements which have a special, sparse form.

VSH-DL-$\mathcal{C}$. The final variant represents the elliptic curve based VSH-DL design as described in Section 4. It works in a prime subgroup of a curve $E(\mathbb{F}_p)$. The parameter p was selected such that $\log_2(p) = 192$. We considered only random curves of the form

$$E : Y^2 = X^3 - 3X + a_6$$

such that special reduction techniques such as those for Mersenne primes were not available; one might expect an incremental improvement in performance by using such a parameterisation. We used Jacobian projective coordinates and a mixed-addition strategy as is common in many point multiplication methods. The form of arithmetic on the curve meant that a dedicated point addition function for addition of small elements did not give any significant benefit.

For each variant, our parameter selection is such that the security of the resulting hash functions is roughly equal. However, we make no attempt to to select values of k that are sensible for the different variants; we use $k = 131$ in all cases. The results in Table 1 detail the performance of our variants; the figures quoted are clock cycles per byte of input measured using the rdtsc instruction. They do not include the cost of initialising the hash function, for example pre-computation of any tables of small elements. We include results, following the nomenclature of Contini et al., for both slow and fast versions: the fast version blocks the input and uses some pre-computation to improve performance.

Analysis. Even considering incremental performance improvements by using, for example, Mersenne primes in the elliptic curve case, our VSH-DL variants perform at least an order or magnitude worse than either the original VSH or VSH-DL proposals by Contini et al. In this respect, their worth in terms of pure performance is untenable. This is exacerbated when one considers that hardware acceleration for modular multiplication as used in VSH is commonplace as a result of use in RSA; one might expect significant performance improvements in a practical setting as a result.

However, one area of advantage which our cyclotomic subgroup variant of VSH-DL gives is memory footprint. That is, the computation of small field elements is essentially free in comparison with the computation of the small primes used in VSH. For the figures in [6][Section 5] one can see that VSH pays a hefty price in terms of memory real-estate to achieve the levels of performance indicated in our results. Although the performance is lower, our cyclotomic subgroup variant of VSH-DL requires far less memory.

A more subtle issue is the selection of the k parameter. By increasing k, which roughly speaking is the block size of the hash function, one can decrease the number of squaring operations in the compression function; the number of multiplications stays the same. The choice of k for VSH is motivated by the need to avoid modular reductions in computing the product P from above. In our schemes we have already highlighted the fact that it is not easy to avoid such reductions; as such we can be more flexible in our choice of k.

6   Conclusion

In this article we have examined in depth the possibility to base VSH-DL on either cyclotomic subgroups of finite fields of extension degree six or on elliptic curves of large prime characteristic. We concluded that for cyclotomic subgroups using CEILIDH we can get a hash function that is about an order of magnitude slower than the original VSH-DL proposal, secure under a slightly different assumption (due to an inevitable redefining of what constitutes a small element), but has a compression factor that is three times as high. Using elliptic curves, we derive a hash function that is significantly slower than the original VSH-DL proposal, but whose compression factor is six times as high and its security can be directly linked to that of the standard ECDLP. In both cases the poor performance is balanced by a potential saving in terms of memory footprint.

We reiterate that our main concern was provable collision resistance under a discrete logarithm-like assumption. Like VSH and VSH-DL on which our constructions are based, our scheme provides only collision resistance and may not be suitable to replace a random oracle in all situations.

References

1. M. Bellare, O. Goldreich and S. Goldwasser. Incremental Cryptography: The Case of Hashing and Signing. In *Advances in Cryptology (CRYPTO)*, Springer-Verlag LNCS 839, 216–233, 1994.
2. M. Bellare and, D. Micciancio. A New Paradigm for Collision-Free Hashing: Incrementality at Reduced Cost. In *Advances in Cryptology (EUROCRYPT)*, Springer-Verlag LNCS 1233, 163–192, 1997.
3. E. Biham and A. Shamir. Differential Cryptanalysis of DES-like Cryptosystems. In *Advances in Cryptology (CRYPTO)*, Springer-Verlag LNCS 537, 2–21, 1990.
4. D. R. L. Brown. Multi-Dimensional Montgomery Ladders for Elliptic Curves. IACR eprint, 2006/220, 2006.
5. M. Brown, D. Hankerson, J. López, and A. Menezes. Software implementation of the NIST elliptic curves over prime fields. In D. Naccache, editor, *CT-RSA'01*, volume 2020 of *Lecture Notes in Computer Science*, pages 250–265. Springer-Verlag, 2001.
6. S. Contini, A. K. Lenstra, and R. Steinfeld. VSH, an efficient and provable collision resistant hash function. In S. Vaudenay, editor, *Advances in Cryptography—Euro'06*, volume 4004 of *Lecture Notes in Computer Science*, pages 165–182. Springer-Verlag, 2006.
7. J. Daemen and V. Rijmen. *The Design of Rijndael*. Springer-Verlag, 2002.
8. I.B. Damgård. Collision Free Hash Functions and Public Key Signature Schemes. In *Advances in Cryptology (EUROCRYPT)*, Springer-Verlag LNCS 304, 203–216, 1987.
9. R. Granger. *On Small Degree Extension Fields in Cryptology*. PhD thesis, University of Bristol, 2005.
10. R. Granger, D. Page, and M. Stam. A comparison of ceilidh and xtr. In D. Buell, editor, *ANTS-VI*, volume 3076 of *Lecture Notes in Computer Science*, pages 235–249. Springer-Verlag, 2004.
11. D. Hankerson, A. Menezes, and S. Vanstone. *Guide to Elliptic Curve Cryptography*. Springer-Verlag, 2004.
12. A. K. Lenstra. Using cyclotomic polynomials to construct efficient discrete logarithm cryptosystems over finite fields. In V. Varadharajan, J. Pieprzyk, and Y. Mu, editors, *ACISP'97*, volume 1270 of *Lecture Notes in Computer Science*, pages 127–138. Springer-Verlag, 1997.
13. A. K. Lenstra and E. R. Verheul. The XTR public key system. In M. Bellare, editor, *Advances in Cryptography—Crypto'00*, volume 1880 of *Lecture Notes in Computer Science*, pages 1–19. Springer-Verlag, 2000.
14. M. Matsui Linear Cryptanalysis Method for DES Cipher. In *Advances in Cryptology (EUROCRYPT)*, Springer-Verlag LNCS 765, 386–397, 1993.
15. R.C. Merkle. A Fast Software One-way Hash Function. *Journal of Cryptology*, **3**, 43–58, 1990.
16. K. Rubin and A. Silverberg. Torus-based cryptography. Technical Report 39, IACR's ePrint Archive, 2003.
17. M.-J. O. Saarinen. Security of VSH in the real world. Technical Report 103, IACR's ePrint Archive, 2006.
18. V. Shoup. NTL: A Library for doing Number Theory. Available from: http://www.shoup.net/ntl/
19. M. Stam and A. K. Lenstra. Efficient subgroup exponentiation in quadratic and sixth degree extensions. In J. Burton S. Kaliski, Ç. Koç, and C. Paar, editors, *CHES'02*, volume 2523 of *Lecture Notes in Computer Science*, pages 318–332. Springer-Verlag, 2003.
20. X. Wang, H. Yu, and Y. L. Yin. Efficient Collision Search Attacks on SHA-0. In *Advances in Cryptology (CRYPTO)*, Springer-Verlag LNCS 3621, 1–16, 2005.
21. X. Wang, Y. Yin, H. Yu. Finding Collisions in the Full SHA-1. In *Advances in Cryptology (CRYPTO)*, Springer-Verlag LNCS 3621, 7–36, 2005.

A   Fast VSH

Recall that for $i \in \mathbb{Z}_{>0}$, we let p_i be the i-th prime number. Let N be an RSA modulus, let k be the block length and let b be the chunking factor. To avoid intermediate reductions, one should ensure that $\prod_{i=1}^{k} p_{(2^b-1)i} < N$. Note that the Merkle-Damgård strengthening listed below might allow collisions on messages of length greater than 2^{bk}, but for reasonable parameter choices of b and k this will not be an issue.

Algorithm 11. VSH compression function [6].

The VSH compression function $\mathcal{H}_{VSH} : \{0,1\}^* \to \mathbb{Z}_N^*$ is defined as follows for an ℓ-bit message m consisting of bits $m_1, m_2, \ldots, m_\ell$.

1. [Padding] Set $\mathcal{L} \leftarrow \lceil \frac{\ell}{bk} \rceil$ and $m_i \leftarrow 0$ for $\ell < i \leq \mathcal{L}bk$.
2. [Radix Conversion] Set $M_i = \sum_{j=0}^{b-1} m_{b(i-1)+j+1} 2^j$ for $0 < i \leq k$.
3. [Merkle-Damgård Strengthening] Let $\ell = \sum_{i=1}^{k} \ell_i 2^{(i-1)b}$ with $\ell_i \in \{0, 2^b - 1\}$ be the 2^b-ary representation of the message length ℓ. Set $M_{\mathcal{L}k+i} \leftarrow \ell_i$ for $0 < i \leq k$.
4. [Initialise Loop] Set $x_0 \leftarrow 1$ and $j \leftarrow 0$.
5. [Padding] Set $m_i \leftarrow 0$ for $\ell < i \leq \mathcal{L}bk$.
6. [Finished] If $j = \mathcal{L} + 1$ terminate with output $x_{\mathcal{L}+1}$.
7. [Prepare product] Set $P \leftarrow \prod_{i=1}^{k} p_{M_i+(i-1)(2^b-1)}$ skipping those i for which $M - i = 0$.
8. [Hash next block] Set $x_{j+1} \leftarrow x_j^2 \times P \pmod{N}$.
9. [Increase j] Increase j by one. Go back to Step 6.

How to Construct Sufficient Conditions for Hash Functions

Yu Sasaki[1], Yusuke Naito[1], Jun Yajima[2], Takeshi Shimoyama[2],
Noboru Kunihiro[1], and Kazuo Ohta[1]

[1] The University of Electro-Communications
Chofugaoka 1-5-1, Chofu-shi, Tokyo, 182-8585, Japan
{yu339,tolucky,kunihiro,ota}@ice.uec.ac.jp
[2] Fujitsu Laboratories Ltd
4-1-1, Kamikodanaka, Nakahara-ku, Kawasaki, 211-8585, Japan
{jyajima,shimo}@labs.fujitsu.com

Abstract. Wang et al. have proposed collision attacks for various hash functions. Their approach is to first construct a differential path, and then determine the conditions (sufficient conditions) that maintain the differential path. If a message that satisfies all sufficient conditions is found, a collision can be generated. Therefore, in order to apply the attack of Wang et al., we need techniques for constructing differential paths and for determining sufficient conditions.

In this paper, we propose the "SC algorithm", an algorithm that can automatically determine the sufficient conditions. The input of the SC algorithm is a differential path, that is, all message differentials and differentials of the chaining variables. The SC algorithm then outputs the sufficient conditions. The computation time of the SC algorithm is within few seconds. In applying the method of Wang et al. to MD5, there are 3 types of sufficient conditions: conditions for controlling the carry length when differentials appear in the chaining variables, conditions for controlling the output differentials of the Boolean function when the input variables of the function have differentials and conditions for controlling the relationship between the carry effect and left rotation operation. Sufficient conditions for SHA-1, SHA-0 and MD4 consist of only Type 1 and Type 2. Type 3 is unique to MD5. The SC algorithm can construct Type 1 and Type 2 conditions; we use the method of Liang et al. to construct Type 3 conditions.

The complexity of the collision attack depends on the number of sufficient conditions needed. The SC algorithm constructs the fewest possible sufficient conditions. To check the feasibility of the SC algorithm, we apply it to the differential path of MD5 given by Wang et al. It is shown to yield 12 fewer conditions than the latest work on MD5. The SC algorithm is applicable to the MD-family and the SHA-family. This paper focuses on the sufficient conditions of MD5, but only as an example.

Keywords: Hash Function, Collision Attack, Differential Path, Sufficient Condition.

P.Q. Nguyen (Ed.): VIETCRYPT 2006, LNCS 4341, pp. 243–259, 2006.

1   Introduction

MD4, MD5, SHA-0 and SHA-1 are hash functions that are in wide use [6,7,4,5]. Since the operations of these hash functions are light, they can be calculated quickly. A hash function $H(\cdot)$ must hold the following 3 properties:

Pre-image Resistance. When y is given, it must be difficult to find M s.t. $H(M) = y$.

Second Pre-image Resistance. When M is given, it must be difficult to find M' s.t. $H(M) = H(M')$ and $M \neq M'$.

Collision Resistance. It must be difficult to find M, M' s.t. $H(M) = H(M')$ and $M \neq M'$

Collision Resistance is more difficult to keep than any other property. This paper uses the term Collision Attack to refer to attacks that break Collision Resistance. Since many collision attacks have been proposed recently, more attention has been paid to the analysis of collision attacks.

Wang et al. proposed collision attacks against various hash functions such as MD5 and SHA-1 [9,10,11,12,13]. We can say that their method is a form of differential attack. Since hash functions perform only numerical addition on the input message, Wang et al. used numerical subtraction as the differential. On the other hand, chaining variables are used for calculating not only numerical addition but also Boolean functions. Therefore, Wang et al. used both numerical subtraction and XOR to express the differentials of chaining variables. In this paper, we call the differentials of numerical subtraction "numerical differentials" and the differentials of XOR "bit differentials". The method of Wang et al. can be divided into 2 phases: pre-computation and collision search. In the pre-computation phase, a differential path is constructed and the conditions for maintaining the differential path are determined. In this paper, we call these conditions "sufficient conditions". The sufficient conditions are constructed to suit the differential path. In the collision search phase, they, together with message modification, enable a message that satisfies all sufficient conditions to be efficiently found. If such a message is found, a collision can be generated. Therefore, in order to apply the attack of Wang et al., we need the following 2 techniques:

1. method of constructing a differential path that can lead to a collision,
2. method of determining the sufficient conditions.

To ensure practicality, these techniques should be automated. This would allow many differential paths that have high probability in terms of generating collision to be automatically constructed by technique #1, and the sufficient conditions of those differential paths to be automatically constructed by technique #2. These techniques also should minimize the number of conditions since the complexity of the collision attack depends on the number of sufficient conditions.

The paper of Wang et al. describes a differential path and the sufficient conditions for maintaining the path. Unfortunately, they did not explain how to construct the sufficient conditions. Furthermore, there was no guarantee that the proposed sufficient conditions were correct. Many researchers have attempted to rectify these omissions. P. Hawkes, M. Paddon and G. G. Rose analyzed how sufficient conditions worked to generate a collision [2], J. Black, M. Cochran and T. Highland studied how to determine the message differential of MD5 [1]. However, these papers did not describe an algorithm for determining the sufficient conditions. M. Schlaffer and E. Oswald analyzed how to construct the differential path and the sufficient conditions, but only for MD4 [8]. These analyses targeted specific hash functions, MD4 or MD5. Therefore, it is unknown whether these analyses can be applied to the SHA-family. In this paper, we propose an algorithm that can automatically determine the sufficient conditions when a differential path is given. For any given differential path, the corresponding sufficient condition sets are not unique. Our algorithm determines the fewest possible sufficient conditions in order to reduce attack complexity. This algorithm is applicable to the SHA-family and the MD-family. In this paper, we explain how to determine the sufficient conditions of MD5 as an example.

Outline of Constructing Sufficient Condition Algorithm. The input of our algorithm is a differential path, that is, all message differentials and the differentials of chaining variables. Our algorithm then outputs the sufficient conditions. In this paper, we call this algorithm "The SC algorithm". In the attack of Wang et al. for MD5, there are 3 types of sufficient conditions.

Type 1: Conditions for controlling the carry length when differentials appear in the chaining variables

Type 2: Conditions for controlling the output differentials of the Boolean function when input variables of the function have differentials

Type 3: Conditions for controlling the relationship between the carry effect and left rotation operation

Sufficient conditions for SHA-1, SHA-0 and MD4 consist of only Type 1 and Type 2. Type 3 is unique to MD5. The SC algorithm can construct Type 1 and Type 2 conditions, so all sufficient conditions for SHA-1, SHA-0 and MD4 can be constructed. However, the SC algorithm cannot, by itself, construct all sufficient conditions for MD5. Although Wang et al. did not describe Type 3 conditions, other research groups have. J. Yajima and T. Shimoyama experimentally found conditions for rotation and carry [14]. J. Liang and X. Lai proposed a method of constructing conditions for rotation and carry, and described the construction of all Type 3 conditions [3].

We need to construct conditions for controlling the carry length for all differentials in the chaining variables. For example, when a chaining variable has a differential of $+2^i$, the number of bits that are changed by the differential is dependent on the value of the chaining variable. If the value of the chaining variable in bit position $i + 1$ is 0, it is changed into 1 by the differential, and the other bits remain unchanged. However, if the value of the chaining variable in bit position $i + 1$ is 1, a carry is triggered in bit position $i + 1$ and the value in bit position $i + 2$ is also changed. Therefore, if we want to stop the carry, we must fix the value in bit position $i + 1$ to 0 by setting a sufficient condition. If we need to trigger a carry, we must fix the value in bit position $i + 1$ to 1. If the carry is long, the number of bit differentials used to calculate the Boolean function becomes big. Therefore, the number of sufficient conditions for controlling the Boolean function increases. The SC algorithm minimizes the sufficient conditions, therefore, the number of bit differentials triggered by a carry must be as small as possible. The SC algorithm, first, sets differentials to prevent carries. If it needs to expand a carry, the SC algorithm tries to make carry length as short as possible.

Conditions for controlling the output differentials of the Boolean function are constructed by bit differentials of input variables of the function. For each bit differential of an input variables we decide whether it should be reflected in the output. The following situation is considered as an example.

- Boolean function ϕ is defined as $\phi(x, y, z) = (x \wedge y) \vee (\neg x \wedge z)$[1].
- The chaining variable z has a differential in bit position i.

In this example, if the value of x in bit position i is 0, the value of ϕ becomes $\phi(0, y, z) = (0 \wedge y) \vee (1 \wedge z) = z$. Therefore, the output of the ϕ has the same differential as z in bit position i. On the other hand, if the value of x in bit position i is 1, the value of ϕ becomes $\phi(1, y, z) = (1 \wedge y) \vee (0 \wedge z) = y$. Therefore, the output of ϕ doesn't have a differential.

To check the feasibility of the proposed algorithm, we applied it to the differential path of MD5 given by Wang et al. [11]. The latest result on the sufficient conditions of the path of Wang et al. was provided by Liang et al. [3]. Although Liang et al. pointed out and corrected some mistakes of the original sufficient conditions of Wang et al., we newly found that the corrected conditions described by Liang et al. still contained unnecessary conditions. Since collision attack complexity depends on the number of sufficient conditions, the SC algorithm yields a more efficient attack. Our sufficient conditions are used to generate an actual collision.

Section 2 explains the structure of the hash functions and notations used in this paper. Section 3 introduces our algorithm for constructing sufficient conditions. This is the main content of this paper. In Section 4, we explain the results of applying the proposed algorithm to the differential path of MD5 given by Wang et al. In Section 5, we explain the application of the proposed algorithm to other hash functions such as MD4, SHA-0, SHA-1. Finally, we summarize this paper.

2   Preliminaries

2.1   Description of Hash Functions: The MD-Family and the SHA-Family

All hash functions attacked by Wang et al. have the Merkle-Damgård structure. They calculate the hash value by repeatedly calculating a compression function.

---

[1] This Boolean function is used in the first round of MD4, MD5, SHA-0 and SHA-1.

In MD4,MD5,SHA-0 and SHA-1, the input message is padded to yield a multiple of 512 bits. Since the padding procedure is not related to the collision attack, we omit an explanation of the padding procedure. Padded message M is divided into 512-bit messages ($M = M_1, M_2, \cdots, M_n, |M_i| = 512$). These divided messages are input to the compression function.

$$h_1 = compress(M_1, IV) \rightarrow h_2 = compress(M_2, h_1) \rightarrow \cdots \rightarrow h_n = compress(M_n, h_{n-1})$$
$$H(M) = h_n$$

In the above expression, IV is a constant defined by the specification of the compression function. In this paper, we call the calculation performed in a single run of the compression function "1 block".

We next explain the structure of the compression function. Although the compression function depends on the hash function, MD4,MD5,SHA-0 and SHA-1 share many similar structures. All calculations in these hash functions are 32-bit. In this paper, we exclude the description of "mod 2^{32}". A message input to compression function M_i is divided into 32 bit messages ($M_i = m_0, m_1, \cdots, m_{15}$). In SHA-0 and SHA-1, m_{16} to m_{79} are calculated by message expansion. However, since the proposed algorithm assumes that message differentials after message expansion are given, we omit details of message expansion. The chaining variables are updated by a certain calculation. In this paper, we call the calculation for updating the chaining variables 1 time the "Step Function". We use the notation Q_i to describe the chaining variable calculated in the i-th step. The step function also differs depending on the hash function used. Since we focus on MD5 in this paper, we explain the step function of MD5. In MD5, the step function is repeated 64 times. Chaining variables $Q_i, 0 \le i \le 63$ are calculated as follows:

$$Q_i = Q_{i-1} + (Q_{i-4} + \phi(Q_{i-1}, Q_{i-2}, Q_{i-3}) + m_i + t_i) \lll j$$

Here, t_i is a constant defined in each step, $\lll j$ denotes left rotation by j bits. m_i is a message. Q_{-1}, Q_{-2}, Q_{-3} and Q_{-4} are the initial values defined by the specification of MD5. Steps 0-15 are called the first round. Steps 16-31, 32-47 and 48-63 are the second, third and fourth rounds, respectively. $\phi(Q_{i-1}, Q_{i-2}, Q_{i-3})$ is a Boolean function defined in each round. Details are as follows:

1st round: $\phi(X, Y, Z) = (X \wedge Y) \vee (\neg X \wedge Z)$
2nd round: $\phi(X, Y, Z) = (X \wedge Z) \vee (Y \wedge \neg Z)$
3rd round: $\phi(X, Y, Z) = X \oplus Y \oplus Z$
4th round: $\phi(X, Y, Z) = (X \vee \neg Z) \oplus Y$

At the last, the compression function outputs $(Q_{-4}+Q_{60})\|(Q_{-3}+Q_{61})\|(Q_{-2}+Q_{62})\|(Q_{-1}+Q_{63})$.

2.2   List of Notations

In this section, we define the notations used in this paper.

- $m_i, 0 \le i \le 63$ denotes a message used in step i.
- $Q_i, 0 \le i \le 63$ denotes a chaining variable calculated in step i. $Q_i, -4 \le i \le -1$ are the initial values.

$$Q_i = Q_{i-1} + (Q_{i-4} + \phi(Q_{i-1}, Q_{i-2}, Q_{i-3}) + m_i + t_i) \lll j$$

- $\phi_i, 0 \le i \le 63$ denotes a Boolean function used in step i.
- $x_{i,j}, 0 \le i \le 63, 1 \le j \le 32$ (x is one of m, Q, ϕ) denotes a value of x in the bit position j in step i.
- Δx (x is one of m, Q, ϕ) denotes the differentials of x. Regarding m_i, ϕ_i, we consider only numerical differentials since they are used only to calculate addition. We describe numerical differential by using the exponentiation of 2.
- $Q_i[\pm j]$ denotes the bit differentials of Q_i. Regarding Q_i, we need to consider not only numerical differentials but also bit differentials since they are used to calculate both addition and Boolean functions. $Q_i[j]$ means that there exists a bit differential whose sign is $+$ in bit position j. $Q_i[-j]$ means that there exists a bit differential whose sign is $-$ in bit position j. For example, the description $Q_3[6, 7, -8, -20, -21, -22, 23]$ means that Q_3 has positive differentials in bit positions 6,7,23 and negative differentials in bit positions 8,20,21,22. If we describe these bit differentials by using numerical differentials, we have $\Delta Q_3 = -2^5 + 2^{19}$. On the other hand, when numerical differentials are given to a chaining variable, we cannot determine the corresponding bit differentials. Unless the values of Q_3 are fixed by sufficient conditions, we cannot determine the bit differentials from numerical differentials. For example, the numerical differential $\Delta Q_3 = -2^5$ can be $Q_3[-6]$, $Q_3[6, -7]$, $Q_3[6, 7, -8]$ and so on.

3 Algorithm for Constructing Sufficient Conditions (The SC Algorithm)

If our algorithm is provided with a differential path, that is, all message differentials and numerical differentials of chaining variables, it outputs the corresponding sufficient conditions. In this paper, we call this algorithm "The SC algorithm". The step function for step i is as follows:

$$Q_i = Q_{i-1} + (Q_{i-4} + \phi(Q_{i-1}, Q_{i-2}, Q_{i-3}) + m_i + t_i) \lll j$$

In the above expression, $\Delta Q_i, \Delta Q_{i-1}, \Delta Q_{i-2}, \Delta Q_{i-3}, \Delta Q_{i-4}$ and Δm_i are given in the SC algorithm. Since ϕ is a Boolean function, ϕ is calculated in each bit position. Therefore, we need to construct sufficient conditions that determine which bits of $\Delta Q_{i-1}, \Delta Q_{i-2}$ and ΔQ_{i-3} have differentials. We need to construct such conditions for all differentials of the chaining variables. All operations except for Boolean function ϕ consist of addition and rotation. Therefore, we cannot control the impact of differentials except for ϕ. However, regarding the differentials of chaining variables that are used as input of ϕ, we can control whether we reflect the differentials of input to the output of ϕ or not by using the property of ϕ. Therefore, we construct sufficient conditions for controlling the output of ϕ for all bits that have differentials. Finally, the SC algorithm constructs 2 types of sufficient conditions.

- Conditions for controlling the carry length when differentials appear in the chaining variables
- Conditions for controlling the output differentials of ϕ when input variables of ϕ have differentials

The SC algorithm goes backward when it constructs sufficient conditions. That is, the SC algorithm constructs conditions from the last step of the last message block to the first step of the first message block. In each step, the SC algorithm uses the following procedure to construct the sufficient conditions.

Process 1: Calculate $\Delta\phi$. We will find some candidates of $\Delta\phi$. We choose the one that has the highest probability in terms of maintaining the differential path (Discussed in Section 3.1).

Process 2: Set the differentials of the chaining variables to prevent carries.

Process 3: Construct sufficient conditions for controlling the output of $\Delta\phi$ from the first bit to the last bit (Discussed in Section 3.2). If the SC algorithm can construct conditions for all bit differentials of all chaining variables, output those sufficient conditions, and stop the algorithm. If a sufficient condition we need contradicts a sufficient condition constructed in a previous step, we run the following procedures to eliminate the contradiction.

    Process 3-1: If we can avoid contradiction within that step, do it. Otherwise, go to process 3-2 (Discussed in Section 3.3).

    Process 3-2: Let the previous step in which the contradicted condition was constructed be step A. First, algorithm resets the conditions which were constructed after step A. After that, algorithm goes back to step A, and choose another sufficient condition that avoids the contradiction. Then, algorithm restarts from step A. If contradiction is not solved, go to process 3-3.

    Process 3-3: We calculate $\Delta\phi$ of this step, and then choose the $\Delta\phi$ that has the next highest probability as $\Delta\phi$ in this step. If we try all $\Delta\phi$ and the contradiction remains, it means that there is no sufficient condition which can hold the given differential path. Therefore, we stop the algorithm.

To show how the SC algorithm works, we offer the following example:

- $\Delta Q_{i-1} = 0, \Delta Q_{i-2} = 0, \Delta Q_{i-3} = 2^5,\ \Delta\phi(Q_{i-1}, Q_{i-2}, Q_{i-3}) = 2^7,$
- $\phi(Q_{i-1}, Q_{i-2}, Q_{i-3}) = (Q_{i-1} \wedge Q_{i-2}) \vee (\neg Q_{i-1} \wedge Q_{i-3}).$

At first, Procedure 1 is executed, so $\Delta\phi$ is calculated. $\Delta\phi(Q_{i-1}, Q_{i-2}, Q_{i-3})$ can be calculated by the following equation (We explain the strict method used to calculate $\Delta\phi$ in Section 3.1):

$$\Delta\phi(Q_{i-1}, Q_{i-2}, Q_{i-3}) = ((\Delta Q_i - \Delta Q_{i-1}) \ggg j) - \Delta Q_{i-4} - \Delta m_i - \Delta t_i$$

Then in Process 2, all differentials of the chaining variables are set so as to prevent carries. After that, in Process 3, we construct sufficient conditions for controlling $\Delta\phi$. In this example, since no input variable of ϕ has a differential in bit position 8, it is impossible to make $\Delta\phi(Q_{i-1}, Q_{i-2}, Q_{i-3}) = 2^7$. This is a contradiction in Process 3. This contradiction can be resolved in Process 3-1 by expanding the carry of differentials in Q_{i-3}. We make the bit differential in bit position 8 by expanding the carry of 2^5 in ΔQ_{i-3}. If the value of Q_{i-3} in bit position 8 changes, we can make the differential 2^7 in the output of ϕ. In order to transmit the carry up to bit position 8, we construct "$Q_{i-3,6} = 1$", "$Q_{i-3,7} = 1$" and "$Q_{i-3,8} = 0$" as sufficient conditions for controlling the carry length. By these conditions, the bit differentials of ΔQ_{i-3} become $Q_{i-3}[-6, -7, 8]$, and we can make $\Delta\phi = 2^7$. Conditions for controlling the carry length are constructed in the same way.

Since the bit differentials of ΔQ_{i-3} change into $Q_{i-3}[-6, -7, 8]$, we have to control all of them in Boolean function ϕ. Regarding bit positions 6 and 7, though Q_{i-3} has differentials, $\Delta\phi$ doesn't have output differentials in those bits. Therefore, we need to construct conditions to guarantee that the bit differentials of Q_{i-3} in bit positions 6 and 7 don't impact the output of $\Delta\phi$. From the property of ϕ, if we fix the value of $Q_{i-1,6}$ and $Q_{i-1,7}$ to be 1, the output of ϕ doesn't have differentials in those bits. Therefore, we construct sufficient conditions "$Q_{i-1,6} = 1$" and "$Q_{i-1,7} = 1$". Then in bit position 8, we make the differential $\Delta\phi(Q_{i-1}, Q_{i-2}, Q_{i-3}) = 2^7$ by using the bit differential of Q_{i-3}. From the property of ϕ, if we fix the value of $Q_{i-1,8}$ to be 0, we can make 2^7 in $\Delta\phi$. This yields sufficient conditions "$Q_{i-1,8} = 0$". Conditions for controlling $\Delta\phi$ are constructed in the same way.

In the SC algorithm, we focus on the input differentials and output differentials of ϕ, and construct sufficient conditions as shown in the example. The SC algorithm goes backward, that is, it constructs conditions from the last step of the last message block to the first step of the first message block. In this section, we give the details of each procedure. In Section 3.1, we explain how to strictly calculate $\Delta\phi$. In Section 3.2, we explain how to construct conditions for controlling the carry length of the chaining variables. In Section 3.3, we explain how to construct conditions for controlling the output of $\Delta\phi$. In Section 3.4, we explain methods that avoid contradictions.

3.1   Calculating $\Delta\phi$

We explain how to calculate $\Delta\phi_i$ when $\Delta Q_i, \Delta Q_{i-1}, \Delta Q_{i-2}, \Delta Q_{i-3}, \Delta Q_{i-4}$ and Δm_i are given. ϕ_i is the Boolean function in step i. It is used to calculate Q_i. The expression for Q_i is as follows:

$$Q_i = Q_{i-1} + (Q_{i-4} + \phi(Q_{i-1}, Q_{i-2}, Q_{i-3}) + m_i + t_i) \lll j.$$

For convenience, in this paper, we introduce notations u_i and v_i.

$$v_i = (Q_{i-4} + \phi(Q_{i-1}, Q_{i-2}, Q_{i-3}) + m_i + t_i)$$
$$u_i = (Q_{i-4} + \phi(Q_{i-1}, Q_{i-2}, Q_{i-3}) + m_i + t_i) \lll j$$

u_i and v_i are also described as $u_i = Q_i - Q_{i-1}$ and $v_i = u_i \ggg j$. We calculate $\Delta\phi_i$ by the following procedure:

Step 1: Calculate Δu_i by using $\Delta u_i = \Delta Q_i - \Delta Q_{i-1}$.

Step 2: Calculate $\Delta v_i = ((u_i + \Delta u_i) \ggg j) - (u_i \ggg j)$ for all u_i. The value of u_i is a 32-bit number. Therefore, we need to repeat this calculation 2^{32} times for each step. As a result of this computation, we will get at most four kinds of Δv_i. In order to raise the probability that a random value that satisfies Δv_i also satisfies Δu_i in this step, we choose the most frequently appearing value as Δv_i. However, if sufficient conditions cannot be constructed for such Δv, it is possible to choose a differential that has smaller probability. The influence of this reduction in probability in the first round can be ignored with regard to overall attack complexity. However, if such a selection must be made in the second or later rounds, it may have a significant influence. In this case, choosing another differential path is required.

Step 3: Calculate $\Delta\phi_i$ by using $\Delta\phi_i = \Delta v_i - \Delta m_{i-1} - \Delta Q_{i-4}$. Δt_i is always 0 since t_i is a constant. Therefore, we don't have to consider it.

3.2 Sufficient Conditions for Controlling the Carry of Chaining Variables

When a differential appears in a chaining variable, we have to decide whether we make a carry or not. In order to control the carry of a differential, we fix the value of the chaining variable where the differential exists by setting a sufficient condition. We construct such conditions by the following branches:

Let chaining variable Q_i have a differential of either $+2^{j-1}$ or -2^{j-1}.

Case 1: The sign of the differential is "+".
  Case 1-1: If we stop the carry, we can construct sufficient condition $Q_{i,j} = 0$. By this condition, the value of $Q_{i,j}$ changes from 0 to 1. No carry is triggered by this differential and no other bits are changed by this differential.
  Case 1-2: If we expand the carry, we can construct sufficient condition $Q_{i,j} = 1$. By this condition, the carry is expanded to the upper bits of $Q_{i,j}$, and $Q_{i,j+1}$ is also changed by this differential.
Case 2: If the sign of the differential is "-"
  Case 2-1: If we stop the carry, we can construct sufficient condition $Q_{i,j} = 1$.
  Case 2-2: If we expand the carry, we can construct sufficient condition $Q_{i,j} = 0$.

If we decide to expand the carry from the i-th bit to the $i{+}1$-th bit, we need to decide whether the carry should be expanded to $i{+}2$-th bit or not. Therefore, we also need to construct a sufficient condition for the expanded carry. By repeating this process, we can make the carry expand in arbitrary length (up to MSB).

We explain this algorithm by using the following example. For example, we assume that a differential of Q_4 is 2^6. In this case, the sign of the differential is

"+", therefore, Case 1 is executed. If we stop the carry, Case 1-1 is executed. Thus, "$Q_{4,7} = 0$" is generated as a sufficient condition. If we expand the carry, Case 1-2 is executed. Thus, "$Q_{4,7} = 1$" is generated as a sufficient condition, and we then decide a condition for $Q_{4,8}$.

If the length of the carry is different, the result of modular integer addition is not changed but the output of the Boolean function ϕ is changed. Generally speaking, if the number of bits that change by differentials increases, the number of sufficient conditions we need also increases (both conditions for controlling carry and conditions for controlling the output of ϕ). Therefore, at first, we construct conditions which stop the carry of differentials. If these conditions raise contradictions when we construct other conditions, we choose the other condition created by expanding the carry of the differential.

3.3 Sufficient Conditions for Controlling the Output of Boolean Function ϕ

The output differentials of Boolean function ϕ can be controlled by constructing sufficient conditions. After we calculate $\Delta\phi$ and decide the carry length of chaining variables, we construct sufficient conditions for controlling $\Delta\phi$.

In order to construct such conditions in step i in bit position j, we apply the following algorithm.

Process 1: If all differentials of the input and output of $\Delta\phi_i$ in bit position j are 0, we don't generate sufficient conditions since the output differential always holds (sufficient conditions are not needed). Otherwise, go to Process 2.

Process 2: For all possible input values, calculate the value of ϕ_i and ϕ'_i, that is, the value of ϕ_i after the differentials are applied.

Process 3: Calculate $\Delta\phi_i = \phi'_i - \phi_i$ for all possible input values.

Process 4: Choose input values whose $\Delta\phi_i$ are the same as the $\Delta\phi_i$ in this step.

Process 5: Extract the smallest common characteristics of the input values chosen in Process 4. Extracted conditions are the sufficient conditions.

By applying this algorithm from $i = 63$ to $i = 0$ and $j = 1$ to $j = 32$, we can construct all sufficient conditions for controlling $\Delta\phi$. To show how the above algorithm works, we give a small example. We assume the following situation.

- We construct sufficient conditions for controlling $\Delta\phi_{63}$. (Input of ϕ_{63} are Q_{62}, Q_{61} and Q_{60}.)
- Differential of ϕ_{63} is $\pm2^{31}$.
- Bit differential of Q_{62} is $Q_{62}[-26, \pm32]$.
- Bit differential of Q_{61} is $Q_{61}[-26, \pm32]$.
- Bit differential of Q_{60} is $Q_{60}[\pm32]$.

We set sufficient conditions for controlling ϕ from the first bit to the last bit.

- Bit positions 1-25: In this example, the input and output of ϕ_{63} don't have any differentials in these bit positions. Therefore, we don't generate any sufficient condition as defined in Process 1.

- Bit position 26: Q_{62} and Q_{61} have differentials in bit position 26, whereas, $\Delta\phi_{63}$ does not have. Therefore, Processes 2-5 are executed. In Process 2, we simulate the value of $\phi_{63,26}$ and $\Delta\phi'_{63,26}$ for all possible input values $Q_{62,26}, Q_{61,26}$ and $Q_{60,26}$. This result is shown in the second and third column of Table 1. In Table 1, x, y, z represents the value of $Q_{62,26}, Q_{61,26}$ and $Q_{60,26}$, respectively. In Process 3, we calculate the value of $\Delta\phi_i = \phi'_i - \phi_i$ for all possible input values. This result is shown in the fourth column of Table 1. In Process 4, we choose $(x, y, z) = (0, 0, 1), (0, 1, 1), (1, 0, 1), (1, 1, 1)$ as possible input values since their $\Delta\phi_i$ are the same as the $\phi_{63,26}$. In Process 5, we extract the smallest common characteristics of the input values chosen in Process 4. We can easily find that "$z = 1$", that is, "$Q_{60,26} = 1$" is the smallest common characteristic of the input values chosen in Process 4. Finally, we construct "$Q_{60,26} = 1$" as a sufficient condition for "$\Delta\phi_{63,26} = 0$".

Table 1. Constructing Sufficient Conditions for "$\Delta\phi_{63,26} = 0$"

x, y, z	$\phi(x, y, z)$	$\phi' = \phi(\neg x, \neg y, z)$	$\Delta\phi(= \phi' - \phi)$
0,0,0	1	0	-1
0,0,1	0	0	0
0,1,0	0	1	1
0,1,1	1	1	0
1,0,0	1	0	-1
1,0,1	1	1	0
1,1,0	0	1	1
1,1,1	0	0	0

- Bit positions 27-31: There are no input and output differentials for ϕ_{63}, so we don't generate any sufficient condition as defined in Process 1.
- Bit position 32: Q_{62}, Q_{61} and Q_{60} have differentials in bit 32, and $\Delta\phi_{63,32} = \pm 2^{31}$. Therefore, Processes 2-5 are executed. In Process 2, we simulate the value of $\phi_{63,32}$ and $\Delta\phi'_{63,32}$ for all possible input values $Q_{62,32}, Q_{61,32}$ and $Q_{60,32}$. This result is shown in the second and third column of Table 2. In Table 2, x, y, z represents the value of $Q_{62,32}, Q_{61,32}$ and $Q_{60,32}$, respectively. In Process 3, we calculate the value of $\Delta\phi_i = \phi'_i - \phi_i$ for all possible input values. This result is shown in the fourth column of Table 2. In Process 4, we choose $(x, y, z) = (0, 0, 0), (0, 1, 0), (1, 0, 1), (1, 1, 1)$ as possible input values since their $\Delta\phi_i$ are one of the $\phi_{63,32}$. In Process 5, we extract the smallest common characteristics of the input values chosen in Process 4. We can easily find that "$x = z$", that is, "$Q_{62,32} = Q_{60,32}$" is the smallest common characteristic of the input values chosen in Process 4. Finally, we construct "$Q_{62,32} = Q_{60,32}$" as a sufficient condition for "$\Delta\phi_{63,32} = \pm 1$".

Table 2. Constructing Sufficient Conditions for "$\Delta\phi_{63,32} = \pm 1$"

x, y, z	$\phi(x, y, z)$	$\phi' = \phi(\neg x, \neg y, \neg z)$	$\Delta\phi(= \phi' - \phi)$
0,0,0	1	0	-1
0,0,1	0	0	0
0,1,0	0	1	1
0,1,1	1	1	0
1,0,0	1	1	0
1,0,1	1	0	-1
1,1,0	0	0	0
1,1,1	0	1	1

Since we now have constructed all sufficient conditions for all bits of ϕ_{63}, the desired differentials of ϕ_{63} can be calculated with probability of 1. By applying this procedure to all steps, we can construct all sufficient conditions for controlling $\Delta\phi$.

3.4 Techniques for Avoiding the Contradiction of Sufficient Condition

The contradiction of a sufficient condition means that a sufficient condition that we need to construct in a certain step is contradicted by a sufficient condition constructed in a previous step. For example, assume that we construct condition "$Q_{30,5} = 0$" to stop the carry of Q_{30}, and then, construct condition "$Q_{30,5} = 1$" for controlling $\Delta\phi_{31}$. In this example, the sufficient condition for Q_{30} is contradicted. When we execute the SC algorithm, such contradictions are frequent, so techniques for avoiding them are very important. This is possible by executing the next procedure.

Process 1: If we can avoid contradiction within that step, do it. Otherwise, go to process 2.

Process 2: Algorithm goes back to the previous step in which the contradicted condition was constructed, and then choose another sufficient condition in order to avoid contradiction. The algorithm restarts from this step. If the contradiction is not resolved, go to process 3.

Process 3: We calculate $\Delta\phi$ of this step, then choose another $\Delta\phi$ that has the next highest probability as the $\Delta\phi$ of this step. (Discussed in 3.1). If we try all $\Delta\phi$ but the contradiction remains, it means there is no sufficient condition that can maintain the given differential path. Therefore, we stop the algorithm.

In this section, we explain the 2 techniques of Process 1.

Expanding Carry

We sometimes need to expand the effect of the chaining variable carry. This situation occurs when $\Delta\phi_i$ in bit position j is not 0, but differentials of all input chaining variables to ϕ_i in bit position j are 0. As long as the differentials of all input chaining variables are 0, it is impossible to make differentials in the output of ϕ. Therefore, we expand the carry of a differential from the next lowest bit position to bit position j. We give an example to explain this technique.

We assume that the value of $\Delta\phi_i$ is 2^{-19}, and bit differentials of chaining variables $Q_{i-1}, Q_{i-2}, Q_{i-3}$ which are inputs of ϕ_i are $Q_{i-1}[5], Q_{i-2}[-16]$, $Q_{i-3}[10, 11, -12]$. Since $\Delta\phi_i$ in bit position 20 is not 0, at least one of the input variables Q_{i-1}, Q_{i-2} and Q_{i-3} need to have a differential in bit position 20. However, no input variable has a differential in bit position 20. Therefore, we need to expand the carry of a chaining variable up to bit position 20. Since the bit differential nearest to 20 is $Q_{i-2}[-16]$, we expand the effect of $Q_{i-2}[-16]$ up to bit position 20 by using the carry. To do this, we use the technique of constructing sufficient conditions for controlling carry length (Discussed in Section 3.2). Finally, we get sufficient conditions $Q_{i-2,16} = 0, Q_{i-2,17} = 0$, $Q_{i-2,18} = 0, Q_{i-2,19} = 0, Q_{i-2,20} = 1$.

Changing the Representative of $\Delta\phi$

The representative of $\Delta\phi$ can be changed. For example, $\Delta\phi = 2^{20}$ can be changed to $\Delta\phi = -2^{20} + 2^{21}$, or $\Delta\phi = -2^{20} - 2^{21} + 2^{22}$ and so on. This technique is useful when $\Delta\phi_i$ in bit position j and input chaining variables of ϕ_i in bit position j and subsequent bit have differentials, but the sign of $\Delta\phi_i$ and the sign of differentials that can be created by input chaining variables in bit position j are opposite. If this situation occurs, we change the expression of $\Delta\phi_i = \pm 2^{j-1}$ to $\Delta\phi_i = \mp 2^{j-1} \pm 2^j$ (the order of the sign must be same). By this change, we can avoid a contradiction in bit position j. Therefore, if sufficient conditions for controlling $\Delta\phi_i$ in bit position $j+1$ are constructed without contradiction, we can solve all problems with regard to $\Delta\phi_i$.

We show the usefulness of this transformation by considering the following situation.

Input chaining variables for ϕ_i are x, y and z.

- The expression of ϕ_i is $\phi_i(x, y, z) = (x \wedge y) \vee (\neg x \wedge z)$
- The value of $\Delta\phi_i$ is 2^{14}.
- Differential of x is 0.
- Differential of y is $y[20]$.
- Differentials of z are $z[-15, -16, -17, -18, -19, -20, 21]$.

In this case, from the property of ϕ, it is impossible to make $\Delta\phi_i = 2^{14}$ by setting sufficient conditions. However, it is possible to make $\Delta\phi_i = -2^{14}$. Therefore, we replace the value of $\Delta\phi_i = 2^{14}$ with $\Delta\phi_i = -2^{14} + 2^{15}$. This allows the construction of sufficient conditions for controlling $\Delta\phi$ in bit position 15. However, there is another problem in bit 16. Since it is impossible to make $\Delta\phi_i = 2^{15}$, we need to further transform the expression of $\Delta\phi$. We continue this discussion until the contradiction of $\Delta\phi_i$ is resolved. In the end, we transform $\Delta\phi_i = -2^{14}$

into $\Delta\phi_i = -2^{14} - 2^{15} - 2^{16} - 2^{17} - 2^{18} + 2^{19}$, and we get sufficient conditions $x_{15} = 0, \ldots, x_{19} = 0, x_{20} = 1$ in order to control the differential of $\Delta\phi_i$.

By using the techniques explained in this section, we can find all sufficient conditions for maintaining the differential path. In the next section, we show the result of applying the SC algorithm to the differential path of MD5 given by Wang et al.

4   Results of Applying the SC Algorithm to the Differential Path of MD5 Given by Wang et al.

To check the feasibility of the SC algorithm, we apply it to the differential path of MD5 given by Wang et al. [11]. Liang et al. corrected some of the mistakes in the original sufficient conditions of Wang et al. in their paper [3]. Our research showed that it was possible to further reduce the number of sufficient conditions. We found that the result of Liang et al. included 11 unnecessary sufficient conditions in the 1st round of the 1st block, and 1 condition in the 4th round of the 2nd block.

Liang et al. pointed out several of the mistakes in the sufficient conditions [3]. For the 1st block, they added the conditions "$Q_{61,26} = 0$" and "$Q_{62,26} = 0$", corrected the condition of $Q_{62,32}$ from "$Q_{62,32} = Q_{61,32}$" to "$Q_{62,32} = Q_{60,32}$" and deleted the condition "$Q_{60,27} = 0$". For the 2nd block, they added the condition "$Q_{59,26} = 0$". In their paper, they showed only these results. They failed to mention how they identified the mistakes of Wang et al. In our research, as a result of applying the SC algorithm to the differential path, we could find the same mistakes and several new ones.

First, the SC algorithm deletes the condition "$Q_{63,26} = 1$" from the 2nd block. This claim can be confirmed by running the SC algorithm for step 63 of the 2nd block. Deleting "$Q_{63,26} = 1$" means that we don't have to wait until "$Q_{63,26} = 1$" is satisfied by random search in the collision search phase. Therefore, the complexity of the 2nd block is reduced by a factor $1/2$.

Second, we found that 11 conditions in the 1st round of the 1st block could be deleted. These conditions are "$Q_{3,21} = Q_{2,21}$", "$Q_{3,22} = Q_{2,22}$", "$Q_{3,23} = Q_{2,23}$", "$Q_{4,21} = 0$", "$Q_{4,22} = 0$", "$Q_{4,23} = 1$", "$Q_{5,21} = 1$", "$Q_{5,22} = 1$", "$Q_{5,23} = 1$", "$Q_{6,22} = 1$", "$Q_{6,23} = 1$". These conditions are related to the carry length of the differential in Q_4. The method of Wang et al. expands the carry of -2^6 in Q_4 to bit position 23. However, the SC algorithm found that the collision paths could hold even if we shorten the carry length to bit position 20. In order to stop the carry in bit position 20, we need to change the condition of $Q_{4,20}$ from "$Q_{4,20} = 0$" to "$Q_{4,20} = 1$". This allows us to delete the conditions for controlling carry length from bit position 21 to 23 and conditions for controlling $\Delta\phi_5, \Delta\phi_6$ and $\Delta\phi_7$ that includes Q_4 as input.

We show a collision that was generated, without the above unnecessary conditions, in Table 3. This collision cannot be generated by using the sufficient conditions given by Wang et al.

Table 3. Generated Collision Messages without Unnecessary Sufficient Conditions

M_0	0x8b075d00f54501bce81f9cab86312f9d3a8bdca58446d56583e9e8365f99ddba 069badd582343c027f16e96793f95b7bdcdbe711c0dc183a6966bb7243c35a00
M_1	0x6c434ce72b9c78834e6f32b8ddfeb19025dc928a2cfc643df71f7512ee2de7d2 117670796628d0b098571b460d91348085075cc9ff33ceb5d8f871a1971f1fe0
M_0'	0x8b075d00f54501bce81f9cab86312f9dba8bdca58446d56583e9e8365f99ddba 069badd582343c027f16e96793f9db7bdcdbe711c0dc183ae966bb7243c35a00
M_1'	0x6c434ce72b9c78834e6f32b8ddfeb190a5dc928a2cfc643df71f7512ee2de7d2 117670796628d0b098571b460d90b48085075cc9ff33ceb558f871a1971f1fe0
Hash value	0x94db011516925a92cb0af8dd07992804

5 Application to Other Hash Functions

The SC algorithm is applicable to not only MD5 but also SHA-1, SHA-0, and MD4. Even if the considered hash function is changed, the input and the output of the SC algorithm does not change. That is, the input of the SC algorithm is all message differentials and the differentials of chaining variables.

In this section, we apply the SC algorithm to SHA-1, and explain how to construct sufficient conditions of SHA-1. In SHA-1, the number of chaining variables is 5 $(a_i, b_i, c_i, d_i, e_i)$. These chaining variables are updated in every step. The number of steps is 80. The step function in step i is as follows:

$$a_i = (a_{i-1} \lll 5) + \phi(b_{i-1}, c_{i-1}, d_{i-1}) + e_{i-1} + m_{i-1} + k_{i-1},$$
$$b_i = a_{i-1},$$
$$c_i = b_{i-1} \lll 30,$$
$$d_i = c_{i-1},$$
$$e_i = d_{i-1}.$$

Here, k_{i-1} is a constant. In the case of SHA-1, the method used to calculate $\Delta\phi$ is different from that for MD5. $\Delta\phi$ can be calculated by transforming the expression for a_i.

$$\Delta\phi = \Delta a_i - (\Delta a_{i-1} \lll 5) - \Delta e_{i-1} - \Delta m_{i-1} - \Delta k_{i-1}$$

Since the expression for $\Delta\phi$ of SHA-1 doesn't include a rotation operation, which exists in the case of MD5, the number of candidates of $\Delta\phi$ for SHA-1 is 1. Therefore, we don't need to search for candidates of $\Delta\phi$ by trying all possible values. The procedure after getting $\Delta\phi$ is the same as that in the case of MD5. However, the SC algorithm cannot execute Procedure 3.3 of MD5, which tries to apply several $\Delta\phi$. Since the structure of the compression functions of SHA-0 and MD4 are similar to that of MD5, the SC algorithm can construct sufficient conditions for SHA-0 and MD4 by using similar procedures.

6   Conclusion

In this paper, we proposed an algorithm, the SC algorithm, that can automatically construct sufficient conditions. The input of the SC algorithm is a differential path, that is, all message differentials and numerical differentials of chaining variables. The SC algorithm outputs the sufficient conditions. In the attack of Wang et al. for MD5, there are 3 types of sufficient conditions.

Type 1: Conditions for controlling the carry length when differentials appear in the chaining variables

Type 2: Conditions for controlling the output differentials of the Boolean function when input variables of the function have differentials

Type 3: Conditions for controlling the relationship between the carry effect and left rotation operation

Sufficient conditions for SHA-1, SHA-0 and MD4 consist of only Type 1 and Type 2. Type 3 is unique to MD5. The SC algorithm can construct Type 1 and Type 2 conditions, so all sufficient conditions for SHA-1, SHA-0 and MD4 can be constructed. However, regarding MD5, it is impossible to construct all sufficient conditions by using only the SC algorithm. Liang et al. proposed a method that constructs Type 3 conditions [3]. Therefore, by running the SC algorithm and the method of Liang et al., all sufficient conditions of MD5 can be constructed.

In this research, in order to show the feasibility of the SC algorithm, we applied it to the differential path of MD5 given by Wang et al. As a result, we could construct more efficient sufficient conditions (11 fewer necessary conditions) compared to the latest work on the same subject.

Acknowledgement

We would like to thank The Telecommunications Advancement Foundation for supporting our research.

References

1. J. Black, M. Cochran and T. Highland. *A Study of the MD5 Attacks.* FSE 2006, Springer-Verlag, 2006.
2. P. Hawkes, M. Paddon and G. G. Rose. *Musings on the Wang et al. MD5 Collision.* Cryptology ePrint Archive, Report 2004/264.
3. J. Liang and X. Lai. *Improved Collision Attack on Hash Function MD5.* Cryptology ePrint Archive, Report 2005/425.
4. NIST. *Secure hash standard.* Federal Information Processing Standard, FIPS-180, May 1993.
5. NIST. *Secure hash standard.* Federal Information Processing Standard, FIPS-180-1, April 1995.
6. R. Rivest. *The MD4 Message Digest Algorithm.* CRYPTO'90 Proceedings, 1992, http://theory.lcs.mit.edu/~rivest/Rivest-MD4.txt

7. R. Rivest. *The MD5 Message Digest Algorithm*. CRYPTO'90 Proceedings, 1992, http://theory.lcs.mit.edu/~rivest/Rivest-MD5.txt
8. M. Schlaffer and E. Oswald. *Searching for Differential Paths in MD4*. FSE 2006, Springer-Verlag, 2006.
9. X. Wang, D. Feng, H. Chen, X. Lai and X. Yu. *Collision for Hash Functions MD4, MD5, HAVAL-128 and RIPEMD*. In Rump Session of CRYPTO 2004 and Cryptology ePrint Archive, Report 2004/199.
10. X. Wang, X. Lai, D. Feng, H. Chen and X. Yu. *Cryptanalysis of the Hash Functions MD4 and RIPEMD*. EUROCRYPT 2005, LNCS 3494, pp1–18, Springer-Verlag, 2005.
11. X. Wang and H. Yu. *How to Break MD5 and Other Hash Functions*. EUROCRYPT 2005, LNCS 3494, pp19–35, Springer-Verlag, 2005.
12. X. Wang, H. Yu and Y. Lisa Yin. *Efficient Collision Search Attack on SHA-0*. CRYPTO 2005, LNCS 3621, pp1–16, Springer-Verlag, 2005.
13. X. Wang, Y. Lisa Yin and H. Yu. *Finding Collisions in the Full SHA-1*. CRYPTO 2005, LNCS 3621, pp17–36, Springer-Verlag, 2005.
14. J. Yajima, T. Shimoyama: On the collision search and the sufficient conditions of MD5, ISEC 2005-78, pp.15-22, 2005.

Improved Fast Correlation Attack
on the Shrinking and Self-shrinking Generators[*]

Kitae Jeong[1], Jaechul Sung[2], Seokhie Hong[1], Sangjin Lee[1],
Jaeheon Kim[3], and Deukjo Hong[1]

[1] Center for Information Security Technologies (CIST),
Korea University, Seoul, Korea
{kite,hsh,sangjin,hongdj}@cist.korea.ac.kr
[2] Department of Mathematics, University of Seoul, Seoul, Korea
jcsung@uos.ac.kr
[3] National Security Research Institute (NSRI),
161 Gajeong-dong, Yuseong-gu, Daejeon 305-350, Korea
jaeheon@etri.re.kr

Abstract. The fast correlation attack on the shrinking generator proposed by Zhang et al. in [8] has a room for improvement that the probability that the guessing bit is incorrect increases in certain case. In this paper, we propose a method to improve Zhang et al.'s attack. Reflecting our idea, the fast correlation attack on the shrinking and self-shrinking generator is more efficient than Zhang et al.'s attack in both data and computational complexities. For the shrinking generator, required keystream bits and computational complexity are reduced about 69% and 27%, respectively; For the self-shrinking generator, required keystream bits and computational complexity are reduced about 46% and 22%, respectively.

Keywords: Clock-controlled generator, Shrinking generator, Self-Shrinking generator, Fast correlation attack.

1 Introduction

The shrinking generator is a clock-controlled generator proposed in [3]. It consists of the generating LFSR and the control LFSR. Both LFSRs are clocked regularly and simultaneously. If a current output bit of the control LFSR is 1, then the corresponding output bit of the generating LFSR is taken as a keystream bit. Otherwise it is discarded. In [8], Zhang et al. showed that a fast correlation attack can be applied to the shrinking generator. They guess the sequence of the generating LFSR by computing the probability that an output bit of the generating LFSR appears in a particular interval of keystream bits and then

[*] "This research was supported by the MIC(Ministry of Information and Communication), Korea, under the ITRC(Information Technology Research Center) support program supervised by the IITA(Institute of Information Technology Advancement)" (IITA-2006-(C1090-0603-0025)).

P.Q. Nguyen (Ed.): VIETCRYPT 2006, LNCS 4341, pp. 260–270, 2006.

recover the initial state of the generating LFSR by applying the fast correlation attack proposed by Chose et al. in [1]. So far, the most efficient attack on the shrinking generator is Zhang et al.'s attack.

The main idea in their guessing the sequence of the generating LFSR is to choose the major bit value between 0 and 1 in an intended interval of the keystream bits. However, their method has a room for improvement that the probability that the guessing bit is incorrect increases as the difference between the number of the 0 bits and 1 bits becomes small.

In this paper, we propose a method to improve Zhang et al.'s attack. We set a threshold of the difference between the number of the 0 bits and 1 bits, and reduce the error probability by guessing the sequence of the generating LFSR only in the interval where the difference between the number of the 0 bits and 1 bits is greater than the threshold. Reflecting our idea, our attack on the shrinking generator is more efficient than Zhang et al.'s attack in both data and computational complexities. The reason is as follows. In the first place, Zhang et al. only consider an interval that includes odd number of integers. We only consider an interval where the difference between the number of the 0 bits and 1 bits is greater than the threshold. Because Zhang et al.'s attack and our attack use refined intervals, the length of the guessed sequence is almost same in two attacks. Secondly, the probability that the guessing bit is correct in our attack is larger than that in Zhang et al.'s attack. So the correlation in our attack is larger than that in Zhang et al.'s attack. Hence our attack is more efficient than Zhang et al.'s attack. Table 1 shows the comparison of Zhang et al.'s attack and our attack on the shrinking generator (the length of the generating LFSR and the control LFSR is respectively 61 and 60) with success probability 99.9%. Zhang et al.'s attack requires $2^{17.1}$ keystream bits and computational complexity of $2^{56.7786}$; Our attack requires $2^{15.43}$ keystream bits and computational complexity of $2^{56.3314}$. In our attack, required keystream bits and computational complexity are reduced about 69% and 27%, respectively.

We also show that our attack works well on the self-shrinking generator. The self-shrinking generator is a modified version of the shrinking generator which is proposed by Meier and Staffelbach in [5]. It requires a single LFSR, whose length will be denoted by L. The selection rule is the same as for the shrinking generator, using even bits as output sequences generated by the control LFSR and odd bits as output sequences generated by the generating LFSR. Thus the selection rule of self-shrinking generator requires a tuple (even bit, odd bit) as input and outputs a odd bit if and only if a even bit is 1. In [5], the initial state of the generator from a short keystream sequence is reconstructed requiring $\mathcal{O}(2^{0.75L})$ steps. In [7], Zenner et al. proposed an attack that reconstructs the initial state of the generator from a short keystream sequence, requiring $\mathcal{O}(2^{0.694L})$ steps. On the other hand, Mihaljevic presented a faster attack that needs a longer part of keystream sequence in [6].

As a simulation, we applied Zhang et al.'s attack and our attack to the self-shrinking generator with the 240-bit internal state. Zhang et al.'s attack requires $2^{46.77}$ keystream bits and computational complexity of $2^{112.768}$. This result is better than previous attacks; Meier et al.'s attack requires $\mathcal{O}(2^{180})$ steps; Zenner et al.'s attack requires $\mathcal{O}(2^{166.56})$ steps; Mihaljevic's attack that is the fastest attack requires $2^{126.91}$ keystream bits and computational complexity of 2^{120}. Complexities of our attack are less than Zhang et al.'s attack in both data and computation. Our attack requires $2^{45.89}$ keystream bits and computational complexity of $2^{112.424}$. In our attack, required keystream bits and computational complexity are reduced about 46% and 22%, respectively. See Table 1.

Table 1. Comparisons of Zhang et al.'s attack and Our attack

		Zhang et al.	Our attack
The	Correlation	0.5098	0.523
shrinking	Data Complexity	$2^{17.1}$	$2^{15.43}$
generator	Computational Complexity	$2^{56.7786}$	$2^{56.3314}$
The	Correlation	0.5098	0.523
self-shrinking	Data Complexity	$2^{46.77}$	$2^{45.89}$
generator	Computational Complexity	$2^{112.768}$	$2^{112.424}$

This paper is organized as follows. Section 2 presents the shrinking and self-shrinking generator. Section 3 provides Zhang et al.'s attack. Section 4 presents our attack. Finally, Section 5 summarizes this paper.

2   The Shrinking and Self-shrinking Generators

The shrinking generator is a clock-controlled generator proposed in [3]. It consists of two LFSRs, say the generating LFSR A and the control LFSR S as shown in the Fig. 1. Both LFSRs are clocked regularly and simultaneously.

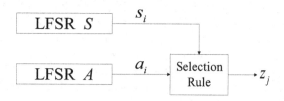

Fig. 1. The shrinking generator

Through this paper, we use notations as follows.

- $a = (a_0, a_1, \cdots)$: the output sequence of LFSR A of which length is L
- $s = (s_0, s_1, \cdots)$: the output sequence of LFSR S
- $z = (z_0, z_1, \cdots)$: the keystream sequence of the shrinking generator
- $\hat{a} = (\hat{a}_{m_0}, \hat{a}_{m_1}, \cdots)$: a guessed sequence associated with sequence a by the relation $P(\hat{a}_i = a_i) = \frac{1}{2} + \varepsilon$ $(\varepsilon > 0)$ where m_j denote indices of a sequence a

If a current output bit of the control LFSR S is 1, then the corresponding output bit of the generating LFSR A is taken as a keystream bit. For simplicity, assume that both LFSR sequences generated by LFSR A and LFSR S are uniformly distributed. The probability that a_r appears as z_k in the shrinking generator is as follows.

$$P(a_r \text{ appears as } z_k) = \binom{r}{k}\left(\frac{1}{2}\right)^{r+1} \quad (k \leq r). \tag{1}$$

On the other hand, if a_r $(r \geq 1)$ appears in the keystream z, we get

$$a_r = z_{\sum_{i=0}^{r-1} s_i}. \tag{2}$$

When r grows large, the distribution of $\sum_{i=0}^{r-1} s_i$ approximates the normal distribution such as (3).

$$\frac{\sum_{i=0}^{r-1} s_i - \frac{r}{2}}{\sqrt{\frac{r}{4}}} \sim N(0, 1). \tag{3}$$

This generator obtains a kind of implicit non-linearity from the shrinking process, i.e. the exact positions of the remaining bits in the generated keystream become uncertain. It is proved that the generated keystream has many merits in cryptographic sense such as a long period, a desirably high linear complexity and good statistical properties.

The self-shrinking generator is a modified version of the shrinking generator and it is proposed by Meier and Staffelbach in [5]. The self-shrinking generator only requires a single LFSR A, whose length will be denoted by L, as shown in the Fig. 2.

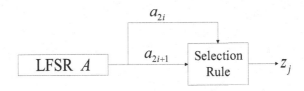

Fig. 2. The self-shrinking generator

LFSR A generates an $(a_i)_{i \geq 0}$ in the usual way. The selection rule is the same as for the shrinking generator, using even bits $(a_0, a_2, \cdots)$ as output sequences generated by LFSR S in the shrinking generator and odd bits $(a_1, a_3, \cdots)$ as output sequences generated by LFSR A in it. Thus the selection rule of the self-shrinking generator requires a tuple (a_{2i}, a_{2i+1}) as input and outputs a_{2i+1} if and only if $a_{2i} = 1$.

It is known that the shrinking generator with registers of lengths $|A|$ and $|S|$ has the same security as the self-shrinking generator of length $L = 2 \cdot (|A| + |S|)$ [5].

3   Zhang et al.'s Attack on the Shrinking Generator

In this section, we present Zhang et al.'s attack on the shrinking generator.

The way to construct a sequence $\hat{a}$ is as follows. An interval $I_{r/2}$ is defined as (4). For arbitrary probability p, there exist α such that whenever a_r appears in keystream z, (5) holds.

$$I_{r/2} = \left[\frac{r}{2} - \alpha \sqrt{\frac{r}{4}}, \frac{r}{2} + \alpha \sqrt{\frac{r}{4}} \right]. \tag{4}$$

$$P \left(\sum_{i=0}^{r-1} s_i \in I_{r/2} \right) = \frac{1}{\sqrt{2\pi}} \int_{-\alpha}^{\alpha} e^{-x^2/2} = p. \tag{5}$$

Definition 1. *W.l.o.g, we assume the interval $I_{r/2}$ includes odd number of integers. Let $S_0 = \{z_i \mid i \in I_{r/2}, z_i = 0\}$, $S_1 = \{z_i \mid i \in I_{r/2}, z_i = 1\}$, the first kind of imbalance of the interval $I_{r/2}$, $Imb_1(I_{r/2})$, is defined as $|S_1| - |S_0|$, where $|\cdot|$ is the cardinality of a set. If $Imb_1(I_{r/2}) \neq 0$, this interval is said to be imbalanced. See the Fig. 3.*

Definition 2. *S_0 and S_1 are the same as those in Definition 1. Let $P_0^{(r)} = \sum_{z_i \in S_0} P(a_r = z_i)$, $P_1^{(r)} = \sum_{z_i \in S_1} P(a_r = z_i)$, the second kind of imbalance of the interval $I_{r/2}$, $Imb_2(I_{r/2})$, is defined as $P_1^{(r)} - P_0^{(r)}$. If $Imb_2(I_{r/2}) \neq 0$, this interval is also said to be imbalanced. See the Fig. 3.*

Using $Imb_1(I_{r/2})$, $Imb_2(I_{r/2})$ of Definition 1 and 2, the method to construct a sequence $\hat{a}$ is as follows.

Method 1. *Following Definition 1, if $Imb_1(I_{r/2}) > 0$, let $\hat{a}_r = 1$. Otherwise, let $\hat{a}_r = 0$.*

Method 2. *Following Definition 2, if $Imb_2(I_{r/2}) \geq 0$, let $\hat{a}_r = 1$. Otherwise, let $\hat{a}_r = 0$.*

Next, we present a brief description of the fast correlation attack proposed by Chose et al. [1]. This attack consists of two stages: pre-processing stage aiming at the construction of parity-check equations of weight k and processing stage in

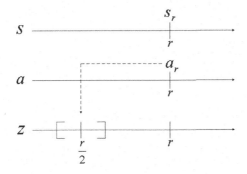

Fig. 3. The interval that a_r probably lies in

which a majority poll is conducted for $D(D > L - B)$ considered bits other than the first B bits $(a_0, a_1, \cdots, a_{B-1})$ of the initial state $(a_0, a_1, \cdots, a_{L-1})$. In pre-processing stage, a match-and-sort algorithm is used to construct parity-check equations of the following form with respect to a given considered bit a_i,

$$a_i = a_{m_1} \oplus \cdots \oplus a_{m_{k-1}} \oplus \sum_{j=0}^{B-1} c_j a_j, \qquad (6)$$

where $m_j(1 \leq j \leq k-1)$ denote indices of output bits and the last sum represents a partial exhaustive search over $(a_0, \cdots, a_{B-1})$ of the initial state $(a_0, \cdots, a_{L-1})$. After regrouping parity-check equations that contain the same pattern of $B - B_1$ initial bits, an application of Walsh transform is suggested to evaluate parity-check equations in processing stage for a given guessed bit $\hat{a}_i$, i.e. when $\omega = [a_{B_1}, a_{B_1+1}, \cdots, a_{B-1}]$, $F_i(\omega) = \sum(-1)^{t_i^1 \oplus t_i^2}$ is just the difference between the number of predicted 0 and the number of predicted 1, where $t_i^1 = \hat{a}_{m_1} \oplus \cdots \oplus \hat{a}_{m_{k-1}} \oplus \sum_{j=0}^{B_1-1} c_j a_j$ and $t_i^2 = \sum_{j=B_1}^{B-1} c_j a_j$. Then for each of D considered bits, if $F_i(\omega) > \theta$, let $a_i = 0$. If $F_i(\omega) < -\theta$, let $a_i = 1$, where θ is the decision threshold. In order to have at least $L - B$ correctly recovered bits among D considered bits, a check procedure is used which requires an exhaustive search on all subsets of size $L - B$ among $L - B + \delta$ bits. The total computational complexity of the processing stage is as follows :

$$\mathcal{O}\left(2^B D \log_2 \omega + (1 + p_{err}(2^B - 1)) \binom{L - B + \delta}{\delta} \frac{1}{\varepsilon^2}\right), \qquad (7)$$

where p_{err} is the probability that a wrong guess results in at least $L - B + \delta$ predicted bits and ω is the expected number of parity-check equations of weight k for each considered bit. For details of these formulae and notations, see [1].

A summary of Zhang et al.'s attack is as follows. Here, an interval $I_{r/2}$ includes odd number of integers. So the number of intervals which they can construct is about $N - \frac{\alpha \cdot \sqrt{N}}{2}$ and the length N' of sequence $\hat{a}$ is also about $N - \frac{\alpha \cdot \sqrt{N}}{2}$.

1. Input: a segment of keystream $z_0, z_1, \cdots, z_{N-1}$.
2. Construct a sequence $\hat{a} = \hat{a}_{m_0}, \cdots, \hat{a}_{m_{N'-1}}$ according to Method 1 or Method 2 from keystream $z_0, z_1, \cdots, z_{N-1}$, where m_j $(0 \leq j \leq N' - 1)$ denote indices of a sequence a.
3. Construct parity check equations such as (6).
4. For each guess of $(a_0, \cdots, a_{B-1})$ and each bit position i $(i = B + 1, B + 2, \cdots, D)$, evaluate parity-check equations using the Walsh transform technique. Select those bits passing the majority poll to recover the initial state of LFSR A using the check procedure.

4 Improved Fast Correlation Attack

In Zhang et al.'s attack, the main idea in their guessing the sequence of the generating LFSR A is to choose the major bit value between 0 and 1 in the keystream bits corresponding to $I_{r/2}$. However, their method has a room for improvement that the probability that the guessing bit is incorrect increases as the difference between the number of the 0 bits and 1 bits becomes small.

Now we propose a method to improve Zhang et al.'s attack. We set a threshold ε' of the difference between the number of the 0 bits and 1 bits, and reduce the error probability by guessing the sequence of the generating LFSR A only in the interval where the ratio of the 0 bits or the 1 bits is greater than $\frac{1}{2} + \varepsilon'$. Reflecting our idea, our attack on the shrinking generator is more efficient than Zhang et al.'s attack in both data and computational complexities. The reason is as follows. In the first place, Zhang et al. only consider $I_{r/2}$ that includes odd number of integers. We only consider $I_{r/2}$ where the ratio of the 0 bits or the 1 bits is greater than $\frac{1}{2} + \varepsilon'$. Because Zhang et al.'s attack and our attack use refined intervals, the length of the guessed sequence $\hat{a}$ is almost same in two attacks. Secondly, the probability that a_r is equal to $\hat{a}_r$ in our attack is larger than that in Zhang et al.'s attack. So the correlation in our attack is larger than that in Zhang et al.'s attack. Hence our attack is more efficient than Zhang et al.'s attack.

4.1 Improved Fast Correlation Attack on the Shrinking Generator

An interval $I_{r/2}$ of length $t = \lfloor 1 + \alpha\sqrt{r} \rfloor$ is defined as (8). Here for arbitrary probability p, there exist α such that whenever a_r appears in keystream z, (9) holds.

$$I_{r/2} = \left[\frac{r}{2} - \alpha\sqrt{\frac{r}{4}}, \ \frac{r}{2} + \alpha\sqrt{\frac{r}{4}} \right]. \tag{8}$$

$$P\left(\sum_{i=0}^{r-1} s_i \in I_{r/2} \right) = \frac{1}{\sqrt{2\pi}} \int_{-\alpha}^{\alpha} e^{-x^2/2} = p. \tag{9}$$

The method to construct a sequence $\hat{a}$ is as follows.

Method 3. *If an interval $I_{r/2}$ in which the ratio of 0 (resp. 1) is more than $\frac{1}{2} + \varepsilon'$ exists, then let $\hat{a}_r = 0$ (resp. 1).*

A summary of our attack on the shrinking generator is as follows. Here $\varepsilon' = 0.02$, 0.03 and 0.04.

1. Input: a segment of keystream $z_0, z_1, \cdots, z_{N-1}$.
2. Construct a guessed sequence $\hat{a} = \hat{a}_{m_0}, \cdots, \hat{a}_{m_{N'-1}}$ according to Method 3 from keystream $z_0, z_1, \cdots, z_{N-1}$, where m_j $(0 \leq j \leq N' - 1)$ denote indices of a sequence a.
3. Construct parity check equations such as (6).
4. For each guess of $(a_0, \cdots, a_{B-1})$ and each bit position i $(i = B + 1, B + 2, \cdots, D)$, evaluate parity-check equations using the Walsh transform technique. Select those bits passing the majority poll to recover the initial state of LFSR A using the check procedure.

Theorem 1. *The probability that a_r is equal to $\hat{a}_r$ is (10). Here, δ means the expected value of the probability of 1 in $I_{r/2}$.*

$$P(a_r = \hat{a}_r) = \frac{1}{2} + \frac{p}{2}\left(\frac{\delta}{t} - \frac{1}{2}\right). \tag{10}$$

Proof.

$$\begin{aligned}
P(a_r = \hat{a}_r) &= P(\hat{a}_r = 1)P(a_r = 1| \hat{a}_r = 1) + P(\hat{a}_r = 0)P(a_r = 0| \hat{a}_r = 0) \\
&= P(\hat{a}_r = 1)P(a_r = 1| \hat{a}_r = 1) \\
&\quad + (1 - P(\hat{a}_r = 1)) P(a_r = 1| \hat{a}_r = 1) \\
&= P(a_r = 1| \hat{a}_r = 1) \\
&= P\left(s_r = 1, \sum_{i=0}^{r-1} s_i \in I_{r/2}, z_{s_r} = 1\right) + P\left(s_r = 0, a_r = 1\right) \\
&\quad + P\left(s_r = 1, \sum_{i=0}^{r-1} s_i \notin I_{r/2}, a_r = 1\right) \\
&= P(s_r = 1)P\left(\sum_{i=0}^{r-1} s_i \in I_{r/2}\right) P(z_{s_r} = 1) + P(s_r = 0)P(a_r = 1) \\
&\quad + P(s_r = 1)P\left(\sum_{i=0}^{r-1} s_i \notin I_{r/2}\right) P(a_r = 1) \\
&= \frac{1}{2} \cdot p \cdot \frac{\delta}{t} + \frac{1}{2} \cdot (1 - p) \cdot \frac{1}{2} + \frac{1}{2} \cdot \frac{1}{2} \\
&= \frac{1}{2} + \frac{p}{2}\left(\frac{\delta}{t} - \frac{1}{2}\right).
\end{aligned}$$

$\square$

Since we only consider the interval in which the ratio of 0 or 1 is more than $\frac{1}{2} + \varepsilon'$, δ is computed as (11). Here $|\cdot|_1$ means the number of 1 in $I_{r/2}$.

$$\delta = \sum_{i=\lceil(\frac{1}{2}+\varepsilon')\cdot t\rceil}^{t} i \cdot P\left(|\cdot|_1 = i \;\middle|\; |\cdot|_1 \geq \left\lceil\left(\frac{1}{2}+\varepsilon'\right)\cdot t\right\rceil\right). \tag{11}$$

The length N' of a guessed sequence $\hat{a}$ guessed is $\beta \cdot (2N - \alpha \cdot \sqrt{N})$ where $2N - \alpha \cdot \sqrt{N}$ is the number of intervals which we can construct and β is computed as follows :

$$\beta = \frac{1}{2^{t-1}}\left(\sum_{i=\lceil(\frac{1}{2}+\varepsilon')\cdot t\rceil}^{t} \binom{t}{i}\right).$$

To compare Zhang et al.'s attack and our attack on the shrinking generator, we use the same parameters as parameters used in [8]. The length of LFSR A is 61 and the length of LFSR S is 60. And $D = 36$, $\delta = 3$, $B = 46$ and $k = 5$. $\alpha = 1.376395$ corresponds to $p = 0.8313$. Table 2 shows the comparison of Zhang et al.'s attack and our attack where $\varepsilon' = 0.02$, 0.03 and 0.04 on the shrinking generator and success probability 99.9%. For $\varepsilon' = 0.04$, required keystream bits and computational complexity are reduced about 69% and 27%, respectively.

Table 2. Comparison of Zhang et al.'s attack and our attack on the shrinking generator

	Zhang et al.	our attack		
		$\varepsilon' = 0.02$	$\varepsilon' = 0.03$	$\varepsilon' = 0.04$
Correlation	0.5098	0.5164	0.5197	0.523
Data Complexity	$2^{17.1}$	$2^{14.89}$	$2^{15.09}$	$2^{15.43}$
Computational Complexity	$2^{56.7786}$	$2^{56.4815}$	$2^{56.4036}$	$2^{56.3314}$

4.2 Improved Fast Correlation Attack on the Self-shrinking Generator

The method to construct a guessed sequence $\hat{a}$ on the self-shrinking generator is very similar to the method on the shrinking generator. The difference is only that we construct a guessed sequence $\hat{a}$ of the following form from keystream z. That is a guessed sequence $\hat{a}$ consists of a tuple $(\hat{a}_{2i}, \hat{a}_{2i+1}) = (1, \hat{a}_{2i+1})$.

$$\hat{a} = (\hat{a}_{m_0}, \hat{a}_{m_1}, \cdots) = (1, \hat{a}_{m_1}, 1, \hat{a}_{m_3}, \cdots, 1, \hat{a}_{m_{2i+1}}, \cdots). \tag{12}$$

An interval I_r is defined as (13). Here for arbitrary probability p, there exist α such that whenever a_{2r} appears in keystream z, (14) holds.

$$I_r = \left[r - \alpha \sqrt{\frac{r}{2}}, \ r + \alpha \sqrt{\frac{r}{2}} \right]. \tag{13}$$

$$P\left(\sum_{i=0}^{r-1} a_{2i} \in I_r \right) = \frac{1}{\sqrt{2\pi}} \int_{-\alpha}^{\alpha} e^{-x^2/2} = p. \tag{14}$$

Theorem 2. *The probability that $\hat{a}_{2r+1}$ is equal to a_{2r+1} is as follows. Here, δ means the expected value of the probability of 1 in I_r and computed as (11).*

$$P(a_{2r+1} = \hat{a}_{2r+1}) = \frac{1}{2} + \frac{p}{2}\left(\frac{\delta}{t} - \frac{1}{2} \right).$$

Proof. It's similar to proof of Theorem 1.   □

As a simulation, we applied Zhang et al.'s attack and our attack to the self-shrinking generator with the 240-bit internal state. Zhang et al.'s attack requires $2^{46.77}$ keystream bits, computational complexity of $2^{112.768}$ and success probability 99.9%. This result is better than previous attacks; Meier et al.'s attack requires $\mathcal{O}(2^{180})$ steps; Zenner et al.'s attack requires $\mathcal{O}(2^{166.56})$ steps; Mihaljevic's attack that is the fastest attack requires $2^{126.91}$ keystream bits and computational complexity of 2^{120}. Complexities of our attack are less than them of Zhang et al.'s attack. Table 3 shows the comparison of Zhang et al.'s attack and our attack where $\varepsilon' = 0.02$, 0.03 and 0.04 on the self-shrinking generator and success probability 99.9%. Here the length of LFSR A is 240 and $D = 150$, $\delta = 3$, $B = 100$ and $k = 5$. $\alpha = 1.376395$ corresponds to $p = 0.8313$. For $\varepsilon' = 0.04$, Our attack requires $2^{45.89}$ keystream bits and computational complexity of $2^{112.424}$. In this case, required keystream bits and computational complexity are reduced about 46% and 22%, respectively.

Table 3. Comparison of Zhang et al.'s attack and our attack on the self-shrinking generator

	Zhang et al.	our attack		
		$\varepsilon' = 0.02$	$\varepsilon' = 0.03$	$\varepsilon' = 0.04$
Correlation	0.5098	0.5164	0.5197	0.523
Data Complexity	$2^{46.77}$	$2^{45.36}$	$2^{45.56}$	$2^{45.89}$
Computational Complexity	$2^{112.768}$	$2^{112.571}$	$2^{112.495}$	$2^{112.424}$

5   Conclusion

In Zhang et al.'s attack, the main idea in their guessing the sequence of the generating LFSR A is to choose the major bit value between 0 and 1 in the keystream bits corresponding to $I_{r/2}$. However, their method has a room for improvement that the probability that the guessing bit is incorrect increases as the difference between the number of the 0 bits and 1 bits becomes small.

We propose a method to improve Zhang et al.'s attack. We set a threshold ε' of the difference between the number of the 0 bits and 1 bits, and reduce the error probability by guessing the sequence of the generating LFSR A only in the interval where the ratio of the 0 bits or the 1 bits is greater than $\frac{1}{2} + \varepsilon'$. Reflecting our idea, our attack on the shrinking and self-shrinking generator is more efficient than Zhang et al.'s attack in both data and computational complexities. Using our attack on the shrinking generator, required keystream bits and computational complexity are reduced about 69% and 27%, respectively. Using Zhang et al.'s attack, we checked that the initial state of generating LFSR (its length is 240) of the self-shrinking generator is recovered faster than previous attacks. Complexities of our attack are less than them of Zhang et al.'s attack in both data and computation. In detail, required keystream bits and computational complexity are reduced about 46% and 22%, respectively using our attack.

References

1. P. Chose, A. Joux, and M. Mitton, *Fast correlation attacks: an algorithmic point of view*, Advances in Cryptology - EUROCRYPT 2002, LNCS 2332, Springer-Verlag, pp. 209–221, 2002.
2. V. V. Chepyzhov, T. Johansson, and B. Smeets, *A simple algorithm for fast correlation attacks on stream ciphers*, Fast Software Encryption - FSE 2000, LNCS 1978, Springer-Verlag, pp. 181–195, 2000.
3. D. Coppersmith, H. Krawczyk, and Y. Mansour, *The Shrinking Generator*, Advances in Cryptologty - CRYPTO 1993, LNCS 773, Springer-Verlag, pp. 22–39, 1994.
4. Ali. A. Kanso, *Clock-Controlled Genrators*, Thesis submitted to the University of London for the degree of Doctor of Philosophy, *www.isg.rhul.ac.uk/alumni/thesis/ kanso_a.pdf*, 1999.
5. W. Meier, and O. Staffelbach, *The Self-Shrinking generator*, Advances in Cryptology - EUROCRYPT 1994, LNCS 950, Springer-Verlag, pp. 205–214, 1995.
6. M. J. Mihaljevic, *A faster cryptanalysis of the self-shrinking generator*, Advances in Cryptology - ACISP 1996, LNCS 1172, Springer-Verlag, pp. 182–189, 1996.
7. E. Zenner, M. Krause and S. Lucks, *Improved Cryptanalysis of the Self-Shrinking Generator*, ACISP 2001, LNCS 2119, Springer-Verlag, pp. 21–35, 2001.
8. B. Zhang, H. Wu, D. Feng and F. Bao, *A Fast Correlation Attack on the Shrinking Generator*, CT-RSA 2005, LNCS 3376, Springer-Verlag, pp. 72–86, 2005.

On the Internal Structure of Alpha-MAC

Jianyong Huang, Jennifer Seberry, and Willy Susilo

School of Information Technology and Computer Science,
University of Wollongong, Wollongong NSW 2522, Australia
{jyh33, jennie, wsusilo}@uow.edu.au

Abstract. Alpha-MAC is a MAC function which uses the building blocks of AES. This paper studies the internal structure of this new design. First, we provide a method to find second preimages based on the assumption that a key or an intermediate value is known. The proposed searching algorithm exploits the algebraic properties of the underlying block cipher and needs to solve eight groups of linear functions to find a second preimage. Second, we show that our idea can also be used to find internal collisions under the same assumption. We do not make any claims that those findings in any way endanger the security of this MAC function. Our contribution is showing how algebraic properties of AES can be used for analysis of this MAC function.

1 Introduction

Hash functions play an important role in many areas of cryptography. The building of hash functions has received extensive work over the years, for example, the design of MD4 [17], MD5 [18], SHA-0 [3] and SHA-1 [2]. On the other hand, the cryptanalysis of hash functions has been carried out by many researchers, for instance, recent attacks on MD4, MD5, SHA-0 and SHA-1 [6,7,10,14,19,20,21,22].

Message Authentication Codes (MACs) are keyed hash functions that provide message integrity by appending a cryptographic checksum to a message which is verifiable only by the intended recipient of the message. Message authentication is one of the most important ways of ensuring the integrity of information, and it has been used in many practical applications. MAC functions take a secret key and a message as input and generate a short digest as output. Many research groups have presented various approaches to construct MAC functions, for example, MAA [13], CBC-MAC [15], UMAC [9], MDx-MAC [16] and HMAC [4,5].

The Alred [11] construction is a new MAC design approach presented at FSE 2005. Alpha-MAC [11] is a specific instance of the Alred construction with AES [1] as the underlying block cipher. The reason why AES was chosen as the underlying block cipher of the Alpha-MAC is because AES is efficient in hardware and software and it has withstood intense public scrutiny since its publication as Rijndael [12].

In this paper, we study the internal structure of the Alpha-MAC by employing the algebraic properties of AES and the structural features of the Alpha-MAC.

P.Q. Nguyen (Ed.): VIETCRYPT 2006, LNCS 4341, pp. 271–285, 2006.

First, we present a method to find second preimages of the ALPHA-MAC by solving eight groups of linear functions, based on the assumption that an authentication key or an intermediate value of this MAC is known. Each of these eight groups of linear functions contains two equations. We divide the second-preimage search algorithm into two steps: the Backwards-aNd-Forwards (BNF) search and the Backwards-aNd-Backwards (BNB) search. The BNF search provides an idea for extending 32-bit collisions to 128-bit collisions[1] by solving four groups of linear functions. Given a key (or an intermediate value) and one four-block message, the BNB search can generate another four-block message such that these two messages produce 32-bit collisions, which are a prerequisite for the BNF search. To do the BNB search, we need to solve another four groups of linear functions. By combining the BNB search with the BNF search, we can find second preimages of ALPHA-MAC. Second, we show that the second-preimage finding method can also be used to generate internal collisions. The proposed collision search method can find two five-block messages such that they produce 128-bit collisions under a selected key (or a selected intermediate value).

This paper is organized as follows: Section 2 provides a description of the ALPHA-MAC, and Section 3 presents the second-preimage search algorithm. Section 4 shows how to generate internal collisions and finally, Section 5 concludes this paper. Appendix A includes our experimental results.

2   A Brief Description of ALPHA-MAC

ALPHA-MAC [11] is a MAC function which uses the building blocks of AES. Similarly to AES, the ALPHA-MAC supports keys of 128, 192 and 256 bits. The word length is 32 bits, and the injection layout places the 4 bytes of each message word $[m_0, m_1, m_2, m_3]$ into a 4×4 array. The format of the injection layout is shown as follows:

$$\begin{bmatrix} m_0 & 0 & m_1 & 0 \\ 0 & 0 & 0 & 0 \\ m_2 & 0 & m_3 & 0 \\ 0 & 0 & 0 & 0 \end{bmatrix}.$$

Like AES, the ALPHA-MAC round function contains SubBytes (SB), ShiftRows (SR), MixColumns (MC) and AddRoundKey (ARK) , and the output of each injection layout acts as the corresponding 128-bit round key. The message padding method appends a single 1 followed by the minimum number of 0 bits such that the length of the result is a multiple of 32. In the initialization, the state is set to all zeros and AES is applied to the state. For every message word, the chaining method carries out an iteration, and each iteration maps the bits of the message word to an injection input. After that, a sequence of AES round functions are applied to the state, with the round keys replaced by the injection input. In the final transformation, AES is applied to the state. The MAC tag is the first l_m bits of the resulting final state. The length of l_m may have any value less than or equal to 128. The ALPHA-MAC function is depicted in Figure 1.

---

[1] Here and in the rest of this paper "collisions" stands for "internal collisions".

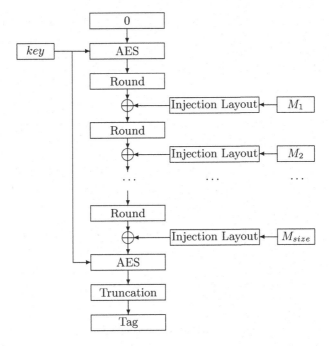

Fig. 1. ALPHA-MAC construction

3   The Second-Preimage Search Algorithm

The proposed second-preimage search algorithm aims to find a five-block second-preimage $\tilde{M}$ for a selected five-block message M, under a selected key (or a selected intermediate value). The assumption of this search is that we know two values: a selected key (or a selected intermediate value) and a selected five-block message M. The result of the search is that M and $\tilde{M}$ generate the same 128-bit value after five rounds of ALPHA-MAC iterations, under the selected key (or the selected intermediate value).

We use Figure 2 to illustrate the second-preimage search. Figure 2 depicts five consecutive rounds of the ALPHA-MAC for two different five-block messages M and $\tilde{M}$. We assume that we are able to select an intermediate value[2, 3] of the Round functions in some round (e.g., in Round $y - 3$), and select five consecutive

---

[2] The intermediate value is:

$$\begin{bmatrix} a_0 & a_4 & a_8 & a_{12} \\ a_1 & a_5 & a_9 & a_{13} \\ a_2 & a_6 & a_{10} & a_{14} \\ a_3 & a_7 & a_{11} & a_{15} \end{bmatrix}.$$

[3] In the case of a selected key, for the sake of simplicity, we assume that $(M_{y-3}, M_{y-2}, M_{y-1}, M_y, M_{y+1})$ are the first five blocks of the selected message. Our search algorithm works without assuming that $(M_{y-3}, M_{y-2}, M_{y-1}, M_y, M_{y+1})$ are the first five blocks of the selected message.

message blocks $M(M_{y-3}, M_{y-2}, M_{y-1}, M_y, M_{y+1})$. Then we can find another five-block message $\tilde{M}(\tilde{M}_{y-3}, \tilde{M}_{y-2}, \tilde{M}_{y-1}, \tilde{M}_y, \tilde{M}_{y+1})$ such that these two five-block messages collide on 128 bits in Round $y + 1$ after ARK.

The second-preimage search algorithm has the following form:

Known: 1. a selected key or a selected intermediate value.
          2. a selected five-block message $M(M_{y-3}, M_{y-2}, M_{y-1}, M_y, M_{y+1})$.
Find: another five-block message $\tilde{M}(\tilde{M}_{y-3}, \tilde{M}_{y-2}, \tilde{M}_{y-1}, \tilde{M}_y, \tilde{M}_{y+1})$ such that M and $\tilde{M}$ collide on 128 bits after ARK in Round $y + 1$.
Method: solve eight groups of linear functions. These eight groups of functions are named as (1), (2), (3), (4), (5), (6), (7) and (8) in this section.

The second-preimage search algorithm consists of two steps: the Backwards-aNd-Forwards search and the Backwards-aNd-Backwards search. The BNF search can extend 32-bit collisions to 128-bit collisions, given two messages M and $\tilde{M}$ which collide on 32 bits, namely Bytes s_4, s_{12}, s_6 and s_{14}, after MC in round y (see Figure 2). Given a key (or an intermediate value) and one four-block message, the BNF search is able to find another four-block message such that these two messages collide on Bytes s_4, s_{12}, s_6 and s_{14} after MC in Round y. The BNB search generates those 32-bit collisions which are required for the BNF search. By merging the BNB search with the BNF search, we can find second preimages of the ALPHA-MAC.

3.1   The Backwards-aNd-Forwards Search

The Backwards-aNd-Forwards search has the following form:

Known: 1. a selected key or a selected intermediate value.
          2. two four-block messages $M(M_{y-3}, M_{y-2}, M_{y-1}, M_y)$ and $\tilde{M}(\tilde{M}_{y-3}, \tilde{M}_{y-2}, \tilde{M}_{y-1}, \tilde{M}_y)$ colliding on 32 bits (Bytes s_4, s_{12}, s_6 and s_{14}) after MC in Round y.
Extend: 32-bit collisions to 128-bit collisions in Round $y + 1$.
Method: solve four groups of linear functions. These four groups of functions are numbered as (1), (2), (3) and (4) in this subsection.

The BNF search assumes that we are able to find two messages M and $\tilde{M}$, which collide on Bytes s_4, s_{12}, s_6 and s_{14} after MC in round y. Based on the algebraic property of the MixColumns transformation and the structure of ALPHA-MAC, we can extend these 32-bit collisions to 128-bit collisions within three rounds by solving four groups of linear equations.

3.1.1   Extending 32-Bit Collisions to 64-Bit Collisions

We use the differential XOR property [8] before and after the MixColumns transformation. In Round y before MC, by XORing those two intermediate values, we get the following result:

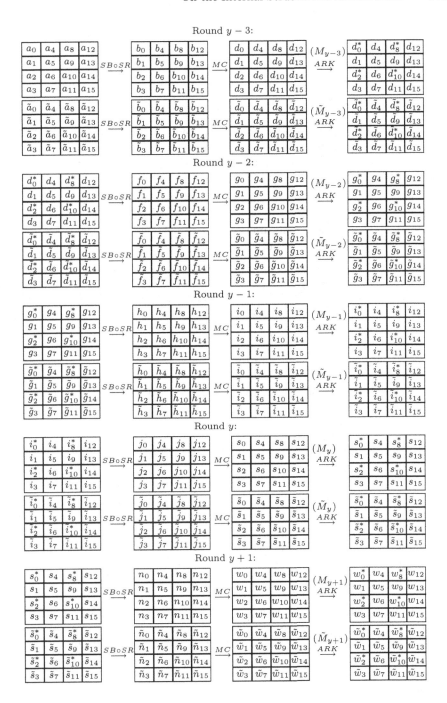

Fig. 2. The five-block collisions

$$\begin{bmatrix} \tilde{j}_0 \oplus j_0 & \tilde{j}_4 \oplus j_4 & \tilde{j}_8 \oplus j_8 & \tilde{j}_{12} \oplus j_{12} \\ \tilde{j}_1 \oplus j_1 & \tilde{j}_5 \oplus j_5 & \tilde{j}_9 \oplus j_9 & \tilde{j}_{13} \oplus j_{13} \\ \tilde{j}_2 \oplus j_2 & \tilde{j}_6 \oplus j_6 & \tilde{j}_{10} \oplus j_{10} & \tilde{j}_{14} \oplus j_{14} \\ \tilde{j}_3 \oplus j_3 & \tilde{j}_7 \oplus j_7 & \tilde{j}_{11} \oplus j_{11} & \tilde{j}_{15} \oplus j_{15} \end{bmatrix} \xrightarrow{MC} \begin{bmatrix} ? & 0 & ? & 0 \\ 0 & \tilde{s}_5 \oplus s_5 & 0 & \tilde{s}_{13} \oplus s_{13} \\ ? & 0 & ? & 0 \\ 0 & \tilde{s}_7 \oplus s_7 & 0 & \tilde{s}_{15} \oplus s_{15} \end{bmatrix}.$$

Here, we use R (to replace $\tilde{j}_0 \oplus j_0$), S (to replace $\tilde{j}_8 \oplus j_8$), T (to replace $\tilde{j}_2 \oplus j_2$) and U (to replace $\tilde{j}_{10} \oplus j_{10}$) so that after the MC transformation in Round y, Bytes $\tilde{s}_1 \oplus s_1$, $\tilde{s}_3 \oplus s_3$, $\tilde{s}_9 \oplus s_9$ and $\tilde{s}_{11} \oplus s_{11}$ become zero. Now the question is "how to decide R, S, T and U". The answer is:

- There exists one and only one pair of (R, T) such that after MC, Bytes $\tilde{s}_1 \oplus s_1$ and $\tilde{s}_3 \oplus s_3$ are both zero.
- There exists one and only one pair of (S, U) such that after MC, Bytes $\tilde{s}_9 \oplus s_9$ and $\tilde{s}_{11} \oplus s_{11}$ are both zero.

According to the MC transformation, we have the following formula:

$$\begin{bmatrix} ? & 0 & ? & 0 \\ 0 & \tilde{s}_5 \oplus s_5 & 0 & \tilde{s}_{13} \oplus s_{13} \\ ? & 0 & ? & 0 \\ 0 & \tilde{s}_7 \oplus s_7 & 0 & \tilde{s}_{15} \oplus s_{15} \end{bmatrix} = \begin{bmatrix} 02\ 03\ 01\ 01 \\ 01\ 02\ 03\ 01 \\ 01\ 01\ 02\ 03 \\ 03\ 01\ 01\ 02 \end{bmatrix} \begin{bmatrix} R & \tilde{j}_4 \oplus j_4 & S & \tilde{j}_{12} \oplus j_{12} \\ \tilde{j}_1 \oplus j_1 & \tilde{j}_5 \oplus j_5 & \tilde{j}_9 \oplus j_9 & \tilde{j}_{13} \oplus j_{13} \\ T & \tilde{j}_6 \oplus j_6 & U & \tilde{j}_{14} \oplus j_{14} \\ \tilde{j}_3 \oplus j_3 & \tilde{j}_7 \oplus j_7 & \tilde{j}_{11} \oplus j_{11} & \tilde{j}_{15} \oplus j_{15} \end{bmatrix}.$$

To find out the values of (R, T) and (S, U), we need to solve the following two groups of equations.

$$(1) \begin{cases} \begin{bmatrix} 01\ 02\ 03\ 01 \end{bmatrix} \begin{bmatrix} R \\ \tilde{j}_1 \oplus j_1 \\ T \\ \tilde{j}_3 \oplus j_3 \end{bmatrix} = 0 \\ \begin{bmatrix} 03\ 01\ 01\ 02 \end{bmatrix} \begin{bmatrix} R \\ \tilde{j}_1 \oplus j_1 \\ T \\ \tilde{j}_3 \oplus j_3 \end{bmatrix} = 0 \end{cases} \qquad (2) \begin{cases} \begin{bmatrix} 01\ 02\ 03\ 01 \end{bmatrix} \begin{bmatrix} S \\ \tilde{j}_9 \oplus j_9 \\ U \\ \tilde{j}_{11} \oplus j_{11} \end{bmatrix} = 0 \\ \begin{bmatrix} 03\ 01\ 01\ 02 \end{bmatrix} \begin{bmatrix} S \\ \tilde{j}_9 \oplus j_9 \\ U \\ \tilde{j}_{11} \oplus j_{11} \end{bmatrix} = 0 \end{cases}$$

In the two equations in (1), there are two variables R and T, and therefore there exists one and only one pair of (R, T) to make these two equations hold simultaneously. Similarly, we can decide the values of S and U by solving the two equations in (2).

Once we get the values of R, S, T and U, message block $\tilde{M}_{y-1}$ can be constructed as follows:

1. Set the values of $\tilde{j}_0^{new}$, $\tilde{j}_8^{new}$, $\tilde{j}_2^{new}$ and $\tilde{j}_{10}^{new}$ as follows: $\tilde{j}_0^{new} = j_0 \oplus R$, $\tilde{j}_8^{new} = j_8 \oplus S$, $\tilde{j}_2^{new} = j_2 \oplus T$, and $\tilde{j}_{10}^{new} = j_{10} \oplus U$. Use $\tilde{j}_0^{new}$ to replace $\tilde{j}_0$, $\tilde{j}_8^{new}$ to replace $\tilde{j}_8$, $\tilde{j}_2^{new}$ to replace $\tilde{j}_2$, and $\tilde{j}_{10}^{new}$ to replace $\tilde{j}_{10}$.

2. Perform SR^{-1} (inverse ShiftRows) and SB^{-1} (inverse SubBytes). As SR^{-1} and SB^{-1} are permutation and substitution, they do not change the properties we have found. Now we have the outputs of ARK in Round $y-1$.
3. Compute the value of $\tilde{M}_{y-1}^{new}$ as follows:

$$\tilde{M}_{y-1}^{new} = (\tilde{j}_0^{new} \oplus \tilde{i}_0)||(\tilde{j}_8^{new} \oplus \tilde{i}_8)||(\tilde{j}_{10}^{new} \oplus \tilde{i}_2)||(\tilde{j}_2^{new} \oplus \tilde{i}_{10}).$$

Use $\tilde{M}_{y-1}^{new}$ to replace $\tilde{M}_{y-1}$.

At this stage, two messages $(M_{y-3}, M_{y-2}, M_{y-1})$ and $(\tilde{M}_{y-3}, \tilde{M}_{y-2}, \tilde{M}_{y-1}^{new})$ collide on 64 bits (Bytes s_4, s_{12}, s_6, s_{14}, s_1, s_9, s_3 and s_{11}) in Round y after MC.

3.1.2   Extending 64-Bit Collisions to 96-Bit Collisions

We only need to focus on Round y and Round $y+1$ to extend 64-bit collisions to 96-bit collisions. The idea is to choose message block $\tilde{M}_y$ to cancel out the differences between Bytes $(s_5, s_{13}, s_7, s_{15})$ and Bytes $(\tilde{s}_5, \tilde{s}_{13}, \tilde{s}_7, \tilde{s}_{15})$ in Round y. The method of choosing $\tilde{M}_y$ is exactly same as the method for constructing $\tilde{M}_{y-1}$ in Section 3.1.1.

By taking the outputs of ARK in Round y, we perform the SB and SR operations, and then XOR the results after SB and SR:

$$\begin{bmatrix} n_0 & n_4 & n_8 & n_{12} \\ n_1 & n_5 & n_9 & n_{13} \\ n_2 & n_6 & n_{10} & n_{14} \\ n_3 & n_7 & n_{11} & n_{15} \end{bmatrix} \oplus \begin{bmatrix} \tilde{n}_0 & n_4 & \tilde{n}_8 & n_{12} \\ \tilde{n}_1 & n_5 & \tilde{n}_9 & n_{13} \\ \tilde{n}_2 & n_6 & \tilde{n}_{10} & n_{14} \\ \tilde{n}_3 & n_7 & \tilde{n}_{11} & n_{15} \end{bmatrix} = \begin{bmatrix} n_0 \oplus \tilde{n}_0 & 0 & n_8 \oplus \tilde{n}_8 & 0 \\ n_1 \oplus \tilde{n}_1 & 0 & n_9 \oplus \tilde{n}_9 & 0 \\ n_2 \oplus \tilde{n}_2 & 0 & n_{10} \oplus \tilde{n}_{10} & 0 \\ n_3 \oplus \tilde{n}_3 & 0 & n_{11} \oplus \tilde{n}_{11} & 0 \end{bmatrix} \xrightarrow{MC} \begin{bmatrix} ? & 0 & ? & 0 \\ 0 & 0 & 0 & 0 \\ ? & 0 & ? & 0 \\ 0 & 0 & 0 & 0 \end{bmatrix}.$$

Here we use π to replace $n_0 \oplus \tilde{n}_0$, ρ to replace $n_8 \oplus \tilde{n}_8$, ϕ to replace $n_2 \oplus \tilde{n}_2$ and ω to replace $n_{10} \oplus \tilde{n}_{10}$ so that after MixColumns in Round $y+1$, Bytes $w_1 \oplus \tilde{w}_1$, $w_9 \oplus \tilde{w}_9$, $w_3 \oplus \tilde{w}_3$ and $w_{11} \oplus \tilde{w}_{11}$ are zero:

$$\begin{bmatrix} \pi & 0 & \rho & 0 \\ n_1 \oplus \tilde{n}_1 & 0 & n_9 \oplus \tilde{n}_9 & 0 \\ \phi & 0 & \omega & 0 \\ n_3 \oplus \tilde{n}_3 & 0 & n_{11} \oplus \tilde{n}_{11} & 0 \end{bmatrix} \xrightarrow{MC} \begin{bmatrix} ? & 0 & ? & 0 \\ 0 & 0 & 0 & 0 \\ ? & 0 & ? & 0 \\ 0 & 0 & 0 & 0 \end{bmatrix}.$$

Now the question is "how to decide π, ρ, ϕ and ω". The answer is:

- There exists one and only one pair of (π, ϕ) such that after MC, Bytes $w_1 \oplus \tilde{w}_1$ and $w_3 \oplus \tilde{w}_3$ are both zero. The values of (π, ϕ) can be decided by solving (3).
- There exists one and only one pair of (ρ, ω) such that after MC, Bytes $w_9 \oplus \tilde{w}_9$ and $w_{11} \oplus \tilde{w}_{11}$ are both zero. By solving (4), we get the values of (ρ, ω).

$$
\left\{
\begin{array}{l}
\begin{bmatrix} 01 & 02 & 03 & 01 \end{bmatrix}
\begin{bmatrix} \pi \\ n_1 \oplus \tilde{n}_1 \\ \phi \\ n_3 \oplus \tilde{n}_3 \end{bmatrix} = 0 \\[2em]
\begin{bmatrix} 03 & 01 & 01 & 02 \end{bmatrix}
\begin{bmatrix} \pi \\ n_1 \oplus \tilde{n}_1 \\ \phi \\ n_3 \oplus \tilde{n}_3 \end{bmatrix} = 0
\end{array}
\right. \quad (3)
\qquad
\left\{
\begin{array}{l}
\begin{bmatrix} 01 & 02 & 03 & 01 \end{bmatrix}
\begin{bmatrix} \rho \\ n_9 \oplus \tilde{n}_9 \\ \omega \\ n_{11} \oplus \tilde{n}_{11} \end{bmatrix} = 0 \\[2em]
\begin{bmatrix} 03 & 01 & 01 & 02 \end{bmatrix}
\begin{bmatrix} \rho \\ n_9 \oplus \tilde{n}_9 \\ \omega \\ n_{11} \oplus \tilde{n}_{11} \end{bmatrix} = 0
\end{array}
\right. \quad (4)
$$

Once we know the values of π, ϕ, ρ and ω, message block $\tilde{M}_y$ can be chosen as follows:

1. Set the values of $\tilde{n}_0^{new}$, $\tilde{n}_8^{new}$, $\tilde{n}_2^{new}$ and $\tilde{n}_{10}^{new}$ as follows: $\tilde{n}_0^{new} = n_0 \oplus \pi$, $\tilde{n}_8^{new} = n_8 \oplus \rho$, $\tilde{n}_2^{new} = n_2 \oplus \phi$, and $\tilde{n}_{10}^{new} = n_{10} \oplus \omega$. Use $\tilde{n}_0^{new}$ to replace $\tilde{n}_0$, $\tilde{n}_8^{new}$ to replace $\tilde{n}_8$, $\tilde{n}_2^{new}$ to replace $\tilde{n}_2$, and $\tilde{n}_{10}^{new}$ to replace $\tilde{n}_{10}$.
2. Perform SR^{-1} and SB^{-1}. Since SR^{-1} and SB^{-1} are permutation and substitution, they do not affect the properties we have found. Now we have the outputs of ARK in Round y.
3. Compute the value of $\tilde{M}_y$ as follows:

$$\tilde{M}_y = (\tilde{n}_0^{new} \oplus \tilde{s}_0) \| (\tilde{n}_8^{new} \oplus \tilde{s}_8) \| (\tilde{n}_{10}^{new} \oplus \tilde{s}_2) \| (\tilde{n}_2^{new} \oplus \tilde{s}_{10}).$$

So far, two messages $(M_{y-3}, M_{y-2}, M_{y-1}, M_y)$ and $(\tilde{M}_{y-3}, \tilde{M}_{y-2}, \tilde{M}_{y-1}^{new}, \tilde{M}_y)$ collide on 96 bits (i.e., Bytes w_1, w_3, w_4, w_5, w_6, w_7, w_9, w_{11}, w_{12}, w_{13}, w_{14} and w_{15}) in Round $y + 1$ after MC transformation.

3.1.3 Extending 96-Bit Collisions to 128-Bit Collisions

This step is straightforward as we can select message M_{y+1} arbitrarily, and construct message $\tilde{M}_{y+1}$ to cancel the differences between Bytes w_0, w_8, w_2 and w_{10}. The construction is provided as follows:

$$\tilde{M}_{y+1} = ((w_0 \oplus \tilde{w}_0) \| (w_8 \oplus \tilde{w}_8) \| (w_2 \oplus \tilde{w}_2) \| (w_{10} \oplus \tilde{w}_{10})) \oplus M_{y+1}.$$

3.2 The Backwards-aNd-Backwards Search

The Backwards-aNd-Backwards search has the following form:

Known:	1. a selected key or a selected intermediate value.
	2. one selected four-block message $M(M_{y-3}, M_{y-2}, M_{y-1}, M_y)$.
Find:	another four-block message $\tilde{M}(\tilde{M}_{y-3}, \tilde{M}_{y-2}, \tilde{M}_{y-1}, \tilde{M}_y)$ such that these two messages collide on 32 bits (Bytes s_4, s_{12}, s_6 and s_{14}) after MC in Round y.
Method:	solve four groups of linear functions. These four groups of functions are named as (5), (6), (7) and (8) in this subsection.

We propose a method to find 32-bit collisions on Bytes s_4, s_{12}, s_6 and s_{14} (see Figure 2) by solving four groups of linear functions. This search assumes that for a selected key (or a selected intermediate value) and a selected four-block message $(M_{y-3}, M_{y-2}, M_{y-1}, M_y)$, we can generate another four-block message $(\tilde{M}_{y-3}, \tilde{M}_{y-2}, \tilde{M}_{y-1}, \tilde{M}_y)$ such that these two messages collide on Bytes s_4, s_{12}, s_6 and s_{14} after MC in Round y. The method used by the BNB search is similar to the idea employed by the BNF search, but works in only one direction (i.e., only backwards).

3.2.1 Deciding Four Values $(\tilde{j}_5, \tilde{j}_7, \tilde{j}_{13} \text{ and } \tilde{j}_{15})$

In the beginning, we choose $(\tilde{M}_{y-3}, \tilde{M}_{y-2}, \tilde{M}_{y-1}, \tilde{M}_y)$ randomly. Assume that the input and the output of MC in Round y are listed as follows:

$$\begin{bmatrix} \tilde{j}_0 & \tilde{j}_4 & \tilde{j}_8 & \tilde{j}_{12} \\ \tilde{j}_1 & \tilde{j}_5^{old} & \tilde{j}_9 & \tilde{j}_{13}^{old} \\ \tilde{j}_2 & \tilde{j}_6 & \tilde{j}_{10} & \tilde{j}_{14} \\ \tilde{j}_3 & \tilde{j}_7^{old} & \tilde{j}_{11} & \tilde{j}_{15}^{old} \end{bmatrix} \xrightarrow{MC} \begin{bmatrix} \tilde{s}_0 & \tilde{s}_4 & \tilde{s}_8 & \tilde{s}_{12} \\ \tilde{s}_1 & \tilde{s}_5 & \tilde{s}_9 & \tilde{s}_{13} \\ \tilde{s}_2 & \tilde{s}_6 & \tilde{s}_{10} & \tilde{s}_{14} \\ \tilde{s}_3 & \tilde{s}_7 & \tilde{s}_{11} & \tilde{s}_{15} \end{bmatrix}.$$

Now we do not use the values of $\tilde{j}_5^{old}$, $\tilde{j}_7^{old}$, $\tilde{j}_{13}^{old}$ or $\tilde{j}_{15}^{old}$. Instead, we use $\tilde{j}_5$ (to replace $\tilde{j}_5^{old}$), $\tilde{j}_7$ (to replace $\tilde{j}_7^{old}$), $\tilde{j}_{13}$ (to replace $\tilde{j}_{13}^{old}$), and $\tilde{j}_{15}$ (to replace $\tilde{j}_{15}^{old}$) such that we get values s_4, s_{12}, s_6, and s_{14} on Bytes $\tilde{s}_4$, $\tilde{s}_{12}$, $\tilde{s}_6$, and $\tilde{s}_{14}$, respectively (illustrated as follows):

$$\begin{bmatrix} \tilde{j}_0 & \tilde{j}_4 & \tilde{j}_8 & \tilde{j}_{12} \\ \tilde{j}_1 & \tilde{j}_5 & \tilde{j}_9 & \tilde{j}_{13} \\ \tilde{j}_2 & \tilde{j}_6 & \tilde{j}_{10} & \tilde{j}_{14} \\ \tilde{j}_3 & \tilde{j}_7 & \tilde{j}_{11} & \tilde{j}_{15} \end{bmatrix} \xrightarrow{MC} \begin{bmatrix} \tilde{s}_0 & s_4 & \tilde{s}_8 & s_{12} \\ \tilde{s}_1 & \tilde{s}_5 & \tilde{s}_9 & \tilde{s}_{13} \\ \tilde{s}_2 & s_6 & \tilde{s}_{10} & s_{14} \\ \tilde{s}_3 & \tilde{s}_7 & \tilde{s}_{11} & \tilde{s}_{15} \end{bmatrix}.$$

Now the question is "how can we make this happen". Our answer is to solve two groups of linear functions. For the values of s_4 and s_6, we have two linear equations in (5) with only two unknown variables ($\tilde{j}_5$ and $\tilde{j}_7$). Therefore, we can solve (5) to obtain the values of $\tilde{j}_5$ and $\tilde{j}_7$.

$$\begin{cases} \begin{bmatrix} 02 & 03 & 01 & 01 \end{bmatrix} \begin{bmatrix} \tilde{j}_4 \\ \tilde{j}_5 \\ \tilde{j}_6 \\ \tilde{j}_7 \end{bmatrix} = s_4 \\ \begin{bmatrix} 01 & 01 & 02 & 03 \end{bmatrix} \begin{bmatrix} \tilde{j}_4 \\ \tilde{j}_5 \\ \tilde{j}_6 \\ \tilde{j}_7 \end{bmatrix} = s_6 \end{cases} \quad (5)$$

$$\begin{cases} \begin{bmatrix} 02 & 03 & 01 & 01 \end{bmatrix} \begin{bmatrix} \tilde{j}_{12} \\ \tilde{j}_{13} \\ \tilde{j}_{14} \\ \tilde{j}_{15} \end{bmatrix} = s_{12} \\ \begin{bmatrix} 01 & 01 & 02 & 03 \end{bmatrix} \begin{bmatrix} \tilde{j}_{12} \\ \tilde{j}_{13} \\ \tilde{j}_{14} \\ \tilde{j}_{15} \end{bmatrix} = s_{14} \end{cases} \quad (6)$$

Similarly, for the values of s_{12} and s_{14}, we have two linear functions in (6) with two unknown variables ($\tilde{j}_{13}$ and $\tilde{j}_{15}$). We can solve (6) to decide the values of $\tilde{j}_{13}$ and $\tilde{j}_{15}$. After getting four values ($\tilde{j}_5$, $\tilde{j}_7$, $\tilde{j}_{13}$, and $\tilde{j}_{15}$) decided, we perform the SR^{-1} and SB^{-1} transformations. As SR^{-1} is permutation and SB^{-1} is substitution, $\tilde{j}_5$, $\tilde{j}_7$, $\tilde{j}_{13}$, and $\tilde{j}_{15}$ are first relocated then substituted by another four values $\tilde{i}_9$, $\tilde{i}_3$, $\tilde{i}_1$, and $\tilde{i}_{11}$, respectively. As the message injection layout does not change the values of $\tilde{i}_9$, $\tilde{i}_3$, $\tilde{i}_1$, and $\tilde{i}_{11}$, these four values are not changed after we do ARK. So, we get four known values ($\tilde{i}_9$, $\tilde{i}_3$, $\tilde{i}_1$, and $\tilde{i}_{11}$) after MC in Round $y - 1$. Our next target is to modify message block $\tilde{M}_{y-2}$ so that we get those four values $\tilde{i}_9$, $\tilde{i}_3$, $\tilde{i}_1$, and $\tilde{i}_{11}$ after MC in Round $y - 1$.

3.2.2 Modifying Message Block $\tilde{M}_{y-2}$

Suppose by using the original message block $\tilde{M}_{y-2}$, we have the following states in Round $y - 1$:

$$
\begin{bmatrix} \tilde{g}_0^{*old} & \tilde{g}_4 & \tilde{g}_8^{*old} & \tilde{g}_{12} \\ \tilde{g}_1 & \tilde{g}_5 & \tilde{g}_9 & \tilde{g}_{13} \\ \tilde{g}_2^{*old} & \tilde{g}_6 & \tilde{g}_{10}^{*old} & \tilde{g}_{14} \\ \tilde{g}_3 & \tilde{g}_7 & \tilde{g}_{11} & \tilde{g}_{15} \end{bmatrix}
\xrightarrow{SB \circ SR}
\begin{bmatrix} \tilde{h}_0^{old} & \tilde{h}_4 & \tilde{h}_8^{old} & \tilde{h}_{12} \\ \tilde{h}_1 & \tilde{h}_5 & \tilde{h}_9 & \tilde{h}_{13} \\ \tilde{h}_2^{old} & \tilde{h}_6 & \tilde{h}_{10}^{old} & \tilde{h}_{14} \\ \tilde{h}_3 & \tilde{h}_7 & \tilde{h}_{11} & \tilde{h}_{15} \end{bmatrix}
\xrightarrow{MC}
\begin{bmatrix} ? & \tilde{i}_4 & ? & \tilde{i}_{12} \\ ? & \tilde{i}_5 & ? & \tilde{i}_{13} \\ ? & \tilde{i}_6 & ? & \tilde{i}_{14} \\ ? & \tilde{i}_7 & ? & \tilde{i}_{15} \end{bmatrix}.
$$

Now we replace values ($\tilde{h}_0^{old}$, $\tilde{h}_2^{old}$, $\tilde{h}_8^{old}$, $\tilde{h}_{10}^{old}$) with ($\tilde{h}_0$, $\tilde{h}_2$, $\tilde{h}_8$, $\tilde{h}_{10}$) and then we get those four values ($\tilde{i}_9$, $\tilde{i}_3$, $\tilde{i}_1$, and $\tilde{i}_{11}$) located as follows:

$$
\begin{bmatrix} \tilde{g}_0^* & \tilde{g}_4 & \tilde{g}_8^* & \tilde{g}_{12} \\ \tilde{g}_1 & \tilde{g}_5 & \tilde{g}_9 & \tilde{g}_{13} \\ \tilde{g}_2^* & \tilde{g}_6 & \tilde{g}_{10}^* & \tilde{g}_{14} \\ \tilde{g}_3 & \tilde{g}_7 & \tilde{g}_{11} & \tilde{g}_{15} \end{bmatrix}
\xrightarrow{SB \circ SR}
\begin{bmatrix} \tilde{h}_0 & \tilde{h}_4 & \tilde{h}_8 & \tilde{h}_{12} \\ \tilde{h}_1 & \tilde{h}_5 & \tilde{h}_9 & \tilde{h}_{13} \\ \tilde{h}_2 & \tilde{h}_6 & \tilde{h}_{10} & \tilde{h}_{14} \\ \tilde{h}_3 & \tilde{h}_7 & \tilde{h}_{11} & \tilde{h}_{15} \end{bmatrix}
\xrightarrow{MC}
\begin{bmatrix} ? & \tilde{i}_4 & ? & \tilde{i}_{12} \\ \tilde{i}_1 & \tilde{i}_5 & \tilde{i}_9 & \tilde{i}_{13} \\ ? & \tilde{i}_6 & ? & \tilde{i}_{14} \\ \tilde{i}_3 & \tilde{i}_7 & \tilde{i}_{11} & \tilde{i}_{15} \end{bmatrix}.
$$

Based on the property of MC transformation, we can form the following two groups of linear functions:

$$
\begin{cases}
\begin{bmatrix} 01 & 02 & 03 & 01 \end{bmatrix} \begin{bmatrix} \tilde{h}_0 \\ \tilde{h}_1 \\ \tilde{h}_2 \\ \tilde{h}_3 \end{bmatrix} = \tilde{i}_1 \\[4ex]
\begin{bmatrix} 03 & 01 & 01 & 02 \end{bmatrix} \begin{bmatrix} \tilde{h}_0 \\ \tilde{h}_1 \\ \tilde{h}_2 \\ \tilde{h}_3 \end{bmatrix} = \tilde{i}_3
\end{cases} \quad (7)
\qquad
\begin{cases}
\begin{bmatrix} 01 & 02 & 03 & 01 \end{bmatrix} \begin{bmatrix} \tilde{h}_8 \\ \tilde{h}_9 \\ \tilde{h}_{10} \\ \tilde{h}_{11} \end{bmatrix} = \tilde{i}_9 \\[4ex]
\begin{bmatrix} 03 & 01 & 01 & 02 \end{bmatrix} \begin{bmatrix} \tilde{h}_8 \\ \tilde{h}_9 \\ \tilde{h}_{10} \\ \tilde{h}_{11} \end{bmatrix} = \tilde{i}_{11}
\end{cases} \quad (8)
$$

We know the values of $\tilde{h}_1$, $\tilde{h}_3$, $\tilde{h}_9$ and $\tilde{h}_{11}$ from the original message block $\tilde{M}_{y-2}$. We can get the values of ($\tilde{h}_0$, $\tilde{h}_2$) by solving (7), and get the values of

$(\tilde{h}_8, \tilde{h}_{10})$ by solving (8). After finding the values of $(\tilde{h}_0, \tilde{h}_2, \tilde{h}_8, \tilde{h}_{10})$, we perform SR^{-1} and SB^{-1}, and obtain the corresponding four values $(\tilde{g}_0^*, \tilde{g}_2^*, \tilde{g}_8^*, \tilde{g}_{10}^*)$. Once we know the values of $(\tilde{g}_0^*, \tilde{g}_2^*, \tilde{g}_8^*, \tilde{g}_{10}^*)$, we replace $\tilde{M}_{y-2}$ with M_{y-2}^{new}. M_{y-2}^{new} is constructed as follows (note that $\tilde{g}_0, \tilde{g}_8, \tilde{g}_2$ and $\tilde{g}_{10}$ are known from the message block $\tilde{M}_{y-3}$ in Round $y-3$):

$$\tilde{M}_{y-2}^{new} = (\tilde{g}_0^* \oplus \tilde{g}_0)||(\tilde{g}_8^* \oplus \tilde{g}_8)||(\tilde{g}_2^* \oplus \tilde{g}_2)||(\tilde{g}_{10}^* \oplus \tilde{g}_{10}).$$

3.3   Combining the BNB Search with the BNF Search

The second-preimage search algorithm combines the BNB search with the BNF search. To search for a second preimage of the ALPHA-MAC, we perform the following steps:

1. Select a key or an intermediate value.
2. Select a five-block message $M(M_{y-3}, M_{y-2}, M_{y-1}, M_y, M_{y+1})$.
3. Generate the second preimage $\tilde{M}(\tilde{M}_{y-3}, \tilde{M}_{y-2}, \tilde{M}_{y-1}, \tilde{M}_y, \tilde{M}_{y+1})$ randomly. We need to guarantee that $\tilde{M}_{y-3}$ is not equal to M_{y-3}.
4. Perform the BNB search to generate 32-bit collisions. The BNB search is done by modifying message block $\tilde{M}_{y-2}$.
5. Use the BNF search to extend those 32-bit collisions to 128-bit collisions. The BNF search is carried out by modifying the values of $\tilde{M}_{y-1}, \tilde{M}_y$, and $\tilde{M}_{y+1}$. Message $\tilde{M}(\tilde{M}_{y-3}, \tilde{M}_{y-2}, \tilde{M}_{y-1}, \tilde{M}_y, \tilde{M}_{y+1})$ is a second preimage of message $M(M_{y-3}, M_{y-2}, M_{y-1}, M_y, M_{y+1})$ under the selected key (or the selected intermediate value).

The routine of finding second preimages is shown in Table 1, and Figure 3 depicts this finding. The name of the BNB search comes from the fact that

Table 1. Second-preimage search = BNB search + BNF search

Search	R	Round $y-2$	Di	Round $y-1$	Di	Round y
BNB	1				$\Leftarrow \tilde{s}_4 \rightharpoonup s_4$, $\tilde{s}_{12} \rightharpoonup s_{12}$, $\tilde{s}_6$ $\rightharpoonup s_6$, $\tilde{s}_{14} \rightharpoonup s_{14}$	
	2		$\Leftarrow \tilde{h}_0^{old} \rightharpoonup \tilde{h}_0$, $\tilde{h}_2^{old} \rightharpoonup \tilde{h}_2$, $\tilde{h}_8^{old} \rightharpoonup \tilde{h}_8$, $\tilde{h}_{10}^{old} \rightharpoonup \tilde{h}_{10}$			
	3	$\tilde{M}_{y-2} \rightharpoonup \tilde{M}_{y-2}^{new}$				
		Round $y-1$	Di	Round y	Di	Round $y+1$
BNF	4	modify $\tilde{M}_{y-1}$	$\Leftarrow$ collisions on s_4, s_{12}, s_6 and s_{14}			
	5		$\Rightarrow$ collisions on s_4, s_{12}, s_6, s_{14}, s_1, s_9, s_3 and s_{11}			
	6		modify $\tilde{M}_y$		$\Rightarrow$ 96-bit collisions	
	7				modify $\tilde{M}_{y+1} \rightarrow$ 128-bit collisions	

Di - Direction

R - Routine

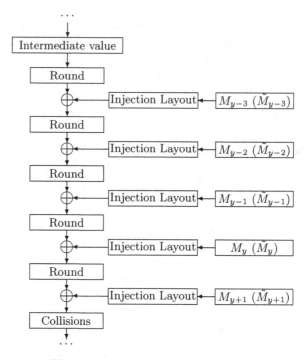

Fig. 3. The second-preimage search

searching for $\tilde{M}_{y-2}$ is carried out by moving backwards and then backwards, and the name of the BNF search comes from the fact that searching for $\tilde{M}_{y-1}$, $\tilde{M}_y$ and $\tilde{M}_{y+1}$ is performed by moving backwards and then forwards (see Table 1). A personal computer takes about 1 second to find a second preimage of the ALPHA-MAC. In Appendix A, we provide a second preimage of a selected key and a selected five-block message.

4   The Collision Search Algorithm

> Known: a selected key or a selected intermediate value.
> Find: two five-block messages M and $\tilde{M}$ such that they collide under the
>   selected key or the intermediate value.
> Method: employ the second-preimage search.

In the second-preimage search, we choose the first five-block message arbitrarily, and once it is decided, we do not modify it. All we need to do is modify the second five-block message so that 128-bit collisions happen. Therefore, the second-preimage search can also be used to find two colliding five-block messages under a selected key (or a selected intermediate value).

5 Conclusions

In this paper, we have presented our analysis on the internal structure of ALPHA-MAC. We proposed a method to find second preimages of the ALPHA-MAC by combining the Backwards-aNd-Forwards search and the Backwards-aNd-Backwards search, based on the assumption that a key or an intermediate value is known. Our method employs the algebraic properties of AES and the structural features of the ALPHA-MAC. To find a second preimage of the ALPHA-MAC, our idea needs to solve eight groups of linear functions. We also showed that the second-preimage finding method can be used to generate internal collisions.

References

1. National Institute of Standards and Technology, U.S. Department of Commerce. *Advanced Encryption Standard (AES)*. Federal Information Processing Standard 197, 2001.
2. National Institute of Standards and Technology, U.S. Department of Commerce. *Secure hash standard*. Federal Information Processing Standard, FIPS-180-1, 1995.
3. National Institute of Standards and Technology, U.S. Department of Commerce. *Secure hash standard*. Federal Information Processing Standard, FIPS-180, 1993.
4. National Institute of Standards and Technology, U.S. Department of Commerce. *The Keyed-Hash Message Authentication Code (HMAC)*. Federal Information Processing Standard 198, 2002.
5. Mihir Bellare, Ran Canetti, and Hugo Krawczyk. *Keying Hash Functions for Message Authentication*. Advances in Cryptology - CRYPTO'96, 16th Annual International Cryptology Conference, Lecture Notes in Computer Science 1109, pp. 1-15, Springer-Verlag, 1996.
6. Eli Biham and Rafi Chen. *Near-Collisions of SHA-0*. Advances in Cryptology - CRYPTO 2004, 24th Annual International Cryptology Conference, Lecture Notes in Computer Science 3152, pp. 290-305, Springer-Verlag, 2004.
7. Eli Biham, Rafi Chen, Antoine Joux, Patrick Carribault, Christophe Lemuet, and William Jalby. *Collisions of SHA-0 and Reduced SHA-1*. Advances in Cryptology - EUROCRYPT 2005, 24th Annual International Conference on the Theory and Applications of Cryptographic Techniques, Lecture Notes in Computer Science 3494, pp. 36-57, Springer-Verlag, 2005.
8. Eli Biham and Adi Shamir. *Differential Cryptanalysis of the Data Encryption Standard*. Springer-Verlag, 1993.
9. John Black, Shai Halevi, Hugo Krawczyk, Ted Krovetz, and Phillip Rogaway. *UMAC: Fast and Secure Message Authentication*. Advances in Cryptology - CRYPTO'99, 19th Annual International Cryptology Conference, Lecture Notes in Computer Science 1666, pp. 216-233, Springer-Verlag, 1999.
10. Florent Chabaud and Antoine Joux. *Differential Collisions in SHA-0*. Advances in Cryptology - CRYPTO'98, 18th Annual International Cryptology Conference, Lecture Notes in Computer Science 1462, pp. 56-71, Springer-Verlag, 1998.

11. Joan Daemen and Vincent Rijmen. *A New MAC Construction ALRED and a Specific Instance ALPHA-MAC.* Fast Software Encryption: 12th International Workshop, FSE 2005, Lecture Notes in Computer Science 3557, pp. 1-17, Springer-Verlag, 2005.

12. Joan Daemen and Vincent Rijmen. *AES Proposal: Rijndael,* AES Round 1 Technical Evaluation CD-1: Documentation, National Institute of Standards and Technology, 1998.

13. Donald Davies. *A Message Authenticator Algorithm Suitable for A Mainframe Computer.* Advances in Cryptology, Proceedings of CRYPTO'84, Lecture Notes in Computer Science 196, pp. 393-400, Springer-Verlag, 1985.

14. Hans Dobbertin. *Cryptanalysis of MD4.* Fast Software Encryption: Third International Workshop, FSE 1996, Lecture Notes in Computer Science 1039, pp. 53-69, Springer-Verlag, 1996.

15. International Organization for Standardization. *ISO/IEC 9797-1, Information technology – Security techniques – Message Authentication Codes (MACs) – Part 1: Mechanisms using a block cipher .* 1999.

16. Bart Preneel and Paul C. van Oorschot. *MDx-MAC and Building Fast MACs from Hash Functions.* Advances in Cryptology - CRYPTO'95, 15th Annual International Cryptology Conference, Lecture Notes in Computer Science 963, pp. 1-14, Springer-Verlag, 1995.

17. Ronald Rivest. The MD4 message-digest algorithm, Request for Comments (RFC) 1320, Internet Activities Board, Internet Privacy Task Force, 1992.

18. Ronald Rivest. The MD5 message-digest algorithm, Request for Comments (RFC) 1320), Internet Activities Board, Internet Privacy Task Force, 1992.

19. Xiaoyun Wang, Xuejia Lai, Dengguo Feng, Hui Chen, and Xiuyuan Yu. *Cryptanalysis of the Hash Functions MD4 and RIPEMD.* Advances in Cryptology - EUROCRYPT 2005, 24th Annual International Conference on the Theory and Applications of Cryptographic Techniques, Lecture Notes in Computer Science 3494, pp. 1-18, Springer-Verlag, 2005.

20. Xiaoyun Wang, Yiqun Lisa Yin, and Hongbo Yu. *Finding Collisions in the Full SHA-1.* Advances in Cryptology - CRYPTO 2005: 25th Annual International Cryptology Conference, Lecture Notes in Computer Science 3621, pp. 17-36, Springer-Verlag, 2005.

21. Xiaoyun Wang and Hongbo Yu. *How to Break MD5 and Other Hash Functions.* Advances in Cryptology - EUROCRYPT 2005, 24th Annual International Conference on the Theory and Applications of Cryptographic Techniques, Lecture Notes in Computer Science 3494, pp. 19-35, Springer-Verlag, 2005.

22. Xiaoyun Wang, Hongbo Yu, and Yiqun Lisa Yin. *Efficient Collision Search Attacks on SHA-0.* Advances in Cryptology - CRYPTO 2005: 25th Annual International Cryptology Conference, Lecture Notes in Computer Science 3621, pp. 1-16, Springer-Verlag, 2005.

A   A Found Second Preimage

For a selected key K (see Table 3) and a selected five-block message M (see Table 2), a second preimage found by our algorithm is $\tilde{M}$ (shown in Table 2). The 128-bit colliding value is listed in Table 4. Note that these two messages are listed after injection layout.

Table 2. Two five-block messages

M (the selected message)																			
M_{y-3}				M_{y-2}				M_{y-1}				M_y				M_{y+1}			
c4	0	8c	0	e6	0	2a	0	77	0	fd	0	ef	0	a1	0	81	0	9f	0
0	0	0	0	0	0	0	0	0	0	0	0	0	0	0	0	0	0	0	0
94	0	f3	0	95	0	04	0	4c	0	37	0	68	0	09	0	25	0	2c	0
0	0	0	0	0	0	0	0	0	0	0	0	0	0	0	0	0	0	0	0
$\tilde{M}$ (the found second preimage)																			
$\tilde{M}_{y-3}$				$\tilde{M}_{y-2}$				$\tilde{M}_{y-1}$				$\tilde{M}_y$				$\tilde{M}_{y+1}$			
1d	0	43	0	22	0	04	0	e4	0	83	0	2f	0	e5	0	69	0	06	0
0	0	0	0	0	0	0	0	0	0	0	0	0	0	0	0	0	0	0	0
1c	0	0d	0	2f	0	30	0	2f	0	9b	0	d4	0	30	0	f4	0	3a	0
0	0	0	0	0	0	0	0	0	0	0	0	0	0	0	0	0	0	0	0

Table 3. The selected key K

83	55	2d	81
88	2c	05	67
c1	63	be	c2
2a	a2	52	a4

Table 4. The 128-bit collisions

7d	69	88	d7
02	cb	1f	af
b9	d8	7b	5e
0e	10	79	21

A Weak Key Class of XTEA for a Related-Key Rectangle Attack*

Eunjin Lee, Deukjo Hong, Donghoon Chang, Seokhie Hong, and Jongin Lim

Center for Information Security Technologies (CIST),
Korea University, Seoul, Korea
{walgadak,hongdj,pointchang,hsh,jilim}@cist.korea.ac.kr

Abstract. XTEA is a block cipher with a very simple structure but there has not been found attack even for half of full round version i.e 32-round version. In this paper we introduce a class of weak keys which makes a 34-round reduced version of XTEA vulnerable to the related-key rectangle attack. The number of such weak keys is about $2^{108.21}$. Our attack on a 34-round reduced version of XTEA under weak key assumption requires 2^{62} chosen plaintexts and $2^{31.94}$ 34-round XTEA encryptions.

Keywords: XTEA algorithm, related-key rectangle attack, weak key class of XTEA.

1 Introduction

In 1994, Wheeler and Needham proposed a simple block cipher TEA [14] that uses exclusive-or, addition, and shift operation. It had a simple round function which looked too weak to give sufficient security, but TEA had large number of rounds of 64 enough to make itself secure against current attacks. Furthermore, since it had a simple structure, its implementation and performance were not bad.

However, Kelsey, Schneier and Wagner proposed a related-key attack on full-round TEA in [9]. They used a differential characteristic for addition mod 2^{32} with probability 1 that can be constructed under a pair of keys with a particular difference. Wheeler and Needham proposed XTEA which was an improved version of TEA [13]. XTEA is 64-bit block cipher using a 128-bit secret key. Until now, the best attack is the related-key truncated differential attack on 27-round XTEA proposed in [11]. Table 1 depicts recent results on XTEA.

In 1993, Biham [2] introduced the related-key attack in which the attacker can choose the relationship between two unknown keys. It depends on a key

---

* This research was supported by the MIC(Ministry of Information and Communication), Korea, under the ITRC(Information Technology Research Center) support program supervised by the IITA(Institute of Information Technology Advancement)(IITA-2006-(C1090-0603-0025)).

P.Q. Nguyen (Ed.): VIETCRYPT 2006, LNCS 4341, pp. 286–297, 2006.

scheduling algorithm and shows that a block cipher with a weak key scheduling algorithm may be vulnerable to this kind of attack. In 1999, Wagner [15] proposed the boomerang attack using chosen plaintexts and adaptively chosen ciphertexts. For a block cipher, it may be that finding a long differential with high probability is difficult but finding a short differential with high probability is easy. In such a block cipher, the boomerang attack is useful since it uses two short differentials with high probability to construct a long-round distinguisher. The boomerang attack was developed into a chosen plaintext attack called the amplified boomerang attack [8]. The transformation to a chosen plaintext attack has price in a much larger data complexity for the identification of the right quartets. After its introduction, Biham, Dunkelman, Keller [3] improved it into the rectangle attack. In 2004, Kim et al. [10] and Biham et al. [5] introduced a combination of the related-key and the rectangle attacks, called the related-key rectangle attack. Recently, Hong et al. [6] and Biham et al. [5] considered four related keys to suggest related-key rectangle attack.

In this paper, we apply the related-key rectangle attack with four related keys to reduced version of XTEA and introduce a class of weak keys for the attack. Although XTEA is an improved version of TEA, it is very interesting that XTEA still has weakness of related-key attack. We show that there exist about $2^{108.21}$ keys which make XTEA not secure against the related-key rectangle attack. Under the assumption of such weak key, we construct a rectangle distinguisher to break a 34-round reduced version of XTEA. This attack requires 2^{62} chosen plaintexts and $2^{31.94}$ 34-round XTEA encryptions.

This paper is organized as follows. In Section 2, we present the XTEA algorithm. In Section 3, we describe a related-key rectangle distinguisher of XTEA with weak key quartets. In Section 4, we present a related-key rectangle attack under weak key assumption on a 34-round reduced version of XTEA. We conclude in Section 5.

Table 1. Various attacks on reduced-round XTEA

Attack method	paper	Rounds	Data complexity	Time Complexity
Impossible Diff.	[12]	14	$2^{62.5}$	2^{85}
Diff.	[7]	15	2^{59}	2^{120}
Truncated Diff.	[7]	23	2^{23}	$2^{120.65}$
R·K Truncated Diff.	[11]	27	$2^{20.5}$	$2^{115.15}$
R·K Rectangle (weak key assumption)	this paper	34	2^{62}	$2^{31.94}$

2   XTEA Algorithm

In this section, we describe several notations used in this paper and briefly describe XTEA.

- $\boxplus$: addition modulo 2^{32}
- $\oplus$: exclusive-or
- $\cdot$: multiplication modulo 2^{32}
- $\ll$ (or $\gg$) : left (or right) shift
- $\|$: concatenation of two binary strings

2.1 Description of XTEA

XTEA is a 64-round block cipher with 64-bit block size and 128-bit key K which is split into four 32-bit words $K = K_0\|K_1\|K_2\|K_3$. Let (L_n, R_n) and (L_{n+1}, R_{n+1}) be the input and the output of the n-th round function, respectively. For an n-th round key S_n, (L_{n+1}, R_{n+1}) is defined as follows.

$$L_{n+1} = R_n,$$
$$R_{n+1} = L_n \boxplus F(R_n, S_n)$$
$$= L_n \boxplus ((G(R_n) \boxplus R_n) \oplus S_n),$$

where $G(x) = (x \ll 4) \oplus (x \gg 5)$ for any 32-bit value x. For $\delta = 9e3779b9_x$ and $1 \le n \le 64$, the n-th round key S_n is generated from K as follows.

$$S_n = \begin{cases} (i-1) \cdot \delta \boxplus K_{((i-1)\cdot\delta \gg 11)\&3} & \text{if } n = 2i - 1, \\ i \cdot \delta \boxplus K_{(i\cdot\delta \gg 11)\&3} & \text{if } n = 2i \end{cases}$$

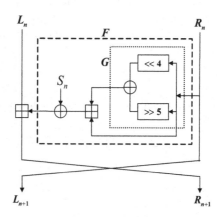

Fig. 1. n-th round function of XTEA

3 Related-Key Rectangle Distinguisher of XTEA Under Weak Key Assumption

In this section we show how to construct a 33-round related-key rectangle distinguisher of XTEA under weak key assumption.

3.1   3-Round Related-Key Differential Characteristic of XTEA

Let $\alpha_1 = 80402010_x$ and $\alpha_2 = 80c02010_x$. We assume two equations $x_1 \boxplus x_2 = x_3$ and $y_1 \boxplus y_2 = y_3$. $\Delta_1 \boxplus \Delta_2 \to \Delta_3$ means $x_1 \oplus y_1 = \Delta_1$(or $x_1 \oplus y_2 = \Delta_1$), $x_2 \oplus y_2 = \Delta_2$(or $x_2 \oplus y_1 = \Delta_2$) and $x_3 \oplus y_3 = \Delta_3$. The probability of $\alpha_1 \boxplus 0 \to \alpha_1$ and the probability of $\alpha_1 \boxplus \alpha_1 \to 0$ are 2^{-3} over the random distribution of (x_1, x_2, y_1, y_2).

We will explain three differential characteristics used in our attack. First, we consider the probability of the 2-round differential characteristic in the case that the input difference is zero and two consecutive round key differences are α_1, namely $\Delta S_n = \Delta S_{n+1} = \alpha_1$. See Fig 2. In the n-th round, the output difference of F is α_1 because $\Delta R_n = 0$ and $\Delta S_n = \alpha_1$. The probability of $0 \boxplus \alpha_1 \to \alpha_1$ is 2^{-3}, so $\Delta R_{n+1} = \alpha_1$ with the probability 2^{-3}. In the $(n+1)$-th round, since $G(\alpha_1) = 0$, the probability of $0 \boxplus \alpha_1 \to \alpha_1$ is 2^{-3} and $\Delta S_{n+1} = \alpha_1$, the output difference of F is zero with the probability 2^{-3}. So, $\Delta R_{n+2} = 0$ when the output difference of F is zero, because $\Delta L_{n+1} = 0$. Consequently, when $\Delta S_n = \Delta S_{n+1} = \alpha_1$, the probability of the differential characteristic $(\Delta L_n, \Delta R_n) \to (\Delta L_{n+2}, \Delta R_{n+2})$ with $(\Delta L_n, \Delta R_n) = (0, 0)$ and $(\Delta L_{n+2}, \Delta R_{n+2}) = (\alpha_1, 0)$ is 2^{-6}. We denote this differential characteristic with ψ_1.

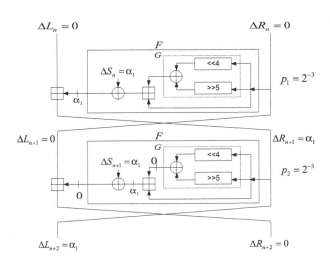

Fig. 2. ψ_1, 2-round related-key differential characteristic of XTEA

ψ_1 can be extended to a 3-round differential characteristic with the probability 2^{-9} by concatenating it to the one-round differential characteristic $(\Delta L_{n+2}, \Delta R_{n+2}) \to (\Delta L_{n+3}, \Delta R_{n+3})$ with $\Delta L_{n+2} = \alpha_1, \Delta R_{n+2} = \Delta L_{n+3} = \Delta R_{n+3} = 0$, and $\Delta S_{n+2} = \alpha_1$. The additionally concatenated one-round differential characteristic has the probability 2^{-3} because the probability of $\alpha_1 \boxplus \alpha_1 \to 0$ is 2^{-3}. We denote this 3-round differential characteristic ψ_2.

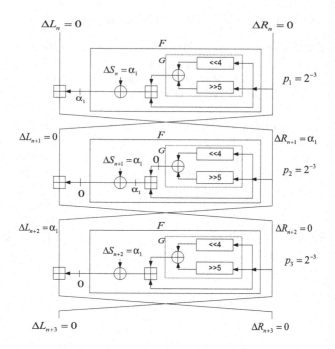

Fig. 3. ψ_2, 3-round related-key differential characteristic of XTEA

ψ_1 can be extended to a 3-round differential characteristic with the probability 2^{-10} by concatenating it to the one-round differential characteristic $(\Delta L_{n+2}, \Delta R_{n+2}) \rightarrow (\Delta L_{n+3}, \Delta R_{n+3})$ with $\Delta L_{n+2} = \alpha_1$, $\Delta R_{n+2} = \Delta L_{n+3} = \Delta R_{n+3} = 0$, and $\Delta S_{n+2} = \alpha_2$. The additionally concatenated one-round differential characteristic has the probability 2^{-4} because the probability of $\alpha_1 \boxplus \alpha_2 \rightarrow 0$ is 2^{-4}. We denote this 3-round differential characteristic ψ_3.

3.2   Related-Key Rectangle Distinguisher of XTEA Under Weak Key Assumption

We use the weakness of the XTEA key schedule to build a 33-round related-key rectangle distinguisher in this subsection. We are interested in the property that round keys inherit the difference from four words of the 128-bit master key. Table 2 shows us the order of four words K_0, K_1, K_2, and K_3 in the master key K of XTEA in generating round keys.

According to Table 2, we can see that K_0 is used from the 8th round to the 10th round and from the 17th round to the 18th round and that K_1 is used from the 26th round to the 28th round. We use this fact and the differential characteristics ψ_1, ψ_2, and ψ_3 to build a 33-round related-key rectangle distinguisher of XTEA.

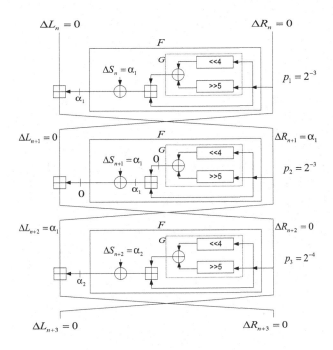

Fig. 4. ψ_3, 3-round related-key differential characteristic of XTEA

Table 2. The order of four words $K_0, K_1, K_2,$ and K_3 in the master key K of XTEA in generating round keys

Round	1	2	3	4	5	6	7	8	9	10	11	12	13	14	15	16
Key	K_0	K_3	K_1	K_2	K_2	K_1	K_3	K_0	K_0	K_0	K_1	K_3	K_2	K_2	K_3	K_1
Round	17	18	19	20	21	22	23	24	25	26	27	28	29	30	31	32
Key	K_0	K_0	K_1	K_0	K_2	K_3	K_3	K_2	K_0	K_1	K_1	K_1	K_2	K_0	K_3	K_3
Round	33	34	35	36	37	38	39	40	41	42	43	44	45	46	47	48
Key	K_0	K_2	K_1	K_1	K_2	K_1	K_3	K_0	K_0	K_3	K_1	K_2	K_2	K_1	K_3	K_1
Round	49	50	51	52	53	54	55	56	57	58	59	60	61	62	63	64
Key	K_0	K_0	K_1	K_3	K_2	K_2	K_3	K_2	K_0	K_1	K_1	K_0	K_2	K_3	K_3	K_2

Let $K1, K2, K3,$ and $K4$ be 128-bit keys. We consider the following quartet of related keys.

$$K1 = K_0||K_1||K_2||K_3, \quad K2 = K_0'||K_1||K_2||K_3,$$
$$K3 = K_0||K_1'||K_2||K_3, \quad K4 = K_0'||K_1'||K_2||K_3.$$

We assume that following conditions are satisfied.

$$(\delta \cdot 4 \boxplus K_0) \oplus (\delta \cdot 4 \boxplus K_0') = \alpha_1, \tag{1}$$
$$(\delta \cdot 8 \boxplus K_0) \oplus (\delta \cdot 8 \boxplus K_0') = \alpha_1, \tag{2}$$

$$(\delta \cdot 9 \boxplus K_0) \oplus (\delta \cdot 9 \boxplus K_0') = \alpha_1, \qquad (3)$$

$$(\delta \cdot 13 \boxplus K_1) \oplus (\delta \cdot 13 \boxplus K_1') = \alpha_1, \qquad (4)$$

$$(\delta \cdot 14 \boxplus K_1) \oplus (\delta \cdot 14 \boxplus K_1') = \alpha_1, \qquad (5)$$

$$(\delta \cdot 5 \boxplus K_0) \oplus (\delta \cdot 5 \boxplus K_0') = \alpha_2. \qquad (6)$$

Let E_0 be the XTEA encryption from the 2nd round to the 19th round and E_1 be the XTEA encryption from the 20th round to the 34th round, and let P_1, and P_2 be 64-bit plaintexts. For $E_0(K1, P_1)$ and $E_0(K2, P_1)$, we can construct a 18-round differential characteristic ψ_4 using the differential characteristics ψ_3 and ψ_1, which is described Table 3. The probability of ψ_4 is $2^{-10} \cdot 2^{-6} = 2^{-16}$. Let $Y_1 = E_0(K1, P_1), Y_2 = E_0(K2, P_1), Y_3 = E_0(K3, P_2)$ and $Y_4 = E_0(K4, P_2)$. We also apply ψ_4 to $E_0(K3, P_2)$ and $E_0(K4, P_2)$ such that $Y_1 \oplus Y_2 = Y_3 \oplus Y_4$. Then the probability of both $Y_1 \oplus Y_2 = Y_3 \oplus Y_4$ and ψ_4 are happened is $\sum_\beta Pr((\Delta 0, \Delta 0) \to (\Delta \beta, \Delta 0))^2 = (2^{-18.38})^2$ where $\Delta \beta$ is a possible value of $\Delta \alpha_1 \boxplus \Delta 0$.

We assume that $Y_1 \oplus Y_3 = Y_2 \oplus Y_4 = (0, 0)$. Then for $E_1(K1, Y_1)$ and $E_1(K3, Y_3)$, we can construct a 15-round differential characteristic ψ_5 using the differential characteristic ψ_2, which is described Table 4. Then the probability of ψ_5 is 2^{-9}. Let $Z_1 = E_1(K1, Y_1), Z_2 = E_1(K2, Y_2), Z_3 = E_1(K3, Y_3)$ and $Z_4 = E_1(K4, Y_4)$. We also apply ψ_5 to $E_1(K3, Y_2)$ and $E_1(K4, Y_4)$ such that $Z_1 \oplus Z_3 = Z_2 \oplus Z_4 = (0, 0)$. Then $Z_1 \oplus Z_3 = Z_2 \oplus Z_4 = (0, 0)$ with the probability $(2^{-9})^2$.

If a quartet (P_1, P_1, P_2, P_2) satisfies above related-key differential characteristics then we call (P_1, P_1, P_2, P_2) a right quartet. That is, right quartet (P_1, P_1, P_2, P_2) satisfies following conditions.

$$Y_1 \oplus Y_2 = Y_3 \oplus Y_4 = (\beta, 0), \qquad (7)$$

$$Y_1 \oplus Y_3 = (0, 0), \qquad (8)$$

$$Z_1 \oplus Z_3 = Z_2 \oplus Z_4 = (0, 0). \qquad (9)$$

Let m be the number of plaintext pairs with input difference $(0,0)$. Then we have about $m^2 \cdot (2^{-18.38})^2$ quartets satisfying (7). If we assume that the intermediate encryption values are distributed uniformly over all possible values, we get $Y_1 \oplus Y_3 = (0, 0)$ with the probability 2^{-64}. This assumption enables us to obtain $m^2 \cdot 2^{-64} \cdot (2^{-18.38})^2$ quartets satisfying (7) and (8). As stated above, (7) and (8) allow us to get $Y_2 \oplus Y_4 = (0, 0)$ with probability 1. Moreover, each of $Y1 \oplus Y3$ and $Y2 \oplus Y4$ satisfies the related-key differential ψ_5 with the probability 2^{-9}. Therefore, the expected number of right quartets is $m^2 \cdot (2^{-18.38})^2 \cdot 2^{-64} \cdot (2^{-9})^2$. For a random permutation the expected number of $Z_1 \oplus Z_3 = Z_2 \oplus Z_4 = (0, 0)$ is $m^2 \cdot (2^{-64})^2$ since there are m^2 possible quartets and each of the $Z_1 \oplus Z_3$ and $Z_2 \oplus Z_4$ satisfies the $(0,0)$ with the probability 2^{-64}. Consequently, our related-key differential characteristics can form a 33-round related-key rectangle distinguisher of XTEA since $m^2 \cdot (2^{-18.38})^2 \cdot 2^{-64} \cdot (2^{-9})^2$ is greater than $m^2 \cdot (2^{-64})^2$.

Table 3. 18-round differential characteristic ψ_4

Round	characteristic	probability
2 ~7	input differences and ΔS_n are all zero	1
8 ~10	This is same to ψ_3	2^{-10}
11~16	input differences and ΔS_n are all zero	1
17~18	This is same to ψ_1	2^{-6}
19	Δinput : (0,0), Δoutput : $(\beta(=\alpha_1 \boxplus 0),0), \Delta S_{19} = 0$	

Table 4. 15-round differential characteristic ψ_5

Round	characteristic	probability
20 ~25	input differences and ΔS_n are all zero	1
26 ~28	This is same to ψ_2	2^{-9}
29~34	input differences and ΔS_n are all zero	1

4   Related-Key Rectangle Attack on 34 Rounds of XTEA Under Weak Key Assumption

We are now ready to show how to exploit the 33-round distinguisher to attack 34-round of XTEA under weak key assumptions. We assume that the 34 round XTEA cipher uses the master key K1 as well as related keys $K2, K3, K4$. The following is an attack procedure of 34 rounds of XTEA.

===

1. Choose 2^{60} plaintext pairs (P_1, P_1) and 2^{60} plaintext pairs (P_2, P_2). $(P_1, P_1,$ $P_2, P_2)$ are encrypted using the keys $(K1, K2, K3, K4)$, respectively, relating the ciphertexts C_1, C_2, C_3, C_4.
2. Check that $C_1 \oplus C_3 = C_2 \oplus C_4 = \Delta(L_{36}, R_{36}) = (\xi, 0)$ where $(\xi, 0)$ is a possible output difference when the input difference is $(0, 0)$. See Table 5 for description of ξ. There are 1080 possible values of ξ.
3. Guess a 32-bit round key quartet (K_1, K_1, K_1', K_1') of the 35th round under weak key assumptions.
   (a) For all ciphertext quartets (C_1, C_2, C_3, C_4) passing Step 2, check that $E_{K1}^{-1}(C_1) \oplus E_{K3}^{-1}(C_3) = (0,0)$ and $E_{K2}^{-1}(C_2) \oplus E_{K4}^{-1}(C_4) = (0,0)$
   (b) If the number of ciphertext quartets passing Step 3(a) is greater than or equal to 2, output the guessed key quartet as the right key quartet of round 35. Otherwise, go to Step 3.

===

This attack requires two pools of 2^{60} plaintext pairs and thus data complexity of attack is 2^{62}. Using two pools of 2^{60} plaintext pairs we can make 2^{120} plaintext quartets. Here, $2^{12.15}(\approx 2^{120} \cdot (2^{-37} \cdot \frac{10}{2^9} \cdot \frac{12}{2^{10}} \cdot \frac{9}{2^8})^2)$ quartets pass Step 2. The

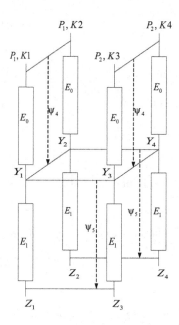

Fig. 5. 33-round related-key rectangle distinguisher of XTEA

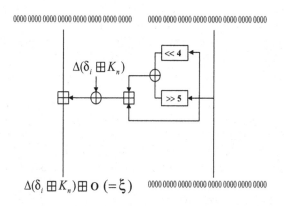

Fig. 6. Description of the 35th round operation

expected number of right quartets is about 2 ($\approx 2^{120} \cdot 2^{-64} \cdot (2^{-18.38} \cdot 2^{-9})^2$). Thus in case that the right key of the 35th round is guessed, the expected number of ciphertext quartets is 2 in Step 3(a). Next, we consider the case that wrong key is guessed. Let ξ_i', ξ_j' for $1 \leq i, j \leq 1080$ be any possible values of ξ'. Then the probability that there exist C_1, C_2, C_3, C_4 such that $C_1 \oplus C_3 = (\xi_i', 0)$ and $C_2 \oplus C_4 = (\xi_j', 0)$ is $\frac{2^{12.15}}{2^{20.15}}$ where $2^{20.15}$ ($\approx 1080 \cdot 1080$). Let $T_{i,j}$ for $1 \leq i, j \leq 1080$ be the probability that $E_{K1}^{-1}(C_1) \oplus E_{K3}^{-1}(C_3) = (0,0)$ and $E_{K2}^{-1}(C_2) \oplus E_{K4}^{-1}(C_4) = (0,0)$ where $C_1 \oplus C_3 = (\xi_i', 0)$ and $C_2 \oplus C_4 = (\xi_j', 0)$.

Table 5. (Possible values of ξ)

i-th bit~j-th bit	possible values of ξ	number of cases
$4 \sim 0$	10000	1
$13 \sim 5$	100000000, 100000001, 100000011, $\cdots$, 111111111, 011111111	10
$23 \sim 14$	1100000000, 1100000001,1100000011, $\cdots$,1111111111,1011111111,0011111111,0111111111	12
$31 \sim 24$	10000000,10000001,10000011, $\cdots$,11111111,01111111	9

Thus for each wrong key quartet the number of ciphertext quartets passing the Step 3(a) is $(\frac{2^{12.15}}{2^{20.15}} \cdot T_{1,1} \cdot 2^{23.87} + \frac{2^{12.15}}{2^{20.15}} \cdot T_{1,2} \cdot 2^{23.87} + \frac{2^{12.15}}{2^{20.15}} \cdot T_{1,3} \cdot 2^{23.87} + \cdots$ $\frac{2^{12.15}}{2^{20.15}} \cdot T_{1080,1080} \cdot 2^{23.87})/2^{23.87} = (\frac{2^{12.15}}{2^{20.15}} \cdot 2^{23.87} \cdot (\sum T_{i,j}))/2^{23.87} \approx 2^{-10}$ where the number of weak keys, $K1_1$'s, is $2^{23.87}$ and $\sum T_{i,j}$ is about 2^{-2}. Therefore we can get the right key used in the 35th round with the above attack procedure. The time complexity of this attack is $2^{12.15} \cdot 2^{23.87} \cdot \frac{1}{34} \cdot \frac{1}{2} \cdot 4 \approx 2^{31.94}$.

5   Conclusion

XTEA is a block cipher with a very simple structure but there was no known attack even for half the total number of rounds version although it has 64 rounds. In this paper, we have presented a related-key rectangle attack on XTEA under weak key assumption. There are $2^{108.21}$ $(= 2^{20.35} \cdot 2^{23.87} \cdot 2^{64})$(a fraction $2^{-19.79}$ of all keys) weak keys which make it possible to construct the above distinguishers. See appendix A for description of weak key quartets. The attack on 34 rounds of XTEA under weak key assumptions requires 2^{62} chosen plaintexts and $2^{31.94}$ 34-round XTEA encryptions.

References

1. E. Biham, A. Shamir, "Differential cryptanalysis of DES-like cryptosystems", *Crypto 1990*, LNCS 537, Springer-Verlag, pp. 2–21, 1991.
2. E. Biham, " New Types of Cryptanalytic Attacks Using Related Keys", *Journal of Cryptology*, LNCS 7, Springer-Verlag, pp. 229–246, 1994
3. E. Biham, O. Dunkelman, N. Keller, "Rectangle Attack-Rectangling the Serpent", *Eurocrypt 2001*, LNCS 2045, Springer-Verlag, pp. 340–357, 2001.
4. E. Biham, O. Dunkelman, N. Keller, "New Results on Boomerang and Rectangle Attacks," *FSE 2002*, LNCS 2365, Springer-Verlag, pp. 1–16, 2002.
5. E. Biham, O. Dunkelman, N. Keller, "Related-Key Boomerang and Rectangle Attacks," *Eurocrypt 2005*, LNCS 3494, Springer-Verlag, pp. 507–525, 2005.
6. S. Hong, J. Kim, S. Lee, B. Preneel, "Related-Key Rectangle Attacks on Reduced Versions of SHACAL-1 and AES-192," *FSE 2005*, LNCS 3557, Springer-Verlag, pp. 368–383, 2005.

7. S. Hong, D. Hong, Y. Ko, D. Chang, W. Lee, S. Lee, "Differential Cryptanalysis of TEA and XTEA," *ICISC 2003*, LNCS 2971, Springer-Verlag, pp. 413–428, 2003.
8. J. Kelsey, T. Kohno, B. Schneier, "Amplified Boomerang Attacks Against Reduced-Round MARS and Serpent," *FSE 2000*, LNCS 1978, Springer-Verlag, pp. 75–93, 2001.
9. J. Kelsey, B. Schneier, D. Wagner, "Related-Key Cryptanalysis of 3-Way, Biham-DES, CAST, DES-X, NewDES, RC2, and TEA," *ICICS 1997*, LNCS 1334, Springer-Verlag, pp. 203–207, 1997.
10. J. Kim, G. Kim, S. Hong, S. Lee and D. Hong, "The Related-key Rectangle Attack Application to SHACAL-1," *ACISP 2004*, LNCS 3108, Springer-Verlag, pp. 123–136, 2005.
11. Y. Ko, S. Hong, W. Lee, S. Lee, J. Kang, "Related Key Differential Attacks on 27 Rounds of XTEA and Full-Round GOST," *FSE 2004*, LNCS 3017, Springer-Verlag, pp. 299–316, 2005.
12. D. Moon, K. Hwang, W. Lee, S. Lee, J. Lim, "Impossible Differential Cryptanalysis of Reduced Round XTEA and TEA," *FSE 2002*, LNCS 2365, Springer-Verlag, pp. 49–60, 2003.
13. R. M. Needham, D. J. Wheeler, "eXtended Tiny Encryption Algorithm," October 1997, Available on http://vader.brad.ac.uk/tea/tea.shtml.
14. D. Wheeler, R. Needham, "TEA,a Tiny Encryption Algorithm," *FSE 1994*, LNCS 1008, Springer-Verlag, pp. 97–110, 1995.
15. D. Wagner, "The Boomerang Attack," *FSE 1998*, LNCS 1636, Springer-Verlag, pp. 156–170, 1999.

A   Description of Weak Key Quartet of XTEA

We describe the form of weak key quartets of XTEA more in detail than $(1){\sim}(6)$. Let $K1 = (K_0||K_1||K_2||K_3)$, $K2 = (K_0'||K_1||K_2||K_3)$, $K3 = (K_0||K_1'||K_2||K_3)$, $K4 = (K_0'||K_1'||K_2||K_3)$ and let $K_j = (K_j^{31}, K_j^{30}, K_j^{29}, \cdots, K_j^1, K_j^0)$ for $0 \le j \le 3$ be 32-bit strings. If $K1$, $K2$, $K3$, $K4$ satisfy $(1.1){\sim}(2.7)$ then $(K1, K2, K3, K4)$ satisfies all conditions of $(1){\sim}(6)$.

$X_i^j(K_0, K_0') \in \{a, b, c\}$ means that $K_0^k = K_0'^k$ have the same value as a or b or c for $i \le k \le j$. $NX_i(K_0, K_0')$ means that $K_0^i = 0$ and $K_0'^i = 1$. Similarly, $NX_i^j(K_0, K_0')$ means that $K_0^k = 0$ and $K_0'^k = 1$ for $i \le k \le j$.

- Let $(K1, K2)$ and $(K3, K4)$ satisfy following $(1.1){\sim}(1.8)$:
    (1.1) $K_0^3 = K_0'^3 = 0$, $X_0^2(K_0, K_0') \in \{011, 100, 101, 110, 111\}$.
    (1.2) $K_0^4 \ne K_0'^4$ i.e if $K_0^4 = 1$ then $K_0'^4 = 0$.
    (1.3) $X_5^{12}(K_0, K_0') \in \{11000100, 11000101, 11000110, 11000111, 11001000\}$
    (1.4) $K_0^{13} \ne K_0'^{13}$
    (1.5) If $K_0^{13} = 0$, $K_0'^{13} = 1$ then K_0^i, $K_0'^i$ for $14 \le i \le 19$ have the same value as follows :
        · $X_{15}^{19}(K_0, K_0') \in \{01000, 01001, 01010, \cdots, 11000\}$, $NX_{14}(K_0, K_0')$ or
        · $X_{16}^{19}(K_0, K_0') \in \{0100, 0101, 0110, 0111, 1000\}$, $NX_{15}(K_0', K_0)$, $NX_{14}(K_0, K_0')$ or
        · $X_{17}^{19}(K_0, K_0') \in \{010, 011\}$, $NX_{16}(K_0', K_0)$, $NX_{14}^{15}(K_0, K_0')$ or
        · $X_{18}^{19}(K_0, K_0') \in \{01\}$, $NX_{17}(K_0', K_0)$, $NX_{14}^{16}(K_0, K_0')$ or
        · $K_0^{19} = K_0'^{19} = 0$, $NX_{18}(K_0', K_0)$, $NX_{14}^{17}(K_0, K_0')$

If $K_0^{13} = 1$, $K_0'^{13} = 0'$ then K_0^i, $K_0'^i$ change the value each other.

(1.6) $K_0^{22} \neq K_0'^{22}$, $K_0^{23} \neq K_0'^{23}$, $K_0^{22} \neq K_0'^{23}$, $K_0^{22} \neq K_0'^{23}$ i.e if $K_0^{22} = 0$ then $K_0'^{22} = 1$, $K_0^{23} = 1$, $K_0'^{23} = 0$

(1.7) $K_0^i = K_0'^i$ for $24 \leq i \leq 30$ and $K_0^{31} \neq K_0'^{31}$

- Let $(K1, K3)$ and $(K2, K4)$ satisfy following $(2.1) \sim (2.7)$:

  (2.1) $X_0^3(K_1, K_1') \in \{0010, 0011, 0100, 0110\}$

  (2.2) $K_1^4 \neq K_1'^4$

  (2.3) $K_1^i, K_1'^i$ for $5 \leq i \leq 12$ have the value same as follows :
  - $X_5^8(K_1, K_1') \in \{0000, 0001, 0010, \cdots, 1010, 1100, \}$, $X_9^{12}(K_1, K_1') \in \{1000\}$

  or
  - $X_5^{10}(K_1, K_1') \in \{010110, 010111, 011000, 011001, 011010, \cdots, 011111, 100000, 100001, \cdots, 111111\}$, $K_1^{11} = K_1'^{11} = 1$, $K_1^{12} = K_1'^{12} = 0$

  (2.4) $K_1^{13} \neq K_1'^{13}$

  (2.5) If $K_1^{13} = 0, K_1'^{13} = 1$ then K_1^i, $K_1'^i$ for $14 \leq i \leq 21$ have the same value as follows :
  - $X_{15}^{20}(K_1, K_1') \in \{011110, 01111, 100000, 100001, \cdots, 101111\}$, $NX_{14}(K_1', K_1)$
  - $X_{16}^{20}(K_1, K_1') \in \{01111, 10000, 10001, \cdots, 10110\}$, $NX_{15}(K_1', K_1)$, $NX_{14}(K_1, K_1')$
  - $X_{17}^{20}(K_1, K_1') \in \{0111, 1000, 1001, \cdots, 1011\}$, $NX_{16}(K_1', K_1)$, $NX_{15}(K_1, K_1')$, $NX_{14}(K_1, K_1')$
  - $X_{18}^{20}(K_1, K_1') \in \{100, 101\}$, $NX_{17}(K_1', K_1)$, $NX_{16}(K_1, K_1')$, $NX_{15}(K_1, K_1')$, $NX_{14}(K_1, K_1')$
  - $X_{19}^{20}(K_1, K_1') \in \{10\}$, $NX_{18}(K_1', K_1)$, $NX_{17}(K_1, K_1')$, $NX_{16}(K_1, K_1')$, $NX_{15}(K_1, K_1')$, $NX_{14}(K_1, K_1')$
  - $K_1^{21} = K_1'^{21} = 1$

  If $K_1^{13} = 1, K_1'^{13} = 0$ then K_1^i, $K_1'^i$ for $14 \leq i \leq 21$ change above values each other.

  (2.6) $K_1^{22} \neq K_1'^{22}$, $K_1^{23} = K_1'^{23} = 1$

  (2.7) $K_1^i = K_1'^i$ for $24 \leq i \leq 30$ and $K_1^{31} \neq K_1'^{31}$

Assume that $K1, K2$ satisfy $(1.1) \sim (1.7)$ and $K1, K3$ satisfy $(2.1) \sim (2.7)$. If $K3, K4$ satisfy $(1.1) \sim (1.7)$ then $K1, K2, K3, K4$ satisfy all conditions of $(1) \sim (6)$. The weak key quartets are computed as follows :

- the number of $K1_0, K2_0$ satisfying $(1.1) \sim (1.7)$

$$2^{11} \cdot 5 \cdot 5 \cdot 26 = 2^{20.34}$$

- the number of $K1_1, K3_1$ satisfying $(2.1) \sim (2.7)$

$$34 \cdot 55 \cdot 2^{13} \approx 2^{23.87}$$

- Thus the total number of weak key quartets is $(2^{20.34}) \cdot (2^{23.87}) \cdot (2^{64}) = 2^{108.21}$

Deniable Group Key Agreement

Jens-Matthias Bohli[1] and Rainer Steinwandt[2]

[1] Institut für Algorithmen und Kognitive Systeme, Fakultät für Informatik,
Am Fasanengarten 5, 76128 Karlsruhe, Germany
bohli@ira.uka.de
[2] Department of Mathematical Sciences, Florida Atlantic University,
777 Glades Road, Boca Raton, FL 33431, USA
rsteinwa@fau.edu

Abstract. Especially for key establishment protocols to be used in internet applications, the (privacy) concern of *deniability* arises: Can a protocol transcript be used—possibly by a participant—to prove the involvement of another party in the protocol? For two party key establishment protocols, a common technique for achieving deniability is the replacement of signature-based message authentication with authentication based on symmetric keys. We explore the question of deniability in the context of group key establishment: Taking into account malicious insiders, using a common symmetric key for authentication is critical, and the question of how to achieve deniability arises.

Building on a model of Bresson et al., we offer a formalization of deniability and present a group key agreement offering provable security in the usual sense, deniability, and security guarantees against malicious insiders. Our approach for achieving deniability through a suitably distributed Schnorr-signature might also be of independent interest.

Keywords: group key agreement, plausible deniability.

1 Introduction

In addition to standard requirements like key secrecy or perfect forward secrecy, often additional conditions are imposed on key establishment protocols. In particular for key establishment protocols geared towards internet applications, e. g., IKEv2 [11] or JFK [1], further issues like DoS resistance and protection of privacy become relevant. One of these (privacy) requirements that has, e. g., been explicitly addressed in a memo on "Features of Proposed Successors to IKE" [12] is the question of *plausible deniability*: Assume we run a key establishment between two principals U_1 and U_2. Then it can be desirable that a transcript of the communication does not allow to prove that indeed U_1 and U_2 have established a key in this session. Going one step further, even for each of the two protocol participants it should be infeasible to prove to a third party that its communication partner has been involved in the key establishment.

In [15] Mao and Paterson put forward (informal) definitions for various degrees of plausible deniability and also key establishment protocols achieving deniability by using identity-based techniques. In joint work with Boyd [5], Mao and

P.Q. Nguyen (Ed.): VIETCRYPT 2006, LNCS 4341, pp. 298–311, 2006.

Paterson discuss how to integrate the design goal of deniability in a two party key establishment with a construction method of Canetti and Krawczyk [8]. A key technique in [5] is the use of public information to derive a shared secret between two principals U_1, U_2 aiming at the establishment of a common key. The general question of deniability, and the question of deniable authentication in particular, has received significant attention in the literature—a very partial list including [10,16,9,18,19]. Nevertheless, for the deniability of key establishment protocols a satisfying formal treatment that integrates with the existing proof frameworks seems to be lacking. The authors of [15] refer to the development of more formal security models and security proofs for deniability as an *important avenue of future research* and raise the question for building proofs for deniability from proofs for the underlying key establishment primitives.

Our contribution. We suggest a definition of deniability in a group key establishment framework along the lines of Bresson et al. [6]. Passing from two parties to a group setting adds qualitatively new problems, e. g., a single malicious participant may be able to impersonate other protocol participants (cf. [7]). While for the two-party case a shared secret seems well-suited for enabling deniability, the group case appears to be more involved. Building on protocols in [13,4], we present an efficient group key agreement protocol with a security analysis in the random oracle model. In addition to provably offering key secrecy and deniability, the suggested protocol offers perfect forward secrecy and security guarantees against malicious insiders.

Organization. After recalling and establishing some theoretical tools for modeling group key establishment in Section 2, we suggest a definition of deniability for group key establishment schemes. Thereafter, we describe a protocol that provably achieves deniability. In addition, we prove the usual key secrecy and perfect forward secrecy requirements as well as security guarantees against malicious insiders. For the proofs, a Computational Diffie Hellman assumption and the random oracle model are used. Our main technical tool for establishing deniability can be interpreted as a Schnorr signature whose computation is distributed to different rounds of the protocol. This construction for achieving deniability might be of independent interest outside the specific protocol discussed here.

2 Group Key Establishment: Modeling Security and Deniability

In this section we summarize the basic components of the framework we use for analyzing the group key establishment protocol proposed in Section 3. We start by recalling a security model of Bresson et al. [6], more precisely we adopt a variant of this framework already used in [4]. For the clarity of exposition, in the formulation of the Send-oracle we introduce an additional role-flag that allows to make explicit different roles taken by protocol participants: While some protocols require all participants to perform identical computations, others exhibit a more asymmetric structure, e.g. in a two-party key establishment, the two participants can play the roles of initiator and responder. Or in a group key transport protocol we may encounter the roles of client and server.

2.1   Modeling Security of a Key Establishment

The modeling of participants, the communication network and adversarial capabilities is fairly standard and—with exception of the indicated modification of the Send-oracle—basically identical to the model in [4]. Because of all adversaries considered being active, we did not include the Execute-oracle in the model: An active adversary can simulate a query to Execute by means of his Send-oracle.

Participants. The set of potential protocol participants is a finite set $\mathcal{U}$ with each U_i being represented as a probabilistic polynomial time (ppt) Turing machine. We allow $\mathcal{U}$ to be of polynomial size in the security parameter k. Each protocol participant $U_i \in \mathcal{P}$ (where $\mathcal{P} \subseteq \mathcal{U}$) is allowed to execute a polynomial number of protocol instances in parallel. We will denote an instance s_i of principal U_i by $\Pi_i^{s_i}$ ($i \in \mathbb{N}$). Each such instance can be interpreted as a process executed by U_i and has assigned seven variables $\mathsf{state}_i^{s_i}$, $\mathsf{sid}_i^{s_i}$, $\mathsf{pid}_i^{s_i}$, $\mathsf{sk}_i^{s_i}$, $\mathsf{term}_i^{s_i}$, $\mathsf{used}_i^{s_i}$ and $\mathsf{acc}_i^{s_i}$:

   $\mathsf{used}_i^{s_i}$ indicates whether this instance is or has been used for a protocol run. The $\mathsf{used}_i^{s_i}$ flag can only be set through a protocol message received by the oracle due to a call to the Send-oracle (see below);

   $\mathsf{state}_i^{s_i}$ keeps the state information during the protocol execution;

   $\mathsf{term}_i^{s_i}$ shows if the execution has terminated;

   $\mathsf{sid}_i^{s_i}$ denotes a non-secret session identifier that can serve as identifier for the session key $\mathsf{sk}_i^{s_i}$—the attacker learns all session identifiers;

   $\mathsf{pid}_i^{s_i}$ stores the set of identities of those principals that $\Pi_i^{s_i}$ aims at establishing a key with—including U_i himself;

   $\mathsf{acc}_i^{s_i}$ indicates if the protocol instance was successful, i.e., the instance $\Pi_i^{s_i}$ accepted the session key;

   $\mathsf{sk}_i^{s_i}$ stores the session key once it is accepted by $\Pi_i^{s_i}$. Before acceptance, it stores a distinguished NULL value.

For more details on the usage of the variables we refer to [2]. We assume that an instance $\Pi_i^{s_i}$ has to accept the session key constructed at the end of the corresponding protocol instance if no deviation from the protocol specification occurs.

Communication network. Arbitrary point-to-point connections among the principals are assumed to be available. The network is considered to be non-private and fully asynchronous. It is controlled by the adversary.

Adversarial model. As just mentioned, the adversary $\mathcal{A}$ has full control of the communication network and may delay, suppress and insert messages at will. To make the adversary's capabilities explicit, the subsequently listed oracles are used, and $\mathcal{A}$ is taken for a ppt Turing machine which may execute any of these.

Send(U_i, s_i, M) This sends the message M to the instance $\Pi_i^{s_i}$ and outputs the reply generated by this instance. If the adversary calls this oracle with an unused instance $\Pi_i^{s_i}$ and $M = (\{U_1, \ldots, U_r\}, \text{role})$, then $\Pi_i^{s_i}$'s $\text{pid}_i^{s_i}$-value is initialized to the value $\text{pid}_i^{s_i} := \{U_1, \ldots, U_r\} \cup \{U_i\}$, the $\text{used}_i^{s_i}$-flag is set, and $\Pi_i^{s_i}$ will act according to the role specified in role. At this, role is just a string over some fixed alphabet to specify a particular function, like initiator, to be played by the instance $\Pi_i^{s_i}$.

If the instance $\Pi_i^{s_i}$ sends a message in the protocol right after receiving M, then Send returns this message to the adversary.

Reveal(U_i, s_i) returns the session key $\text{sk}_i^{s_i}$.

Corrupt(U_i) reveals the long term secret key SK_i of U_i to the adversary. Given a concrete protocol run, involving instances $\Pi_i^{s_i}$ of principals $U_1, \ldots, U_k$ we say that principal $U_{i_0} \in \{U_1, \ldots, U_k\}$ is *honest* if and only if no query of the form Corrupt(U_{i_0}) has been made by the adversary.

Test(U_i, s_i) Only one query of this form is allowed for an active adversary $\mathcal{A}$. Provided that $\text{sk}_i^{s_i}$ is defined, (i. e. $\text{acc}_i^{s_i} = \text{true}$ and $\text{sk}_i^{s_i} \neq \text{NULL}$), $\mathcal{A}$ can execute this oracle query at any time when being activated. Then with probability $1/2$ the session key $\text{sk}_i^{s_i}$ and with probability $1/2$ a uniformly chosen random session key is returned.

Initialization. Before the actual key establishment protocol is executed for the first time, an initialization phase takes place where for each principal $U_i \in \mathcal{P}$ a public key/secret key pair (SK_i, PK_i) is generated. The value SK_i is revealed to U_i only, and PK_i is given to all principals. In the protocol below, (SK_i, PK_i) will just be a pair (α_i, g^{α_i}) with g a generator of a suitable cyclic group.

For the sake of simplicity, we assume all key pairs (SK_i, PK_i) to be generated by a trusted party which also takes care of distributing the PK_i-values. We do not address the issue of malicious principals who try to generate incorrect key pairs or adversaries that can influence the initialization phase.

Correctness. This property basically expresses that the protocol will establish a good key without adversarial interference and allows us to exclude "useless" protocols. We take a group key establishment protocol for *correct* if in the absence of attacks indeed a common key along with a common identifier is established:

Definition 1. *A group key establishment protocol $\mathcal{P}$ is called* correct *if upon honest delivery of all messages a single execution of the protocol for establishing a key among $U_1, \ldots, U_r$ involves r instances $\Pi_1^{s_1}, \ldots, \Pi_r^{s_r}$ and ensures that with overwhelming probability all instances:*

- *accept, i. e.,* $\text{acc}_1^{s_1} = \cdots = \text{acc}_r^{s_r} = \text{true}$.
- *obtain a common session identifier* $\text{sid}_1^{s_1} = \cdots = \text{sid}_r^{s_r}$ *which is globally unique.*
- *have accepted the same session key* $\text{sk}_1^{s_1} = \cdots = \text{sk}_r^{s_r} \neq \text{NULL}$ *associated with the common session identifier* $\text{sid}_1^{s_1}$.
- *know their partners* $\text{pid}_1^{s_1} = \text{pid}_2^{s_2} = \cdots = \text{pid}_r^{s_r}$ *and it is* $\text{pid}_1^{s_1} = \{U_1, \ldots U_r\}$.

Partnering. For detailing the security definition, we will have to specify under which conditions a Test-query may be executed. To do so, we follow the same idea as in [4].

Definition 2. *Two instances* $\Pi_i^{s_i}$, $\Pi_j^{s_j}$ *are* partnered *if* $\mathsf{sid}_i^{s_i} = \mathsf{sid}_j^{s_j}$, $\mathsf{acc}_i^{s_i} = \mathsf{acc}_j^{s_j} = \mathsf{true}$ *and* $\mathsf{pid}_i^{s_i} = \mathsf{pid}_j^{s_j}$.

Freshness. A Test-query should only be allowed to those instances holding a key that is not for trivial reasons known to the adversary. To this aim, an instance $\Pi_i^{s_i}$ is called *fresh* if none of the following two conditions hold:

- For some $U_j \in \mathsf{pid}_i^{s_i}$ a Corrupt(U_j) query was executed before a query of the form Send($U_k, s_k, *$) has taken place where $U_k \in \mathsf{pid}_i^{s_i}$.
- The adversary queried Reveal(U_j, s_j) with $\Pi_i^{s_i}$ and $\Pi_j^{s_j}$ being partnered.

The idea here is that revealing a session key from an instance $\Pi_i^{s_i}$ trivially yields the session key of all instances partnered with $\Pi_i^{s_i}$, and hence this kind of "attack" will be excluded in the security definition.

Security (key secrecy). The security definition of [6] can be summarized as follows. As a function of the security parameter k we define the advantage $\mathsf{Adv}_{\mathcal{A}}(k)$ of a ppt adversary $\mathcal{A}$ in attacking protocol P as

$$\mathsf{Adv}_{\mathcal{A}} := |2 \cdot \mathsf{Succ} - 1|$$

where Succ is the probability that the adversary queries Test on a fresh instance $\Pi_i^{s_i}$ and guesses correctly the bit b used by the Test oracle in a moment when $\Pi_i^{s_i}$ is still fresh.

Definition 3. *We call the group key establishment protocol* P *secure if for any ppt adversary* $\mathcal{A}$ *the function* $\mathsf{Adv}_{\mathcal{A}} = \mathsf{Adv}_{\mathcal{A}}(k)$ *is negligible.*

2.2   Modeling Deniability in a Group Key Establishment Protocol

To introduce a definition of deniability for group key establishment schemes, we build on the model outlined in the previous section. Before stating the definition, we quickly review the notion of *plausible deniability* for the SIGMA protocol [14] which serves as example for plausible deniability in [15,5].

The SIGMA protocol. In the SIGMA protocol, both participants sign the pair of ephemeral public keys of a Diffie-Hellman key exchange (g^x, g^y) instead of a message including identities of the participants. However, that two principals A and B signed a message that includes (g^x, g^y) and did not establish the key with each other certainly would only happen with a negligible probability for honest participants. The plausible explanation for A, being confronted with a transcript as above, is to argue that a corrupted B could have intentionally signed the tuple (g^x, g^y) she caught from one of A's former protocol runs with a different partner.

For a group key establishment, which in general involves more than two participants, new questions come up:

- One may argue to what extent a plausible explanation of protocol data may impose a maliciously acting collusion of several other protocol participants. A straightforward application of SIGMA's method to a group key establishment protocol—not signing identities, but rather nonces and ephemeral keys of the participants—would require the denying party to argue that all of his presumable partners actually colluded to produce the transcript, this could be seen no longer to be plausible.
- One may argue which former protocol runs are accepted as an excuse. If A wants to deny a key establishment with, say, B and C. Would an actual protocol run between A, B, C and D (a strict superset of $\{A, B, C\}$) be accepted as a plausible excuse? Depending on the application context, different views can be adopted here.

Another weakness of plausible deniability as in SIGMA certainly is the undeniability of a protocol execution itself. If n different tuples (g^x, g^y) signed by A are found, there is no plausible way for A to deny that he executed n protocol runs. He might only repudiate his respective partners. To overcome these problems we directly aim at a stronger form of deniability for group key establishment, following the goals of complete deniability in [15]. Principals should be able to deny involvement in any protocol run. This should also hold in a situation where the adversary is even willing to disclose internal state information, possibly including long term keys, in order to provide evidence for the involvement of some principal in a key establishment.

Deniability for group key establishment. We consider an adversary $\mathcal{A}_d$ that tries to break deniability in a group key establishment protocol. More specifically, we take $\mathcal{A}_d$ for a ppt algorithm expecting as input the security parameter k and the initial public keys PK_i of all potential protocol participants $U_i \in \mathcal{U}$ as well as a bound $q_c \in \mathbb{N}_0$ on the number of possibly dishonest (corrupted) principals. Having received this input, $\mathcal{A}_d$ interacts with the instances $\Pi_i^{s_i}$ of the principals U_i by querying the oracles Corrupt, Send and Reveal. Access to the Test oracle is not granted and the Corrupt oracle may be queried at most q_c times. Finally, $\mathcal{A}_d$ outputs a protocol transcript $T_{\mathcal{A}_d}(k, q_c, \{PK_i\}_i)$, which from a formal point of view can be an arbitrary bitstring, and intuitively represents evidence for the involvement of a certain principal in a particular key establishment.

Let $T_{\mathcal{A}_d}(k, q_c)$ be the random variable that describes $T_{\mathcal{A}_d}(k, q_c, \{PK_i\}_i)$ with uniformly chosen randomness for the adversary, the oracles and the key generation in the initialization phase.

The idea is now to introduce a simulator $\mathcal{S}_d$ that accepts the same input and can impose the same number of corrupted principles as $\mathcal{A}_d$ does, but must not invoke any uncorrupted principal. This means, $\mathcal{S}_d$ may execute up to q_c queries to the Corrupt-oracle, but has *no* access to Send and Reveal. Analogously as above, we define a transcript $T_{\mathcal{S}_d}(k, q_c, \{PK_i\}_i)$ and a random variable $T_{\mathcal{S}_d}(k, q_c)$.

Definition 4. *We call a group key establishment protocol* deniable *if for each adversary $\mathcal{A}_d$ as specified above and for all inputs $k \in \mathbb{N}$, $q_c \in \mathbb{N}_0$ a ppt simulator $\mathcal{S}_d$ as specified above exists such that $T_{\mathcal{A}_d}(k, q_c)$ and $T_{\mathcal{S}_d}(k, q_c)$ are computationally indistinguishable, i. e. no ppt algorithm D can distinguish them with non-negligible probability.*

In the next section we present a four round group key agreement protocol that, under a Computational Diffie Hellman assumption and in the random oracle model, offers deniability along with other security guarantees that are common in group key establishment.

3   A Deniable Group Key Agreement Protocol

We present a group key establishment protocol that achieves deniability in the sense of Definition 4. Our protocol builds on protocols in [13,4] and from these inherits features like being contributory, perfect forward secrecy and offering resistance against malicious insiders. The system parameters are a cyclic group G of prime order q with generator g, such that the Computational Diffie Hellman problem in G is hard. Also, we make use of the random oracle model. In the initialization phase, all principals U_i obtain a secret key $SK_i := \alpha_i$ chosen uniformly at random from $\mathbb{Z}_q$, and the corresponding public keys $PK_i := g^{\alpha_i}$ are distributed to all principals.

3.1   Protocol Description and Design Rationale

For authentication we will use a protocol which lies in-between Schnorr's zero-knowledge identification scheme and signature scheme [20]. Unlike as in the signature scheme, the verifiers' challenge will not depend on the prover's first random value—only on the message to be authenticated. This means that also the message may not depend on the prover's random value, thus must be determined before. For the protocol to be sound, the verifiers must be convinced, that the message to be authenticated is not known to the prover at the time he sends the first random value. This fact, that a message is determined but not yet known, restricts the usability of the deniable authentication protocol for general use. Key establishment protocols, however, will lead to a *fresh* and previously unknown key so that the authentication scheme is particularly well-suited for key establishment protocols.

The proposed protocol is summarized in Figure 1 with $H(\cdot)$ denoting a random oracle. All protocol participants perform identical computations, i. e., play an identical role participant. So we can restrict to specifying the computations of an instance $\Pi_i^{s_i}$ initialized with $\mathsf{pid}_i^{s_i} = \{U_1, \ldots, U_n\}$. As there is no risk of confusion, for the sake of readability, we omitted the upper index s_i in the protocol description for instance $\Pi_i^{s_i}$. Also, we note that in Figure 1, for the computation of c_i in Round 4, the bitstring output by $H(\cdot)$ is interpreted as binary representation of a non-negative integer, and when writing y_{i+1} resp. y_{i-1}, indices are to be understood mod n, i. e., $y_{n+1} = y_1$ and $y_0 = y_n$. Finally, as usual $\cdot\xleftarrow{R}\cdot$ denotes a random choice with uniform distribution. Before proving properties of the protocol, some comments on the underlying basic ideas are in order:

- The first round is essentially the same as in [13,4]. In our protocol all principals will broadcast $H(k_i)$, which acts as a commitment to their nonce k_i. Thus, after the first round, the session identifier and also the session key are determined, though the session key is not yet known to any participant.
- The second round prepares the deniable authentication. Each principal U_i chooses a value $z_i = g^{r_i}$. It is important, that the session key was fixed beforehand and does not depend on g^{r_i} to obtain deniability. Choosing his value k_j after knowing g^{r_i} would allow a malicious participant U_j to obtain an undeniable Schnorr signature of U_i. Further on, the value z_i has to be fixed before U_i learns the the session key (and therewith the key confirmation message that is to be authenticated). Otherwise the authentication is not convincing. Hence, only in the third round the participants reveal their nonces k_i. Deviating from the previous protocols, for the sake of symmetry, in the protocol below all participants do this encrypted.
- Finally, in the fourth round all principals know the session key and can provide an "a posteriori authentication" for the session. If the final verification in Round 4 succeeds, too, an instance accepts the session key.

Verifying Correctness of this protocol is straightforward—the only possibly non-obvious step is the decryption of the k_j-values in Round 3. One easily checks, however, that the T_j-values received in Round 2 enable U_i to iteratively recover all needed t_j^R-values, starting with a neighbor in the "circle of protocol participants".

3.2   Security Analysis

Deniability. We start by an analysis of the deniability feature. As the authentication is based on Schnorr's zero-knowledge identification scheme, anyone can simulate a transcript of the authentication protocol. This fulfills our definition of complete deniability.

Proposition 1. *The protocol in Figure 1 is deniable in the sense of Definition 4.*

Proof. For constructing the required simulator $\mathcal{S}_d$, we use the adversary $\mathcal{A}_d$ as black-box. Namely, $\mathcal{S}_d$ will initiate $\mathcal{A}_d$ with $(k, q_c, \{PK_i\}_i)$ and will simulate the instances of all protocol participants and the oracles Send and Reveal. If $\mathcal{A}_d$ queries Corrupt(U_j), $\mathcal{S}_d$ will do likewise, learn the secret key SK_j and hand it over to $\mathcal{A}_d$.

Once $\mathcal{S}_d$ needs to access the secret key $SK_i = \alpha_i$ of any uncorrupted protocol participant to compute $M_i^4 = d_i = r_i - c_i\alpha_i \bmod q$, the simulator stops the execution and rewinds $\mathcal{A}_d$ such that $\mathcal{S}_d$ can choose a different z_i for U_i in this session in Round 2. The simulator chooses first at random a value $d \xleftarrow{R} \mathbb{Z}_q$ for use in this session and computes then $z_i = g^d(PK_i)^{c_i}$ with $c_i = H(\mathsf{sid}_i \| \mathsf{sconf}_i)$.

	Protocol for instance Π_i of principal U_i
Round 1: Compute	$k_i \xleftarrow{R} \{0,1\}^k$, $x_i \xleftarrow{R} \mathbb{Z}_q$, $y_i = g^{x_i}$
Broadcast	$M_i^1 = (H(k_i), y_i, U_i)$
Round 2: Compute	$\mathsf{sid}_i = H(\mathsf{pid}_i \| H(k_1) \| \dots \| H(k_n))$, $r_i \xleftarrow{R} \mathbb{Z}_q$, $z_i = g^{r_i}$
Broadcast	$M_i^2 = (\mathsf{sid}_i, z_i, U_i)$.
Round 3: Compute	$t_i^L = H(y_{i-1}^{x_i})$, $t_i^R = H(y_{i+1}^{x_i})$, $T_i = t_i^L \oplus t_i^R$
Broadcast	$M_i^3 = (k_i \oplus t_i^R, T_i, U_i)$
Round 4: Verify	$T_1 \oplus \dots \oplus T_n = 0$, and for all decrypted k_j, $H(k_j)$ equals
	the 1st component of M_j^1 ($j \in \{1, \dots, n\} \setminus \{i\}$)
Session Key	$\mathsf{sk}_i = H(\mathsf{pid}_i \| k_1 \| \dots \| k_n)$
Session Confirmation	$\mathsf{sconf}_i = H((y_1, k_1) \| \dots \| (y_n, k_n))$
Compute	$c_i = H(\mathsf{sid}_i \| \mathsf{sconf}_i) \bmod q$, $d_i = r_i - c_i \alpha_i \bmod q$
Broadcast	$M_i^4 = (d_i, U_i)$
Verify	$g^{d_j}(PK_j)^{c_i} = z_j$ for all $j \in \{1, \dots, n\} \setminus \{i\}$

Fig. 1. A deniable group key agreement protocol

The element z_i will be uniformly distributed in G, perfectly indistinguishable from the honest choice as g^{r_i} with $r_i \xleftarrow{R} \mathbb{Z}_q$.

From this point, the adversary received a different message than in the former protocol run and will generally deviate. However, unless $\mathcal{A}_d$ finds a collision of the hash function being able to reveal another value k_i, $\mathcal{A}_d$ cannot anymore influence the values sid and sconf and therewith c_i from that moment. Thus, $\mathcal{S}_d$ will be able to return d as a valid authentication in Round 4 of this session.

Once $\mathcal{A}_d$ outputs a transcript, $\mathcal{S}_d$ uses it as its output. Because $\mathcal{A}_d$ was used with an indistinguishable simulation of the instances it interacted with, the output of $\mathcal{A}_d$ in this experiment—thus the output of $\mathcal{S}_d$—must be indistinguishable from $\mathcal{A}_d$'s output in interaction with real instances. □

Key secrecy. For proving key secrecy, it is important that the confirmation messages are authenticated to all participants. Thus, the proof of security begins by understanding that the value d computed in Round 4 indeed authenticates the session key to the protocol participants.

Lemma 1. *Suppose the discrete logarithm problem in $G = \langle g \rangle$ is hard. Then, with message M_a^4, a principal $U_a \in \mathcal{P}$ unforgeably authenticates the session identifier sid_a and the key confirmation message sconf_a to all participants.*

Proof. An adversary $\mathcal{A}$ who is able to produce with non-negligible probability a valid message M_a^4 for an uncorrupted protocol participant U_a can be used as a black-box to solve the dlog-problem in G. A given instance of the dlog problem $y \in G$ is assigned to U_a as his public key PK_a. Signing queries of $\mathcal{A}$, i.e., a Send-query to an instance of U_a requiring to compute a message of Round 4, can be answered as before by rewinding the adversary.

Assume now that $\mathcal{A}$ outputs with non-negligible probability a message $M_a^4 = (d_a, U_a)$ such that for a certain instance with $c_i = H(\mathsf{sid}_i \| \mathsf{sconf}_i)$ the verification $z_a = g^{d_a}(PK_a)^{c_i}$ holds. The confirmation message sconf includes the nonces k_i

that will be released in the third round after z_a is already fixed. For any honest participant who does not publish his nonce k_i before he knows z_a, the signature of U_a can only be computed in the order $(z_a, \text{sconf}, H(\text{sid}\|\text{sconf}), d_a)$. Then, by the forking lemma [17], $\mathcal{A}$ can be restarted given the same random tape and the same inputs, except that the random oracle H will deviate from the old answers from a certain point such that now $c_i' = H'(\text{sid}_i\|\text{sconf}_i)$. The adversary $\mathcal{A}$ will now with non-negligible probability output a message $M_a^{4'} = (d_a', U_a)$ for the same instance, and it holds that $z_a = g^{d_a'}(PK_a)^{c_i'}$. Then one can compute $SK_a = \log_g PK_a = (d_a - d_a')/(c_i' - c_i) \bmod q$. $\qquad\square$

Proposition 2. *If the CDH problem in G is hard, the protocol in Figure 1 is a secure authenticated key establishment protocol.*

Proof. Intuitively the secrecy follows from the secrecy of the original protocol: The message of the first round is authenticated, the messages of Rounds 2 and 4 constitute the authentication. Moreover, modifying the message of the third round cannot give any information to the adversary: The partnering bases on the session identifier, that is already defined after Round 1—thus, Reveal is of no use—and no participant would accept a weak key, because the correctness of the key can be checked via the commitments given in Round 1—the Test-session cannot be influenced. A more detailed proof is given in the appendix. $\qquad\square$

Perfect forward secrecy. Perfect forward secrecy is implied by the standard argument that the long term secret keys are used for message authentication exclusively.

Agreement property and protection against malicious insiders. Due to the computation of the session key with the random oracle involving a nonce from each participant, the protocol is certainly contributory. For the same reason, if at least one participant is honest, the resulting session key will be chosen uniformly at random and cannot be predicted by malicious insiders.

Due to the construction of the session identifier that allows to verify the session key, the proof for integrity and entity authentication is analogous to [3,4].

4   Conclusions

The above discussion illustrates that the concept of deniable key establishment becomes qualitatively more involved, when passing from the two party case to a group setting. On the constructive side, the suggested protocol shows that a rather strong form of deniability can provably be achieved with reasonabe efficiency and without having to sacrifice other security features offered by a group key establishment protocol. The chosen approach to enable deniability through a suitably distributed Schnorr signature might be of independent interest, when trying to augment other group key establishment protocols with deniability.

References

1. William Aiello, Steven M. Bellovin, Matt Blaze, Ran Canetti, John Ioannidis, Angelos D. Keromytis, and Omer Reingold. Just Fast Keying: Key Agreement In A Hostile Internet. *ACM Transactions on Information and System Security*, 7(2):1–30, 2004.

2. Mihir Bellare, David Pointcheval, and Phillip Rogaway. Authenticated Key Exchange Secure Against Dictionary Attacks. In Bart Preneel, editor, *Advances in Cryptology — EUROCRYPT'00*, volume 1807 of *Lecture Notes in Computer Science*, pages 139–155. Springer, 2000.

3. Jens-Matthias Bohli. A Framework for Robust Group Key Agreement. In *Computational Science and Its Applications – ICCSA 2006*, volume 3982 of *Lecture Notes in Computer Science*, pages 355–364. Springer, 2006.

4. Jens-Matthias Bohli, María Isabel González Vasco, and Rainer Steinwandt. Secure Group Key Establishment Revisited. Cryptology ePrint Archive, Report 2005/395, 2005. http://eprint.iacr.org/2005/395/.

5. Colin Boyd, Wenbo Mao, and Kenneth G. Paterson. Deniable Authenticated Key Establishment for Internet Protocols. In Bruce Christianson, Bruno Crispo, James A. Malcolm, and Michael Roe, editors, *Security Protocols: 11th International Workshop*, volume 3364 of *Lecture Notes in Computer Science*, pages 255–271. Springer, 2003.

6. Emmanuel Bresson, Olivier Chevassut, David Pointcheval, and Jean-Jacques Quisquater. Provably Authenticated Group Diffie-Hellman Key Exchange. In Pierangela Samarati, editor, *Proceedings of the 8th ACM Conference on Computer and Communications Security (CCS-8)*, pages 255–264. ACM, 2001.

7. Daniel R. L. Brown. Deniable Authentication with RSA and Multicasting. Cryptology ePrint Archive, Report 2005/056/, 2005. http://eprint.iacr.org/2005/056/.

8. Ran Canetti and Hugo Krawczyk. Analysis of Key-Exchange Protocols and Their Use for Building Secure Channels. In *Advances in Cryptology – EUROCRYPT 2001*, volume 2045 of *Lecture Notes in Computer Science*, pages 453–474. Springer, 2001.

9. Tianjie Cao, Dongdai Lin, and Rui Xue. An Efficient ID-Based Deniable Authentication Protocol from Pairings. In *19th International Conference on Advanced Information Networking and Applications (AINA '05)*, volume 1 (AINA papers), pages 388–391. IEEE, 2005.

10. Cynthia Dwork, Moni Naor, and Amit Sahai. Concurrent Zero-Knowledge. In *Proceedings of the 30th ACM Symposium on Theory of Computing, STOC 98*, pages 409–418. ACM, 1998.

11. Charlie Kaufman (editor). Internet Key Exchange (IKEv2) Protocol. Network Working Group Request for Comments: 4306, December 2005. See http://www.ietf.org/rfc/rfc4306.txt.

12. Paul Hoffman (editor). Internet Draft draft-ietf-ipsec-soi-features-01.txt, May 2002. See http://www3.ietf.org/proceedings/03mar/I-D/draft-ietf-ipsec-soi-feature%s-01.txt

13. Hyun-Jeong Kim, Su-Mi Lee, and Dong Hoon Lee. Constant-Round Authenticated Group Key Exchange for Dynamic Groups. In Pil Joong Lee, editor, *Advances in Cryptology – ASIACRYPT 2004*, volume 3329 of *Lecture Notes in Computer Science*, pages 245–259. Springer, 2004.

14. Hugo Krawczyk. SIGMA: The 'SIGn-and-MAc' Approach to Authenticated Diffie-Hellman and Its Use in the IKE Protocols. In Dan Boneh, editor, *Advances in Cryptology – CRYPTO 2003*, volume 2729 of *Lecture Notes in Computer Science*, pages 400–425. Springer, 2003.

15. Wenbo Mao and Kenneth G. Paterson. On The Plausible Deniability Feature of Internet Protocols. See http://isg.rhul.ac.uk/~kp/IKE.ps.

16. Moni Naor. Deniable Ring Authentication. In Moti Yung, editor, *Advances in Cryptology – CRYPTO 2002*, volume 2442, pages 481–498. Springer, 2002.

17. David Pointcheval and Jacques Stern. Security Proofs for Signature Schemes. In Ueli Maurer, editor, *Advances in Cryptology – EUROCRYPT '96*, volume 1070 of *Lecture Notes in Computer Science*, pages 387–399. Springer, 1996.

18. Haifeng Qian, Zhenfu Cao, Lichen Wang, and Qingshui Xue. Efficient Non-interactive Deniable Authentication Protocols. In *The Fifth International Conference on Computer and Information Technology (CIT'05)*, pages 673–679. IEEE, 2005.

19. Mario Di Raimondo and Rosario Gennaro. New Approaches for Deniable Authentication. In Vijay Atluri, Catherine Meadows, and Ari Juels, editors, *Proceedings of the 12th ACM Conference on Computer and Communications Security, CCS 2005*, pages 112–121. ACM, 2005.

20. Claus P. Schnorr. Efficient Identification and Signatures for Smart Cards. In Gilles Brassard, editor, *Advances in Cryptology – CRYPTO '89*, volume 435 of *Lecture Notes in Computer Science*. Springer, 1990.

A Proof of Proposition 2

Proof. Let $\mathcal{A}$ be an adversary that is allowed at most q_s, q_{ro} queries to the Send respectively random oracle. Moreover, let Adv_{CDH}, Adv_{Auth} be the by assumption negligible probabilities to solve the CDH-problem in G respectively break the authentication scheme.

Let Forge be the event that the adversary succeeds in forging the message of Round 4 for an uncorrupted participant U_i such that it is accepted by an instance of any honest user U_j. Lemma 1 guarantees that Forge only occurs with negligible probability.

Let Collision be the event that the random oracle produces a collision. The random oracle can be queried from one Send-query at most $(n + 2)$ times in Round 4 or directly by the adversary. The number of queries to the random oracle is bounded by $(n + 2) \cdot q_s + q_{ro}$, the probability that a collision of the random oracle occurs is then

$$P(\mathsf{Collision}) \leq \frac{((n + 2) \cdot q_s + q_{ro})^2}{2^k}.$$

Let Repeat be the event that an instance chooses a nonce k_i that was previously used by any other instance of any principal. There are at most q_s used oracles that may have chosen a nonce k_i and thus Repeat can only happen with a probability

$$P(\mathsf{Repeat}) \leq \frac{(q_s)^2}{2^k}.$$

Let $\mathsf{Succ} := (\mathsf{Adv}_\mathcal{A} + 1)/2$ be the success probability of adversary $\mathcal{A}$ to win the Test-experiment. Now we connect $\mathcal{A}$ to a simulator Sim that simulates the oracles and instances. We consider a sequence of games and bound the difference of the success probability for the adversary between the games.

In Game 0 the simulator Sim simulates the oracles and principals' instances faithfully. Thus, there is no difference for the adversary and denoting $\mathcal{A}$'s success probability in Game i by $\mathsf{Succ}_{\mathrm{Game}\ i}$, we have $\mathsf{Succ}_{\mathrm{Game}\ 0} = \mathsf{Succ}$.

In Game 1 the simulator stops the simulation as soon as one of the events Forge, Collision or Repeat occurs.

$$|\mathsf{Succ}_{\mathrm{Game}\ 1} - \mathsf{Succ}_{\mathrm{Game}\ 0}| \le P(\mathsf{Forge}) + P(\mathsf{Collision}) + P(\mathsf{Repeat}).$$

In Game 2 the simulation of the Send oracle is modified. On a $\mathsf{Send}(U_i, s_i, M_j^1)$ query, which delivers the last message of Round 1 to $\Pi_i^{s_i}$ and executes Round 2 for this oracle starting with computing the session identifier $\mathrm{sid}_i^{s_i}$, the simulator checks if all users in $\mathrm{pid}_i^{s_i}$ are uncorrupted and all messages $(H(k_j), y_j, U_j)$ were unmodified delivered to $\Pi_i^{s_i}$, i.e. the simulator itself generated the messages in the name of an instance of principal U_j. In this case, instead of querying the random oracle, the simulator simulates an own random oracle and chooses random values $t_i^L, t_i^R \in \{0,1\}^k$. The simulator keeps a list of the mappings $y_{i\pm1}^{x_i} \rightarrow t_i^{L/R}$ for consistency in the protocol and will in further rounds first check if the value exists already in the list.

If the condition on corrupted users is not fulfilled the simulator checks his list and returns the corresponding value if available. However, if the value is not available the simulator queries the random oracle $H(\cdot)$ and inserts the result in his list. The simulator does not generate an own random element, because it could be known to the adversary. This procedure guarantees consistency if some users get the messages delivered honestly but others in the same session do not. We will see later that such a session does not qualify as Test-session.

The success probabilities can only differ, if $\mathcal{A}$ queries one of the Diffie-Hellman keys $y_{i-1}^{x_i}, y_{i+1}^{x_i}$ to the random oracle and detects the difference. Denoting this event by Random, we have

$$|\mathsf{Succ}_{\mathrm{Game}\ 2} - \mathsf{Succ}_{\mathrm{Game}\ 1}| \le \Pr(\mathsf{Random}).$$

Lemma 2. *The probability* $\Pr(\mathsf{Random})$ *of the event* Random *to occur is negligible if CDH in G is hard.*

Proof. Given a Diffie-Hellman challenge (g^a, g^b) the adversary that reaches Random can be used to obtain g^{ab}. The simulator will use the challenge in the first message of two instances randomly selected from the set $\{\Pi_i^{s_i} | i \in \{1,\ldots,n\}, s_i \in \{1,\ldots,q_s\}\}$. Then the simulator will pick a random element from the adversary's Random Oracle queries and give it as answer to the CDH instance. The probability to be right is

$$\mathsf{Adv}_{CDH} \geq \frac{1}{q_s^2 q_{ro}} \Pr(\mathsf{Random}),$$

thus

$$\Pr(\mathsf{Random}) \leq q_s^2 q_{ro} \mathsf{Adv}_{CDH}.$$

$\square$

Because all users authenticate in sconf_i all Round 1 messages it follows that if any participant would have received a different message, the verification of this user's authentication message fails for all participants. Therefore the Test-session must only consist of instances among which all Round 1 messages were delivered honestly.

Now it is clear, having random values XORed on the nonces k_i, that the transcript provides no information about the key and the adversary's success probability is $\frac{1}{2}$. In the Test-session, the adversary cannot modify any Round 1 message.

Putting it all together we obtain

$$\mathsf{Adv}_{\mathcal{A}} = |\mathsf{Succ} - 1/2| \leq \frac{(q_s)^2}{2^k} + \frac{((n+2) \cdot q_s + q_{ro})^2}{2^k} + \mathsf{Adv}_{Auth} + q_s^2 q_{ro} \mathsf{Adv}_{CDH}$$

$\square$

An Ideal and Robust Threshold RSA

Hossein Ghodosi[1] and Josef Pieprzyk[2]

[1] School of Mathematics, Physics and Information Technology
James Cook University, Townsville, Qld 4811, Australia
hossein@cs.jcu.edu.au
[2] Department of Computing
Center for Advanced Computing – Algorithms and Cryptography
Macquarie University, Sydney, NSW 2109 Australia
josef@ics.mq.edu.au

Abstract. We present a novel implementation of the threshold RSA. Our solution is conceptually simple, and leads to an *easy* design of the system. The signing key is shared in additive form, which is desirable for collaboratively performing cryptographic transformations, and its size, at *all times*, is $\log n$, where n is the RSA modulus. That is, the system is *ideal*.

Keywords: Threshold RSA, Robust Systems, Ideal Secret Sharing Schemes.

1 Introduction

Society-oriented cryptography [4] requires that cryptographic transformations to be performed by a group of users, rather than just an individual. A particularly interesting class of society-oriented cryptographic transformations is threshold cryptosystems. In a threshold cryptosystem, the power to perform a cryptographic operation is distributed among ℓ users, such that the following conditions are satisfied:

- any set of more than k ($k < \ell$) users can successfully perform the required cryptographic operation;
- any set of k or fewer users fail to perform the required cryptographic operation successfully;
- neither the group secret key nor the shares of users from the group secret key can be derived from the partial cryptographic results.

An early implementation of a proper threshold cryptosystem is due to Desmedt and Frankel [6]. Their proposed threshold decryption is based on the ElGamal [8] cryptosystem. The main concern in implementing the threshold RSA cryptosystem is how to distribute the secret key over $\mathbb{Z}_{\phi(n)}$, when $\phi(n)$ is kept secret (here n is the RSA modulus). In [7], Desmedt and Frankel have demonstrated a method which solves this problem. The drawback of their solution, however,

P.Q. Nguyen (Ed.): VIETCRYPT 2006, LNCS 4341, pp. 312–321, 2006.

is that the key associated with each user increases in size with respect to the number of users who take part in the cryptographic transformation.

Implementation of an efficient threshold RSA system has been the subject of extensive investigation [5,11,1,12,15,2]. Due to large size of the keys associated to each signatories in signature generation phase, some implementations of the threshold RSA systems are not practical. For example, in [3], for a $(4, 10)$ threshold RSA system, the size of the key associated to each user for signature generation is $\ell \log n$, and in [9], it is $10^6 \log n$. In [12], the authors claim that the size of the key in their scheme is comparable to that of [10], and is $2 \log (\ell n)^1$. The most efficient threshold RSA system was invented by Shoup [15], in which the key size of each shareholder is bound by a constant multiplied by the size of the RSA module. Shoup's scheme also requires that primes p and q have a special form. Although this assumption is removed by Damgård and Dupont [2], the sise of the key still is bound by a constant multiplied by the size of the RSA module. That is, no *ideal* threshold RSA signatures has ever been presented in the literature[2].

In this paper, we present a novel technique for implementing an *ideal* threshold RSA signature. The size of the key for each user is equivalent to the size of the secret itself (i.e., $\log n$), *at all times*. Furthermore, our solution is conceptually simple that leads to an easy implementation. The organization of this paper is as follows. In Section 2, we will give an overview of our scheme. In Section 3, we will give an implementation of our basic system, in which the adversary is passive. In Section 4, we will discuss the robustness of our system in the presence of active adversary. Section 5 is devoted to the security and efficiency consideration. Finally, the paper concludes in Section 6.

2   Overview of Our System

The RSA system [13] uses a composite integer n, which is the product of two large primes p and q, i.e., $n = pq$. The public key e is chosen such that $\gcd(e, \phi(n)) = 1$, where $\phi(n) = (p - 1)(q - 1)$. The secret key is an integer d that satisfies the equation $e \times d \equiv 1 \mod \phi(n)$. The signature on a message m is $\sigma = m^d \mod n$. The signature is accepted to be genuine, if $m = \sigma^e \mod n$ is satisfied.

A common technique in the design of a (k, ℓ) threshold RSA signature is to distribute the secret key, d, among a set of ℓ users in such a way that for any authorized set, $\mathcal{A}$, $(|\mathcal{A}| > k)$ the set of modified shares, d_i, corresponding to user u_i, satisfies the following equation:

$$\sum_{u_i \in \mathcal{A}} d_i = d \quad (\mod \phi(n)).$$

---

[1] Considering their *backup shares* description, the memory requirement for each user is $2\ell \log(\ell n)$.

[2] A threshold signature scheme is ideal if the length of the modified shares that each participant uses, for generating its partial signature, is the same as the group's secret key.

An authorized set of users, $\mathcal{A}$, can generate the signature on a message m, since each user generates its partial signature on message m (i.e., $\sigma_i = m^{d_i}$), and the signature on m can be calculated by multiplication of these partial signatures,

$$\sigma = \prod_{u_i \in \mathcal{A}} \sigma_i = m^{\sum_{u_i \in \mathcal{A}} d_i} = m^d \pmod{n}.$$

The problem with this design is that the computation of modified shares, d_i cannot be done over modulo $\phi(n)$, since $\phi(n)$ is unknown to the participants. In order to overcome this problem, computations are performed over integers. This solution, however, yields to schemes in which the size of the modified shares d_i is larger than the secret.

2.1   A Novel Implementation of Threshold RSA

Our implementation utilizes the RSA modulus n for all computations (i.e., in the underlying secret sharing scheme and/or in the performance of cryptographic operations). That is, we distribute the secret key, d, among a set of ℓ users in such a way that for any authorized set, $\mathcal{A}$, ($|\mathcal{A}| > k$) the set of modified shares, d_i, corresponding to users $u_i \in \mathcal{A}$, satisfies the following equation

$$\sum_{u_i \in \mathcal{A}} d_i = d \pmod{n}.$$

An advantage of this technique is that each user can compute their modified share in modular arithmetic environment, since the modulus n is public. Hence, the size of each modified share, d_i is bound by $\log n$, which is the sise of the secret.

In the signature generation phase, each participant of an authorized set, $\mathcal{A}$, generates its partial signature on message m, i.e., $\sigma_i = m^{d_i}$. Multiplication of these partial signatures is not a correct signature. However, as will will see shortly, the correct signature can be obtained easily.

3   Basic Scheme

In this scheme we assume that the adversary is passive. That is, it can corrupt up to k users, and thus, learns all the information held by the corrupted users. However, it has no control on the behavior of users and/or on their information (i.e., all users follow the protocol appropriately).

Let $\mathcal{U} = \{u_1, \ldots, u_\ell\}$ be the set of users and k ($k < \ell$) be the threshold parameter –the maximum number of users that can be corrupted by the adversary.

3.1   Initialization

This is a one-time protocol, and can be run by a trusted dealer or any of the known distributed RSA key generation protocols. It accepts system parameters as input, and generates the RSA modulus $n = pq$, where p and q are distinct primes of requested size. It also chooses the public key e, such that $\gcd(e, \phi(n)) = 1$, and computes the secret key $d = 1/e \pmod{\phi(n)}$. It utilizes the following secret sharing scheme, in order to distribute the secret key among all users:

1. Secretly chooses, independently at random, k elements of $\mathbb{Z}_n$, denoted $a_1, \ldots, a_k$ and forms a polynomial

$$f(x) = d + a_1 x + a_2 x^2 + \cdots + a_k x^k.$$

   Note that, $a_k \neq 0$, i.e., $f(x)$ is a polynomial of degree k.
2. Computes $s_i = f(x_i)$, for $1 \leq i \leq \ell$. Since x_is are public, without loss of generality, we let $x_i = i$ and thus, $s_i = f(i) \pmod{n}$.
3. Gives (in private) share s_i to user u_i.

This threshold secret sharing scheme is due to Shamir [14] and it has been proven to be information theoretically secure, i.e., any subset of up to k shareholders, collaboratively, cannot get any useful information about the secret. On the other hand, any subset $\mathcal{A}$ ($|\mathcal{A}| > k$) is an authorized subset and they can collaboratively reconstruct the associated polynomial, $f(x)$, using Lagrange interpolation formula,

$$f(x) = \Sigma_{x_i \in \mathcal{A}} f(x_i) \Pi_{\substack{x_i \in \mathcal{A} \\ x_i \neq x_j}} \frac{(x - x_j)}{(x_i - x_j)}, \tag{1}$$

and thus, uniquely determine the secret.

3.2   Signature Generation

Let m ($0 \leq m < n$) be the hash value of the message that is requesting a signature. Given a message m, the signature of the message is $\sigma = m^d \bmod n$. The verification of the signature utilizes the public key, e, and the signature is accepted as genuine if it satisfies the equation $m = \sigma^e \bmod n$.

In threshold signatures, the group's secret key is not known to any user. The generation of the signature, however, can be carried out by collaboration of every authorized set $\mathcal{A}$, which can reconstruct the group's secret. In our proposed threshold RSA system, the secret key d can be obtained according to the formula given in equation (1), since $d = f(0)$:

$$d = \Sigma_{x_i \in \mathcal{A}} f(x_i) \Pi_{\substack{x_i \in \mathcal{A} \\ x_i \neq x_j}} \frac{(0 - x_j)}{(x_i - x_j)}.$$

That is, each user u_i, of an authorized subset, $\mathcal{A}$, ($|\mathcal{A}| > k$) calculates its modified share, using

$$d_i = s_i \prod_{\substack{u_j \in \mathcal{A} \\ j \neq i}} \frac{j}{j - i} \pmod{n} \tag{2}$$

such that,

$$d = \sum_{u_i \in \mathcal{A}} d_i \pmod{n}. \tag{3}$$

For security reasons, however, the participants do not recover the secret key d, otherwise the secret key will be known to single participants. This would then enable them to sign any messages individually. Instead, the participants take part in a protocol that outputs the group's signature on the message, without compromising the group's secret key. Our signing protocol works as follows.

1. Each user, $u_i \in \mathcal{A}$ computes his partial signature $\sigma_i = m^{d_i}$.
2. After collecting all partial signatures from participants of the active group $\mathcal{A}$, the combiner computes

$$\sigma' = \prod_{u_i \in \mathcal{A}} \sigma_i = m^{\sum_{u_i \in \mathcal{A}} d_i \bmod n} = m^{d + I_{\mathcal{A}} \times n} = \sigma \times m^{I_{\mathcal{A}} \times n} \quad (\bmod \; n).$$

The required signature σ, can be obtained if the above result is multiplied by $m^{-I_{\mathcal{A}} \times n}$. We call $I_{\mathcal{A}}$ the *index* of the active subset $\mathcal{A}$. It is not difficult to see that $I_{\mathcal{A}}$ is approximately $|\mathcal{A}|/2$, since each d_i is an element of $\mathbb{Z}_n$. A naive algorithm requires one exponentiation to compute m^{-n}, and approximately $|\mathcal{A}|/2$ multiplications, in order to obtain the required result. In Section 4, we will provide a direct and efficient method for deriving the correct signature from collected partial signatures.

4   Robust Scheme

Up to this stage we have assumed that adversary is passive. i.e, all users appropriately follow the signing procedure. In this section we consider an active adversary. That is, the adversary not only learns all the information held by the corrupted users, it also controls the behavior of all corrupted users. So, it may force corrupted users to not follow the protocol.

A desirable characteristic of a threshold signature is that the participants must be able to generate the signature, even if unauthorized subsets want to prevent the signing protocol. That is, the system must, to an extent, tolerate deceptive users who do not cooperate properly in the signature generation protocol. We will show that in our scheme, a signature generation can be carried out successfully if majority of participants follow the protocol appropriately. We set the threshold parameter $k < \ell/2$. Hence, the system tolerates up to k corrupted or deceptive users, who do not cooperate properly in the signature generation protocol.

In order to prevent deceptive users from interfering with the signature generation protocol, the system must possess a facility to distinguish faulty partial results from correct results. This requires the normal protocols to be armed with verification facilities.

4.1   Initialization

1. Secretly chooses, independently at random, k elements of $\mathbb{Z}_n$, denoted $a_1, \ldots, a_k$ and forms a polynomial $f(x) = d + a_1 x + a_2 x^2 + \cdots + a_k x^k$, where $a_k \neq 0$. That is, $f(x)$ is a polynomial of degree k.

2. Computes $s_i = f(i)$, for $1 \le i \le \ell$.
3. Gives (in private) share s_i to user u_i, and broadcast $g_0 = g^d$, $g_1 = g^{a_1}$, $g_2 = g^{a_2}, \ldots, g_k = g^{a_k}$ (mod n), where g is an element of high order in $\mathbb{Z}_n^*$.
4. User u_i verifies that $g^{s_i} = \Pi_{j=0}^{k} g_j^{i^j}$ (mod n). If the equality does not hold, u_i publishes s_i. If more than k users complain, the dealer fails.
5. All users can check that the set of following public values,

$$w_0 = g^{0 \times n} = 1, \quad w_1 = g^{1 \times n}, \quad w_2 = g^{2 \times n}, \ldots, w_\ell = g^{\ell \times n}$$

are computed correctly.

Note that giving away the signature of g is not a security problem, because finding a message that its hash value (with proper padding) is equal to g is an intractable problem –assuming that the underlying hash function is collision resistant.

4.2 Signature Generation

Signing a message m ($0 \le m < n$) is more or less the same as in the basic scheme, but armed with partial signatures verification that eliminates corrupted users.

1. Each user $u_i \in \mathcal{A}$ ($|\mathcal{A}| > k$), computes their modified share d_i, according to equation (2).
2. Each user $u_i \in \mathcal{A}$ computes their partial signature $\sigma_i = m^{d_i}$ and a verification value g^{d_i}. Our partial signature verification is similar to that of [11], and works as follows:
3. After combiner received all verified pairs (σ_i, g^{d_i}) from all active participants, it computes

$$w_{I_\mathcal{A}} = \frac{\Pi_{u_i \in \mathcal{A}} g^{d_i}}{g^d}, \tag{4}$$

and obtains $I_\mathcal{A}$, which is the index of an element in the set of public values $w_0, w_1, \ldots, w_\ell$.

Input
Secret: modified shares $d_i \in \mathbb{Z}_n$
Common: g, n, m, partial signatures σ_i, and verification values g^{d_i}

1.   The verifier, V, chooses $a, b \in_R \mathbb{Z}_n$ and computes $x = g^a m^b \bmod n$, which is sent to the prover, $u_i \in \mathcal{A}$.
2. u_i computes $y = x^{d_i} \bmod n$ and sends it to the verifier, V.
3. V verifies that $y = (g^{d_i})^a \sigma_i^b \bmod n$.
      If equality holds, then the verifier accepts the partial result, σ_i, as a genuine partial signature; otherwise, it is rejected.

Fig. 1. Verification of partial signatures

4. The combiner computes the signature of m using

$$\sigma = \frac{\prod_{u_i \in \mathcal{A}} \sigma_i}{m^{I_{\mathcal{A}} n}}. \tag{5}$$

Theorem 1. *The above protocol generates a correct RSA signature.*

Proof. Since each d_i is an integer smaller than n, equation (3) can be rewritten as $\sum_{u_i \in \mathcal{A}} d_i = d + I_{\mathcal{A}} n$ for some integer $0 \leq I_{\mathcal{A}} < |\mathcal{A}|$. Therefore,

$$\frac{\prod_{u_i \in \mathcal{A}} g^{d_i}}{g^d} = \frac{g^{d + I_{\mathcal{A}} n}}{g^d} = g^{I_{\mathcal{A}} n} = w_{I_{\mathcal{A}}}$$

That is, $w_{I_{\mathcal{A}}}$ must be one of the elements $w_0, w_1, \ldots, w_{|\mathcal{A}|}$. Knowing $I_{\mathcal{A}}$, the signature on the message m can be obtained from equation (5).

5 Evaluation

5.1 Security

The underlying secret sharing scheme employed for the share distribution protocol is Shamir's threshold scheme, which is believed to be information-theoretically secure. In signature generation protocol, however, one might be able to learn the constant integer $I_{\mathcal{A}}$, associated to the active signing group $\mathcal{A}$, where

$$\Sigma_{i \in \mathcal{A}} d_i = \Sigma_{i \in \mathcal{A}} s_i \prod_{\substack{u_j \in \mathcal{A} \\ j \neq i}} \frac{j}{j - i} = I_{\mathcal{A}} \times n + d.$$

From the point of view of an honest but curious user no information (neither about the shares s_i, nor about the group secret d) leaks from the index $I_{\mathcal{A}}$. So, let us consider the scenario in which an adversary has corrupted k users, and has thus learnt k shares s_i (w.l.o.g. let the set of corrupted users is $u_1, u_2, \ldots, u_k$). We want to see whether or not this adversary can learn any useful information about other shares and/or the group secret d.

We assume that the adversary (who knows k shares $s_1, s_2, \ldots, s_k$) participates (along with all k corrupted signatories) in a signature generation. That is, in a signature generation there is only one user u_x ($k < x \leq \ell$), in which the adversary does not know the respective secret value. In this setting, the adversary easily can determine whether $\Sigma_{i=1}^k d_i$ (mod n) is smaller or larger than the group secret, d (obviously if the number of non-corrupted users participating in the signature generation is more than one, the adversary cannot determine whether $\Sigma_{i=1}^k d_i$ (mod n) is larger or smaller than d). Furthermore, if the system is made one-time system or the number of users is small, then the adversary will cannot get any useful information that enables him a successful attack to the system.

Considering the fact that all shares in the Shamir scheme are indistinguishable from random values, the modified shares d_i, and thus their summation, are random values in the interval $[0, n[$. That is, after signing p messages (with the assumption that each message is signed by a group, consisting of all corrupted users and a honest user) the adversary has a list of p values, some of them which are smaller than d, and the rest which are larger than d. In order to see how likely/unlikely it is that the adversary can learn any useful information about the group secret d, let the RSA module have a moderate size (e.g., n is a 1024-bit integer). After signing 2^{24} messages[3], which will never happens in practice, the average distance between any two of these numbers is approximately 2^{1000} (due to uniform distribution of random values). Although this provides some extra information to the adversary, we are not aware of any method in which this information enables the adversary to launch a successful attack to the system.

5.2 Efficiency

The proposed threshold RSA signature scheme requires each signatories to sign the message using their modified shares, which is of size $\log n$. This is the most efficient way that a shared generation signature can be designed (note that in a secret sharing scheme, if shares are smaller than the secret, it leaks some information about the secret). It is worth mentioning that in the most efficient existing schemes the size of the share of each participants is $\log(2\ell!n)$.

In order to calculated the correct signature, however, we need to know the index of the active group (which is an integer smaller than the number of co-signers). Considering the facts that:

- The index of each group of collaborating servers/users is a constant integer, and therefore does not need to be calculated more than once.
- In order to increase the speed of the signature generation protocol, groups of users who prefer to work together can calculate their group's index prior to signing procedure.

The cost of our threshold RSA signature scheme is just one exponentiation by each user and the combiner, where the size of the exponent is the same as the secret key.

6 Conclusions

We have presented a novel technique for implementing the first *ideal* threshold RSA system. The proposed scheme has the following advantage:

- The share of each user from the group's secret key, *at all times*, is not larger than the size of the group's secret key itself.
- In the signature generation phase, each user performs only one exponentiation, where the exponent is not larger than the size of the secret key of the underlying RSA system.

---

[3] This implies that there must be at least 2^{24} honest users in the system.

- The combining process requires only one exponentiation and a few multiplications.
- The scheme is robust, i.e., the signature generation process cannot be prevented by k or less corrupted users.

Acknowledgment. The authors would like to thank Yvo Desmedt for a productive discussion and unanimous referees for their helpful comments.

References

1. D. Boneh and M. Franklin, "Efficient Generation of Shared RSA Keys," in *Advances in Cryptology - Proceedings of CRYPTO '97* (S. Burton and J. Kaliski, eds.), vol. 1294 of *Lecture Notes in Computer Science*, pp. 425–439, Springer-Verlag, 1997.
2. I. Damgård and K. Dupont, "Efficient Threshold RSA Signatures with General Moduli and No Extra Assumptions," in *Proceedings of the 8th International Workshop on Practice and Theory in Public Key cryptography (PKC 2005)* (S. Vaudeny, ed.), vol. 3386 of *Lecture Notes in Computer Science*, pp. 346–361, Springer-Verlag, 2005.
3. A. De Santis, Y. Desmedt, Y. Frankel, and M. Yung, "How to Share a Function Securely," in *26th Annual ACM Symp. on the Theory of Computing*, pp. 522–533, 1994.
4. Y. Desmedt, "Society and group oriented cryptography: A new concept," in *Advances in Cryptology - Proceedings of CRYPTO '87* (C. Pomerance, ed.), vol. 293 of *Lecture Notes in Computer Science*, pp. 120–127, Springer-Verlag, 1988.
5. Y. Desmedt, "Threshold Cryptosystems," in *Advances in Cryptology - Proceedings of AUSCRYPT '92* (J. Seberry and Y. Zheng, eds.), vol. 718 of *Lecture Notes in Computer Science*, pp. 3–14, Springer-Verlag, 1993.
6. Y. Desmedt and Y. Frankel, "Threshold cryptosystems," in *Advances in Cryptology - Proceedings of CRYPTO '89* (G. Brassard, ed.), vol. 435 of *Lecture Notes in Computer Science*, pp. 307–315, Springer-Verlag, 1990.
7. Y. Desmedt and Y. Frankel, "Shared generation of authenticators and signatures," in *Advances in Cryptology - Proceedings of CRYPTO '91* (J. Feigenbaum, ed.), vol. 576 of *Lecture Notes in Computer Science*, pp. 457–469, Springer-Verlag, 1992.
8. T. ElGamal, "A Public Key Cryptosystem and a Signature Scheme Based on Discrete Logarithms," *IEEE Trans. on Inform. Theory*, vol. IT-31, pp. 469–472, July 1985.
9. Y. Frankel, P. Gemmell, P. MacKenzie, and M. Yung, "Proactive RSA," in *Advances in Cryptology - Proceedings of CRYPTO '97* (S. Burton and J. Kaliski, eds.), vol. 1294 of *Lecture Notes in Computer Science*, pp. 440–454, Springer-Verlag, 1997.
10. Y. Frankel, P. Gemmell, P. Mackenzie, and M. Yung, "Optimal Resilience Proactive Public-key Cryptosystems," in *38th Annual Symp. on Foundations of Computer Science (FOCS)*, pp. 384–393, IEEE, 1997.
11. R. Gennaro, S. Jarecki, H. Krawczyk, and T. Rabin, "Robust and Efficient Sharing of RSA Functions," in *Advances in Cryptology - Proceedings of CRYPTO '96* (S. Burton and J. Kaliski, eds.), vol. 1109 of *Lecture Notes in Computer Science*, pp. 157–172, Springer-Verlag, 1996.
12. T. Rabin, "A Simplified Approach to Threshold and Proactive RSA," in *Advances in Cryptology - Proceedings of CRYPTO '98* (H. Krawczyk, ed.), vol. 1462 of *Lecture Notes in Computer Science*, pp. 89–104, Springer-Verlag, 1998.

13. R. Rivest, A. Shamir, and L. Adleman, "A Method for Obtaining Digital Signatures and Public-Key Cryptosystems," *Communications of the ACM*, vol. 21, pp. 120–126, Feb 1978.
14. A. Shamir, "How to Share a Secret," *Communications of the ACM*, vol. 22, pp. 612–613, Nov. 1979.
15. V. Shoup, "Practical Threshold Signatures," in *Advances in Cryptology - Proceedings of EUROCRYPT 2000* (B. Preneel, ed.), vol. 1807 of *Lecture Notes in Computer Science*, pp. 207–220, Springer-Verlag, 2000.

Towards Provably Secure Group Key Agreement Building on Group Theory

Jens-Matthias Bohli[1], Benjamin Glas[2], and Rainer Steinwandt[3]

[1] Institut für Algorithmen und Kognitive Systeme, Fakultät für Informatik,
Am Fasanengarten 5, 76128 Karlsruhe, Germany
bohli@ira.uka.de
[2] Institut für Technik der Informationsverarbeitung,
Fakultät für Elektrotechnik & Informationstechnik,
Engesserstraße 5, 76128 Karlsruhe, Germany
glas@itiv.uka.de
[3] Department of Mathematical Sciences, Florida Atlantic University,
777 Glades Road, Boca Raton, FL 33431, USA
rsteinwa@fau.edu

Abstract. Known proposals for key establishment schemes based on combinatorial group theory are often formulated in a rather informal manner. Typically, issues like the choice of a session identifier and parallel protocol executions are not addressed, and no security proof in an established model is provided. Successful attacks against proposed parameter sets for braid groups further decreased the attractivity of combinatorial group theory as a candidate platform for cryptography.

We present a 2-round group key agreement protocol that can be proven secure in the random oracle model if a certain group-theoretical problem is hard. The security proof builds on a framework of Bresson et al., and explicitly addresses some issues concerning malicious insiders and also forward secrecy. While being designed as a tool for basing group key agreement on non-abelian groups, our framework also yields a 2-round group key agreement basing on a Computational Diffie-Hellman assumption.

Keywords: group key establishment, provable security, conjugacy problem, automorphisms of groups.

1 Introduction

While in recent years cryptographic proposals building on combinatorial group theory, in particular braid groups, proliferated, repeated cryptanalytic success also diminished the initial optimism on the subject significantly. Dehornoy's paper [15] gives a good survey on the state of the subject, and evidently significant research is still needed to reach a definite conclusion on the cryptographic potential of braid groups. As far as key establishment is concerned, especially an idea of Anshel et al. [2,1] received a lot of attention (e. g., [16,19,27]). Several further

P.Q. Nguyen (Ed.): VIETCRYPT 2006, LNCS 4341, pp. 322–336, 2006.

ideas for deriving a key establishment scheme from combinatorial group theory have been put forward, including the work in [22,23,26,28,25]. Unfortunately, to the best of our knowledge for none of these proposals a modern security analysis in an established cryptographic framework like [5,3,24,4,11,12] has been carried out. It should be mentioned, however, that the 2-party construction considered by Catalano et al. in [13] seems suitable for a non-abelian setting, but no further exploration in this direction is known to us.

One approach to build a key establishment protocol on non-abelian groups is to prove a scheme secure against passive adversaries, followed by applying a generic compiler that establishes stronger security guarantees (cf. [21], for instance). In this contribution we focus on group[1] key establishment. So far, the only known proposals for basing a group key establishment on non-abelian groups we are aware of are due to Lee et al. [23] and Grigoriev and Ponomarenko [18]. Unfortunately the former builds on ideas from a two-party protocol presented in [22], so Cheon and Jun's polynomial time solution to the Braid Diffie-Hellman conjugacy problem [14] impairs the attractivity of Lee et al.'s scheme. Grigoriev and Ponomarenko use a different approach to build a group key establishment. They build on ideas from [2,1] and make repeated use of a 2-party protocol.

As an intermediate step, we build a key encapsulation mechanism from a group theoretic problem and then construct a group key establishment protocol on top. We do not follow the indicated 2-step approach (proving security in the passive case, followed by, say, applying the compiler of Katz and Yung [21])—aiming at a 2-round solution, a direct design approach appears to be no less attractive. While we cannot give a concrete non-abelian instance of our scheme, a concrete protocol can be derived from a Computational Diffie-Hellman (CDH) assumption in a cyclic group. In a sense, our approach can be seen in the spirit of [17], which takes a similar effort to identify requirements on finite non-abelian groups that allow to implement a provably IND-CCA secure public key encryption scheme.

Our security proof makes use of an existentially unforgeable signature scheme and the random oracle, and it is fair to ask whether there is really a need for another protocol in such a setting. For instance, in terms of communication complexity Boyd and Nieto's 1-round protocol from [9] certainly can be seen as superior to the 2-round proposal below. However, the latter protocol lacks forward secrecy, and to our knowledge it is not known whether a one-round protocol achieving forward secrecy can be constructed at all [8]. Moreover, we aim at a 2-round protocol offering security guarantees in the presence of malicious insiders. Currently the treatment of malicious insiders in group key establishment receives increasing attention—including the work in [7,20,6]. At the current state of the art, to design a 2-round protocol with security guarantees against malicious insiders and offering forward secrecy, a "one step" design strategy building on a random oracle and a signature scheme still appears to be fair.

---

[1] Unfortunately, in this paper the term *group* has to express two different meanings; here it refers to a set of principals.

2   Preliminaries

Giving a general introduction to the existing models for group key establishment is beyond the scope of this paper, and we refer to the standard reference [10] for this. Moreover, the background in group theory needed for describing our protocol is extremely modest. Hence, we restrict to recalling some details of the cryptographic proof model used for the security proof below. A more detailed discussion of this model can be found in [4,11,7].

2.1   Security Model

Participants. A finite set $\mathcal{P}$ of probabilistic polynomial time (ppt) Turing machines U_i models the users that constitute the (potential) protocol participants. Each user $U_i \in \mathcal{P}$ may execute a polynomial number of protocol instances in parallel. We denote the instance $s \in \mathbb{N}$ of principal $U_i \in \mathcal{P}$ by Π_i^s. Each instance Π_i^s may be taken for a process executed by U_i and has assigned seven variables state_i^s, sid_i^s, pid_i^s, sk_i^s, term_i^s, used_i^s and acc_i^s:

used_i^s is initialized with false and set to true as soon as the instance begins a protocol run triggered by a call to the Execute-oracle or a call to the Send-oracle (see below);

state_i^s stores the state information during the protocol execution;

term_i^s is initialized with false and set to true when the execution has terminated;

sid_i^s holds the (non-secret) session identifier that serves as identifier for the session key sk_i^s and is initialized with a distinguished NULL value—the adversary has access to all sid_i^s-values;

pid_i^s stores the set of user identities that Π_i^s aims at establishing a key with—it also includes U_i itself;

acc_i^s is initialized with false and set to true if the protocol execution terminated successfully (i. e., the principal accepted the session key for use with users pid_i^s in session sid_i^s);

sk_i^s contains the session key after the execution is accepted by instance Π_i^s. Before acceptance, it stores a distinguished NULL value.

Initialization. In a one-time initialization phase, before the first execution of the key establishment protocol, for each user $U_i \in \mathcal{P}$ a secret key/public key pair (SK_i, PK_i) is generated. The secret key SK_i is only revealed to U_i, the corresponding public key PK_i is given to all users[2].

Communication network. We assume arbitrary point-to-point connections to be available between the users. However, the connections are insecure and fully asynchronous, modeled by an active adversary with full control over the network (cf. the adversarial model below).

---

[2] We assume these keys to be generated and distributed honestly by a trusted party.

Adversarial model. The adversary $\mathcal{A}$ interacts with the user instances via a set of oracles Execute, Send, Reveal, Corrupt and Test. We call the adversary *passive* if no access to the Send- and Corrupt-oracle is granted.

Execute($\{U_1, U_2, \ldots, U_r\}$) This query executes a protocol run between unused instances Π_i^s of the specified users and returns a transcript of all messages sent during the protocol execution.

Send(U_i, s, M) This query sends the message M to instance Π_i^s and returns the reply generated by this instance. A special message $M = \{U_1, \ldots, U_r\}$ sent to an unused instance will set $\mathsf{pid}_i^s := M$, $\mathsf{used}_i^s := \mathsf{true}$ and provoke Π_i^s to begin with the protocol execution.

Reveal(U_i, s) returns the session key sk_i^s.

Corrupt(U_i) returns the long-term secret key SK_i that U_i holds. We will refer to a user U_i as *honest* if no query of the form Corrupt(U_i) was made.

Test(U_i, s) The adversary is allowed to use this query only once. Provided that $\mathsf{sk}_i^s \neq \mathrm{NULL}$, a random bit b is drawn and depending on b with probability $1/2$ the session key sk_i^s and with probability $1/2$ a uniformly chosen random session key is returned. The adversary is allowed to query other oracles after its Test-query, but no query that would repeal the freshness of Π_i^s is allowed.

Correctness. To exclude "useless" protocols, we take a group key establishment protocol P for *correct* if in the presence of a passive adversary a single execution of P among arbitrary participants $U_1, \ldots, U_r$ involves r instances $\Pi_1^{s_1}, \ldots, \Pi_r^{s_r}$ and ensures that with overwhelming probability all instances accept a matching session key with a common partner identifier and a common and unique session identifier. More formally, with overwhelming probability the following conditions have to hold:

- $\mathsf{used}_1^{s_1} = \cdots = \mathsf{used}_r^{s_r} = \mathsf{true}$;
- $\mathsf{acc}_1^{s_1} = \cdots = \mathsf{acc}_r^{s_r} = \mathsf{true}$;
- $\mathsf{sk}_1^{s_1} = \cdots = \mathsf{sk}_r^{s_r}$;
- $\mathsf{sid}_1^{s_1} = \cdots = \mathsf{sid}_r^{s_r}$ globally unique;
- $\mathsf{pid}_1^{s_1} = \mathsf{pid}_2^{s_2} = \cdots = \mathsf{pid}_r^{s_r} = \{U_1, \ldots, U_r\}$.

Freshness. For the security definition, we have to specify which instances are fresh, i.e., hold a session key that should be unknown to the adversary. As a first step we define the notion of partnering.

Definition 1 (Partnering). *Two instances $\Pi_i^{s_i}$, $\Pi_j^{s_j}$ are partnered if* $\mathsf{sid}_i^{s_i} = \mathsf{sid}_j^{s_j}$, $\mathsf{pid}_i^{s_i} = \mathsf{pid}_j^{s_j}$ *and* $\mathsf{acc}_i^{s_i} = \mathsf{acc}_j^{s_j} = \mathsf{true}$.

Now the freshness of an instance is defined as follows.

Definition 2. *We call a user instance $\Pi_i^{s_i}$ that has accepted, i.e., $\mathsf{acc}_i^{s_i} = \mathsf{true}$, fresh if none of the following two conditions holds:*

- *For a $U_j \in \mathsf{pid}_i^{s_i}$ a Corrupt(U_j) query was executed before a query of the form Send(U_ℓ, s_ℓ, M) with $U_\ell \in \mathsf{pid}_i^s$ has taken place.*
- *A Reveal(U_j, s_j) was executed where $\Pi_i^{s_i}$ and $\Pi_j^{s_j}$ are partnered.*

We say that an adversary $\mathcal{A}$ was successful if $\mathcal{A}$, after interacting with the oracles including one $\mathsf{Test}(\Pi_i^{s_i})$ query for a fresh oracle $\Pi_i^{s_i}$, outputs a bit d and it holds that $d = b$ for the bit b used by the Test-oracle. We denote this probability by Succ and define $\mathcal{A}$'s advantage to be

$$\mathsf{Adv}_{\mathcal{A}} := |2 \cdot \mathsf{Succ} - 1|.$$

Definition 3 (Key secrecy/(basic) security). *We call the group key establishment protocol* P *secure if for all ppt adversaries* $\mathcal{A}$ *the function* $\mathsf{Adv}_{\mathcal{A}} = \mathsf{Adv}_{\mathcal{A}}(k)$ *is negligible in the security parameter* k.

Forward secrecy is addressed in the usual manner:

Definition 4 (Forward secrecy). *We say the group key establishment protocol* P *fulfills* forward secrecy, *if the disclosure of the private long-term keys used in the protocol execution does not compromise earlier derived session keys.*

The following extended security properties aim at avoiding further attacks imposed by malicious participants:

Definition 5 (Strong entity authentication). Strong entity authentication *to an oracle* $\Pi_i^{s_i}$ *is provided if both* $\mathsf{acc}_i^{s_i} = \mathsf{true}$ *and for all honest* $U_j \in \mathsf{pid}_i^{s_i}$ *with overwhelming probability there exists an oracle* $\Pi_j^{s_j}$ *with* $\mathsf{sid}_j^{s_j} = \mathsf{sid}_i^{s_i}$ *and* $U_i \in \mathsf{pid}_j^{s_j}$.

Definition 6 (Integrity). *We say a correct group key establishment protocol fulfills* integrity *if with overwhelming probability all oracles of honest principals that have accepted with the same session identifier* $\mathsf{sid}_j^{s_j}$ *hold identical session keys* $\mathsf{sk}_j^{s_j}$, *and associate this key with the same principals* $\mathsf{pid}_j^{s_j}$.

2.2 Assumptions on the Underlying Group

For the security proof of our protocol, the underlying group G (resp. family of groups $G = G(k)$, indexed by the security parameter) has to satisfy certain requirements. In particular, we assume products and inverses of group elements to be computable by ppt algorithms. For the sake of simplicity, we also assume that G allows a ppt computable canonical representation of elements, so that we can identify group elements with their canonical representation. To generate the group elements needed in a protocol execution, we rely on the existence of three algorithms, that capture the problem of creating "good instances":

– DomPar denotes a (stateless) ppt *domain parameter generation* algorithm that upon input of the security parameter 1^k outputs a finite sequence S of elements in G. The subgroup $\langle S \rangle$ of G spanned by S will be publicly known. For the special case of applying our framework to a CDH-assumption, S specifies a public generator of a cyclic group.

- SamAut denotes a (stateless) ppt *automorphism group sampling* algorithm that upon input of the security parameter 1^k and a sequence S output by DomPar returns a description of an automorphism ϕ on the subgroup $\langle S \rangle$, so that both ϕ and ϕ^{-1} can be evaluated efficiently. E. g., for a cyclic group, ϕ could be given as an exponent, or for an inner automorphism the conjugating group element could be specified.
- SamSub denotes a (stateless) ppt *subgroup sampling* algorithm that upon input of the security parameter 1^k and a sequence S output by DomPar returns a word $x(S)$ in the generators S (and their inverses) representing an element $x \in \langle S \rangle$. Intuitively, SamSub chooses a random $x \in \langle S \rangle$, so that it is hard to recognize x if we know elements of x's orbit under $\mathrm{Aut}(G)$. Our protocol needs an explicit representation of x in terms of the generators S.

With this notation, we can define a computational problem of parallel automorphism application, where $o \leftarrow \mathsf{A}(i)$ denotes that algorithm A outputs o upon receiving input i:

Definition 7 (Parallel automorphism application). *Let $r \in \mathbb{N}_{>0}$ be a natural number. By the* problem of r-fold parallel automorphism application $(r\text{-PAA})$ *w. r. t. the quadruple $(G, \mathsf{DomPar}, \mathsf{SamAut}, \mathsf{SamSub})$ we mean the task of finding an algorithm $\mathcal{A}$ which on input of S, $\phi_i(S) := (\phi_i(s))_{s \in S}$ for $i = 1, \ldots, r$ and $\phi_1(x), \ldots, \phi_r(x)$ outputs the group element x represented by the word $x(S)$, where*

- $S \leftarrow \mathsf{DomGen}(1^k)$,
- $x(S) \leftarrow \mathsf{SamSub}(1^k, S)$,
- $(\phi_i, \phi_i^{-1}) \leftarrow \mathsf{SamAut}(1^k, S)$ $(i = 1, \ldots, r)$.

To capture the assumption needed in the security proof below, we also define the advantage of an adversary in solving the above problem:

Definition 8 (r-PAA advantage). *For an algorithm $\mathcal{A}$ trying to solve r-PAA, we denote its advantage as a function in the security parameter k and its runtime t by $\mathsf{Adv}_{\mathcal{A}}^{r-\mathrm{PAA}} = \mathsf{Adv}_{\mathcal{A}}^{r-\mathrm{PAA}}(k, t) =$*

$$\Pr\left(x \leftarrow \mathcal{A}(S, (\phi_i(S), \phi_i(x))_{1 \leq i \leq r}) \;\middle|\; \begin{array}{l} S \leftarrow \mathsf{DomGen}(1^k), x(S) \leftarrow \mathsf{SamSub}(1^k, S), \\ (\phi_i, \phi_i^{-1}) \leftarrow \mathsf{SamAut}(1^k, S) \ (i = 1, \ldots, r) \end{array} \right).$$

Our security proof builds on the assumption that for any ppt adversary $\mathcal{A}$ the advantage $\mathsf{Adv}_{\mathcal{A}}^{r-\mathrm{PAA}}$ is negligible. For the case of ϕ being an inner automorphism, r-PAA expresses a kind of parallel conjugacy problem. Note however, that instead of looking for concrete instances building on a non-abelian group, we may apply our framework to an "ordinary" Computational Diffie-Hellman (CDH) setting, too:

Example 1 (Basing on CDH). Let G be a cyclic group and choose for $S := \langle g \rangle$ an element $g \in G$ of prime order q. Now let SamSub choose uniformly at random an exponent $x \in \{1, \ldots, q\}$. Similarly, we specify SamAut to choose uniformly at random an exponent $\phi \in \{1, \ldots, q - 1\}$. Then r-PAA is polynomial time equivalent to the CDH-problem in $\langle g \rangle$:

"CDH solution" $\Rightarrow$ r-PAA solution": A CDH-oracle allows to find g^x from a single pair $(g^\phi, g^{x\phi})$ as follows. First compute $g^{\phi^{-1} \bmod q}$ by using the CDH-oracle to multiply the exponents of (g, g) with $g = \left(g^\phi\right)^{\phi^{-1}}$ taken for a power of the group generator g^ϕ. Next we can obtain g^x by applying the CDH-oracle to $g^{\phi^{-1} \bmod q}$ and $g^{x\phi}$.

"CDH solution" $\Leftarrow$ r-PAA solution": Given g^{ϕ_1}, g^v ($\phi_1, v \in \{1, \ldots, q-1\}$), we can use an oracle solving r-PAA to compute $g^{v\phi_1^{-1}}$: We can interpret v as having the form $v = x \cdot \phi_1 \bmod q$, and by raising g^{ϕ_1} and g^v to uniformly at random chosen powers $\phi_1^{-1}\phi_i \in \{1, \ldots, q-1\}$ ($i = 2, \ldots, r$), we obtain the input needed by an oracle solving r-PAA to compute $g^x = g^{v\phi_1^{-1}}$. Hence, given a pair (g^u, g^v) we can compute g^{uv} as follows:

1. Apply the above method to (g^u, g^v), yielding $g^{vu^{-1}}$.
2. Apply the above method to $(g^v, g^{vu^{-1}})$, yielding $g^{u^{-1}}$.
3. Apply the above method to $(g^{u^{-1}}, g^v)$, yielding g^{uv}.

2.3   Groundwork of the Protocol

The r-PAA assumption is in quintessence a variant of a *key encapsulation mechanism* (KEM). A KEM provides the public key algorithm that is needed for constructing a hybrid encryption system. In contrast to public key encryption it is not necessary to be able to encrypt arbitrary messages, but only random messages, which don't need to be given as input to the algorithm.

Smart [29] extends KEM to mKEM which captures key encapsulation to multiple parties. An mKEM consists of Algorithms Gen, Enc and Dec. At this Gen will take the domain parameters as input and output a public/private key pair (pk, sk). The algorithm Enc takes as input a list of public keys $(pk_1, \ldots, pk_n)$ and outputs a pair consisting of a key K and a ciphertext C. Finally, Dec takes as input a ciphertext C and a private key sk_i and outputs the key K.

The assumption about the group in Section 2.2 resembles an mKEM. However, for a KEM, the key space will consist of a finite set, such that K is indistinguishable from an element chosen uniformly at random. The value x, sampled by the algorithm SamSub will generally not offer this property, so a randomness extraction has to be applied on x to build a KEM. We now give the interpretation of the PAA as an mKEM, using a random oracle H as a randomness extractor. Note that the protocol in Section 3 will not need the randomness extraction for an individual x, but only for the collection of the x-values of all participants.

After having generated domain parameters with DomPar, SamAut produces the automorphism ϕ on the subgroup $\langle S \rangle$. The images $(\phi_i(t))_{t \in S}$ of the generators S will act as public key and ϕ_i^{-1} as private key. This will provide the algorithm Gen. Given the subgroup generators S, SamSub returns a word $x(S)$ in the generators S. Then, given any number of public keys $\phi_i(S)$ for the subset S, the ciphertext $\phi_i(x(S))$ can be computed. Thus, the combination of SamSub and application of ϕ_i can be seen as providing Enc. Again, Dec is only given implicitly, as the application of ϕ^{-1} to $\phi_i(x(S))$ is straightforward.

Domain parameter $\mathbf{D}$: $S \leftarrow \mathsf{DomGen}(1^k)$

$(pk, sk) \leftarrow \mathsf{Gen}(\mathbf{D})$: $(\phi, \phi^{-1}) \leftarrow \mathsf{SamAut}(1^k, S)$
$\phantom{(pk, sk) \leftarrow \mathsf{Gen}(\mathbf{D})}$ $pk = (\phi(t))_{t \in S}$
$\phantom{(pk, sk) \leftarrow \mathsf{Gen}(\mathbf{D})}$ $sk = \phi^{-1}$

$(K, C) \leftarrow \mathsf{Enc}(pk_1, \dots, pk_n)$: $x(S) \leftarrow \mathsf{SamSub}(1^k, S)$
$\phantom{(K, C) \leftarrow \mathsf{Enc}(pk_1, \dots, pk_n)}$ $K = H(x(S))$
$\phantom{(K, C) \leftarrow \mathsf{Enc}(pk_1, \dots, pk_n)}$ $C = (\phi_1(x(S)), \dots, \phi_n(x(S)))$

$K \leftarrow \mathsf{Dec}(C, sk_i, (pk_1, \dots, pk_n))$: $K = H(\phi_i^{-1}(\phi_i(x(S))))$

Security of r-PAA as an mKEM. With the interpretation as above, intuitively r-PAA is secure as an mKEM. Using the random oracle to derive the key K, transforms the indistinguishability of keys in the mKEM into the problem to compute the preimage, as the r-PAA advantage. However, Smart [29] defines an r out of n security where the adversary is offered n public keys and can chose a set of r on which he will mount his attack. In this respect, the above r-PAA problem yields an r out of r secure mKEM. Though, the weaker requirements for a secure r-PAA might help the construction of concrete instances.

On Burmester-Desmedt style key agreements. The Burmester-Desmedt principle constructs a group key by arranging the participants in a circle, establishing keys between neighbors and broadcasting information, that allows anyone who knows one key in the circle, to compute all other keys. Having in mind the construction of a 2-round protocol, the key establishment should be possible in one round. However, forward secrecy requires ephemeral public keys, such that in order to establish a key, first U_i has to execute Gen and send the result to U_j who has to execute Enc and return the ciphertext to U_i. As this requires already 2 rounds, we have chosen a different approach, which is similar to [9] but guarantees forward secrecy in addition.

Protocol idea. The idea for the protocol is now as follows: In the first round, all participants will generate an ephemeral key, what will be necessary to achieve forward secrecy. In the next round, each participant U_i will use the encryption algorithm of the KEM to obtain a key contribution K_i and a ciphertext C, and broadcast the ciphertext C. Finally, the participants compute a group key from the contributions U_j, $j = 1, \dots, n$.

3   A 2-Round Protocol for Group Key Agreement

To discuss our group key agreement protocol we adopt the common assumption that, from some protocol-external context, the set of protocol participants $\mathcal{U} \subseteq \mathcal{P}$ is known to all $U_i \in \mathcal{U}$. To simplify notation, w.l.o.g. we assume $\mathcal{U} = \{U_1, \dots, U_r\}$. Moreover, we assume that an asymmetric signature scheme is available that is existentially unforgeable under adaptive chosen message attacks. The respective signing and verification keys are to be fixed and distributed throughout the initialization phase mentioned in Section 2.1, and we denote a signature of a protocol participant U_i on a message M by $\mathsf{Sig}_i(M)$.

3.1 Description of the Protocol

Having fixed the security parameter k, first we have to run $\mathsf{DomGen}(1^k)$ to generate the public subgroup generators S. Hereafter, for an instance $\Pi_i^{s_i}$ of a protocol participant U_i a single protocol run can be described as shown in Figure 1. At this, *Broadcast: M* means that message M is sent to all other participants $U_j \in \mathcal{U}$ over *point-to-point* connections, i.e., the adversary is allowed to delay, suppress or modify some or all of the transmitted messages. In contrast to the idea in the last section, it is possible to separate portions for each participant $\phi_i(x(S))$ and instead of broadcasting all ciphertexts, every participant gets only the necessary part. Moreover, the randomness extraction is only applied on the list $x_1, \ldots, x_n$, instead on every x_i. Finally, H denotes a cryptographic hash function which will be modeled as a random oracle.

Round 1: Initialization Set $\mathsf{pid}_i^{s_i} := \mathcal{U}$, $\mathsf{used}_i^{s_i} := \mathsf{true}$.
  Choose $(\phi_i^{s_i}, (\phi_i^{s_i})^{-1}) \leftarrow \mathsf{SamAut}(1^k)$, $x_i^{s_i}(S) \leftarrow \mathsf{SamSub}(1^k, S)$, and compute
  the message $m_1^{s_i}(U_i) := (U_i, (\phi_i^{s_i}(t))_{t \in S}, H(x_i^{s_i}))$.
  Broadcast: $m_1^{s_i}(U_i)$.
Round 2: Key Exchange Set $\mathsf{sid}_i^{s_i} := H\left(m_1^{s_1}(U_1), \ldots, m_1^{s_r}(U_r), \mathsf{pid}_i^{s_i}\right)$.
  Compute and send $m_2^{s_i}(U_i, U_j) := \left(U_i, \phi_j^{s_j}(x_i^{s_i}), \mathsf{Sig}_i(\mathsf{sid}_i^{s_i})\right)$ to each participant $U_j \in \mathsf{pid}_i^{s_i}$, $j \neq i$. (To compute $\phi_j^{s_j}(x_i^{s_i})$ use the representation of
  $x_i^{s_i} = x_i^{s_i}(S)$ in terms of the generators S.)
Key Generation Compute from $\phi_i^{s_i}(x_j^{s_j})$ the original $x_j^{s_j}$ for all $j \neq i$ by applying the inverse of $\phi_i^{s_i}$.
  Compute the common session key $K := H\left(x_1^{s_1}, \ldots, x_r^{s_r}, \mathsf{pid}_i^{s_i}\right)$.
Verification Check for all $U_j \in \mathsf{pid}_i^{s_i}$ if $\mathsf{Sig}_j(\mathsf{sid}_j^{s_j})$ is a valid signature for $\mathsf{sid}_i^{s_i}$
  and if for $x_j^{s_j}$ the received hash value $H(x_j^{s_j})$ in $m_1^{s_j}(U_j)$ was correct.
  If true, set $\mathsf{acc}_i^{s_i} := \mathsf{term}_i^{s_i} := \mathsf{true}$, and $\mathsf{sk}_i^{s_i} := K$.
  Else set $\mathsf{acc}_i^{s_i} := \mathsf{false}$, $\mathsf{term}_i^{s_i} := \mathsf{true}$.

Fig. 1. A 2-round group key agreement protocol basing on r-PAA

At first glance, Round 1 of the protocol may look suspicious: The message is not signed and hence an attacker may tamper with this message. The underlying idea here is, that any tampering with the message in Round 1 will result in a failed signature verification in Round 2, because the session identifier $\mathsf{sid}_i^{s_i}$ computed and signed by $\Pi_i^{s_i}$ involves the correct message from Round 1. Further on, one may wonder whether one shouldn't simply fix the $\phi_i^{s_i}(S)$-values and include them in the public key of user U_i. Effectively, the latter would render $\phi_i^{s_i}$ a long-term secret and the protocol would no longer achieve forward secrecy.

Remark 1. Having in mind instances of the protocol in Figure 1 where the ϕ_i are inner automorphisms, it is worth noting that the protocol is symmetric in the sense that all participants perform the same steps: Differing from Anshel et al.'s 2-party construction, the key computation for the initiator is the same as for the other protocol participants.

3.2 Security Analysis

Correctness of the protocol in Figure 1 is immediate. To prove its security, we first observe that the constructed session identifier is with overwhelming probability globally unique:

Lemma 1. *If for all ppt adversaries $\mathcal{A}$ the advantage $\mathsf{Adv}_{\mathcal{A}}$ in solving r-PAA is negligible, then the session identifier $\mathsf{sid}_i^{s_i}$ constructed in the above protocol is with overwhelming probability globally unique.*

Proof. The assumption of the lemma implies in particular that the probability of SamSub outputting twice the same value in a ppt number of executions is negligible. Thus the collision-freeness of H yields the desired uniqueness of the session identifier. $\qquad\square$

Next, before looking at (basic) security, we note that the above protocol also offers strong entity authentication and integrity:

Proposition 1. *The protocol provides strong entity authentication according to Definition 5 and integrity according to Definition 6.*

Proof. Strong entity authentication. Consider an arbitrary instance $\Pi_i^{s_i}$ of an uncorrupted participant U_i that has accepted with session identifier $\mathsf{sid}_i^{s_i}$. Let $U_j \in \mathsf{pid}_i^{s_i}$ be some other uncorrupted participant. Instance $\Pi_i^{s_i}$ must have received a message of U_j with a signature on U_j's session identifier $\mathsf{sid}_j^{s_j}$. By unforgeability of the signature scheme, uniqueness of the session identifier $\mathsf{sid}_j^{s_j} = \mathsf{sid}_i^{s_i}$, and the collision resistance of the hash function we obtain $\mathsf{pid}_j^{s_j} = \mathsf{pid}_i^{s_i}$ with overwhelming probability.

Integrity. Consider any two instances $\Pi_i^{s_i}$ and $\Pi_j^{s_j}$ that both have accepted with $\mathsf{sid} = \mathsf{sid}_i^{s_i} = \mathsf{sid}_j^{s_j}$ and where the participants U_i and U_j are honest. By unforgeability of the signature scheme, uniqueness of the session identifier sid, and the collision resistance of the hash function, with overwhelming probability we get $\mathsf{pid}_i^{s_i} = \mathsf{pid}_j^{s_j}$ and the equivalence of the messages $m_1^{s_\ell}(U_\ell)$ they received in Round 1. Those messages include hash values $H(x_\ell^{s_\ell})$ from all protocol participants and before accepting, all participants check if the computed values $x_\ell^{s_\ell}$ in Round 2 are consistent with the $H(x_\ell^{s_\ell})$. Unless a collision of H occurs they compute the same key. $\qquad\square$

For the ease of presentation, in the proof of the basic security property we imagine the protocol without the hash value $H(x_i^{s_i})$ in Round 1. This simplification can be justified with a standard random oracle argument as in the proof of Lemma 3.

Proposition 2. *Denote the maximal number of protocol participants by $n = |\mathcal{P}|$, and let $\mathcal{A}$ be an adversary that is allowed at most $q_{\mathrm{s}}, q_{\mathrm{ex}}, q_H$ queries to the Send, Execute and random oracle H, respectively. Moreover, let $\mathsf{Adv}^{(n-1)-\mathrm{PAA}}$ resp. $\mathsf{Adv}^{\mathrm{Sig}}$ be the maximum advantage of a ppt algorithm solving $(n-1)$-PAA resp. achieving an existential forgery in running time t. Then*

$$\mathsf{Adv}_{\mathcal{A}} = |\mathsf{Succ} - 1/2| \le n \cdot (q_{\mathrm{s}} + q_{\mathrm{ex}})^n \cdot q_H \cdot \mathsf{Adv}^{(n-1)-\mathrm{PAA}} + n \cdot \mathsf{Adv}^{\mathrm{Sig}} + \mathrm{negl}(k)$$

where $\mathrm{negl}(k)$ is negligible in k.

Proof. Let $\mathsf{Succ} := (\mathsf{Adv}_{\mathcal{A}} + 1)/2$ be the success probability of adversary $\mathcal{A}$ to win the experiment. Imagine $\mathcal{A}$ now to be connected to a simulator Sim that simulates the oracles. We consider a sequence of games and bound the difference of the adversary's success probability between subsequent games.

In Game 0 the simulator Sim simulates the oracles and principals' instances faithfully. Thus, there is no difference for the adversary and denoting $\mathcal{A}$'s success probability in Game i by $\mathsf{Succ}_{\mathrm{Game}\ i}$, we have $\mathsf{Succ}_{\mathrm{Game}\ 0} = \mathsf{Succ}$.

In Game 1 the simulator will keep a list with an entry $(i, \mathsf{sid}_i^{s_i})$ for every session identifier $\mathsf{sid}_i^{s_i}$ the simulator signs with the secret key of user U_i and returns it in a Round 2 message to $\mathcal{A}$ following an Execute-query or on a Send-query. We define the event Forge to occur, if $\mathcal{A}$ comes up with a query $\mathsf{Send}(*, *, M)$ where M includes a signature $\mathsf{Sig}(\mathsf{sid}_i^{s_i})$, signed by an uncorrupted principal U_i and $(i, \mathsf{sid}_i^{s_i})$, does not appear in the simulator's list. In this case we abort the experiment and count it as success for the adversary. Thus we have:

$$|\mathsf{Succ}_{\mathrm{Game}\ 1} - \mathsf{Succ}_{\mathrm{Game}\ 0}| \le \Pr(\mathsf{Forge}).$$

Lemma 2. *If the signature scheme is existentially unforgeable, the probability of* Forge *is negligible. Formally:*

$$\Pr(\mathsf{Forge}) \le n \cdot \mathsf{Adv}^{\mathrm{Sig}}$$

Proof. The simulator can use an adversary that can reach Forge with a non-negligible probability as black box to forge a signature from the underlying signature scheme.

The simulator is given a public key PK and a signing oracle. In the initialization it will uniformly choose one user U_i and assign the key PK as PK_i to U_i. If in the following simulation Sim has to generate a signed message for U_i it will use the signing oracle to sign the message. If $\mathcal{A}$ will send a message $(*, \sigma)$, where σ is a signature of a session identifier $\mathsf{sid}_i^{s_i}$ that is not in the simulator's list, the simulator will return $(\mathsf{sid}_i^{s_i}, \sigma)$ as existential forgery. Otherwise the simulator returns $\bot$. As i was chosen uniformly the simulator will succeed with a probability of $1/n \cdot \Pr(\mathsf{Forge})$, thus $\Pr(\mathsf{Forge}) \le n \cdot \mathsf{Adv}^{\mathrm{Sig}}$.   □

In Game 2 the simulator will keep a list with entries

$$(m_1^{s_1}(U_1), \ldots, m_1^{s_n}(U_n), H(m_1^{s_1}(U_1), \ldots, m_1^{s_n}(U_n)))$$

for every computation of a session identifier invoked by an Execute-query or Send-query and all entries $(M, H(M))$ where $\mathcal{A}$ queried the random oracle directly. We define the event Collision to occur, if the simulator computes a session identifier $H(m_1^{s_1}(U_1), \ldots, m_1^{s_n}(U_n))$ which equals a session identifier that $\mathcal{A}$ obtained previously with *non-identical* messages. In this case we abort the experiment and count it as success for the adversary. From H being a random oracle, we conclude that $|\mathsf{Succ}_{\mathrm{Game}\ 2} - \mathsf{Succ}_{\mathrm{Game}\ 1}|$ is negligible.

In Game 3 the simulation of the Test oracle is modified. On a query $\mathsf{Test}(U_i, s_i)$, the simulator checks if $\Pi_i^{s_i}$ is fresh. If so, then *Sim* will not query the random oracle, but return a random value in any case. As now no information about the Test-oracle's secret bit b is given to $\mathcal{A}$ in Game 3, the success probability is $\mathsf{Succ}_{\text{Game 3}} = 1/2$.

Now we have to determine the difference in the adversary's success probability between Game 2 and Game 3. For $\mathcal{A}$, a random value and the random oracle's answer are indistinguishable as long as $\mathcal{A}$ does not know the actual query to the random oracle. The success probabilities can only differ, if $\mathcal{A}$ queries $H(x_1^{s_1}, \ldots, x_r^{s_r}, \mathsf{pid}_i^{s_i})$ to the random oracle. Denoting this event by Random, we have

$$|\mathsf{Succ}_{\text{Game 3}} - \mathsf{Succ}_{\text{Game 2}}| \leq \Pr(\mathsf{Random}).$$

Lemma 3. *The probability* $\Pr(\mathsf{Random})$ *of the event* Random *to occur is negligible if* n *is constant and* $\mathsf{Adv}^{(n-1)-\text{PAA}}$ *is negligible.*

Proof. The simulator is given an instance $(S, (\phi_i(S), \phi_i(x))_{1 \leq i \leq n-1})$ of the $(n-1)$-PAA problem. In the initialization phase, the simulator will give S as parameter to $\mathcal{A}$ and uniformly choose n random numbers $\alpha_i \in \{1, q_s + q_{\text{ex}}\}$ $(i = 1, \ldots, n)$ to point to the instances $\Pi_i^{\alpha_i}$. The simulator will choose uniformly at random $\beta \in \{1, \ldots, n\}$ to select one distinguished instance $\Pi_\beta^{\alpha_\beta}$ among them.

When the simulator has to process Round 1 for one instance $\Pi_i^{\alpha_i}, i = 1, \ldots, n$, $i \neq \beta$, *Sim* will use the given $\phi_i(S)$ instead of computing a new ϕ_i with SamAut. For instance $\Pi_\beta^{\alpha_\beta}$ the simulator will use the given $x(S)$. If instance $\Pi_\beta^{\alpha_\beta}$ does not only get messages containing $\phi_i(S), (i = 1, \ldots, n, i \neq \beta)$ the simulator aborts and outputs $\perp$. Also, if the simulator ever has to apply a ϕ_i^{-1} it aborts and outputs $\perp$ (this will only happen from a Reveal-query).

Because $\Pi_\beta^{\alpha_\beta}$ is uniformly selected out of a set of $n \cdot (q_s + q_{\text{ex}})$ potential instances, it will be used in the Test-query with a probability of $(n \cdot (q_s + q_{\text{ex}}))^{-1}$. To be able to apply the Test-query the adversary has to let $\Pi_\beta^{\alpha_\beta}$ accept. All $U_i \in \mathsf{pid}_\beta^{\alpha_\beta}$ have to be uncorrupted. Then by uniqueness of the session identifier (Lemma 1) the messages $\Pi_\beta^{\alpha_\beta}$ must have got in Round 1 were generated by the same instances as the messages $\Pi_\beta^{\alpha_\beta}$ received in Round 2. These have to be the distinguished oracles $\Pi_i^{\alpha_i}$ for $U_i \in \mathsf{pid}_\beta^{\alpha_\beta}$. For the principals $U_j \notin \mathsf{pid}_\beta^{\alpha_\beta}$, $\Pi_j^{\alpha_j}$ must be an oracle that was not revealed. There must be at least one potential instance that is not used for each $U_j \notin \mathsf{pid}_\beta^{\alpha_\beta}$. Consequently, with a probability of $1/n \cdot (q_s + q_{\text{ex}})^n$ the principals are distributed as needed.

If $\mathcal{A}$ halts, the simulator chooses uniformly one of the at most q_H queries to the random oracle, extracts x_β (assuming it is of the form $H(x_1^{s_1}, \ldots, x_r^{s_r}, \mathsf{pid}_i^{s_i})$) and answers this to the $(n-1)$-PAA challenge. The probability to pick the correct query is $1/q_H \cdot \Pr(\mathsf{Random})$. With a probability of at least

$$\Pr((n-1)-\mathsf{PAAsolved}) \geq \frac{1}{n \cdot (q_s + q_{\text{ex}})^n \cdot q_H} \cdot \Pr(\mathsf{Random})$$

the simulator solves the challenge. □

Putting it all together we see that $\mathsf{Adv}_{\mathcal{A}} = |\mathsf{Succ} - 1/2|$ is smaller or equal than

$$n \cdot (q_{\mathrm{s}} + q_{\mathrm{ex}})^n \cdot q_H \cdot \mathsf{Adv}^{(n-1)-\mathrm{PAA}} + n \cdot \mathsf{Adv}^{\mathrm{Sig}} + \mathrm{negl}(k) \quad .$$

$\square$

Remark 2. If instead of a constant number n of potential protocol participants, we want to allow a size $\mathcal{P}$ of polynomial size, the bound in Proposition 2 is in general no longer negligible. However, if we base on a CDH-assumption as in Example 1, we can allow for a set $\mathcal{P}$ of polynomial size: With the argument given in Example 1 we see that in this case solving 1-PAA is equivalent to solving r-PAA for an arbitrary r of polynomial size. In the security proof, this reduction allows us to replace the exponent n by the constant 2.

Also, an r-PAA that is secure even if the adversary can choose the r public keys out of a polynomial sized set will help. Because the definition of an mKEM takes such a choice into account, the protocol allows a polynomial sized set of users, if it bases on such an mKEM.

Finally, forward secrecy follows with the standard argument that the long-term keys are used for message authentication exclusively, and we obtain:

Proposition 3. *The protocol in Figure 1 fulfills forward secrecy in the sense of Definition 4.*

4   Conclusion

In this contribution we have described a 2-round group key agreement and showed it to be secure under the assumption that certain group-theoretical tools are available. In addition to the "standard" security requirement, the proposed protocol also offers strong entity authentication and integrity. While our framework is primarily geared towards building a provably secure group key agreement on non-abelian groups, it also allows to derive a 2-round group key agreement from a CDH assumption.

Acknowledgment

We are indebted to Dennis Hofheinz and Jonathan Katz for valuable discussions and comments.

References

1. Iris Anshel, Michael Anshel, Benji Fisher, and Dorian Goldfeld. New Key Agreement Protocols in Braid Group Cryptography. In David Naccache, editor, *Topics in Cryptology, Proceedings of CT-RSA 2001*, number 2020 in Lecture Notes in Computer Science, pages 13–27. Springer-Verlag, 2001.

2. Iris Anshel, Michael Anshel, and Dorian Goldfeld. An Algebraic Method for Public-Key Cryptography. *Mathematical Research Letters*, 6:287–291, 1999.

3. Mihir Bellare, Ran Canetti, and Hugo Krawczyk. A Modular Approach to the Design and Analysis of Authentication and Key Exchange Protocols. In *Proceedings of the 30th Annual ACM Symposium on Theory of Computing STOC*, pages 319–428, 1998.

4. Mihir Bellare, David Pointcheval, and Phillip Rogaway. Authenticated Key Exchange Secure Against Dictionary Attacks. In Bart Preneel, editor, *Advances in Cryptology – EUROCRYPT 2000*, volume 1807 of *Lecture Notes in Computer Science*, pages 139–155. Springer, 2000.

5. Mihir Bellare and Phillip Rogaway. Entitiy Authentication and Key Distribution. In Douglas R. Stinson, editor, *Advances in Cryptology – CRYPTO '93*, volume 773 of *Lecture Notes in Computer Science*, pages 232–249. Springer, 1994.

6. Jens-Matthias Bohli. A Framework for Robust Group Key Agreement. In Marina L. Gavrilova, Osvaldo Gervasi, Vipin Kumar, Chih Jeng Kenneth Tan, Antonio Laganà David Taniar, Youngsong Mun, and Hyunseung Choo, editors, *Computational Science and Its Applications – ICCSA 2006*, volume 3982 of *Lecture Notes in Computer Science*, pages 355–364. Springer, 2006.

7. Jens-Matthias Bohli, María Isabel González Vasco, and Rainer Steinwandt. Secure Group Key Establishment Revisited. Cryptology ePrint Archive, Report 2005/395, 2005. `http://eprint.iacr.org/2005/395/`.

8. Dan Boneh and Alice Silverberg. Applications of Multilinear Forms to Cryptography. *Contemporary Mathematics*, 324:71–90, 2003.

9. Colin Boyd and Juan Manuel González Nieto. Round-Optimal Contributory Conference Key Agreement. In Yvo Desmedt, editor, *Public Key Cryptography, Proceedings of PKC 2003*, volume 2567 of *Lecture Notes in Computer Science*, pages 161–174. Springer, 2002.

10. Colin Boyd and Anish Mathuria. *Protocols for Authentication and Key Establishment*. Information Security and Cryptography; Texts and Monographs. Springer, 2003.

11. Emmanuel Bresson, Olivier Chevassut, David Pointcheval, and Jean-Jacques Quisquater. Provably Authenticated Group Diffie-Hellman Key Exchange. In Pierangela Samarati, editor, *Proceedings of the 8th ACM Conference on Computer and Communications Security*, pages 255–264. ACM Press, 2001.

12. Ran Canetti and Hugo Krawczyk. Analysis of Key-Exchange Protocols and Their Use for Building Secure Channels. In *Advances in Cryptology – EUROCRYPT 2001*, volume 2045 of *Lecture Notes in Computer Science*, pages 453–474. Springer, 2001.

13. Dario Catalano, David Pointcheval, and Thomas Pornin. IPAKE: Isomorphisms for Password-based Authenticated Key Exchange. In Matthew K. Franklin, editor, *Advances in Cryptology – CRYPTO 2004*, volume 3152 of *Lecture Notes in Computer Science*, pages 477–493. Springer, 2004.

14. Jung Hee Cheon and Byungheup Jun. A Polynomial Time Algorithm for the Braid Diffie-Hellman Conjugacy Problem. In Dan Boneh, editor, *Advances in Cryptology – CRYPTO 2003*, volume 2729 of *Lecture Notes in Computer Science*, pages 212–225. Springer, 2003.

15. Patrick Dehornoy. Braid-based cryptography. In Alexei G. Myasnikov, editor, *Group Theory, Statistics, and Cryptography*, number 360 in Contemporary Mathematics, pages 5–33. ACM Press, 2004. Online available at `http://www.math.unicaen.fr/ dehornoy/Surveys/Dgw.ps`.

16. David Gerber, Shmuel Kaplan, Mina Teicher, Boaz Tsaban, and Uzi Vishne. Probabilistic solutions of equations in the braid group. *Advances in Applied Mathematics*, 35(3):323–334, 2005.
17. María Isabel González Vasco, Consuelo Martínez, Rainer Steinwandt, and Jorge L. Villar. A new Cramer-Shoup like methodology for group based provably secure schemes. In Joe Kilian, editor, *Proceedings of the 2nd Theory of Cryptography Conference TCC 2005*, volume 3378 of *Lecture Notes in Computer Science*, pages 495–509. Springer, 2005.
18. Dimitri Grigoriev and Ilia Ponomarenko. Constructions in public-key cryptography over matrix groups. arXiv preprint, 2005. Online available at http://arxiv.org/abs/math.GR/0506180.
19. Dennis Hofheinz and Rainer Steinwandt. A Practical Attack on Some Braid Group Based Cryptographic Primitives. In Yvo Desmedt, editor, *Public Key Cryptography, Proceedings of PKC 2003*, number 2567 in Lecture Notes in Computer Science, pages 187–198. Springer-Verlag, 2002.
20. Jonathan Katz and Ji Sun Shin. Modeling Insider Attacks on Group Key-Exchange Protocols. In *12th ACM Conference on Computer and Communications Security*, pages 180–189. ACM Press, 2005.
21. Jonathan Katz and Moti Yung. Scalable Protocols for Authenticated Group Key Exchange. In Dan Boneh, editor, *Advances in Cryptology – CRYPTO 2003*, volume 2729 of *Lecture Notes in Computer Science*, pages 110–125. Springer, 2003.
22. Ki Hyoung Ko, Sang Jin Lee, Jung Hee Cheon, Jae Woo Han, Ju sung Kang, and Choonsik Park. New Public-Key Cryptosystem Using Braid Groups. In Mihir Bellare, editor, *Advances in Cryptology – CRYPTO 2000*, volume 1880 of *Lecture Notes in Computer Science*, pages 166–183. Springer, 2000.
23. Ho-Kyu Lee, Hyang-Sook Lee, and Young-Ran Lee. Cryptology ePrint Archive: Report 2003/018, 2003. http://eprint.iacr.org/2003/018.
24. Victor Shoup. On Formal Models for Secure Key Exchange (version 4). Revision of IBM Research Report RZ 3120 (April 1999), November 1999. Online available at http://www.shoup.net/papers/skey.pdf.
25. Vladimir Shpilrain and Alexander Ushakov. A new key exchange protocol based on the decomposition problem. Cryptology ePrint Archive: Report 2005/447, 2005. http://eprint.iacr.org/2005/447.
26. Vladimir Shpilrain and Alexander Ushakov. Thompson's Group and Public Key Cryptography. In John Ioannidis, Angelos Keromytis, and Moti Yung, editors, *Applied Cryptography and Network Security: Third International Conference, ACNS 2005*, volume 3531 of *Lecture Notes in Computer Science*, pages 151–163, 2005.
27. Vladimir Shpilrain and Alexander Ushakov. The conjugacy search problem in public key cryptography: unnecessary and insufficient. *Applicable Algebra in Engineering, Communication and Computing*, to appear. Online available at http://www.sci.ccny.cuny.edu/~shpil/csp.pdf.
28. Vladimir Shpilrain and Gabriel Zapata. Combinatorial group theory and public key cryptography. *Applicable Algebra in Engineering, Communication and Computing*, to appear. Online available at http://www.sci.ccny.cuny.edu/~shpil/pkc.pdf.
29. Nigel Smart. Efficient Key Encapsulation to Multiple Parties. In Carlo Blundo and Stelvio Cimato, editors, *Security in Communication Networks – SCN 2004*, volume 3352 of *Lecture Notes in Computer Science*, pages 208–219. Springer, 2005.

Universally Composable
Identity-Based Encryption

Ryo Nishimaki[1], Yoshifumi Manabe[1,2], and Tatsuaki Okamoto[1,2]

[1] Graduate School of Infomatics, Kyoto University,
Yoshida-honmachi, Kyoto, 606-8501 Japan
nisimaki@ai.soc.i.kyoto-u.ac.jp
[2] NTT Laboratories, Nippon Telegraph and Telephone Corporation,
1-1 Hikari-no-oka, Yokosuka, 239-0847 Japan
{manabe.yoshifumi, okamoto.tatsuaki}@lab.ntt.co.jp

Abstract. Identity-based encryption (IBE) is one of the most important primitives in cryptography, and various security notions of IBE (e.g., IND-ID-CCA2, NM-ID-CCA2, IND-sID-CPA etc.) have been introduced. The relations among them have been clarified recently. This paper, for the first time, investigates the security of IBE in the universally composable (UC) framework. This paper first defines the UC-security of IBE, i.e., we define the ideal functionality of IBE, $\mathcal{F}_{\text{IBE}}$. We then show that UC-secure IBE is equivalent to conventionally-secure (IND-ID-CCA2-secure) IBE.

Keywords: identity-based encryption, IND-ID-CCA2, universal composition.

1 Introduction

1.1 Background

The concept of identity-based encryption (IBE), introduced by Shamir [22], is a variant of public-key encryption (PKE), where the identity of a user is employed in place of the user's public-key.

Boneh and Franklin [6] defined IND-ID-CCA2 (indistinguishability against adaptive chosen-ciphertext attacks under chosen identity attacks) as the desirable security of IBE schemes. Canetti, Halevi, and Katz [11,12] defined a weaker notion of security in which the adversary, ahead of time, commits to the challenge identity it will attack. We refer to this notion as *selective identity* (sID) adaptive chosen-ciphertext secure IBE (IND-sID-CCA2). In addition, they also defined a weaker security notion of IBE, selective-identity *chosen-plaintext* (CPA) secure IBE (IND-sID-CPA). Attrapadung, Cui, Galindo, Hanaoka, Hasuo, Imai, Matsuura, Yang and Zhang [1] introduced *non-malleability* (NM) and *semantic security* (SS) to the set of security notions of IBE. Thus, the security definitions considered up to now in the literature are: G-A1-A2, where G ∈ {IND, NM, SS}, A1 ∈ {ID, sID}, ID denotes chosen identity attacks, and A2 ∈ {CPA, CCA1, CCA2}.

P.Q. Nguyen (Ed.): VIETCRYPT 2006, LNCS 4341, pp. 337–353, 2006.
© Springer-Verlag Berlin Heidelberg 2006

Attrapadung, Cui, Galindo, Hanaoka, Hasuo, Imai, Matsuura, Yang and Zhang [1] clarified the relationship among these notions, and showed that IND-ID-CCA2 is equivalent to the strongest security notion among them, NM-ID-CCA2.

Since Canetti introduced universal composability (UC) as a framework for analyzing the security of cryptographic primitives/protocols [8], investigating the relation between UC-secure primitives/protocols and conventionally-secure primitives/protocols has been a significant topic in cryptography [2,3,9,10,13,14,17,21]. Since UC represents stronger security requirements, a lot of conventionally-secure protocols fail to meet UC security requirements. For example, we cannot design secure two party protocols in the UC framework with no setup assumption [8,10,15,16,18,19], while there are conventionally-secure two party protocols (e.g., commitment and zero-knowledge proofs) with no setup assumption.

We do know, however, that the conventional security notions are equivalent to UC security notions for a few cryptographic primitives. For example, UC-secure PKE is equivalent to conventionally-secure (IND-CCA2-secure) PKE [8], UC-secure signatures are equivalent to conventionally-secure (existentially unforgeable against chosen message attacks: EUF-CMA-secure) signatures [9] and UC-secure Key Encapsulation Mechanism (KEM) is equivalent to conventionally-secure (IND-CCA2-KEM-secure) KEM [20].

IBE is a more complex cryptographic primitive than PKE or signatures, so it is not clear whether conventionally-secure (i.e., IND-ID-CCA2-secure) IBE is equivalent to UC-secure IBE or not. Since IBE is one of the most significant primitives [4,5,6,7,23] like PKE and signatures in cryptography, it is important to clarify the relationship between UC security and the conventional security notions of IBE. The UC security of IBE, however, has not been investigated.

That is, we have the following problems:

1. What is the security definition of IBE in the UC framework (i.e., how to define an ideal functionality of IBE)?
2. Is UC-secure IBE equivalent to IND-ID-CCA2-secure IBE?

1.2 Our Results

This paper answers the above problems:

1. This paper defines the UC-security of IBE, i.e., we define the ideal functionality of IBE, $\mathcal{F}_{\text{IBE}}$.
2. We show that UC-secure IBE is equivalent to conventionally-secure (IND-ID-CCA2-secure) IBE.

2 Preliminaries

2.1 Notations

We describe probabilistic algorithms and experiments using standard notations and conventions. For probabilistic algorithm A, $A(x_1, x_2, ...; r)$ denotes the random variable of A's output on inputs $x_1, x_2, ...$ and coins r. We let $y \xleftarrow{\text{R}} A(x_1, x_2, ...)$

denote that y is randomly selected from $A(x_1, x_2, ...; r)$ according to its distribution. If S is a finite set, then $x \overset{U}{\leftarrow} S$ denotes that x is uniformly selected from S. If α is neither an algorithm nor a set, then $x \leftarrow \alpha$ indicates that we assign α to x.

We say that function $f : \mathbb{N} \to \mathbb{R}$ is negligible in security parameter k if for every constant $c \in \mathbb{N}$, there exists $k_c \in \mathbb{N}$ such that $f(k) < k^{-c}$ for any $k > k_c$. Hereafter, we often use $f < \epsilon(k)$ to mean that f is negligible in k. On the other hand, we use $f > \nu(k)$ to mean that f is non-negligible in k. i.e., function $f : \mathbb{N} \to \mathbb{R}$ is non-negligible in k, if there exists a constant $c \in \mathbb{N}$ such that for every $k_c \in \mathbb{N}$, there exists $k > k_c$ such that $f(k) > k^{-c}$. A distribution ensemble $X = \{X(k, z)\}_{k \in \mathbb{N}, z \in \{0,1\}^*}$ is an infinite set of probability distributions, where a distribution $X(k, z)$ is associated with each $k \in \mathbb{N}$ and $z \in \{0,1\}^*$. The ensembles considered in this paper describe outputs of computations where the parameter z represents input, and the parameter k is taken to be the security parameter. Two *binary* distribution ensembles X and Y are statistically indistinguishable (written $X \approx Y$) if for any $c, d \in \mathbb{N}$ there exists $k_c \in \mathbb{N}$ such that for any $k > k_c$ and any $z \in \cup_{\kappa \leq k^d} \{0,1\}^\kappa$ we have:

$$| \Pr[X(k, z) = 1] - \Pr[Y(k, z) = 1]| < k^{-c}$$

2.2 Identity-Based Encryption

Identity-based encryption scheme Σ is specified by four algorithms: $\mathcal{S}$, $\mathcal{X}$, $\mathcal{E}$, $\mathcal{D}$:

Setup: $\mathcal{S}$ takes security parameter k and returns PK (system parameters) and MK (master-key). The system parameters include a description of a finite message space $\mathcal{M}$, and a description of a finite ciphertext space $\mathcal{C}$. Intuitively, the system parameters should be publicly known, while MK is known only by the setup party.

Extract: $\mathcal{X}$ takes as input PK, MK, and an arbitrary $ID \in \{0,1\}^*$, and returns private key dk. Here ID is an arbitrary string that will be used as a public key, and dk is the corresponding private decryption key. The extract algorithm extracts a private key from the given public key.

Encrypt: $\mathcal{E}$ takes as input PK, ID, and $m \in \mathcal{M}$. It returns ciphertext $c \in \mathcal{C}$.

Decrypt: $\mathcal{D}$ takes as input PK, $c \in \mathcal{C}$, and private key dk. It returns $m \in \mathcal{M}$.

These algorithms must satisfy the standard consistency constraint, namely,
$$\forall m \in \mathcal{M} : \mathcal{D}(PK, c, dk) = m \text{ where } c = \mathcal{E}(PK, ID, m), dk = \mathcal{X}(PK, MK, ID)$$

2.3 Definitions of Security Notions for IBE Schemes

Let $\mathcal{A} = (\mathcal{A}_1, \mathcal{A}_2)$ be an adversary; we say $\mathcal{A}$ is polynomial time if both probabilistic algorithms $\mathcal{A}_1$ and $\mathcal{A}_2$ are polynomial time. In the first stage, given the system parameters, the adversary computes and outputs challenge template τ. $\mathcal{A}_1$ can output some information, s, which will be transferred to $\mathcal{A}_2$. In the second stage, the adversary is challenged with target ciphertext c^* generated from τ by a probabilistic function, in a manner depending on the goal. We say adversary $\mathcal{A}$ successfully breaks the scheme if she achieves her goal. We consider a security goal, IND [1], and three attack models, ID-CPA, ID-CCA, ID-CCA2, listed in order of increasing strength. The difference among the models is whether or not $\mathcal{A}_1$ or $\mathcal{A}_2$ is granted access to decryption oracles.

We describe in Table 1 and Table 2 the ability with which the adversary can, in the different attack models, access the Extraction Oracle $\mathcal{X}(PK, MK, \cdot)$, the Encryption Oracle $\mathcal{E}(PK, ID, \cdot)$ and the Decryption Oracle $\mathcal{D}(PK, \cdot, dk)$. When we say $\mathcal{O}_i = \{\mathcal{X}\mathcal{O}_i, \mathcal{E}\mathcal{O}_i, \mathcal{D}\mathcal{O}_i\} = \{\mathcal{X}(PK, MK, \cdot), \mathcal{E}(PK, ID, \cdot), \perp\}$, where $i \in \{1, 2\}$, we mean that no decryption oracle can be used.

Table 1. Oracle Set $\mathcal{O}_1$ in the Definitions of the Notions for IBE

	$\mathcal{O}_1 = \{\mathcal{X}\mathcal{O}_1, \mathcal{E}\mathcal{O}_1, \mathcal{D}\mathcal{O}_1\}$
ID-CPA	$\{\mathcal{X}(PK, MK, \cdot), \mathcal{E}(PK, ID, \cdot), \perp\}$
ID-CCA	$\{\mathcal{X}(PK, MK, \cdot), \mathcal{E}(PK, ID, \cdot), \mathcal{D}(PK, \cdot, dk)\}$
ID-CCA2	$\{\mathcal{X}(PK, MK, \cdot), \mathcal{E}(PK, ID, \cdot), \mathcal{D}(PK, \cdot, dk)\}$

Table 2. Oracle Set $\mathcal{O}_2$ in the Definitions of the Notions for IBE

	$\mathcal{O}_2 = \{\mathcal{X}\mathcal{O}_2, \mathcal{E}\mathcal{O}_2, \mathcal{D}\mathcal{O}_2\}$
ID-CPA	$\{\mathcal{X}(PK, MK, \cdot), \mathcal{E}(PK, ID, \cdot), \perp\}$
ID-CCA	$\{\mathcal{X}(PK, MK, \cdot), \mathcal{E}(PK, ID, \cdot), \perp\}$
ID-CCA2	$\{\mathcal{X}(PK, MK, \cdot), \mathcal{E}(PK, ID, \cdot), \mathcal{D}(PK, \cdot, dk)\}$

Let $\Sigma = (\mathcal{S}, \mathcal{X}, \mathcal{E}, \mathcal{D})$ be an identity based encryption scheme and let $\mathcal{A} = (\mathcal{A}_1, \mathcal{A}_2)$ be an adversary. For atk $\in \{\text{id-cpa, id-cca, id-cca2}\}$ and $k \in N$ let,

$$\mathbf{Adv}_{\Sigma,\mathcal{A}}^{\text{ind-atk}}(k) = \Pr[\mathbf{Exp}_{\Sigma,\mathcal{A}}^{\text{ind-atk-1}}(k) = 1] - \Pr[\mathbf{Exp}_{\Sigma,\mathcal{A}}^{\text{ind-atk-0}}(k) = 1]$$

where for $b, d \in \{0, 1\}$ and $|m_0| = |m_1|$,

$$\text{Experiment } \mathbf{Exp}_{\Sigma,\mathcal{A}}^{\text{ind-atk-b}}(k)$$
$$(PK, MK) \xleftarrow{\text{R}} \mathcal{S}(k);$$
$$(m_0, m_1, s, ID) \xleftarrow{\text{R}} \mathcal{A}_1^{\mathcal{O}_1}(PK);$$
$$c^* \xleftarrow{\text{R}} \mathcal{E}(PK, ID, m_b);$$
$$d \xleftarrow{\text{R}} \mathcal{A}_2^{\mathcal{O}_2}(m_0, m_1, s, c^*, ID);$$
$$\text{return } d$$

We say that Σ is secure in the sense of IND-ATK, if $\mathbf{Adv}_{\Sigma,\mathcal{A}}^{\text{ind-atk}}(k)$ is negligible for any $\mathcal{A}$.

2.4 Universal Composability

The universally composable security framework allows the security properties of cryptographic tasks to be defined such that security is maintained under a general composition with an unbounded number of instances of arbitrary protocols running concurrently. Security in this framework is called universally composable (UC) security. Informally, we describe this framework as follows: (See [8] for more details.)

We consider the real life world, the ideal process world, and environment $\mathcal{Z}$ that tries to distinguish these two worlds.

The real life world. In this world, there are adversary $\mathcal{A}$ and protocol π which realizes a functionality among some parties. Let $\text{REAL}_{\pi,\mathcal{A},\mathcal{Z}}(k, z, \mathbf{r})$ denote the output of environment $\mathcal{Z}$ when interacting with adversary $\mathcal{A}$ and parties $P_1, ..., P_n$ running protocol π (hereafter denoted as $(\mathcal{A}, \pi)$) on security parameter k, auxiliary input z and random input $\mathbf{r} = (r_{\mathcal{Z}}, r_{\mathcal{A}}, r_1, ..., r_n)$ (z and $r_{\mathcal{Z}}$ for $\mathcal{Z}$, $r_{\mathcal{A}}$ for $\mathcal{A}$, r_i for party P_i). Let $\text{REAL}_{\pi,\mathcal{A},\mathcal{Z}}(k, z)$ denote the random variable describing $\text{REAL}_{\pi,\mathcal{A},\mathcal{Z}}(k, z, \mathbf{r})$ when $\mathbf{r}$ is uniformly chosen.

The ideal process world. In this world, there are a simulator $\mathcal{S}$ that simulates the real life world, an ideal functionality $\mathcal{F}$, and dummy parties. Let $\text{IDEAL}_{\mathcal{F},\mathcal{S},\mathcal{Z}}(k, z, \mathbf{r})$ denote the output of environment $\mathcal{Z}$ when interacting with adversary $\mathcal{S}$ and ideal functionality $\mathcal{F}$ (hereafter denoted as $(\mathcal{S}, \mathcal{F})$) on security parameter k, auxiliary input z and random input $\mathbf{r} = (r_{\mathcal{Z}}, r_{\mathcal{S}}, r_{\mathcal{F}})$ (z and $r_{\mathcal{Z}}$ for $\mathcal{Z}$, $r_{\mathcal{S}}$ for $\mathcal{S}$, $r_{\mathcal{F}}$ for party $\mathcal{F}$). Let $\text{IDEAL}_{\mathcal{F},\mathcal{S},\mathcal{Z}}(k, z)$ denote the random variable describing $\text{IDEAL}_{\mathcal{F},\mathcal{S},\mathcal{Z}}(k, z, \mathbf{r})$ when $\mathbf{r}$ is uniformly chosen.

Let $\mathcal{F}$ be an ideal functionality and let π be a protocol. We say that π UC-realizes $\mathcal{F}$, if for any adversary $\mathcal{A}$, there exists simulator $\mathcal{S}$, such that for any environment $\mathcal{Z}$ we have: (See Section 2.1 for $\approx$.)

$$\text{IDEAL}_{\mathcal{F},\mathcal{S},\mathcal{Z}} \approx \text{REAL}_{\pi,\mathcal{A},\mathcal{Z}}$$

where $\mathcal{A}$, $\mathcal{S}$ and $\mathcal{Z}$ are probabilistic polynomial-time interactive Turing machines.

3   UC-Secure IBE Is Equivalent to IND-ID-CCA2-Secure IBE

3.1   The Identity-Based Encryption Functionality $\mathcal{F}_{\text{IBE}}$

We define IBE functionality $\mathcal{F}_{\text{IBE}}$ in Fig.1. Our definition of $\mathcal{F}_{\text{IBE}}$ follows the one for $\mathcal{F}_{\text{PKE}}$ of regular public-key encryption schemes given by Canetti [8]. The idea of $\mathcal{F}_{\text{IBE}}$ is to allow parties to obtain idealized (information theoretically secure) ciphertexts for messages by using their IDs, such that private keys do not appear in the interface, but at the same time the designated decryptor can retrieve the plaintexts. There may be multiple designated decryptors who let the setup party extract their private keys from their IDs.

$\mathcal{F}_{\text{IBE}}$ should be defined as follows: If no party is corrupted, setup, extract, encrypt and decrypt are information theoretically securely executed. $\mathcal{F}_{\text{IBE}}$ realizes such idealized setup, extract, encrypt and decrypt by recording IDs, ciphertexts and plaintexts. (i.e., $\mathcal{F}_{\text{IBE}}$ plays the role of the centralized database of encrypted messages and the corresponding ciphertexts and IDs used to encrypt.) $\mathcal{F}_{\text{IBE}}$ is

---

Functionality $\mathcal{F}_{IBE}$

$\mathcal{F}_{IBE}$ proceeds as follows, given domain $\mathcal{M}$ of plaintexts and domain N of ID. Let $\mu \in \mathcal{M}$ be a fixed message.

Setup

In the first activation, expect to receive a value (**Setup**, sid, T) from some party T. Then do:

1. Hand (**Setup**, sid, T) to adversary $\mathcal{S}$.
2. Upon receiving value (**Algorithms**, sid, x, e, d) from the adversary, where x, e, d are descriptions of PPT ITMs, output (**Encryption Algorithm**, sid, e) to T.
3. Record (T, x, e, d).

Extract

Upon receiving value (**Extract**, sid, ID, D, e') from party T, proceed as follows:

1. If $ID \notin N$ or (ID, P) is recorded in ID-Reg for some $P(\neq D)$, then output an error message to T.
2. If e' is not recorded, ignore the request. Else, record (ID, D) in ID-Reg and output (**Extracted**, sid, D) to T.

Encrypt

Upon receiving value (**Encrypt**, sid, m, ID, e') from some party E, proceed as follows:

1. If $m \notin \mathcal{M}$ or $ID \notin N$ then output an error message to E. Else, if $e' = e$, the setup party is uncorrupted, (ID, P) is recorded in ID-Reg for some P and decryptor P is uncorrupted, then let $c = e'(ID, \mu)$ and record (m, c, ID) in Plain-Cipher. Else, let $c = e'(ID, m)$.
2. Output (**Ciphertext**, sid, c) to E

Decrypt

Upon receiving value (**Decrypt**, sid, c, ID) from D, proceed as follows:

1. If the following two conditions are satisfied then hand (**Plaintext**, sid, m) to D.
   (a) (ID, D) is recorded in ID-Reg.
   (b) (m, c, ID) is stored in Plain-Cipher.
2. If (ID, D) is not recorded in ID-Reg then hand **not-recorded** to D.
3. Otherwise, return (**Plaintext**, $sid, d(c, x(ID))$) to D.

Fig. 1. The ideal identity-based encryption functionality, $\mathcal{F}_{IBE}$

written in a way that can be realized by protocols that have only local operations (setup, extract, encrypt, decrypt). All communication is left to the protocols that call $\mathcal{F}_{IBE}$. The important difference between PKE and IBE is that IBE schemes have the extract algorithm. Users need to extract private keys corresponding to their IDs to decrypt ciphertexts. They cannot locally generate private keys. The setup party generates user's private keys. $\mathcal{F}_{\text{IBE}}$ takes four types of input: setup, extract, encrypt, and decrypt.

Upon receiving a setup request from party T (the setup party), $\mathcal{F}_{\text{IBE}}$ asks the adversary to provide three descriptions of PPT algorithms: Extract algorithm x, encryption algorithm e, and decryption algorithm d. (Note that x, e, and d can be probabilistic.) It then outputs to T the description of encryption algorithm e. While the encryption algorithm is public and given to the environment (via T), the extract algorithm and decryption algorithm do not appear in the interface between $\mathcal{F}_{\text{IBE}}$ and T. Encryption algorithm e also plays the role of system parameters.

Upon receiving a request from setup party T to extract a private key with an ID, D (party ID) and encryption algorithm e' (as system parameters), $\mathcal{F}_{\text{IBE}}$ proceeds as follows. If ID is not in domain N or (ID, P) is recorded in ID-Reg for some P, then $\mathcal{F}_{\text{IBE}}$ outputs an error message to T. If e' is not recorded, $\mathcal{F}_{\text{IBE}}$ ignores the request. Else, $\mathcal{F}_{\text{IBE}}$ records pair (ID, D) in ID-Reg and outputs message "extracted" to T. (Notice that one party may extract multiple private keys.) $\mathcal{F}_{\text{IBE}}$ only records the correspondence between parties and IDs. $\mathcal{Z}$ may obtain private keys of some parties only when $\mathcal{Z}$ corrupts them. Thus $\mathcal{F}_{\text{IBE}}$ need not output private keys in the interface of extract.

Upon receiving a request from some arbitrary party E to encrypt message m with ID and encryption algorithm e', $\mathcal{F}_{\text{IBE}}$ proceeds as follows. If m is not in domain $\mathcal{M}$ or ID is not in domain N, $\mathcal{F}_{\text{IBE}}$ outputs an error message to E. Else, $\mathcal{F}_{\text{IBE}}$ outputs formal ciphertext c to E, where c is computed as follows. If $e' = e$, the setup party is uncorrupted, (ID, P) is recorded in ID-Reg for some P and decryptor P is uncorrupted, then $c = e(ID, \mu)$, where $\mu \in \mathcal{M}$ is some fixed message. In this case, (m, c, ID) is recorded for future decryption. Else, $c = e'(ID, m)$. In this case, no secrecy is guaranteed, since c may depend on m in arbitrary ways. Notice that if $\mathcal{F}_{\text{IBE}}$ receives (Encrypt, sid, m, ID, e) before receiving (Extract, sid, ID, P_h, e) for any party P_h, then c is not information theoretically secure, ($c = e(ID, m)$) even if (Extract, sid, ID, P_h, e) arrives later for uncorrupted party P_h. However, this does not influence the UC security of IBE, because $c = e(ID, m)$ in the ideal world is the same as $c = e(ID, m)$ in the real world. (i.e., $\mathcal{Z}$ cannot distinguish two worlds.)

Upon receiving a request from party D to decrypt ciphertext c encrypted for ID ID, $\mathcal{F}_{\text{IBE}}$ first checks if there are records (ID, D) in ID-Reg (i.e., D is the decryptor) and (m, c, ID) in Plain-Cipher for some m. If so, then it returns m as the decrypted value. This guarantees perfectly correct decryption for messages that were encrypted via this instance of $\mathcal{F}_{\text{IBE}}$. If (ID, D) is not recorded in ID-Reg, this means that D has not extracted a private key for ID. Accordingly, $\mathcal{F}_{\text{IBE}}$ returns an error message. If no (m, c, ID) record exists for any m, this means that c was not generated legitimately via this instance of $\mathcal{F}_{\text{IBE}}$, so no correctness guarantee is provided, and $\mathcal{F}_{\text{IBE}}$ returns the value $d(c, x(ID))$. In IBE, multiple users may extract their private keys from a single master key, so single instance of $\mathcal{F}_{\text{IBE}}$ should deal with multiple decryptors.

3.2 UC-Secure IBE Is Equivalent to IND-ID-CCA2-Secure IBE

Next, we present a protocol that UC-realizes $\mathcal{F}_{\text{IBE}}$.

Let $\Sigma = (\mathcal{S}, \mathcal{X}, \mathcal{E}, \mathcal{D})$ be an identity based encryption scheme. We define protocol π_Σ that is constructed from Σ and has the same interface with the environment as $\mathcal{F}_{\text{IBE}}$.

protocol π_Σ

Setup: Upon input (Setup, sid, T) within some setup party T, T obtains the system parameters PK and master-key MK by running algorithm $\mathcal{S}()$ and sets $x = \mathcal{X}(PK, MK, \cdot)$, $e = \mathcal{E}(PK, \cdot, \cdot)$, $d = \mathcal{D}(PK, \cdot, \cdot)$. It then outputs (Encryption Algorithm, sid, e).

Extract: Upon input (Extract, sid, ID, D, e') within setup party T, if $ID \notin N$ or T has already obtained $dk_{ID} = x(ID)$, then T outputs an error message. Else if $e' \neq e$, T ignores the request. Else, T obtains private key $dk_{ID} = x(ID)$. T outputs (Extracted, sid, D) and pair (ID, dk_{ID}) is immediately transferred to D. (See the remark below for how to transfer.)

Encrypt: Upon input (Encrypt, sid, m, ID, e') within some party E, if $m \notin \mathcal{M}$ or $ID \notin N$, E outputs an error message. Else, E obtains ciphertext $c = e'(ID, m)$ and outputs (Ciphertext, sid, c). (Note that it does not necessarily hold that ID is E's)

Decrypt: Upon input (Decrypt, sid, c, ID') within D, if $ID' \neq ID$ or D does not have correct private key dk_{ID} yet, outputs not-recorded. (Notice that D received her ID when she received her private key.) Else, D obtains $m = d(c, dk_{ID})$ and outputs (Plaintext, sid, m).

Remark (On the communication between the setup party and the decryptor). IBE is specified by four algorithms which are locally executed. The procedure to send and receive keys is outside of the definition of the IBE scheme. In order to realize a secure communication mechanism based on IBE, some transmission protocol must be used with IBE. The security of the communication mechanism depends also on the security of the transmission protocol. Our $\mathcal{F}_{\text{IBE}}$ and Σ definitions do not describe procedures to transfer keys, because the aim of our paper is investigating the security of IBE, not the communication mechanism. When (Extracted, sid, D) is output, the private key is immediately transferred from the setup party to D.

Security against adaptive adversaries. Recall that, even in the case of $\mathcal{F}_{\text{PKE}}$, when the adversary is allowed to corrupt parties during the course of the computation, and obtain their internal state, realizing $\mathcal{F}_{\text{PKE}}$ is a very hard problem [8]. The reason is as follows: If $\mathcal{Z}$ is allowed to corrupt adaptively, $\mathcal{Z}$ makes uncorrupted party E generate ciphertext c of message m for ID ID whose decryptor D is uncorrupted. $\mathcal{Z}$ then corrupts D and can distinguish whether $c = e(ID, \mu)$ or $c = e(ID, m)$ (($\mathcal{S}, \mathcal{F}_{\text{IBE}}$) or ($\mathcal{A}, \pi_\Sigma$)) by obtaining corrupted D's internal states.

Theorem 1. π_Σ UC-realizes $\mathcal{F}_{\text{IBE}}$ with respect to non-adaptive adversaries if and only if IBE scheme Σ is IND-ID-CCA2-secure.

Proof
("only if" part). We prove that if Σ is not IND-ID-CCA2-secure, then π_Σ does not UC-realize $\mathcal{F}_{\mathrm{IBE}}$. In more detail, assuming that there exists adversary G that can break Σ in the sense of IND-ID-CCA2 with non-negligible probability (i.e., $\mathbf{Adv}_{\Sigma,G}^{\mathrm{ind\text{-}id\text{-}cca2}} > \nu(k)$), we prove that we can construct environment $\mathcal{Z}$ and real life adversary $\mathcal{A}$ such that for any ideal process adversary (simulator) $\mathcal{S}$, $\mathcal{Z}$ can tell with non-negligible probability whether $(\mathcal{S}, \mathcal{F}_{\mathrm{IBE}})$ or $(\mathcal{A}, \pi_\Sigma)$ by using adversary G. $\mathcal{Z}$ proceeds as follows:

1. Activates party T with $(\mathtt{Setup}, sid, T)$ and obtains encryption algorithm (and system parameters) e.
2. Hands e to G and plays the role of $\mathcal{XO}_1$ (the extraction oracle as in Preliminaries) and $\mathcal{DO}_1$ (the decryption oracle) for adversary G in the IND-ID-CCA2 game.
3. Obtains (ID^*, m_0, m_1) from G. ID^* is the ID G attacks.
4. If $\mathcal{Z}$ has not activated T with $(\mathtt{Extract}, sid, ID^*, D, e)$ yet, it does so and obtains $(\mathtt{Extracted}, sid, D)$.
5. Chooses random bit $b \xleftarrow{\mathsf{U}} \{0, 1\}$, selects an arbitrary party $E(\neq D)$, activates E with $(\mathtt{Encrypt}, sid, m_b, ID^*, e)$ and obtains c^*.
6. Hands c^* to G as the target ciphertext.
7. Plays the role of $\mathcal{XO}_2$ and $\mathcal{DO}_2$ for adversary G in the IND-ID-CCA2 game, and obtains guess $b' \in \{0, 1\}$.
8. Outputs 1 if $b = b'$, otherwise outputs 0 and halts.

Notice that we consider non-adaptive adversary case. The corrupted parties are denoted $\tilde{P}_1, ..., \tilde{P}_t$.

In step 3, the adversary issues queries $q_1, ..., q_m$ where query q_l is one of:

1. Extraction query $\langle ID_l \rangle$. If this is the x-th extraction, $\mathcal{Z}$ activates T with $(\mathtt{Extract}, sid, ID_l, \tilde{P}_x, e)$ to obtain private key dk_{ID_l} corresponding to public key ID_l from corrupted $\tilde{P}_x$. When $(\mathtt{Extracted}, sid, \tilde{P}_x)$ is output, private key dk_{ID_l} is transferred to $\tilde{P}_x$ in the real world. So $\mathcal{Z}$ can obtain dk_{ID_l} from corrupted $\tilde{P}_x$. In the ideal world, $\mathcal{Z}$ can do so as in the real world, because simulator $\mathcal{S}$ generates master-key and extraction algorithm x when T is activated with $(\mathtt{Setup}, sid, T)$ and uses simulated copy of real life adversary $\mathcal{A}$. In both cases, $\mathcal{Z}$ can hand correct private key dk_{ID_l} to G.
2. Decryption query $\langle ID_l, c_l \rangle$. If this is the first decryption query for ID_l, $\mathcal{Z}$ selects a new uncorrupted party, P_y, and activates T with $(\mathtt{Extract}, sid, ID_l, P_y, e)$ and then activates P_y with $(\mathtt{Decrypt}, sid, c_l, ID_l)$. Otherwise $\mathcal{Z}$ activates P_y' with $(\mathtt{Decrypt}, sid, c_l, ID_l)$, where P_y' is the process $\mathcal{Z}$ activated T with $(\mathtt{Extract}, sid, ID_l, P_y', e)$. When $\mathcal{Z}$ receives $(\mathtt{Plaintext}, sid, v_l)$, it hands v_l to G.

These queries may be asked adaptively, that is, each query q_l may depend on the replies to $q_1, ..., q_{l-1}$.

In step 8, the adversary issues more queries $q_{m+1}, ..., q_n$ where query q_l is one of:

1. Extraction query $\langle ID_l \rangle$ where $ID_l \neq ID^*$. $\mathcal{Z}$ responds as in step 3.
2. Decryption query $\langle ID_l, c_l \rangle \neq \langle ID^*, c^* \rangle$. $\mathcal{Z}$ responds as in step 3.

These queries may be asked adaptively as in step 3.

When $\mathcal{Z}$ interacts with $\mathcal{A}$ and π_Σ, $\mathcal{Z}$ obtains $c^* = \mathcal{E}(PK, ID^*, m_b)$ in Step 6. G can break IND-ID-CCA2 security with non-negligible advantage $\mathbf{Adv}_{\Sigma,G}^{\text{ind-id-cca2}} > \nu(k)$. $\Pr[\mathcal{Z} \to 1 | \mathcal{Z} \leftrightarrow \text{REAL}]$ denotes the probability that $\mathcal{Z}$ outputs 1 when $\mathcal{Z}$ interacts with $\mathcal{A}$ and π_Σ.

$$
\begin{aligned}
\Pr[\mathcal{Z} \to 1 | \mathcal{Z} \leftrightarrow \text{REAL}] &= \Pr[m_b = m_0] \Pr[b' = 0 | c^* = \mathcal{E}(PK, ID^*, m_0)] \\
&\quad + \Pr[m_b = m_1] \Pr[b' = 1 | c^* = \mathcal{E}(PK, ID^*, m_1)] \\
&= \frac{1}{2}(1 - \Pr[b' = 1 | c^* = \mathcal{E}(PK, ID^*, m_0)]) \\
&\quad + \frac{1}{2} \Pr[b' = 1 | c^* = \mathcal{E}(PK, ID^*, m_1)] \\
&= \frac{1}{2} + \frac{1}{2}(\Pr[\mathbf{Exp}_{\Sigma,G}^{\text{ind-id-cca2-1}}(k) = 1] - \Pr[\mathbf{Exp}_{\Sigma,G}^{\text{ind-id-cca2-0}}(k) = 1]) \\
&> \frac{1}{2} + \frac{1}{2}\nu(k)
\end{aligned}
$$

In contrast, when $\mathcal{Z}$ interacts with the ideal process for $\mathcal{F}_{\text{IBE}}$ and any adversary, the view of the instance of G within $\mathcal{Z}$ is statistically independent of b, thus in this case $b = b'$ with probability exactly one half. To see why G's view is independent of b, recall that the view of G consists of the target ciphertext c^* and the decryptions of all ciphertexts generated by G (except for the decryption of c^*). However, $c^* = e(ID^*, \mu)$ for fixed message μ is independent of b. Furthermore, all ciphertexts c_l generated by G are independent of b, thus decryption $d(c_l, dk_{ID_l})$ is independent of b.

$\Pr[\mathcal{Z} \to 1 | \mathcal{Z} \leftrightarrow \text{IDEAL}]$ denotes the probability that $\mathcal{Z}$ outputs 1 when $\mathcal{Z}$ interacts with $\mathcal{S}$ in the ideal process for $\mathcal{F}_{\text{IBE}}$.

$$
\begin{aligned}
\Pr[\mathcal{Z} \to 1 | \mathcal{Z} \leftrightarrow \text{IDEAL}] &= \Pr[m_b = m_0] \Pr[b' = 0 | c^* = e(ID^*, \mu)] \\
&\quad + \Pr[m_b = m_1] \Pr[b' = 1 | c^* = e(ID^*, \mu)] \\
&= \frac{1}{2}(1 - \Pr[b' = 1 | c^* = e(ID^*, \mu)] + \Pr[b' = 1 | c^* = e(ID^*, \mu)]) \\
&= \frac{1}{2}
\end{aligned}
$$

Thus, $\Pr[\mathcal{Z} \to 1 | \mathcal{Z} \leftrightarrow \text{REAL}] - \Pr[\mathcal{Z} \to 1 | \mathcal{Z} \leftrightarrow \text{IDEAL}] > \frac{1}{2}\nu(k)$. Therefore, $\mathcal{Z}$ can tell whether $(\mathcal{S}, \mathcal{F}_{\text{IBE}})$ or $(\mathcal{A}, \pi_\Sigma)$ with non-negligible probability.

("if" part). We show that if π_Σ does not UC-realize $\mathcal{F}_{\text{IBE}}$, then Σ is not IND-ID-CCA2-secure. In more detail, we assume for contradiction that there is real life adversary $\mathcal{A}$ such that for any ideal process adversary $\mathcal{S}$ there exists environment $\mathcal{Z}$ that can tell whether $(\mathcal{S}, \mathcal{F}_{\mathrm{IBE}})$ or $(\mathcal{A}, \pi_{\Sigma})$. We then show that there exists an IND-ID-CCA2 attacker G against Σ using $\mathcal{Z}$.

First, we show that $\mathcal{Z}$ can distinguish whether $(\mathcal{S}, \mathcal{F}_{\mathrm{IBE}})$ or $(\mathcal{A}, \pi_{\Sigma})$ only when setup party T, some encryptor E and some decryptor D are not corrupted. Since we are dealing with non-adaptive adversaries, there are seven cases; Case 1: setup party T is corrupted (throughout the protocol), Case 2: Encryptor E is corrupted (throughout the protocol), Case 3: Decryptor D is corrupted (throughout the protocol), Case 4: T and E are corrupted (throughout the protocol), Case 5: T and D are corrupted (throughout the protocol), Case 6: D and E are corrupted (throughout the protocol), Case 7: T, E and D are uncorrupted.

In Case 1, we can construct simulator $\mathcal{S}$ such that no $\mathcal{Z}$ can distinguish whether$(\mathcal{S}, \mathcal{F}_{\mathrm{IBE}})$ or $(\mathcal{A}, \pi_{\Sigma})$ as follows:

1. When $\mathcal{Z}$ sends $(\texttt{Setup}, sid, T)$ to corrupted party T (i.e., $\mathcal{S}$), $\mathcal{S}$ receives the message and sends it to $\mathcal{F}_{\mathrm{IBE}}$ on behalf of T and the simulated copy of $\mathcal{A}$, which returns a reply message (which may be $\perp$) to $\mathcal{S}$. When $\mathcal{S}$ receives $(\texttt{Setup}, sid, T)$ from $\mathcal{F}_{\mathrm{IBE}}$, $\mathcal{S}$ sends $\mathcal{A}$'s reply to $\mathcal{F}_{\mathrm{IBE}}$. $\mathcal{S}$ sends $\mathcal{A}$'s reply to $\mathcal{Z}$.

2. When $\mathcal{Z}$ sends $(\texttt{Extract}, sid, ID, D, e)$ to corrupted party T (i.e., $\mathcal{S}$), $\mathcal{S}$ receives the message and sends it to the simulated copy of $\mathcal{A}$, which returns a reply message (which may be $\perp$) to $\mathcal{S}$. If $\mathcal{A}$'s reply is the correct private key, $\mathcal{S}$ sends $(\texttt{Extract}, sid, ID, D, e)$ to $\mathcal{F}_{\mathrm{IBE}}$ on behalf of T. Lastly, $\mathcal{S}$ sends $\mathcal{A}$'s reply to $\mathcal{Z}$.

3. When $\mathcal{Z}$ sends $(\texttt{Encrypt}, sid, m, ID, e)$ to E, E forwards it to $\mathcal{F}_{\mathrm{IBE}}$. $\mathcal{F}_{\mathrm{IBE}}$ generates $c = e(ID, m)$ and returns $(\texttt{Ciphertext}, sid, c)$ to E, since T is corrupted.

4. When $\mathcal{Z}$ sends $(\texttt{Decrypt}, sid, c, ID)$ to D, D forwards it to $\mathcal{F}_{\mathrm{IBE}}$. If $(\texttt{Extract}, sid, ID, D, e)$ was sent to T and the simulated copy of $\mathcal{A}$ output the correct key in step 2, $\mathcal{F}_{\mathrm{IBE}}$ returns $(\texttt{Plaintext}, sid, d(c, x(ID)))$. Otherwise, $\mathcal{F}_{\mathrm{IBE}}$ outputs $\texttt{not-recorded}$, because (ID, D) is not recorded in ID-Reg.

In this case, $\mathcal{Z}$ cannot distinguish whether $(\mathcal{S}, \mathcal{F}_{\mathrm{IBE}})$ or $(\mathcal{A}, \pi_{\Sigma})$, because the message returned by $\mathcal{S}$ (using $\mathcal{A}$) as T in the ideal world is the same as that returned by $\mathcal{A}$ as T in the real world, and $(\texttt{Ciphertext}, sid, c)$ returned by $\mathcal{F}_{\mathrm{IBE}}$ is exactly the same as that returned by E in the real world, and $\texttt{not-recorded}$ or $(\texttt{Plaintext}, sid, d(c, x(ID)))$ returned by $\mathcal{F}_{\mathrm{IBE}}$ is exactly the same as that returned by D in the real world.

In Case 2, we can construct simulator $\mathcal{S}$ such that no $\mathcal{Z}$ can distinguish whether $(\mathcal{S}, \mathcal{F}_{\mathrm{IBE}})$ or $(\mathcal{A}, \pi_{\Sigma})$ as follows:

1. When $\mathcal{Z}$ sends $(\texttt{Setup}, sid, T)$ to T, T forwards it to $\mathcal{F}_{\mathrm{IBE}}$. $\mathcal{F}_{\mathrm{IBE}}$ sends $(\texttt{Setup}, sid, T)$ to $\mathcal{S}$, $\mathcal{S}$ computes (PK, MK) by running algorithm $\mathcal{S}()$, and generates x, e and d, where $x = \mathcal{X}(PK, MK, \cdot)$, $e = \mathcal{E}(PK, \cdot, \cdot)$ and $d = \mathcal{D}(PK, \cdot, \cdot)$. $\mathcal{S}$ returns $(\texttt{Algorithms}, sid, x, e, d)$ to $\mathcal{F}_{\mathrm{IBE}}$.

2. When $\mathcal{Z}$ sends $(\texttt{Extract}, sid, ID, D, e)$ to T, T forwards it to $\mathcal{F}_{\mathrm{IBE}}$. $\mathcal{F}_{\mathrm{IBE}}$ records (ID, D) and returns $(\texttt{Extracted}, sid, D)$.

3. When $\mathcal{Z}$ sends (Encrypt, sid, m, ID, e) to corrupted party E (i.e., $\mathcal{S}$), $\mathcal{S}$ receives the message and sends it to the simulated copy of $\mathcal{A}$, which replies to $\mathcal{S}$. $\mathcal{S}$ then returns $\mathcal{A}$'s reply (which may be $\perp$) to $\mathcal{Z}$.
4. When $\mathcal{Z}$ sends (Decrypt, sid, c, ID) to D, D forwards it to $\mathcal{F}_{\mathrm{IBE}}$. $\mathcal{F}_{\mathrm{IBE}}$ then returns (Plaintext, $sid, d(c, x(ID))$), since E (i.e., $\mathcal{S}$) sent no (Encrypt, sid, m, ID, e) to $\mathcal{F}_{\mathrm{IBE}}$, which records nothing as (m, c, ID).

In this case, $\mathcal{Z}$ cannot distinguish whether $(\mathcal{S}, \mathcal{F}_{\mathrm{IBE}})$ or $(\mathcal{A}, \pi_\Sigma)$, because the message returned by $\mathcal{S}$ (using $\mathcal{A}$) as E in the ideal world is the same as that returned by $\mathcal{A}$ as E in the real world, and (Encryption Algorithm, sid, e) returned by $\mathcal{F}_{\mathrm{IBE}}$ is exactly the same as that returned by T in the real world, (Extracted, sid, D) returned by $\mathcal{F}_{\mathrm{IBE}}$ is exactly the same as that returned by T in the real world, and (Plaintext, $sid, d(c, x(ID))$) returned by $\mathcal{F}_{\mathrm{IBE}}$ is exactly the same as that returned by D in the real world.

In Case 3, we can construct simulator $\mathcal{S}$ such that no $\mathcal{Z}$ can distinguish whether $(\mathcal{S}, \mathcal{F}_{\mathrm{IBE}})$ or $(\mathcal{A}, \pi_\Sigma)$ as follows:

1. When $\mathcal{Z}$ sends (Setup, sid, T) to T, T forwards it to $\mathcal{F}_{\mathrm{IBE}}$. $\mathcal{F}_{\mathrm{IBE}}$ sends (Setup, sid, T) to $\mathcal{S}$, $\mathcal{S}$ computes (PK, MK) by running algorithm $\mathcal{S}()$, and generates x, e and d, where $x = \mathcal{X}(PK, MK, \cdot)$, $e = \mathcal{E}(PK, \cdot, \cdot)$ and $d = \mathcal{D}(PK, \cdot, \cdot)$. $\mathcal{S}$ returns (Algorithms, sid, x, e, d) to $\mathcal{F}_{\mathrm{IBE}}$.
2. When $\mathcal{Z}$ sends (Extract, sid, ID, D, e) to T, T forwards it to $\mathcal{F}_{\mathrm{IBE}}$. $\mathcal{F}_{\mathrm{IBE}}$ records (ID, D) and returns (Extracted, sid, D).
3. When $\mathcal{Z}$ sends (Encrypt, sid, m, ID, e) to E, E forwards it to $\mathcal{F}_{\mathrm{IBE}}$. $\mathcal{F}_{\mathrm{IBE}}$ generates $c = e(ID, m)$ and returns (Ciphertext, sid, c) to E, since D is corrupted.
4. When $\mathcal{Z}$ sends (Decrypt, sid, c, ID) to corrupted party D (i.e., $\mathcal{S}$), $\mathcal{S}$ sends (Decrypt, sid, c, ID) to $\mathcal{A}$. $\mathcal{A}$ returns a reply (which may be $\perp$) to $\mathcal{S}$, which forwards $\mathcal{A}$'s reply to $\mathcal{Z}$.

In this case, $\mathcal{Z}$ cannot distinguish whether $(\mathcal{S}, \mathcal{F}_{\mathrm{IBE}})$ or $(\mathcal{A}, \pi_\Sigma)$, because the message returned by $\mathcal{S}$ (using $\mathcal{A}$) as D in the ideal world is the same as that returned by $\mathcal{A}$ as D in the real world, (Encryption Algorithm, sid, e) returned by $\mathcal{F}_{\mathrm{IBE}}$ is exactly the same as that returned by T in the real world, and (Extracted, sid, D) returned by $\mathcal{F}_{\mathrm{IBE}}$ is exactly the same as that returned by T in the real world, and (Ciphertext, sid, c) returned by $\mathcal{F}_{\mathrm{IBE}}$ is exactly the same as that returned by E in the real world.

In Case 4, we can construct simulator $\mathcal{S}$ such that no $\mathcal{Z}$ can distinguish whether $(\mathcal{S}, \mathcal{F}_{\mathrm{IBE}})$ or $(\mathcal{A}, \pi_\Sigma)$ as follows:

1. When $\mathcal{Z}$ sends (Setup, sid, T) to corrupted party T (i.e., $\mathcal{S}$), $\mathcal{S}$ receives the message and sends it to $\mathcal{F}_{\mathrm{IBE}}$ on behalf of T and the simulated copy of $\mathcal{A}$, which returns a reply message (which may be $\perp$) to $\mathcal{S}$. When $\mathcal{S}$ receives (Setup, sid, T) from $\mathcal{F}_{\mathrm{IBE}}$, $\mathcal{S}$ sends $\mathcal{A}$'s reply to $\mathcal{F}_{\mathrm{IBE}}$. $\mathcal{S}$ sends $\mathcal{A}$'s reply to $\mathcal{Z}$.
2. When $\mathcal{Z}$ sends (Extract, sid, ID, D, e) to corrupted party T (i.e., $\mathcal{S}$), $\mathcal{S}$ receives the message and sends it to the simulated copy of $\mathcal{A}$, which returns

a reply message (which may be $\perp$) to $\mathcal{S}$. If $\mathcal{A}$'s reply is the correct private key, $\mathcal{S}$ sends (Extract, sid, ID, D, e) to $\mathcal{F}_{\text{IBE}}$ on behalf of T. Lastly, $\mathcal{S}$ sends $\mathcal{A}$'s reply to $\mathcal{Z}$.

3. When $\mathcal{Z}$ sends (Encrypt, sid, m, ID, e) to corrupted party E (i.e., $\mathcal{S}$), $\mathcal{S}$ receives the message and sends it to the simulated copy of $\mathcal{A}$, which replies to $\mathcal{S}$. $\mathcal{S}$ then returns $\mathcal{A}$'s reply (which may be $\perp$) to $\mathcal{Z}$.

4. When $\mathcal{Z}$ sends (Decrypt, sid, c, ID) to D, D forwards it to $\mathcal{F}_{\text{IBE}}$. If (Extract, sid, ID, D, e) was sent to T and the simulated copy of $\mathcal{A}$ output the correct key in step 2, $\mathcal{F}_{\text{IBE}}$ returns (Plaintext, $sid, d(c, x(ID))$). Otherwise, $\mathcal{F}_{\text{IBE}}$ outputs not-recorded, because (ID, D) is not recorded in ID-Reg.

In this case, $\mathcal{Z}$ cannot distinguish whether $(\mathcal{S}, \mathcal{F}_{\text{IBE}})$ or $(\mathcal{A}, \pi_\Sigma)$, because the message returned by $\mathcal{S}$ (using $\mathcal{A}$) as T and E in the ideal world is the same as that returned by $\mathcal{A}$ as T and E in the real world, and not-recorded or (Plaintext, $sid, d(c, x(ID))$) returned by $\mathcal{F}_{\text{IBE}}$ is exactly the same as that returned by D in the real world.

In Case 5, we can construct simulator $\mathcal{S}$ such that no $\mathcal{Z}$ can distinguish whether $(\mathcal{S}, \mathcal{F}_{\text{IBE}})$ or $(\mathcal{A}, \pi_\Sigma)$ as follows:

1. When $\mathcal{Z}$ sends (Setup, sid, T) to corrupted party T (i.e., $\mathcal{S}$), $\mathcal{S}$ receives the message and sends it to the simulated copy of $\mathcal{A}$, which returns a reply message (which may be $\perp$) to $\mathcal{S}$. $\mathcal{S}$ sends it to $\mathcal{Z}$.

2. When $\mathcal{Z}$ sends (Extract, sid, ID, D, e) to corrupted party T (i.e., $\mathcal{S}$), $\mathcal{S}$ receives the message and sends it to the simulated copy of $\mathcal{A}$, which returns a reply message (which may be $\perp$) to $\mathcal{S}$. $\mathcal{S}$ sends it to $\mathcal{Z}$.

3. When $\mathcal{Z}$ sends (Encrypt, sid, m, ID, e) to E, E forwards it to $\mathcal{F}_{\text{IBE}}$. $\mathcal{F}_{\text{IBE}}$ generates $c = e(ID, m)$ and returns (Ciphertext, sid, c) to E, since T (i.e., $\mathcal{S}$) sent no (Setup, sid, T) to $\mathcal{F}_{\text{IBE}}$, which records nothing as encryption algorithm e.

4. When $\mathcal{Z}$ sends (Decrypt, sid, c, ID) to corrupted party D (i.e., $\mathcal{S}$), $\mathcal{S}$ sends (Decrypt, sid, c, ID) to $\mathcal{A}$. $\mathcal{A}$ returns a reply (which may be $\perp$) to $\mathcal{S}$, which forwards $\mathcal{A}$'s reply to $\mathcal{Z}$.

In this case, $\mathcal{Z}$ cannot distinguish whether $(\mathcal{S}, \mathcal{F}_{\text{IBE}})$ or $(\mathcal{A}, \pi_\Sigma)$, because the message returned by $\mathcal{S}$ (using $\mathcal{A}$) as T and D in the ideal world is the same as that returned by $\mathcal{A}$ as T and D in the real world, and (Ciphertext, sid, c) returned by $\mathcal{F}_{\text{IBE}}$ is exactly the same as that returned by E in the real world.

In Case 6, we can construct simulator $\mathcal{S}$ such that no $\mathcal{Z}$ can distinguish whether $(\mathcal{S}, \mathcal{F}_{\text{IBE}})$ or $(\mathcal{A}, \pi_\Sigma)$ as follows:

1. When $\mathcal{Z}$ sends (Setup, sid, T) to T, T forwards it to $\mathcal{F}_{\text{IBE}}$. $\mathcal{F}_{\text{IBE}}$ sends (Setup, sid, T) to $\mathcal{S}$, $\mathcal{S}$ computes (PK, MK) by running algorithm $\mathcal{S}$, and generates x, e and d, where $x = \mathcal{X}(PK, MK, \cdot)$, $e = \mathcal{E}(PK, \cdot, \cdot)$ and $d = \mathcal{D}(PK, \cdot, \cdot)$. $\mathcal{S}$ returns (Algorithms, sid, x, e, d) to $\mathcal{F}_{\text{IBE}}$.

2. When $\mathcal{Z}$ sends (Extract, sid, ID, D, e) to T, T forwards it to $\mathcal{F}_{\text{IBE}}$. $\mathcal{F}_{\text{IBE}}$ records (ID, D) and returns (Extracted, sid, D).

3. When $\mathcal{Z}$ sends $(\texttt{Encrypt}, sid, m, ID, e)$ to corrupted party E (i.e., $\mathcal{S}$), $\mathcal{S}$ receives the message and sends it to the simulated copy of $\mathcal{A}$, which replies to $\mathcal{S}$. $\mathcal{S}$ then returns $\mathcal{A}$'s reply (which may be $\perp$) to $\mathcal{Z}$.
4. When $\mathcal{Z}$ sends $(\texttt{Decrypt}, sid, c, ID)$ to corrupted party D (i.e., $\mathcal{S}$), $\mathcal{S}$ sends $(\texttt{Decrypt}, sid, c, ID)$ to $\mathcal{A}$. $\mathcal{A}$ returns a reply (which may be $\perp$) to $\mathcal{S}$, which forwards $\mathcal{A}$'s reply to $\mathcal{Z}$.

In this case, $\mathcal{Z}$ cannot distinguish whether $(\mathcal{S}, \mathcal{F}_{\text{IBE}})$ or $(\mathcal{A}, \pi_\Sigma)$, because the message returned by $\mathcal{S}$ (using $\mathcal{A}$) as E and D in the ideal world is the same as that returned by $\mathcal{A}$ as E and D in the real world, $(\texttt{Encryption Algorithm}, sid, e)$ returned by $\mathcal{F}_{\text{IBE}}$ is exactly the same as that returned by T in the real world, and $(\texttt{Extracted}, sid, D)$ returned by $\mathcal{F}_{\text{IBE}}$ is exactly the same as that returned by T in the real world.

Thus, $\mathcal{Z}$ cannot distinguish $(\mathcal{S}, \mathcal{F}_{\text{IBE}})$ or $(\mathcal{A}, \pi_\Sigma)$ in Cases $1, 2, 3, 4, 5$, and 6. Hereafter, we consider only Case 7.

Recall that $\mathcal{A}$ takes three types of messages from $\mathcal{Z}$: either to corrupt parties, or to report on messages sent in the protocol, or to deliver some messages. There are no party corruption instructions, since we are dealing with non-adaptive adversaries. However, $\mathcal{Z}$ may request some corrupted parties to reveal their private keys, so $\mathcal{A}$ need report private keys to $\mathcal{Z}$.

Thus, the activity of $\mathcal{S}$ is to provide the algorithms to $\mathcal{F}_{\text{IBE}}$ and to report private keys. Since $\mathcal{Z}$ succeeds in distinguishing for any $\mathcal{S}$, it also succeeds for the following specific $\mathcal{S}$. Simulator $\mathcal{S}$ acts as follows:

When $\mathcal{S}$ receives message $(\texttt{Setup}, sid, T)$ from $\mathcal{F}_{\text{IBE}}$, it runs setup algorithm S, obtains system parameters PK and master-key MK, and returns $x = \mathcal{X}(PK, MK, \cdot)$, $e = \mathcal{E}(PK, \cdot, \cdot)$ and $d = \mathcal{D}(PK, \cdot, \cdot)$ to $\mathcal{F}_{\text{IBE}}$.

When $\mathcal{Z}$ requests private keys, $\mathcal{S}$ relays the request to simulated copy of $\mathcal{A}$ and returns the message from $\mathcal{A}$ to $\mathcal{Z}$ by using extraction algorithm $x = \mathcal{X}(PK, MK, \cdot)$.

We consider the case where setup party T, encryptor E and decryptor D are uncorrupted and assume for contradiction that there is environment $\mathcal{Z}$ that can distinguish whether $(\mathcal{S}, \mathcal{F}_{\text{IBE}})$ or $(\mathcal{A}, \pi_\Sigma)$. We now prove that we can construct adversary G that breaks IND-ID-CCA2 security by using environment $\mathcal{Z}$. More precisely, we assume that there is real life adversary $\mathcal{A}$ such that for any ideal process adversary $\mathcal{S}$, there exists environment $\mathcal{Z}$ such that for fixed value k of security parameter and fixed input z for $\mathcal{Z}$,

$$|\text{IDEAL}_{\mathcal{F}_{\text{IBE}}, \mathcal{S}, \mathcal{Z}}(k, z) - \text{REAL}_{\pi_\Sigma, \mathcal{A}, \mathcal{Z}}(k, z)| > \nu(k)$$

We then show that there exists G_h whose advantage $\mathbf{Adv}_{\Sigma, G_h}^{\text{ind-id-cca2}}(k) > \nu(k)/l$ in the IND-ID-CCA2 game, where l is the total number of messages that were encrypted by uncorrupted party's ID (Extract has already executed) throughout the running of the system and $h \in \{1, ..., l\}$. G_h is given system parameters PK, and is allowed to query $\mathcal{XO}_i$, $\mathcal{EO}_i$ and $\mathcal{DO}_i$ (as in Preliminaries). G_h runs $\mathcal{Z}$ on the following simulated interaction with a system running $\pi_\Sigma / \mathcal{F}_{\text{IBE}}$. Let

(m_j, ID_j) denote the jth pair of message and ID that $\mathcal{Z}$ activates some party with $(\text{Encrypt}, sid, m_j, ID_j, e)$ in this simulation.

1. When $\mathcal{Z}$ activates some party T with input (Setup, sid, T), G_h lets T output value e calculated from PK.
2. When $\mathcal{Z}$ activates some party T with input $(\text{Extract}, sid, ID, P, e)$, G_h lets T output message $(\text{Extracted}, sid, P)$ from G_h's input. If P is corrupted and $\mathcal{Z}$ requests P's private key, then G_h queries $\mathcal{XO}_i$ on ID, obtains value u and lets P return u to $\mathcal{Z}$. This is perfect simulation, so $\mathcal{Z}$ cannot distinguish $(\mathcal{S}, \mathcal{F}_{\text{IBE}})$ or $(\mathcal{A}, \pi_\Sigma)$ in this step.
3. For the first $h - 1$ times that $\mathcal{Z}$ asks encryptor E to encrypt some message, m_j, G_h lets E return $c_j = e(ID_j, m_j)$.
4. The h-th time that $\mathcal{Z}$ asks E to encrypt message, m_h by ID^*, G_h queries encryption oracle $\mathcal{EO}_i$ with the pair of messages (m_h, μ), where $\mu \in \mathcal{M}$ is the fixed message, and obtains target ciphertext c_h. It then hands c_h to $\mathcal{Z}$ as the encryption of m_h. That is, $c_h = \mathcal{E}(PK, ID^*, m_h)$ $(b = 0)$ or $c_h = \mathcal{E}(PK, ID^*, \mu)$ $(b = 1)$.
5. For the remaining $l - h$ times that $\mathcal{Z}$ asks E to encrypt some message, m_j, G_h lets E return $c_j = \mathcal{E}(PK, ID_j, \mu)$.
6. Whenever decryptor D is activated with input $(\text{Decrypt}, sid, c, ID)$ where $c = c_j$ and $ID = ID_j$ for some j, G_h lets D return the corresponding plaintext m_j. If c is different from all c_j's and ID_j is extracted, G_h queries $\mathcal{DO}_i$ on (ID, c), obtains value v, and lets D return v to $\mathcal{Z}$. If c is different from all c_j's and ID_j is not extracted, G_h lets D output not-recorded. This is perfect simulation, so $\mathcal{Z}$ cannot distinguish $(\mathcal{S}, \mathcal{F}_{\text{IBE}})$ or $(\mathcal{A}, \pi_\Sigma)$ in this step.
7. When $\mathcal{Z}$ halts, G_h outputs whatever $\mathcal{Z}$ outputs and halts.

Notice that $\mathcal{Z}$ cannot distinguish $(\mathcal{S}, \mathcal{F}_{\text{IBE}})$ or $(\mathcal{A}, \pi_\Sigma)$ by activating E with $(\text{Encrypt}, sid, m, ID, e)$ before activating T with $(\text{Extract}, sid, ID, P_h, e)$, because in this case, $c = e(ID, m)$ in both the real and the ideal world.

We apply a standard hybrid argument for analyzing the success probability of G_h. For $j \in \{0, ..., l\}$, let Env_j be an event that $\mathcal{Z}$ interacts with $\mathcal{S}$ in the ideal process, with the exception that the first j ciphertexts are computed as an encryption of the real plaintexts, rather than encryptions of μ. The replies to $\mathcal{Z}$ from setup party T and decryptor D are the same as those shown in step $1, 2$ and 6 above. Let H_j be $\Pr[\mathcal{Z} \rightarrow 1 | \text{Env}_j]$.

Notice that in steps 2 and 6, $\mathcal{Z}$ cannot tell whether it is interacting with $\mathcal{A}$ and π_Σ or with $\mathcal{S}$ in the ideal process for $\mathcal{F}_{\text{IBE}}$, because G_h offers perfect simulation.

It is easy to see that H_0 is identical to the probability that $\mathcal{Z}$ outputs 1 in the ideal process, and that H_l is identical to the probability that $\mathcal{Z}$ outputs 1 in the real life model. Furthermore, in a run of G_h, if value c_h that G_h obtains from its encryption oracle is encryption m_h, the probability that $\mathcal{Z}$ outputs 1 is identical to H_{h-1}. If c_h is an encryption of μ, the probability that $\mathcal{Z}$ outputs 1 is identical to H_h. Details follow:

$$H_0 = \text{IDEAL}_{\mathcal{F}_{\text{IBE}},\mathcal{S},\mathcal{Z}}(k,z)$$
$$H_1 = \text{REAL}_{\pi_\Sigma,\mathcal{A},\mathcal{Z}}(k,z)$$
$$H_h = \Pr[G_h \to 1|c_h = \mathcal{E}(PK,ID^*,\mu)]$$
$$H_{h-1} = \Pr[G_h \to 1|c_h = \mathcal{E}(PK,ID^*,m_h)]$$

$$\sum_{i=1}^{l}|H_{i-1} - H_i| \geq |\sum_{i=1}^{l}(H_{i-1} - H_i)|$$
$$= |H_0 - H_l|$$
$$= |\text{IDEAL}_{\mathcal{F}_{\text{IBE}},\mathcal{S},\mathcal{Z}}(k,z) - \text{REAL}_{\pi_\Sigma,\mathcal{A},\mathcal{Z}}(k,z)|$$
$$> \nu(k)$$

Therefore, there exists some $h \in \{1,...,l\}$ such that $|H_{h-1} - H_h| > \nu(k)/l$. Here, w.l.o.g, let $H_{h-1} - H_h > \nu(k)/l$, since if $H_h - H_{h-1} > \nu(k)/l$ for $\mathcal{Z}$, we can obtain $H_{h-1} - H_h > \nu(k)/l$ for $\mathcal{Z}^*$, where $\mathcal{Z}^*$ outputs the opposite of $\mathcal{Z}$'s output bit.

We have the advantage of adversary G_h as follows:

$$\textbf{Adv}_{\Sigma,G_h}^{\text{ind-id-cca2}}(k) = \Pr[\textbf{Exp}_{\Sigma,\mathcal{A}}^{\text{ind-id-cca2-1}}(k) = 1] - \Pr[\textbf{Exp}_{\Sigma,\mathcal{A}}^{\text{ind-id-cca2-0}}(k) = 1]$$
$$= \Pr[G_h \to 1|c_h = \mathcal{E}(PK,ID^*,\mu)] - \Pr[G_h \to 1|c_h = \mathcal{E}(PK,ID^*,m_h)]$$
$$= H_h - H_{h-1} > \nu(k)/l$$

That is, G has non-negligible advantage in k since l is polynomially bounded in k. □

Acknowledgements

The authors would like to thank anonymous reviewers of VietCrypt 2006 for their invaluable comments and suggestions.

References

1. N. Attrapadung, Y. Cui, D. Galindo, G. Hanaoka, I. Hasuo, H. Imai, K. Matsuura, P. Yang, and R. Zhang. Relations Among Notions of Security for Identity Based Encryption Schemes. *In proceedings of LATIN'06*, 3887 of LNCS, 2006.
2. B. Barak, R. Canetti, Y. Lindell, R. Pass, and T. Rabin. Secure Computation Without Authentication. *In proceedings of CRYPTO'05*, 2005.
3. B. Barak, R. Canetti, J. B. Nielsen, and R. Pass. Universally Composable Protocols with Relaxed Set-up Assumptions. *In proceedings of FOCS'04*, 2004.
4. D. Boneh and X. Boyen. Efficient Selective-ID Secure Identity Based Encryption Without Random Oracles. *In proceedings of EUROCRYPT'04*, 3027 of LNCS, 2004.
5. D. Boneh and X. Boyen. Secure Identity Based Encryption Without Random Oracles. *In proceedings of CRYPTO'04*, 3152 of LNCS, 2004.
6. D. Boneh and M. Franklin. Identity-based encryption from the Weil pairing. *In proceedings of CRYPTO'01*, 2139 of LNCS, 2001.

7. D. Boneh and J. Katz. Improved Efficiency for CCA-Secure Cryptosystems Built Using Identity Based Encryption. *In proceedings of RSA-CT'05*, 2005.

8. R. Canetti. Universally Composable Security: A New Paradigm for Cryptograpic Protocols. *In proceedings of FOCS'01*, 2001. Current Full Version Available at Cryptology ePrint Archive, Report 2000/067 http://eprint.iacr.org/.

9. R. Canetti. Universally Composable Signatures, Certification, and Authenticated Communication. *In proceedings of 17th Computer Security Foundations Workshop*, 2004.

10. R. Canetti and M. Fischlin. Universally Composable Commitments. *In proceedings of CRYPTO'01*, 2139 of LNCS, 2001.

11. R. Canetti, S. Halevi, and J. Katz. A Forward-Secure Public-Key Encryption Scheme. *In proceedings of EUROCRYPT'03*, 2656 of LNCS, 2003.

12. R. Canetti, S. Halevi, and J. Katz. Chosen-Ciphertext Security from Identity-based Encryption. *In proceedings of EUROCRYPT'04*, 3027 of LNCS, 2004.

13. R. Canetti and H. Krawczyk. Universally Composable Key Exchange and Secure Channels. *In proceedings of EUROCRYPT'02*, 2332 of LNCS, 2002.

14. R. Canetti, H. Krawczyk, and J. B. Nielsen. Relaxing Chosen-Ciphertext Security. *In proceedings of CRYPTO'03*, 2729 of LNCS, 2003.

15. R. Canetti, E. Kushilevitz, and Y. Lindell. On the Limitations of Universally Composable Two-Party Computation Without Set-up Assumptions. *In proceedings of EUROCRYPT'03*, 2656 of LNCS, 2003.

16. R. Canetti, Y. Lindell, R. Ostrovsky, and A. Sahai. Universally Composable Two-Party and Multi-Party Secure Computaion. *In proceedings of STOC'02*, 2002.

17. R. Canetti and T. Rabin. Universal Composition with Joint State. *In proceedings of CRYPTO'03*, 2729 of LNCS, 2003.

18. I. Damgård and J. B. Nielsen. Perfect Hiding and Perfect Binding Universally Composable Commitment Schemes with Constant Expansion Factor. *In proceedings of CRYPTO'02*, 2442 of LNCS, 2002.

19. A. Datta, A. Derek, J. C. Mitchell, A. Ramanathan, and A. Scedrov. Games and the Impossibility of Realizable Ideal Functionality. *In proceedings of TCC'06*, 3876 of LNCS, 2006.

20. W. Nagao, Y. Manabe, and T. Okamoto. On the Equivalence of Several Security Notions of Key Encapsulation Mechanism. Cryptology ePrint Archive, Report 2006/268, 2006. http://eprint.iacr.org/.

21. M. Prabhakaran and A. Sahai. New Notions of Security: Achieving Universal Composability without Trusted Setup. *In proceedings of STOC'04*, 2004.

22. A. Shamir. Identity-based Cryptosystems and Signature Schemes. *In proceedings of CRYPTO'84*, 196 of LNCS, 1984.

23. B. Waters. Efficient Identity-Based Encryption Without Random Oracles. *In proceedings of EUROCRYPT'05*, 3494 of LNCS, 2005.

Traitor Tracing for Stateful Pirate Decoders with Constant Ciphertext Rate

Duong Hieu Phan

University College London
h.phan@adastral.ucl.ac.uk

Abstract. Stateful pirate decoders are history recording and abrupt pirate decoders. These decoders can keep states between decryptions to detect whether they are being traced and are then able to take some counter-actions against the tracing process, such as "shutting down" or erasing all internal information. We propose the first constant ciphertext rate scheme which copes with such pirate decoders. Our scheme moreover supports black-box public traceability.

1 Introduction

In the secure distribution of digital content, there are two main types of schemes: *broadcast encryption* schemes, which enable a center to prevent a set of users from recovering the broadcasted information; and *traitor tracing* schemes, which enable the center to trace users who collude to produce pirate decoders. This paper focuses on the traceability property for pirate decoders in the strongest model (according to the hierarchy established by Kiayias and Yung [7]). In [7], the authors described various categories of pirate decoders which are resumed below:

Stateless pirate decoders. These decoders are *resettable* and *available*. A resettable decoder can be reset to its initial state by the tracer at any time. This gives the tracer the advantage of making independent tests during the tracing process. An available pirate decoder is a device that does not take any counter-action against the tracing process and thus is always available for the tracer.

Stateful pirate decoders. In contrast to stateless pirate decoders, these are *history recording* and *abrupt* pirate decoders. A history recording pirate decoder can remember previous queries made by the tracer in order to detect if it is being traced. Abrupt pirate decoders can take some counter-actions against the tracing process such as the "shutting down" mechanism, a process by which pirate decoders erase all internal key information and thus defeat the tracing process. The history recording capability along with abrupt capability can be used by pirate decoders to evade tracing.

Kiayias and Yung [7] also showed an interesting method to convert some types of tracing systems for stateless pirate decoders into tracing systems for stateful ones by embedding robust watermarks in the content. However, previous tracing systems for stateful decoders are inefficient in terms of ciphertext rate.

P.Q. Nguyen (Ed.): VIETCRYPT 2006, LNCS 4341, pp. 354–365, 2006.

Schemes with constant transmission rate. These schemes are well-suited to encrypt large messages. An interesting property of these scheme is the efficient black-box traceability, *i.e.* the tracing procedure does not have to open the pirate decoder, but only interacts with it. However, the constant transmission rate is asymptotically achieved and for large plaintexts (due to the use of collision-secure codes and codes with identifiable parent property). We note that the constant ciphertext rate schemes [7,5,10] were only designed for stateless pirate decoders.

Public Traceability. Chabanne *et al.* [5] introduced the notion of public traceability where tracing is a black-box and publicly computable procedure which allows an untrusted party to trace pirate decoders. Phan *et al.* [10] introduced the first constant ciphertext rate traitor tracing scheme with public-traceability.

1.1 Contribution

We propose the first traitor tracing scheme for stateful decoders with constant ciphertext rate. Furthermore, our scheme still keeps the desirable properties of previous traitor tracing with constant ciphertext rate, namely black-box traceability and public traceability.

We first propose a basic scheme for stateful pirate decoders by employing watermarking technique [7] in the Phan, *et al.*'s basic scheme [10]. We then introduce an efficient generalization of the basic scheme to obtain an efficient general scheme. Although our basic scheme is significantly less efficient than Phan *et al.*'s basic scheme, the general scheme is almost as efficient as theirs. We moreover point out that the latter cannot deal with stateful pirate decoders.

2 Preliminaries

In this section, we first recall definitions of Public Key Encryption (PKE) and of Data Encapsulation Mechanism (DEM) which will be used in our constructions. We then review the definition of traitor tracing systems with public traceability.

2.1 Public-Key Encryption

A public-key encryption scheme PKE is defined by the three following algorithms:

- The *key generation algorithm* Gen. On input 1^λ, where λ is the security parameter, the algorithm Gen produces a pair (pk, sk) of matching public and private keys.
- The *encryption algorithm* Enc. Given a message m (in the space of plaintexts $\mathcal{M}$) and a public key pk, $\mathsf{Enc_{pk}}(m)$ produces a ciphertext c (in the space of ciphertexts $\mathcal{C}$) of m. This algorithm may be probabilistic (involving random coins $r \in \mathcal{R}$), it is then denoted $\mathsf{Enc_{pk}}(m; r)$.
- The *decryption algorithm* Dec. Given a ciphertext $c \in \mathcal{C}$ and the secret key sk, $\mathsf{Dec_{sk}}(c)$ gives back the plaintext $m \in \mathcal{M}$.

2.2   Data Encapsulation Mechanism (DEM)

A DEM is a symmetric encryption scheme that consists of the following algorithms:

- *Setup algorithm* DEM.Setup$(1^\lambda) \to \mathcal{K}_D$: an algorithm that specifies the symmetric key space $\mathcal{K}_D$.
- *Encryption algorithm* DEM.Enc$_{dk}(m) \to \tau$: a deterministic, polynomial-time algorithm that encrypts m into τ, using a symmetric-key $dk \in \mathcal{K}_D$.
- *Decryption algorithm* DEM.Dec$_{dk}(\tau) \to m$: a deterministic, polynomial-time algorithm that decrypts τ to m, using a symmetric-key $dk \in \mathcal{K}_D$.

2.3   Traitor Tracing System with Public Traceability

Definition 1 (Pirate Decoder). *A pirate decoder D is defined as a probabilistic circuit that takes as input a ciphertext C and outputs some message M or $\perp$.*

Definition 2 (Traitor Tracing System with Public Traceability). *A Traitor Tracing system with public traceability consists of the following four algorithms:*

Setup(N, λ) *akes as input N, the number of users in the system, and λ, the security parameter. The algorithm runs in polynomial time in λ and outputs a public key* pk *and private keys $K_1, ..., K_N$, where K_u is given to user u.*

Encrypt$($pk$, M)$ *encrypts M using the public broadcasting key* pk *and outputs ciphertext C.*

Decrypt$(j, K_j, C,$ pk$)$ *decrypts C using the private key K_j of user j. The algorithm outputs a message M or $\perp$.*

Trace$(\mathcal{D},$ pk$)$ *is an oracle algorithm that is only given as input the public key* pk *and a pirate decoder $\mathcal{D}$. The tracing algorithm queries the pirate decoder $\mathcal{D}$ as a black-box oracle, as defined above. It outputs at least a user in $1, 2, \ldots, N$*

The system, called a $(N, t)-$ TTS, must satisfy the following properties:

Correctness property: *for all user $i \in \{1, \ldots, N\}$, and all messages M:*

$$\text{Decrypt}(j, K_j, \text{Encrypt}(\text{pk}, M), \text{pk}) = M$$

Traceability property: *from a pirate decoder $\mathcal{D}$ produced from a collusion of up to t users, the above tracing algorithm should be able to correctly return at least a user in the collusion producing $\mathcal{D}$.*

3   PST Basic Scheme [10]

Below, we briefly review the basic scheme of Phan, Safavi-Naini, Tonien [10] (called PST scheme) which will take part in our construction of general scheme. Their basic scheme, called PSTBasic(N, λ), is described as follows.

Primitives: a public-key encryption PKE and a data encapsulation mechanism DEM.

Setup(N, λ): *Algorithm* PSTBasic.Gen($1^\lambda, N$) $\rightarrow$ (pk, $sk_1, \ldots, sk_N$).

The algorithm PSTBasic.Gen(.) takes as input the number of users N and a security parameter λ. It simply runs the setup algorithm of the public key encryption to define its public key pk and its private keys $sk_1, \ldots, sk_N$:

- For each $i = 1, \ldots, N$, call PKE.Gen(1^λ) $\rightarrow (pk_i, sk_i)$.
- Set pk $= (pk_1, \ldots, pk_N)$.

Encrypt(pk, m): *Algorithm* PSTBasic.Enc(pk, m) $\rightarrow c$.

The algorithm PSTBasic.Enc(.) takes as input the public key pk, a message m and outputs a ciphertext c. It uses DEM to encrypt the message m and PKE to encrypt the key used in DEM:

- Choose a random dk
- Call DEM.Enc$_{dk}(m) \rightarrow \tau$
- Compute $h = H(\tau)$ and for each $i = 1, \ldots, N$, call PKE.Enc$_{pk_i}(dk\|h) \rightarrow \sigma_i$.
- Define $c = (\sigma_1, \ldots, \sigma_N, \tau)$.

Decrypt(i, sk_i, c, pk): *Algorithm* PSTBasic.Dec(sk_i, c) $\rightarrow m$ or $\perp$.

The algorithm PSTBasic.Dec(.) takes as input a secret key sk_i and a cipher-text $c = (\sigma_1, \ldots, \sigma_N, \tau)$ and outputs a message m ot $\perp$:

- Call PKE.Dec$_{sk_i}(\sigma_i) \rightarrow dk\|h$
- If $h \neq H(\tau)$, return $\perp$
- Otherwise, call DEM.Dec$_{dk}(\tau) \rightarrow m$ and output m

Trace($\mathcal{D}$, pk):

Algorithm PSTBasic.Public-Trace(pk, $\mathcal{D}$) $\rightarrow t$.

The algorithm PSTBasic.Public-Trace(.) takes as input the public key pk and a pirate decoder $\mathcal{D}$ and outputs a traitor t as follows:

- Choose randoms dk, m, then call PSTBasic.Enc$_{pk}(m) \rightarrow (\sigma_1, \ldots, \sigma_N, \tau)$.
- For each $i = 1, \ldots, N$, choose random $d'_i \neq dk\|h$ such that d'_i has the same length as $dk\|h$.
- Call PKE.Enc$_{pk_i}(d'_i) \rightarrow \sigma'_i$.
- Calculate the following probabilities:
  - $p_0 = Pr[\mathcal{D}(\sigma_1, \sigma_2, \ldots, \sigma_N, \tau) = m]$
  - $p_1 = Pr[\mathcal{D}(\sigma'_1, \sigma_2, \ldots, \sigma_N, \tau) = m]$
  - $\ldots$
  - $p_n = Pr[\mathcal{D}(\sigma'_1, \sigma'_2, \ldots, \sigma'_N, \tau) = m]$.
- If $|p_i - p_{i-1}|$ is not negligible, output $t = i$ as a traitor.

We first show that the above scheme can not be used for stateful pirate decoders.

Proposition 3. *A stateful pirate decoder can defeat the above tracing algorithm.*

Proof. Assume that user 1 and user N collude to produce a pirate decoder $\mathcal{D}$, whenever $\mathcal{D}$ receives a ciphertext $(\sigma'_1, \sigma_2, \ldots, \sigma_N, \tau)$, it can detect whether a tracing procedure has been applied. Therefore, by applying a standard delaying technique (such as the one used in [7]), the stateful pirate decoder can defeat the tracing algorithm: upon detecting tracing, the decoder might continue to work by returning the message m (by decrypting the ciphertext σ_N) for a random number of trials and then start returning a random message $m^\star$. By this strategy, any user $2, \ldots, N$ can be claimed to be guilty.

4 Basic Scheme Against Stateful Pirate Decoders

We now transform the PSTBasic scheme for stateless decoders to a scheme for stateful decoders. In our construction, inspired by Kiayias and Yung's method, we employ a watermarking scheme. The basic scheme is therefore quite inefficient. However, in the next section, we will show how to use this basic scheme to construct an efficient general scheme.

Adequate presentation of a message. In almost all applications of traitor tracing, a slight modification of data does not affect the user. For example, users are not affected by a slight modification of a pixel in a figure or small changes in spaces between words in text display. For a data M, we call "adequate" presentations of M varied copies which can replace M without affecting the users. A decoder is usable if, from a ciphertext of M, it returns an adequate presentation of M.

Formally, as in [7], we restrict ourselves to plaintext spaces for which the following watermarking assumption is true:

Assumption 4 (Watermarking Assumption). *For some t, h, there is a probabilistic algorithm (t, h)-$\mathcal{W}$ such that, given any $M \in \mathcal{M}$, it produces h "versions" of M, $M_1, M_2, \ldots, M_h$, such that the following are true:*

(i) M_i are "adequate" presentations of M
(ii) there is an algorithm $\mathcal{W}'$ such that for any algorithm $\mathcal{A}$ that generates a M' given $M_{j_1}, \ldots, M_{j_k}$, $\mathcal{W}'$ given M' traces back to one of the M_{j_l} , provided that M' is an adequate presentation of M, and that k is below a certain threshold t.

This assumption has been used in [6,7] and can be achieved in most audio or video streams. We can thus make use of watermarking techniques as those of [9,2] to support the tracing process.

We remark that a $(t, h)-$watermarking scheme is also a $(t', h')-$ watermarking scheme, for all $t' \leq t, h' \leq h$. Therefore, possessing a $(t, h)-$watermarking scheme, we can use it as a $(N, N)-$watermarking scheme, for $N \leq t$ and $N \leq h$.

Basic scheme. Our basic scheme, called StatefulBasic(N, λ) is described as follows:

Primitives: a public-key encryption PKE, a data encapsulation mechanism DEM and a watermarking algorithm (N, N)-$\mathcal{W}$.
Setup(N, λ): It simply runs PSTBasic(N, λ) with the same parameters, and outputs pk as the public encryption key and $sk_1, \ldots, sk_N$ as private keys.

Encrypt(pk, M): *Algorithm* StatefulBasic.Enc(pk, m) $\rightarrow c$.

The algorithm StatefulBasic.Enc(.) takes as input the public key pk, a message m and outputs a ciphertext c. It uses $\mathcal{W}$ to produce "adequate" presentations of m, DEM to encrypt the messages and PKE to encrypt the keys used in DEM:

- Call $\mathcal{W}(m) \rightarrow m_1, \ldots, m_N$.
- Choose random keys $dk_1, \ldots, dk_N$
- For each $i = 1, \ldots, N$, call DEM.Enc$_{dk_i}(m_i) \rightarrow \tau_i$
- Compute $h_i = H(\tau_i)$ and for each $i = 1, \ldots, N$, call PKE.Enc$_{pk_i}(dk_i \| h_i) \rightarrow \sigma_i$.
- Define $c = (\sigma_1, \ldots, \sigma_N, \tau_1, \ldots, \tau_N)$.

Decrypt(j, K_j, C, pk): *Algorithm* StatefulBasic.Dec(sk_i, c) $\rightarrow m_i$ or $\perp$.

The algorithm StatefulBasic.Dec(.) takes as input a secret key sk_i and a ciphertext
$c = (\sigma_1, \ldots, \sigma_N, \tau_1, \ldots, \tau_N)$, it outputs a message m_i or $\perp$:

- Call PKE.Dec$_{sk_i}(\sigma_i) \rightarrow dk_i \| h_i$
- If $h_i \neq H(\tau_i)$, return $\perp$
- Otherwise, call DEM.Dec$_{dk_i}(\tau_i) \rightarrow m_i$ and output m_i

Trace($\mathcal{D}$, pk):

Algorithm StatefulBasic.Public-Trace(pk, $\mathcal{D}$) $\rightarrow t$.

The algorithm StatefulBasic.Public-Trace(.) takes as input the public key pk and a pirate decoder $\mathcal{D}$ and outputs a traitor t as follows:

- Choose random dk, m, and call PSTBasic.Enc(pk, m) $\rightarrow (\sigma_1, \ldots, \sigma_N, \tau_1, \ldots, \tau_N)$. Recall that $\tau_i = $ DEM.Enc$_{dk_i}(m_i)$ and $m_1, \ldots, m_N$ are outputted by $\mathcal{W}(m)$.
- Give $(\sigma_1, \ldots, \sigma_N, \tau_1, \ldots, \tau_N)$ to $\mathcal{D}$.
- Suppose that $\mathcal{D}(\sigma_1, \ldots, \sigma_N, \tau_1, \ldots, \tau_N) \rightarrow m^\star$
- Call $\mathcal{W}'(m^\star) \rightarrow m_t$ and output t as a traitor.

Traceability. We first consider the traceability of the above system.

Theorem 5. *If W is a (N, N)−watermarking scheme, PKE and DEM are semantically secure, the above scheme is a fully collusion resistant traitor tracing scheme against stateful pirate decoders.*

Proof. We remark that the above tracing only makes one query to the pirate decoder. Moreover, this query is a valid ciphertext. Therefore, if a pirate decoder is usable, it should return an "adequate" presentation $m^\star$ of the original message m.

Let $(\sigma_1, \ldots, \sigma_N, \tau_1, \ldots, \tau_N)$ be the query to the pirate decoder and $m_1, \ldots, m_N$ be the N underlying messages of $\sigma_1, \ldots, \sigma_N$. We note that $m_1, \ldots, m_N$ are outputted by $\mathcal{W}(m)$.

Suppose that the pirate decoder $\mathcal{D}$ is produced from a collusion of k users $u_{j_1}, \ldots, u_{j_k}$, we show that all the information the pirate decoder knows from the unique query $(\sigma_1, \ldots, \sigma_N, \tau_1, \ldots, \tau_N)$ are the k messages $m_{j_1}, \ldots, m_{j_k}$. For this, we use a hybrid argument as follows.

Denote by $i_1, \ldots, i_{N-k}$ the $N - k$ indexes that are not in $\{j_1 \ldots j_k\}$. At each qth step (q runs from 1 to $N - K$), we replace σ_{i_q} by $\sigma_{i_q} = \mathsf{PKE.Enc}_{pk_{i_q}}(dk'_{i_q} \| h_{i_q})$, where dk'_{i_q} is randomly chosen in the key space, and τ_{i_q} by $\tau_{i_q} = \mathsf{DEM}_{dk'_{i_q}}(m'_{i_q})$, where m'_{i_q} is randomly chosen in the message space. The pirate decoder has thus, after the qth step, no information about m_{i_q}. Because PKE and DEM are semantically secure, the pirate decoder has a negligible advantage to distinguish between two successive steps.

The message $m^\star$ is thus produced from $m_{j_1}, \ldots, m_{j_k}$. At this stage, we can use the algorithm $\mathcal{W}'$ to reveal one of $m_{j_t} \in \{m_{j_1} \ldots m_{j_k}\}$ and can correctly return the user j_t as a traitor. $\qquad\square$

Security of Encryption. The $\mathsf{PSTBasic}$ scheme can be considered as a particular case of our $\mathsf{StatefulBasic}$ scheme with $dk_1 = \ldots = dk_N = dk$ and $m_1 = \ldots = m_N = m$. Using the argument in the security proof of $\mathsf{PSTBasic}$ [10], our scheme achieves the same security level, *i.e.* semantic security against "Replayable" CCA adversaries.

Remarks. The above basic scheme is inefficient in comparison with $\mathsf{PSTBasic}$ scheme because of its linear ciphertext rate. However, its main advantage is that it can be used for stateful pirate decoders and that the tracing procedure is very efficient with only one query to each pirate decoder. More interestingly, this basic scheme, although inefficient, helps us to construct an efficient general scheme which is almost as efficient as the PST general scheme [10].

The construction of the general scheme is described in the next section.

5   General Scheme for Stateful Pirate Decoders

5.1   IPP Codes

Let $\mathcal{Q}$ be an alphabet containing q symbols. If $C = \{w_1, w_2, \ldots, w_N\} \subset \mathcal{Q}^\ell$, then C is called a q-ary code of size N and length ℓ. Each $w_i \in C$ is called a codeword and we write $w_i = (w_{i,1}, w_{i,2}, \ldots, w_{i,\ell})$ where $w_{i,j} \in \mathcal{Q}$ is called the jth component of the codeword w_i.

We define *descendants* of a subset of codewords as follows. Let $X \subset C$ and $u = (u_1, u_2, \ldots, u_\ell) \in \mathcal{Q}^\ell$. The word u is called a descendant of X if for any $1 \leq j \leq \ell$, the jth component u_j of u is equal to a jth component of a codeword in X. In this case, codewords in X are called *parent codewords* of u. For example, $(3, 2, 1, 3)$ is a descendant of the three codewords $(3, 1, 1, 2)$, $(1, 2, 1, 3)$ and $(2, 2, 2, 2)$.

$$
\begin{array}{l}
\mathbf{3\ 1\ 1\ 2} \\
\mathbf{1\ 2\ 1\ 3} \quad \longleftarrow \quad \text{parent codewords} \\
\mathbf{2\ 2\ 2\ 2} \\
\hline
\mathbf{3\ 2\ 1\ 3} \quad \longleftarrow \quad \text{a descendant}
\end{array}
$$

We denote by $\mathsf{Desc}(X)$ the set of all descendants of X. For a positive integer c, denote by $\mathsf{Desc}_c(C)$ the set of all descendants of subsets of up to c codewords. Codes with identifiable parent property (IPP codes) are defined as follows.

Definition 6. *A code C is called c-IPP if, for any $u \in \text{Desc}_c(C)$, there exists a $w \in C$ such that for any $X \subset C$, if $|X| \leq c$ and $u \in \text{Desc}(X)$, then $w \in X$.*

In a c-IPP code, given a descendant $u \in \text{Desc}_c(C)$, we can always identify at least one of its parent codewords. Binary c-IPP codes (with more than two codewords) do not exists, thus in any c-IPP code, the alphabet size $q \geq 3$. Some typical constructions are in [12]. The best known algorithms construct c-IPP codes and c-collusion secure codes [2] with logarithmic length in number of codewords.

5.2 Framework for Combination of Basic Schemes

In [8,5,10], the authors constructed general schemes from combinations of basic schemes by using collusion secure codes or IPP codes. We resume this framework, in the case of using IPP codes, as follows:

Primitives: $BasicSch_1, \ldots, BasicSch_\ell$ are basic schemes of q users. C is a q-ary c-IPP code of length ℓ.

Step 1: Each user is associated to a codeword $w = w_1 \ldots w_\ell$ $(w_i \in \{1, \ldots, q\})$ of C. This codeword determines the user key which contains ℓ sub-keys corresponding to ℓ basic schemes $BasicSch_1, \ldots, BasicSch_\ell$: the w_i^{th} key in scheme $BasicSch_i$, for each $i = 1, 2, \ldots, n$.

Step 2: The general scheme is combined from ℓ basic schemes:

$$GeneralSch = (BasicSch_1, \ldots, BasicSch_\ell)$$

Step 3: For tracing, we first find out the codeword $w^\star$ associated to the pirate decoder $\mathcal{D}$ by finding out each $w_i^\star$ as follows:
- create valid ciphertexts in $BasicSch_1, \ldots, BasicSch_{i-1}, BasicSch_{i+1}, \ldots, BasicSch_\ell$:
  $c_1, \ldots, c_{i-1}, c_{i+1}, \ldots, c_\ell$
- create probe ciphertext (ciphertext for tracing) in $BasicSch_i$: $c_i^\star$
- give the combined ciphertext $\mathsf{c} = (c_1, \ldots, c_{i-1}, c_i^\star, ci+1, \ldots, c_\ell)$ to the pirate decoder $\mathcal{D}$. On feedback of $\mathcal{D}$, by using the tracing algorithm in $BasicSch_i$, find out $w_i^\star$.

Step 4: from $w^\star = w_1^\star \ldots w_\ell^\star$, the codeword associated to the pirate decoder, thanks to the identifiable parent property of codes, we can trace back one traitor.

In the PST general scheme, in Step 2, $BasicSch_1, \ldots, BasicSch_\ell$ are independent instances of the PSTBasic. We can replace $BasicSch_1, \ldots, BasicSch_\ell$ by our basic schemes StatefulBasic. However, in this case, we lost a q factor of efficiency in comparison with the PST general scheme, as each of our StatefulBasic scheme loses a linear factor in comparison with PSTBasic scheme.

By using 3-ary IPP codes ($q = 3$), we only loose a small constant factor of 3. We would like nevertheless to improve the efficiency to make it comparable to the PST scheme. The idea is to include a StatefulBasic scheme in a sequence of PSTBasic schemes. More precisely, in Step 2, we replace one of ℓ PSTBasic schemes by one StatefulBasic scheme. Due to the independency between basic schemes, such a combination works well for encryption and decryption. We are only worrying about the tracing capability. Below, we present such a combination with traceability. We note however that this technique of combination cannot be used for constructions where basic schemes share common data. In [5], in order to improve the efficiency, the general scheme combines basic schemes which share some common data. Our technique is thus not suitable to make the scheme in [5] resistant against stateful pirate decoders.

5.3   General Scheme

Let $C = \{\omega_1, \ldots, \omega_N\}$ be a q-ary c-IPP code that allows collusion of up to c users. The N-user general scheme is a combination of ℓ basic schemes PSTBasic $S_1, S_2, \ldots, S_\ell$ and a basic scheme StatefulBasic, each basic scheme supports q users:

Setup: Given security parameters λ and c, the algorithm works as follows:
- For each $j = 1, \ldots, \ell$, call the algorithm PSTBasic.Gen$(1^\lambda, q)$ to generate an encryption key pk_j and q decryption keys $sk_{j,1}, \ldots, sk_{j,q}$ for the q-user system S_j.
- For each $j = 1, \ldots, \ell$, call the algorithm StatefulBasic.Gen$(1^\lambda, q)$ to generate an encryption key $\overline{pk_j}$ and q decryption keys $\overline{sk}_{j,1}, \ldots, \overline{sk}_{j,q}$ for the q-user system $\overline{S}_j$.
- Public key pk is defined by the tuple $(pk_i)_{i=1,\ldots,\ell}$, $(\overline{pk_i})_{i=1,\ldots,\ell}$ and the code C.
- Private key K_i of each user i (for $i = 1, \ldots, N$) contains a codeword $w_i \in C$, $\overline{sk}_i$, the ℓ-tuple key $sk_{1,w_{i,1}}, sk_{2,w_{i,2}}, \ldots, sk_{\ell,w_{i,\ell}}$ and the ℓ-tuple key $\overline{sk}_{1,w_{i,1}}, \overline{sk}_{2,w_{i,2}}, \ldots, \overline{sk}_{\ell,w_{i,\ell}}$, where $w_{i,j} \in \mathcal{Q} = \{1, 2, \ldots, q\}$ is the symbol at the jth position of the codeword w_i.

$$K_i \doteq (w_i, sk_{1,w_{i,1}}, sk_{2,w_{i,2}}, \ldots, sk_{\ell,w_{i,\ell}}, \overline{sk}_{1,w_{i,1}}, \overline{sk}_{2,w_{i,2}}, \ldots, \overline{sk}_{\ell,w_{i,\ell}})$$

Encrypt(pk, m): The plaintext space of the ℓ-key system is $\mathcal{M}^\ell$. On input $m = (m_1, m_2, \ldots, m_\ell)$, the encryption algorithm works as follows:
- an index k is randomly chosen $k \xleftarrow{R} i = 1, \ldots, \ell$.
- the kth component c_k is encrypted by StatefulBasic $\overline{S}_k$:

$$c_k = \text{StatefulBasic.Enc}_{\overline{pk_k}}(m_k) = (\sigma_{k,1}, \ldots, \sigma_{k,q}, \tau_{k,1}, \ldots, \tau_{k,q})$$

- for $j = 1, \ldots, \ell$, $j \neq k$, the jth component c_k is encrypted by PSTBasic S_j:

$$c_j = \text{PSTBasic.Enc}_{pk_j}(m_j) = (\sigma_{j,1}, \ldots, \sigma_{j,q}, \tau_j)$$

- the ciphertext $\mathsf{c} \doteq (k, c_1, c_2, \ldots, c_\ell)$

Decrypt$(j, K_j, \mathsf{c}, \mathsf{pk})$: On the ciphertext $(k, c_1, c_2, \ldots, c_\ell)$, user i uses his secret key to compute:

$$m_k = \mathsf{StatefulBasic.Dec}_{\overline{sk}_{k,w_{i,k}}}(c_k)$$

$$m_j = \mathsf{PSTBasic}_{sk_{j,w_{i,j}}}(c_j), \text{ for } j = 1, \ldots, \ell, j \neq k$$

Trace$(\mathcal{D}, \mathsf{pk})$: Let $E = \{1, \ldots \ell\}$. Repeat the following steps until $E = \varnothing$.

- randomly choose an index $k \xleftarrow{R} i = 1, \ldots, \ell$, $E = E \backslash \{k\}$
- choose a message $m = (m_1, m_2, \ldots, m_\ell)$
- create components: $c_j = \mathsf{PSTBasic.Enc}_{pk_j}(m_j)$, for $j = 1, \ldots, \ell, j \neq k$
- create component $c_k = \mathsf{StatefulBasic.Enc}_{\overline{pk}_k}(m_k)$
- define $\mathsf{c} = (k, c_1, c_2, \ldots, c_\ell)$ and give c to the pirate decoder $\mathcal{D}$
- on feedback $m' = (m_1', m_2', \ldots, m_\ell')$, extract the component m_k' and use the same argument used in the tracing algorithm of StatefulBasic scheme to get w_k.
- from the descendant codeword $w = (w_1 || \ldots || w_\ell) \in \mathcal{Q}^\ell$, identify one of its parent codewords. The user associated with this codeword is a traitor.

Traceability. We first consider the traceability of the above system.

Proposition 7. *Suppose that $\mathcal{D}$ is a stateful pirate decoder, the above tracing algorithm can correctly trace back a traitor.*

Proof. We remark that, in the above tracing algorithm, query ciphertexts are identical to valid ciphertexts. Therefore, a pirate decoder cannot detect whether it is being traced. Consequently, a stateful pirate decoder always decrypts correctly as it would do. In this case, a stateful pirate decoder is not more powerful than a stateless one.

Using the same arguments from Theorem 5, the tracing procedure can reveals each w_k as the tracing procedure in StatefulBasic can correctly reveal a traitor. Therefore, the tracing can correctly associate a descendant codeword $w = (w_1 || \ldots || w_\ell)$ of the set of codewords corresponding to the users in the collusion. By the property of IPP codes, the tracing algorithm can thus identify at least one traitor.

Security of Encryption. We now consider the security of the encryption. For this, one could use the following assumption, used in [8,5,10]:

Assumption 8 (threshold assumption). *A pirate-decoder that only returns correctly a fraction p of a plaintext of length λ where $1 - p$ is a non-negligible function in λ, is useless.*

We emphasis that, as already mentioned in [8], by employing an all-or-nothing transform [11,4], this assumption is not necessary.

Proposition 9. *In the general scheme, a collusion of users in the $(\ell - 1)$ basic schemes does not affect the security of the remaining basic scheme.*

Proof. The proof follows the one in [5]. Assume there is an adversary $\mathcal{A}$ that, having information I of the targeted system, called $Basic_1$, and also all the information for ℓ remained systems, called $Basic_2, \ldots, Basic_n$ can get an advantage ε for breaking the system $Basic_1$ (for some goal G). We can then construct an algorithm $\mathcal{B}$ that, having only the information I of the system $Basic_1$, can break the system $Basic_1$ (for the goal G) with advantage ε. Indeed, the algorithm $\mathcal{B}$ can perfectly simulate all the information about $Basic_2, \ldots, Basic_n$ systems by generating itself all parameters for $Basic_2, \ldots, Basic_n$. Because the $Basic_1, \ldots, Basic_n$ systems are totally independent that they do not have any common information, this simulation of $\mathcal{B}$ is perfect.   □

This proposition shows that the security of the general scheme is at least the same as the security of each basic scheme. Therefore, the encryption in the above general system is secure.

Efficiency. The ciphertext contains two parts: ciphertext body $(\tau_1, \ldots, \tau_\ell)$ and ciphertext header $(\sigma_{j,1}, \ldots, \sigma_{j,q})_{j=1,\ldots,\ell}$. Between ℓ sub-ciphertext bodies $(\tau_1, \ldots, \tau_\ell)$, $\ell-1$ sub-ciphertext bodies correspond to the basic schemes PSTBasic and only one sub-ciphertext bodie corresponds to the basic scheme StatefulBasic. As in PSTBasic scheme, the ciphertext body approximately has the same size as the plaintext, and as ℓ is large, the ciphertext body in our general scheme approximately has the same size as the original message. Concerning the ciphertext header, as we use the hybrid framework, each $\sigma_{j,k}$ is significantly smaller than τ_j. Therefore, the header's size is small compared to the message size and the ciphertext rate is almost optimal (rate =1). Without the hybrid argument, the header rate is also small (this rate $= q = 3$) and therefore, the ciphertext rate still remains constant.

6   Conclusion

We proposed the first constant ciphertext rate traitor tracing for stateful pirate decoders. Our scheme moreover supports black-box public traceability. However, due to the use of IPP codes, our scheme does not support full collusion as schemes for stateless pirate decoders in [1,3]. We raise thus the open question of constructing a fully collusion resistant scheme for stateful pirate decoders with constant ciphertext rate.

Acknowledgments

We would like to thank Antoine Joux and Louis Goubin for helpful discussions and suggestions.

References

1. Dan Boneh, Amit Sahai, and Brent Waters. Fully Collusion Resistant Traitor Tracing with Short Ciphertexts and Private Keys. In Serge Vaudenay, editor, *EUROCRYPT 2006*, volume 4004 of LNCS, pages 573–592, St. Petersburg, Russia, May 2006. Springer-Verlag, Berlin, Germany.

2. Dan Boneh and James Shaw. Collusion-secure fingerprinting for digital data. In Don Coppersmith, editor, *CRYPTO'95*, volume 963 of *LNCS*, pages 452–465, Santa Barbara, CA, USA, August 27–31, 1995. Springer-Verlag, Berlin, Germany.
3. Dan Boneh and Brent Waters. A Fully Collusion Resistant Broadcast, Trace and Revoke System, 2006. (To Appear in) Proceedings of 13th ACM Conference on Computer and Communications Security (*ACM CCS 2006*). http://crypto.stanford.edu/ dabo/papers/tr.pdf.
4. Ran Canetti, Yevgeniy Dodis, Shai Halevi, Eyal Kushilevitz and Amit Sahai. Exposure-Resilient Functions and All-or-Nothing Transforms. In Bart Preneel, editor, *EUROCRYPT 2000*, volume 1807 of *LNCS*, pages 453–469, Bruges, Belgium, May 14–18, 2000. Springer-Verlag, Berlin, Germany.
5. Hervé Chabanne, Duong Hieu Phan, and David Pointcheval. Public Traceability in Traitor Tracing Schemes. In Ronald Cramer, editor, *EUROCRYPT 2005*, volume 3494 of *LNCS*, pages 542–558, Aarhus, Denmark, May 22–26, 2005. Springer-Verlag, Berlin, Germany.
6. Amos Fiat and Tamir Tassa. Dynamic Traitor Traicing. In Michael J. Wiener, editor, *CRYPTO'99*, volume 1666 of *LNCS*, pages 354–371, Santa Barbara, CA, USA, August 15–19, 1999. Springer-Verlag, Berlin, Germany.
7. Aggelos Kiayias and Moti Yung. On Crafty Pirates and Foxy Tracers. In T. Sander, editor, *ACM Workshop in Digital Rights Management – DRM 2001*, volume LNCS 2320, pages 22–39. Springer, 2001.
8. Aggelos Kiayias and Moti Yung. Traitor Tracing with Constant Transmission Rate. In Lars R. Knudsen, editor, *EUROCRYPT 2002*, volume 2332 of *LNCS*, pages 450–465, Amsterdam, The Netherlands, April 28 – May 2, 2002. Springer-Verlag, Berlin, Germany.
9. Joe Kilian, F. Thompson Leighton, Lesley R. Matheson, Talal G. Shamoon, Robert E. Tarjan, and Francis Zane. Resistance of digital watermarks to collusive attacks. In *Proceedings of the 1998 IEEE International Symposium on Information Theory*, page 271, 1998.
10. Duong Hieu Phan, Reihaneh Safavi-Naini, and Dongvu Tonien. Generic Construction of Hybrid Public Key Traitor Tracing with Full-Public-Traceability. In M. Bugliesi, editor, *33rd ICALP*, volume 4052 of *LNCS*, pages 630–648, Venice - Italy, July 9–16, 2006. Springer-Verlag, Berlin, Germany.
11. Ronald L. Rivest. All-or-Nothing Encryption and the Package Transform. In *Proceedings of the 4th FSE*, LNCS 1267. Springer-Verlag, Berlin, 1997.
12. Tran Van Trung and Sosina Martirosyan. New Constructions for IPP Codes. *Des. Codes Cryptography*, 35(2):227–239, 2005.

Reducing the Spread of Damage of Key Exposures in Key-Insulated Encryption

Thi Lan Anh Phan[1], Yumiko Hanaoka[2], Goichiro Hanaoka[3], Kanta Matsuura[1], and Hideki Imai[3]

[1] The University of Tokyo, Japan
{phananh,kanta}@iis.u-tokyo.ac.jp
[2] NTT DoCoMo, Inc.
claudia@nttdocomo.co.jp
[3] National Institute of Advanced Industrial Science and Technology, Japan
{hanaoka-goichiro,h-imai}@aist.go.jp

Abstract. A proposal for key exposure resilient cryptography called, *key-insulated public key encryption* (KIPE), has been proposed by Dodis, Katz, Xu, and Yung [6] in which the secret key is changed over time so that the exposure of current key minimizes the damage overall. We take this idea further toward betterment by introducing new schemes with improved helper key security: in our schemes, we introduce an *auxiliary helper key* to update the secret key less frequently than the main helper key (and only one of these keys is used at each key updates,) as a result, this gives added protection to the system, by occasional auxiliary key updates, reducing the spread of further harm that may be caused by key exposure when compared to the original KIPE. Our proposed schemes are proven to be semantically secure in the random oracle model.

1 Introduction

Background. Advances in mobile and wireless technology are getting more and more complex each day and so are the information handling and processing, which are also becoming more mobile and diversified than ever, creating higher risks than in the past for security breaches to occur. Mobile devices are especially vulnerable to data loss and secret key exposure, and such incidents are said to be one of the most serious attacks in real life. Laptops are the obvious examples of these devices, but PDAs, cell phones, USB drives, and even iPods can store large amounts of data, making them tempting targets for thieves. Some incidents are malicious, some are accidents, and some are the result of sloppy management of personal information by the users, but whatever the cause may be, protecting our assets, nevertheless, is essential. Also, security is becoming much more under the control of end users than before, and most end users aren't aware of the risks or how to properly protect against privacy and security incidents.

Dodis et al.[6] proposed a new paradigm called *key-insulated public key cryptosystem* (KIPE) which gives "resilience" to key exposure by changing (or "evolving") the secret key over time so that exposure of current key mitigates the

P.Q. Nguyen (Ed.): VIETCRYPT 2006, LNCS 4341, pp. 366–384, 2006.

overall damage: even if the current key is exposed, all the rest of (unexposed) keys remain unaffected, and both future and past security are guaranteed. KIPE performs key updates with the aid of a "helper key". Helper key is stored inside an external device which is connected to the user device (therefore, the network,) only at occasions the secret key needs to be updated. Hence, the chance of helper key exposure to occur is quite small, although, it is not to say definitely that such event will never occur: for instance, if we try to enhance the security of the secret key by updating the key frequently, the helper key will in turn be more vulnerable to the network hackers (as it will be connected to the network more often).

Our Contribution. In this paper, we propose new KIPE schemes with improved helper key security. More specifically, in our schemes, we introduce an *auxiliary helper key* which is used to update a secret key less frequently than the main helper key. Suppose if you update the secret key once each day: basically, you only need to pay attention to the main helper key which you will use to update the secret key just like in the original KIPE, and use the auxiliary helper key only occasionally, at an interval longer, e.g. the first day of each month, on which the auxiliary key is used instead of the main key (and the main key will not be used at all on that day). It turns out that the auxiliary key updates reduce further spread of harm that may be caused by key exposure, i.e. by one month, if ever a secret key and the main helper key are compromised. Also, since the auxiliary key is used less frequently than the main helper key, it can be stored at some place much safer and managed differently than the main key, as an effect, further decreases the likelihood of both helper keys to be compromised at the same time, increasing the overall security of the system considerably. The encryption procedure in our schemes is carried out using public key and time, and exposure of secret key(s) will still guarantee the security of all unexposed time periods, just like in the original KIPE. Our proposed schemes are proven to be semantically secure in the random oracle model.

Initialization in our scheme involves providing main helper device H_{main} and auxiliary helper device H_{aux} with a main helper key mk and an auxiliary helper key ak, respectively, and the user's terminal with a *stage 0 user secret key* usk_0. Similarly to the original KIPE, user's public encryption key pk is treated like that of an ordinary encryption scheme with regard to certification, but its lifetime is divided into stages $i = 1, 2, ..., N(= n \cdot \ell)$ with encryption in stage i performed as a function of pk, i and the plaintext, and decryption in stage i performed by the user using a *stage i user secret key* usk_i obtained by the following key-update process performed at the beginning of stage i:

- If $i \neq k \cdot \ell$ for $k \in \{1, 2, ..., n\}$, H_{main} sends to the user's terminal over a secure channel a *stage i helper key* hsk_i computed as a function of mk and i
- If $i = k \cdot \ell$ for $k \in \{1, 2, ..., n\}$, similarly to the above, H_{aux} sends hsk_i computed as a function of ak and i,

the user computes usk_i as a function of usk_{i-1} and hsk_i, and erases usk_{i-1}. Like the original KIPE, our schemes also address random access key update [6] in which the user can compute an arbitrary stage user secret key (that could also be a past key). Note that it is reasonable to assume that mk and ak will not be exposed simultaneously as they can be managed separately.

The security intentions are:

1. Similarly to the original KIPE, if none of the helpers is compromised, then exposure of any of user secret keys does not compromise the security of the non-exposed stages,
2. even if one of H_{main} and H_{aux} is compromised, security is still guaranteed unless other secret information is exposed as well,
3. if mk and usk_i are compromised for some i $(k \cdot \ell \leq i \leq (k+1) \cdot \ell - 1)$, then security of stages $k \cdot \ell, ..., (k+1) \cdot \ell - 1$ are compromised,
4. if ak and usk_i are compromised for some i $(k \cdot \ell + 1 \leq i \leq (k+1) \cdot \ell - 2)$, then security of stage i is compromised, and
5. if ak and usk_i are compromised for some $i(= k \cdot \ell - 1$ or $k \cdot \ell)$, then security of stages $k \cdot \ell - 1$ and $k \cdot \ell$ are compromised.

Similar to the original KIPE, we can further address the case when all of the helper keys are exposed:

6. Even if both helpers H_{main} and H_{aux} are compromised, security of all stages remain secure as long as user secret key (of any one stage) is not compromised as well.

Application. Many of us are familiar with the following setting: a user with his portable device, such a device can be a laptop computer or a cell phone, either way, a portable device which he carries around with him daily where all his secret transactions such as decryption take place; needless to say, risk of leakage of sensitive data inside his device whether by accident or malicious intent is always an issue to him. As an application of our schemes, we can let the laptop be the main helper H_{main} where he stores the main helper key mk, and a dedicated smart card or the auxiliary helper H_{aux} in which the auxiliary helper key ak is stored and also managed securely (preferably at somewhere reasonably safe like home) when it is not in use. Laptop, i.e. H_{main}, is the one mainly being used to update his secret key just like in the original KIPE, and only occasionally, his smart card, i.e. H_{aux}, and by doing so can prevent further spreading of the damage that may be caused by key exposure even if H_{main} is ever compromised. To make things more clear, let us consider the next example: daily key updating is carried out on his laptop PC, and "safety guard" updates with his smart card at the first day of each month. As you can see, even if both the master helper key mk and a user secret key (let say, 6/12 user secret key) are exposed at the same time, still, the damage is kept to the minimum by losing only the security of a month of June and the rest remain secure. Also, even if the auxiliary key ak is exposed, the harm cause is merely for a day. End users are ultimately responsible for securing information more than ever before and this is a simple and effective way for the users to give added proof to their system.

Related Works. With the aim of enhancing the security of KIPE against helper key exposure, Dodis, Katz, Xu and Yung also proposed schemes called, *strongly key insulated public key encryption* (sKIPE) [6], here, exposure of the helper key will not compromise the security of the system right away, but this only holds if not one of secret keys is exposed at the same time. The work here was followed by Dodis, Franklin, Katz, Miyaji and Yung. In their scheme, *intrusion-resilient public key encryption* (IRPKE) [7], forward security is strengthened, so even if both the helper key and the secret key are exposed simultaneously, it compromises future periods but will protect the past periods. The security for IRPKE is improved from sKIPE, only, it became less convenient, as it no longer allow random access key update. Hanaoka, Hanaoka, Shikata, and Imai [12] proposed a scheme that also enhances the security of sKIPE by hierarchically structuring the helper key with added identity-based property. Also, in another work, Hanaoka, Hanaoka, and Imai [11] increased the security against helper key exposure by alternatively using two helper keys to update a secret key. [11] can be viewed as a special case of our proposing scheme. On the other hand, as an encryption scheme that allows key update, we have the KIPE, and also, *forward secure public key encryption* (FSPKE). FSPKE was introduced by Anderson [1] and the first efficient construction, by Canetti, Halevi and Katz [5]. Dodis, Franklin, Katz, Miyaji and Yung showed that by using FSPKE with certain homomorphic property, gives a generic IRPKE [8].

2   Definitions

First, we give the model of KIPE with an auxiliary helper key (KwAH) and the security notion. We follow by showing some of the characteristics of bilinear maps. We then briefly review the related computational assumptions.

2.1   Model: KIPE with an Auxiliary Helper Key

A KwAH scheme $\mathcal{E}$ consists of five efficient algorithms (**KeyGen, Δ-Gen, Update, Encrypt, Decrypt**).

KeyGen: Takes a security parameter k and returns mk, ak, usk_0 and pk. Public key pk includes a description of finite message space $\mathcal{M}$, and description of finite ciphertext space $\mathcal{C}$.

Δ-Gen: Takes as inputs, mk and i, and returns stage i helper key hsk_i if $\ell \nmid i$, or $\perp$ otherwise, and takes as inputs, ak and i, and returns stage i helper key hsk_i if $\ell | i$, or $\perp$ otherwise

Update: Takes as inputs, usk_{i-1}, hsk_i and i, and returns stage i user secret key usk_i.

Encrypt: Takes as inputs, pk, i and $M \in \mathcal{M}$, and returns ciphertext $C \in \mathcal{C}$.

Decrypt: Takes as inputs, pk, usk_i and $C \in \mathcal{C}$, and returns $M \in \mathcal{M}$ or $\perp$.

These algorithms must satisfy the standard consistency constraint, namely,

$$\forall i \in \{1, 2, ..., N\}, \ \forall M \in \mathcal{M} : \textbf{Decrypt}(pk, usk_i, C) = M$$
$$\text{where } C = \textbf{Encrypt}(pk, i, M).$$

2.2 Security Notion

Here, we define the notion of semantic security for KwAH. This is based on the security definition in the original KIPE [6,4]. It should be noticed that the definition in [4] looks simpler than in [6] but they are essentially the same.

We say that a KwAH scheme $\mathcal{E}$ is *semantically secure against an adaptive chosen ciphertext attack under an adaptive chosen key exposure attack* (IND-KE-CCA) if no polynomially bounded adversary $\mathcal{A}$ has a non-negligible advantage against the challenger in the following IND-KE-CCA game:

Setup: The challenger takes a security parameter k and runs the **KeyGen** algorithm. He gives the adversary the public key pk and keeps usk_0, mk and ak to himself.

Phase 1: The adversary issues several queries $q_1, \cdots, q_m$ where each of the queries q_i is one of:
- Exposure query $\langle j, \mathtt{class} \rangle$: If $\mathtt{class}$ = "user", the challenger responds by running the algorithms Δ-**Gen** and **Update** to generate usk_j and sends it to the adversary. If $\mathtt{class}$ = "main helper" or "auxiliary helper", the challenger sends mk or ak to the adversary, respectively.
- Decryption query $\langle j, C \rangle$: The challenger responds by running the algorithms Δ-**Gen** and **Update** to generate usk_j. He then runs **Decrypt** to decrypt the ciphertext C using usk_j and sends the result to the adversary.

These queries may be asked adaptively, that is, each query q_i may depend on the replies to $q_1, \cdots, q_{i-1}$.

Challenge: Once the adversary decides that Phase 1 is over, she outputs two equal length plaintexts $M_0, M_1 \in \mathcal{M}$ and $j^* \in \{1, 2, ..., N\}$ on which she wishes to be challenged. The challenger picks a random bit $\beta \in \{0, 1\}$ and sets $C^* = \mathbf{Encrypt}(pk, j^*, M_\beta)$. The challenger sends C^* as the challenge to the adversary.

Phase 2: The adversary issues additional queries $q_{m+1}, \cdots, q_{max}$ where each of the queries is one of:
- Exposure query $\langle j, \mathtt{class} \rangle$: Challenger responds as in Phase 1.
- Decryption query $\langle j, C \rangle$: Challenger responds as in Phase 1.

These queries may be asked adaptively as in Phase 1.

Guess: Finally, the adversary outputs her guess $\beta' \in \{0, 1\}$. She wins the game if $\beta' = \beta$ and

1. $\langle j^*, C^* \rangle$ does not appear in Decryption queries,
2. $\langle j^*, \text{"user"} \rangle$ does not appear in Exposure queries,
3. both $\langle j, \text{"user"} \rangle$, such that $m \cdot \ell \le j^* \le (m+1) \cdot \ell - 1$ and $m \cdot \ell \le j \le (m+1) \cdot \ell - 1$ for some m $(0 \le m \le n - 1)$, and $\langle \cdot, \text{"main helper"} \rangle$ do not simultaneously appear in Exposure queries,
4. both $\langle j, \text{"user"} \rangle$, such that $j^* = (m+1) \cdot \ell - 1$ or $(m+1) \cdot \ell$ and $j = (m+1) \cdot \ell - 1$ or $(m+1) \cdot \ell$ for some m $(0 \le m \le n-1)$, and $\langle \cdot, \text{"auxiliary helper"} \rangle$ do not simultaneously appear in Exposure queries,
5. both $\langle \cdot, \text{"main helper"} \rangle$ and $\langle \cdot, \text{"auxiliary helper"} \rangle$ do not simultaneously appear in Exposure queries.

We refer to such an adversary $\mathcal{A}$ as an IND-KE-CCA adversary. We define adversary $\mathcal{A}$'s advantage in attacking the scheme $\mathcal{E}$ as:

$$Adv_{\mathcal{E},\mathcal{A}} = \Pr[\beta' = \beta] - 1/2.$$

The probability is over the random bits used by the challenger and the adversary.

Definition 1. We say that a KwAH system $\mathcal{E}$ is (t, ϵ)-*adaptive chosen ciphertext secure under adaptive chosen key exposure attacks* if for any t-time IND-KE-CCA adversary $\mathcal{A}$, we have $Adv_{\mathcal{E},\mathcal{A}} < \epsilon$. As shorthand, we say that $\mathcal{E}$ is IND-KE-CCA *secure*.

As usual, we can define chosen plaintext security similarly to the game above except that the adversary is not allowed to issue any Decryption queries. The adversary still can adaptively issue Exposure queries. We call this adversary IND-KE-CPA *adversary*.

Definition 2. We say that a KwAH system $\mathcal{E}$ is (t, ϵ)-*adaptive chosen plaintext secure under adaptive chosen key exposure attacks* if for any t-time IND-KE-CPA adversary $\mathcal{A}$, we have $Adv_{\mathcal{E},\mathcal{A}} < \epsilon$. As shorthand, we say that $\mathcal{E}$ is IND-KE-CPA *secure*.

IND-KE-CCA is already a strong security notion, but its security can be enhanced further to cover the compromise of both the helper keys. Concretely, as a constraint on the above adversary's Exposure query, we can modify 5. so that:

5′. $\langle \cdot$, "main helper"$\rangle$, $\langle \cdot$, "auxiliary helper"$\rangle$, and $\langle j$, "user"$\rangle$ do not simultaneously appear in Exposure queries for any $j \in \{1, 2, ..., N\}$.

Such modification allows the adversary $\mathcal{A}$ to obtain both mk and ak if $\mathcal{A}$ doesn't ask any of user secret keys. Let this adversary be a *strong* IND-KE-CCA adversary.

Definition 3. We say that a KwAH system $\mathcal{E}$ is (t, ϵ)-*adaptive chosen ciphertext secure under strongly adaptive chosen key exposure attacks* if for any t-time strong IND-KE-CCA adversary $\mathcal{A}$, we have $Adv_{\mathcal{E},\mathcal{A}} < \epsilon$. As shorthand, we say that $\mathcal{E}$ is *strongly* IND-KE-CCA *secure*.

Similarly, we can define *strong* IND-KE-CPA *adversary*, and here as well, she is not allowed to issue any Decryption queries.

Definition 4. We say that a KwAH system $\mathcal{E}$ is (t, ϵ)-*adaptive chosen plaintext secure under strongly adaptive chosen key exposure attacks* if for any t-time strong IND-KE-CPA adversary $\mathcal{A}$, we have $Adv_{\mathcal{E},\mathcal{A}} < \epsilon$. As shorthand, we say that $\mathcal{E}$ is *strongly* IND-KE-CPA *secure*.

A Remark on the Security Notion: Exposure of the Helper Keys. In the discussion we had so far, it may seem like we may have overlooked the exposure of stage i helper key, but actually, we haven't. It is obvious that if hsk_i can be computed from usk_{i-1} and usk_i for any stage i, then exposure of hsk_i can be emulated by using the responses to the Exposure queries. So, the security definition so far given is sufficient as it is even against exposure of stage i helper keys for any i, if we assume that such property holds. As a matter of fact, all of our constructions satisfy this property.

2.3   Bilinear Maps

We give brief review of the bilinear maps. Throughout this paper, we let $\mathbb{G}_1$ and $\mathbb{G}_2$ be two multiplicative cyclic groups of prime order q, and g be a generator of $\mathbb{G}_1$. A *bilinear map* $e : \mathbb{G}_1 \times \mathbb{G}_1 \rightarrow \mathbb{G}_2$ satisfies the following properties:

1. Bilinearity: For all $u, v \in \mathbb{G}_1$ and $a, b \in \mathbb{Z}$, $e(u^a, v^b) = e(u, v)^{ab}$.
2. Non-degeneracy: $e(g, g) \neq 1$.
3. Computability: There is an efficient algorithm to compute $e(u, v)$ for all $u, v \in \mathbb{G}_1$.

Note that a bilinear map is symmetric since $e(g^a, g^b) = e(g^b, g^a) = e(g, g)^{ab}$.

2.4   Complexity Assumptions

Here, we consider two complexity assumptions related to bilinear maps: the Computational Bilinear Diffie-Hellman (CBDH) assumption and the Gap Bilinear Diffie-Hellman (GBDH) assumption.

CBDH Assumption. The CBDH problem [2] in $\langle \mathbb{G}_1, \mathbb{G}_2, e \rangle$ is as follows: given a tuple $(g, g^a, g^b, g^c) \in (\mathbb{G}_1)^4$ as input, output $e(g, g)^{abc} \in \mathbb{G}_2$. An algorithm $\mathcal{A}_{cbdh}$ solves CBDH problem in $\langle \mathbb{G}_1, \mathbb{G}_2, e \rangle$ with the probability ϵ_{cbdh} if

$$\Pr[\mathcal{A}_{cbdh}(g, g^a, g^b, g^c) = e(g, g)^{abc}] \geq \epsilon_{cbdh},$$

where the probability is over the random choice of generator $g \in \mathbb{G}_1 \backslash \{1\}$, the random choice of $a, b, c \in \mathbb{Z}_q$ and random coins consumed by $\mathcal{A}_{cbdh}$.

Definition 5. We say that the $(t_{cbdh}, \epsilon_{cbdh})$-*CBDH assumption* holds in $\langle \mathbb{G}_1, \mathbb{G}_2, e \rangle$ if no t_{cbdh}-time algorithm has advantage of at least ϵ_{cbdh} in solving the CBDH problem in $\langle \mathbb{G}_1, \mathbb{G}_2, e \rangle$.

GBDH Assumption. The GBDH problem in $\langle \mathbb{G}_1, \mathbb{G}_2, e \rangle$ is as follows: given a tuple $(g, g^a, g^b, g^c) \in (\mathbb{G}_1)^4$ as input, output $e(g, g)^{abc} \in \mathbb{G}_2$ with the help of a decision BDH oracle $\mathcal{O}$ which for given $(g, g^a, g^b, g^c, T) \in (\mathbb{G}_1)^4 \times \mathbb{G}_2$, answers *"true"* if $T = e(g, g)^{abc}$, or *"false"* otherwise [13]. An algorithm $\mathcal{A}_{gbdh}$ solves GBDH problem in $\langle \mathbb{G}_1, \mathbb{G}_2, e \rangle$ with the probability ϵ_{gbdh} if

$$\Pr[\mathcal{A}_{gbdh}^{\mathcal{O}}(g, g^a, g^b, g^c) = e(g, g)^{abc}] \geq \epsilon_{gbdh},$$

where the probability is over the random choice of generator $g \in \mathbb{G}_1 \backslash \{1\}$, the random choice of $a, b, c \in \mathbb{Z}_q$ and random coins consumed by $\mathcal{A}_{gbdh}$.

Definition 6. We say that the $(t_{gbdh}, \epsilon_{gbdh})$-*GBDH assumption* holds in $\langle \mathbb{G}_1, \mathbb{G}_2, e \rangle$ if no t_{gbdh}-time algorithm has advantage at least ϵ_{gbdh} in solving the GBDH problem in $\langle \mathbb{G}_1, \mathbb{G}_2, e \rangle$.

3   Proposed Schemes

In this section, we propose our KwAH schemes and prove its security under CBDH assumption in the random oracle model. Our schemes are reasonably efficient since efficiency of our KwAH schemes can said to be comparable to [2]. In our schemes, we let $N = O(\mathsf{poly}(k))$.

3.1   IND-ID-CPA Scheme

Let $\mathbb{G}_1$ and $\mathbb{G}_2$ be two groups of order q of size k, and g be a generator of $\mathbb{G}_1$. Let $e : \mathbb{G}_1 \times \mathbb{G}_1 \to \mathbb{G}_2$ be a bilinear map. Let G, H be cryptographic hash functions $G : \mathbb{G}_2 \to \{0,1\}^n$ for some n, $H : \{0,1\}^* \to \mathbb{G}_1$, respectively. The message space is $\mathcal{M} = \{0,1\}^n$. The KwAH1 scheme consists of the following algorithms:

KwAH1: IND-KE-CPA CONSTRUCTION

KeyGen: Given a security parameter k, **KeyGen** algorithm:
1. generates $\mathbb{G}_1, \mathbb{G}_2, g$ and e.
2. picks $s_1, s_2 \in \mathbb{Z}_q^*$ uniformly at random, and sets $h_1 = g^{s_1}$ and $h_2 = g^{s_2}$,
3. chooses cryptographic hash functions G and H,
4. computes $d_{-1} = H(-1)^{s_1}$ and $d_0 = H(0)^{s_2}$,
5. outputs $pk = \langle q, \mathbb{G}_1, \mathbb{G}_2, e, n, g, h_1, h_2, G, H \rangle$, $mk = s_1$, $ak = s_2$ and $usk_0 = d_{-1} \cdot d_0$.

Δ-Gen: For given mk and $i \in \{1, 2, ..., N\}$, **Δ-Gen** algorithm:
1. outputs $\bot$ if $i = 0 \bmod \ell$,
2. outputs $hsk_i = H(i-1)^{-s_1} \cdot H(i)^{s_1}$ if $i \neq m \cdot \ell + 1$ for some m ($0 \leq m \leq n-1$),
3. outputs $hsk_i = H(i-2)^{-s_1} \cdot H(i)^{s_1}$ if $i = m \cdot \ell + 1$ for some m ($0 \leq m \leq n-1$).

For given ak and $i \in \{1, 2, ..., N\}$, **Δ-Gen** algorithm:
1. outputs $\bot$ if $i \neq 0 \bmod \ell$,
2. outputs $hsk_i = H(i - \ell)^{-s_2} \cdot H(i)^{s_2}$ otherwise.

Update: For given usk_{i-1}, hsk_i and i, **Update** algorithm:
1. computes $usk_i = usk_{i-1} \cdot hsk_i$,
2. deletes usk_{i-1} and hsk_i,
3. outputs usk_i.

Encrypt: For given pk, i, and a message $M \in \{0,1\}^n$, assuming that $m \cdot \ell + 1 \leq i \leq (m+1) \cdot \ell$ for some m ($0 \leq m \leq n-1$), **Encrypt** algorithm:
1. chooses random $r \in \mathbb{Z}_q^*$,
2. computes $W = (e(h_1, H(i)) \cdot e(h_2, H(m \cdot \ell)))^r$ if $i \neq (m+1) \cdot \ell$,
3. computes $W = (e(h_1, H(i-1)) \cdot e(h_2, H((m+1) \cdot \ell)))^r$ if $i = (m+1) \cdot \ell$,
4. sets $C = \langle i, g^r, G(W) \oplus M \rangle$,
5. outputs C as a ciphertext.

Decrypt: For given pk, usk_i and $C = \langle i, c_0, c_1 \rangle$, **Decrypt** algorithm:
1. computes $W' = e(c_0, usk_i)$,
2. computes $M' = c_1 \oplus G(W')$,
3. outputs M' as a plaintext.

3.2 Security

Now, we prove that KwAH1 is IND-KE-CPA under the CBDH assumption. Here, we briefly mention the technical hurdle for the security proof. Since we consider adaptively chosen key exposure adversary, the simulator has to deal with various types of key exposures, i.e. mixture of mk, ak, and user secret keys, and moreover, it does not know the adversary's strategy before the simulation. Nevertheless, the simulator must provide successful simulation. This makes the proof complicated.

Theorem 1. *Suppose* $(t_{cbdh}, \epsilon_{cbdh})$*-CBDH assumption holds in* $\langle \mathbb{G}_1, \mathbb{G}_2, e \rangle$ *and hash functions G and H are random oracles. Then,* KwAH1 *is* $(t_{kwah}, \epsilon_{kwah})$*-*IND-KE-CPA *secure as long as:*

$$\epsilon_{kwah} \leq \frac{3q_G N}{2} \epsilon_{cbdh}$$
$$t_{kwah} \leq t_{cbdh} + \Theta(\tau(2q_H + 3q_E)),$$

where IND-KE-CPA *adversary $\mathcal{A}_{kwah}$ issues at most q_H H-queries and q_E Exposure queries. Here, τ is the maximum time for computing an exponentiation in* $\mathbb{G}_1, \mathbb{G}_2$*, and pairing e.*

The proof of Theorem 1 is given in Appendix. Security of KwAH1 can also be proven under GBDH assumption with a tighter security reduction.

Theorem 2. *Suppose* $(t_{gbdh}, \epsilon_{gbdh})$*-GBDH assumption holds in* $\langle \mathbb{G}_1, \mathbb{G}_2, e \rangle$ *and hash functions G and H are random oracles. Then,* KwAH1 *is* $(t_{kwah}, \epsilon_{kwah})$*-*IND-KE-CPA *secure as long as*

$$\epsilon_{kwah} \leq \frac{3N}{2} \epsilon_{gbdh}$$
$$t_{kwah} \leq t_{gbdh} + \Theta(\tau(2q_H + 3q_E)),$$

where IND-KE-CPA *adversary $\mathcal{A}_{kwah}$ issues at most q_H H-queries and q_E Exposure queries. Here, τ is the maximum time for computing an exponentiation in* $\mathbb{G}_1, \mathbb{G}_2$*, and pairing e.*

The proof of the theorem is similar to Theorem 1.

3.3 Strongly IND-KE-CPA Scheme

We can build a construction of a strongly IND-KE-CPA scheme KwAH2 by only slightly modifying KwAH1. The KwAH2 scheme consists of the following algorithms:

KwAH2: STRONGLY IND-KE-CPA CONSTRUCTION

KeyGen: Given a security parameter k, **KeyGen** algorithm does the same as that of KwAH1 except that it:
   2. picks random, $s_1, s_2, s_3 \in \mathbb{Z}_q^*$, and sets $h_1 = g^{s_1 s_3}$ and $h_2 = g^{s_2 s_3}$,
   6. outputs $pk = \langle q, \mathbb{G}_1, \mathbb{G}_2, e, n, g, h_1, h_2, G, H \rangle$, $mk = s_1$, $ak = s_2$ and $usk_0 = \langle d_{-1}^{s_3} \cdot d_0^{s_3}, s_3 \rangle$.

Δ-Gen: Same as in KwAH1.

Update: For given $usk_{i-1} = \langle usk'_{i-1}, s_3 \rangle$, hsk_i and i, **Update** algorithm:
   1. computes $usk'_i = usk'_{i-1} \cdot hsk_i^{s_3}$,
   2. deletes usk'_{i-1} and hsk_i,
   3. outputs $usk_i = \langle usk'_i, s_3 \rangle$.

Encrypt: Same as in KwAH1.

Decrypt: For given $usk_i = \langle usk'_i, s_3 \rangle$ and $C = \langle i, c_0, c_1 \rangle$, **Decrypt** algorithm does the same as that of KwAH1 except that it:
   1. computes $W' = e(c_0, usk'_i)$.

The security proof of KwAH2 can be done similarly to KwAH1. Here we briefly explain why both master keys, mk and ak, can be exposed and still guarantee security. Since the plaintext M is perfectly hidden by $G(e(g^r, usk'_i))$, it is necessary to compute $e(g^r, usk'_i)$ to compromise the semantic security of KwAH2. However, this is almost as difficult as the CBDH problem without knowing s_3 even if the adversary knows both mk and ak. Hence, KwAH2 is more secure than KwAH1 against exposure of master helper keys.

3.4 Chosen Ciphertext Secure Schemes

We can also construct chosen ciphertext secure KwAH schemes by extending KwAH1 and KwAH2 with Fujisaki-Okamoto padding [9,10]. Especially, by applying it to KwAH2, we have a strongly IND-KE-CCA scheme which is the strongest scheme amongst all our proposed KwAH schemes.

4   Conclusion

We presented in this paper a new KIPE schemes with improved helper key security. Our schemes use two helper keys: the auxiliary helper key and the main helper key. The auxiliary helper key is used to update less frequently while it is stored safer than the main helper key. By using these two kinds of helper keys, the system can be added more careful protection and the damage can be reduced when the key exposure happens. Our schemes are proven to be semantically secure in the random oracle model.

References

1. R. Anderson, "Two remarks on public key cryptology," Invited Lecture, ACM CCCS'97, available at http://www.cl.cam.ac.uk/users/rja14/.
2. D. Boneh and M. Franklin, "Identity-based encryption from the Weil pairing," Proc. of Crypto'01, LNCS 2139, Springer-Verlag, pp.213-229, 2001.
3. D. Boneh and M. Franklin, "Identity-based encryption from the Weil pairing," SIAM J. of Computing, vol. 32, no. 3, pp.586-615, 2003 (full version of [2]).
4. M. Bellare and A. Palacio, "Protecting against key exposure: strongly key-insulated encryption with optimal threshold," available at http://eprint.iacr.org/2002/064/.
5. R. Canetti, S. Halevi and J. Katz, "A forward secure public key encryption scheme," Proc. of Eurocrypt'03, LNCS 2656, Springer-Verlag, pp.255-271, 2003.
6. Y. Dodis, J. Katz, S. Xu and M. Yung, "Key-insulated public key cryptosystems," Proc. of Eurocrypt'02, LNCS 2332, Springer-Verlag, pp.65-82, 2002.
7. Y. Dodis, M. Franklin, J. Katz, A. Miyaji and M. Yung, "Intrusion-resilient public-key encryption," Proc. of CT-RSA'03, LNCS 2612, Springer-Verlag, pp.19-32, 2003.
8. Y. Dodis, M. Franklin, J. Katz, A. Miyaji and M. Yung, "A generic construction for intrusion-resilient public-key encryption," Proc. of CT-RSA'04, LNCS 2964, Springer-Verlag, pp.81-98, 2004.
9. E. Fujisaki and T. Okamoto, "How to enhance the security of public-key encryption at minimum cost," Proc. of PKC'99, LNCS 1560, Springer-Verlag, pp.53-68, 1999.
10. E. Fujisaki and T. Okamoto, "Secure integration of asymmetric and symmetric encryption schemes," Proc. of Crypto'99, LNCS 1666, Springer-Verlag, pp.537-554, 1999.
11. G. Hanaoka, Y. Hanaoka and H. Imai, "Parallel key-insulated public key encryption," Proc. of PKC'06, LNCS 3958, Springer-Verlag, pp.105-122, 2006.
12. Y. Hanaoka, G. Hanaoka, J. Shikata and H. Imai, "Identity-based hierarchical strongly key-insulated encryption and its application" Proc. of Asiacrypt'05, LNCS 3958, Springer-Verlag, pp.495-514, 2005.
13. T. Okamoto and D. Pointcheval, "The gap-problems: a new class of problems for the security of cryptographic schemes," Proc. of PKC'01, LNCS 1992, Springer-Verlag, pp.104-118, 2001.

Appendix: Proof of Theorem 1

We show that we can construct an algorithm $\mathcal{A}_{cbdh}$ that can solve the CBDH problem in $\langle \mathbb{G}_1, \mathbb{G}_2, e \rangle$ by using an adversary $\mathcal{A}_{kwah}$ that breaks IND-KE-CPA security of our scheme. The algorithm $\mathcal{A}_{cbdh}$ is given an instance $\langle g, g^a, g^b, g^c \rangle$ in $\mathbb{G}_1$ from the challenger and tries to output $e(g,g)^{abc}$ using $\mathcal{A}_{kwah}$. Let $g_1 = g^a, g_2 = g^b, g_3 = g^c$. The algorithm $\mathcal{A}_{cbdh}$ works by interacting with $\mathcal{A}_{kwah}$ in an IND-KE-CPA game as follows:

Before we start the simulation, we let $\mathcal{A}_{cbdh}$ flip a coin $\mathcal{COIN} \in \{0,1\}$ such that we have $\Pr[\mathcal{COIN} = 0] = \delta$ for some δ which we will determine later. If $\mathcal{COIN} = 0$, $\mathcal{A}_{cbdh}$ simulates the responses to $\mathcal{A}_{kwah}$'s queries expecting that $\mathcal{A}_{kwah}$ will never submit $\langle \cdot, \text{"main helper"} \rangle$ nor $\langle \cdot, \text{"auxiliary helper"} \rangle$ as Exposure query. If $\mathcal{COIN} = 1$, $\mathcal{A}_{cbdh}$ carries out the simulation expecting that $\mathcal{A}_{kwah}$ will submit $\langle \cdot, \text{"main helper"} \rangle$ or $\langle \cdot, \text{"auxiliary helper"} \rangle$.

If $\mathcal{COIN} = 0$, $\mathcal{A}_{cbdh}$ responses to $\mathcal{A}_{kwah}$'s queries will be as follows:

Setup: $\mathcal{A}_{cbdh}$ picks a random $s \in \mathbb{Z}_q^*$. Also, $\mathcal{A}_{cbdh}$ gives $\mathcal{A}_{kwah}$ the system parameter

$$pk = \langle q, \mathbb{G}_1, \mathbb{G}_2, e, n, g, h_1, h_2, G, H \rangle,$$

where $h_1 = g_1$ and $h_2 = g_1^s$, and random oracles G, H are controlled by $\mathcal{A}_{cbdh}$ as described below.

G-queries: $\mathcal{A}_{kwah}$ issues up to q_G queries to the random oracle G. To respond to these queries algorithm, $\mathcal{A}_{cbdh}$ forms a list of tuples $\langle W, x \rangle$ as explained below. We call this list G_{list}. The list is initially empty. When $\mathcal{A}_{kwah}$ gives $\mathcal{A}_{cbdh}$ a query W to the oracle G, $\mathcal{A}_{cbdh}$ responds as follows:

1. If the query W already appears on the G_{list} in a tuple $\langle W, x \rangle$, then outputs $G(W) = x$.
2. $\mathcal{A}_{cbdh}$ chooses a random $x \in \{0,1\}^n$.
3. $\mathcal{A}_{cbdh}$ adds the tuple $\langle W, x \rangle$ to the G_{list} and outputs $G(W) = x$.

H-queries: $\mathcal{A}_{cbdh}$ picks a random $\alpha \in \{1, ..., N\}$ in advance. $\mathcal{A}_{kwah}$ issues up to q_H queries to the random oracle H. To respond to these queries algorithm, $\mathcal{A}_{cbdh}$ forms a list of tuples $\langle i, u_i, r_i \rangle$ as explained below. We call the list H_{list}. The list is initially empty. When $\mathcal{A}_{kwah}$ gives $\mathcal{A}_{cbdh}$ a query i to the oracle H, $\mathcal{A}_{cbdh}$ responds as follows:

1. If the query i already appears on the H_{list} in a tuple $\langle i, u_i, r_i \rangle$, then outputs $H(i) = u_i$.
2. If $i = \alpha$, $\mathcal{A}_{cbdh}$ sets $u_i = g_2$ and $r_\alpha = 0$.
3. If $i < \alpha$, $\mathcal{A}_{cbdh}$ chooses a random $r_i \in \mathbb{Z}_q^*$ and sets $u_i = g^{r_i}$.
4. If $i > \alpha$, $\mathcal{A}_{cbdh}$ chooses a random $r_i \in \mathbb{Z}_q^*$ and sets $u_i = g_2^z \cdot g^{r_i}$, where
   - $z = 1$ if $\alpha = 0 \bmod \ell$ and $i = 0 \bmod \ell$,
   - $z = -s$ if $\alpha = 0 \bmod \ell$ and $i \neq 0 \bmod \ell$,
   - $z = 1$ if $\alpha = -1 \bmod \ell$ and $i \neq 0 \bmod \ell$,
   - $z = -s^{-1}$ (s^{-1} is the inverse of $s \bmod q$) if $\alpha = -1 \bmod \ell$ and $i = 0 \bmod \ell$,
   - $z = 0$ otherwise,
5. $\mathcal{A}_{cbdh}$ adds the tuple $\langle i, u_i, r_i \rangle$ to the H_{list} and outputs $H(i) = u_i$.

Challenge: Once algorithm $\mathcal{A}_{kwah}$ decides that Phase 1 is over, it outputs a target stage i^* and two messages M_0, M_1 on which it wishes to be challenged. Algorithm $\mathcal{A}_{cbdh}$ responds as follows:

1. $\mathcal{A}_{cbdh}$ sets $C^* = \langle i^*, c_0^*, c_1^* \rangle$ as $c_0^* = g_3$ and $c_1^* = \mu$ where $\mu \in_R \{0,1\}^n$.
2. $\mathcal{A}_{cbdh}$ gives $C^* = \langle i^*, c_0^*, c_1^* \rangle$ as the challenge ciphertext to $\mathcal{A}_{kwah}$.

Exposure queries: $\mathcal{A}_{kwah}$ issues up to q_E Exposure queries. When $\mathcal{A}_{kwah}$ gives a query $\langle i, \mathtt{class} \rangle$, $\mathcal{A}_{cbdh}$ responds as follows:

1. If $\mathtt{class} =$ "main helper" or "auxiliary helper", $\mathcal{A}_{cbdh}$ aborts the simulation.
2. If $i = \alpha$, $\mathcal{A}_{cbdh}$ aborts the simulation.
3. $\mathcal{A}_{cbdh}$ runs the algorithm for responding to H-queries to obtain $\langle i, u_i, r_i \rangle$ and $\langle j, u_j, r_j \rangle$, where $j = i - 1$ if $i = 0 \bmod \ell$, or $j = L$ such that $i - \ell < L < i$, $L = 0 \bmod \ell$ otherwise.

4. $\mathcal{A}_{cbdh}$ sets $usk_i = h_1^{r_{i-1}} \cdot h_2^{r_i}$ if $i = 0 \bmod \ell$, or $usk_i = h_1^{r_i} \cdot h_2^{r_L}$ otherwise. Observe that usk_i is the user secret key corresponding to the stage i. Especially, when $i > \alpha$,

$$u_{i-1}^{\log_g h_1} \cdot u_i^{\log_g h_2} = (g_2^{-s} \cdot g^{r_{i-1}})^a \cdot (g_2 \cdot g^{r_i})^{s \cdot a} = g^{a \cdot r_{i-1}} \cdot g^{r_i \cdot s \cdot a} = h_1^{r_{i-1}} h_2^{r_i}$$
$$\text{(if } i = 0 \bmod \ell, \ \alpha = 0 \bmod \ell)$$

$$= (g_2 \cdot g^{r_{i-1}})^a \cdot (g_2^{-s^{-1}} \cdot g^{r_i})^{s \cdot a} = g^{a \cdot r_{i-1}} \cdot g^{r_i \cdot s \cdot a} = h_1^{r_{i-1}} h_2^{r_i}$$
$$\text{(if } i = 0 \bmod \ell, \ \alpha = -1 \bmod \ell)$$

$$u_L^{\log_g h_2} \cdot u_i^{\log_g h_1} = (g_2 \cdot g^{r_L})^{s \cdot a} \cdot (g_2^{-s} \cdot g^{r_i})^a = g^{s \cdot a \cdot r_L} \cdot g^{r_i \cdot a} = h_2^{r_L} h_1^{r_i}$$
$$\text{(if } i \neq 0 \bmod \ell, \ \alpha = 0 \bmod \ell)$$

$$= (g_2^{-s^{-1}} \cdot g^{r_L})^{s \cdot a} \cdot (g_2 \cdot g^{r_i})^a = g^{s \cdot a \cdot r_L} \cdot g^{r_i \cdot a} = h_2^{r_L} h_1^{r_i}$$
$$\text{(if } i \neq 0 \bmod \ell, \ \alpha = -1 \bmod \ell)$$

5. $\mathcal{A}_{cbdh}$ outputs usk_i to $\mathcal{A}_{kwah}$.

Guess: When $\mathcal{A}_{kwah}$ decides that Phase 2 is over, $\mathcal{A}_{kwah}$ outputs its guess bit $\beta' \in \{0, 1\}$. At the same time, algorithm $\mathcal{A}_{cbdh}$ terminates the simulation. Then, $\mathcal{A}_{cbdh}$ picks a tuple $\langle W, x \rangle$ uniformly at random from the G_{list}, and computes

$$T = \left(\frac{W}{e(g_1, g_3)^{r_{\alpha-1}}}\right)^{s^{-1}} \quad \text{if } \alpha = 0 \bmod \ell,$$

$$= \left(\frac{W}{e(g_1, g_3)^{s \cdot r_A}}\right), \ (\alpha - \ell < A < \alpha, \ A = 0 \bmod \ell) \quad \text{otherwise.}$$

Finally, $\mathcal{A}_{cbdh}$ outputs T.

Claim 1. *If $i^* = \alpha$ and $\mathcal{A}_{cbdh}$ does not abort, then $\mathcal{A}_{kwah}$'s view is identical to its view in the real attack until $\mathcal{A}_{kwah}$ submits W^* as a G-query, where*

$$W^* = e(g_1, g_3)^{r_{\alpha-1}} \cdot e(g, g)^{s \cdot abc} \quad \text{if } \alpha = 0 \bmod \ell,$$
$$= e(g_1, g_3)^{s \cdot r_A} \cdot e(g, g)^{abc} \quad \text{otherwise.}$$

We note that if $i^* = \alpha$,

$$e(g, g)^{abc} = \left(\frac{W^*}{e(g_1, g_3)^{r_{\alpha-1}}}\right)^{s^{-1}} \quad \text{if } \alpha = 0 \bmod \ell,$$

$$= \left(\frac{W^*}{e(g_1, g_3)^{s \cdot r_A}}\right) \quad \text{otherwise.}$$

Proof. It is obvious that the responses to G are perfect. The responses to H are also as in the real attack since each response is uniformly and independently distributed in $\mathbb{G}_1$. Interestingly, the responses to Exposure queries are perfect if $\mathcal{A}_{cbdh}$ does not abort. Finally, we show that the response to Challenge is indistinguishable from the real attack until $\mathcal{A}_{kwah}$ submits W^*. Let the response to Challenge be $C^* = \langle \alpha, c_0^*, c_1^* \rangle$. Then, c_0^* is uniformly distributed in $\mathbb{G}_1$ due to random $\log_g g_3 (= c)$, and therefore are as in the real attack. Also, since

$c_1^* = M_\beta \oplus G(W^*)$, it is information-theoretically impossible to obtain any information on M_β unless $\mathcal{A}_{kwah}$ asks $G(W^*)$.

Next, let us define by E_1, an event assigned to be true if and only if $i^* = \alpha$. Similarly, let us define by E_2, an event assigned to be true if and only if a G-query coincides with W^*, and by E_{msk}, an event assigned to be true if and only if an Exposure query coincides with $\langle \cdot, \text{"main helper"} \rangle$ or $\langle \cdot, \text{"auxiliary helper"} \rangle$.

Claim 2. We have that $\Pr[\beta' = \beta | E_1, \neg E_{msk}] \geq \Pr[\beta' = \beta | \neg E_{msk}]$.

Proof. It is clear that

$$\sum_{i \in \{1,...,N\}} \Pr[\beta' = \beta | i^* = i, \neg E_{msk}] \Pr[i^* = i | \neg E_{msk}] = \Pr[\beta' = \beta | \neg E_{msk}].$$

Since α is uniformly chosen from $\{1, ..., N\}$ at random, we have

$$\Pr[\beta' = \beta | i^* = \alpha, \neg E_{msk}]] \Pr[i^* = \alpha | \neg E_{msk}] \geq \frac{1}{N} \Pr[\beta' = \beta | \neg E_{msk}].$$

Therefore, we have $\Pr[\beta' = \beta | E_1, \neg E_{msk}] \geq \Pr[\beta' = \beta | \neg E_{msk}]$, which proves the claim.

Claim 3. We have that $\Pr[\beta' = \beta | E_1, \neg E_2, \neg E_{msk}] = 1/2$.

Proof. Let the response to Challenge be $C^* = \langle \alpha, c_0^*, c_1^* \rangle$. Since $c_1^* = M_\beta \oplus G(W^*)$, it is information-theoretically impossible to obtain any information on M_β without submitting W^* as a G-query. This implies that $\mathcal{A}_{kwah}$'s best strategy becomes a random guess if E_2 is false. Hence, the claim is proven.

Claim 4. We have that

$$\Pr[\mathcal{A}_{cbdh}(g, g^a, g^b, g^c) = e(g,g)^{abc} | \mathcal{COIN} = 0]$$
$$\geq \frac{1}{q_G N} \cdot \Pr[E_2 | E_1, \neg E_{msk}] \Pr[\neg E_{msk}].$$

Proof. If $i^* = \alpha$, then $e(g,g)^{abc}$ can easily be calculated from W^*, and W^* appears in G_{list} with probability $\Pr[E_2]$. Obviously, $\Pr[E_2] \geq \Pr[E_2, E_1, \neg E_{msk}] = \Pr[E_2 | E_1, \neg E_{msk}] \Pr[E_1 | \neg E_{msk}] \Pr[\neg E_{msk}]$ and $\Pr[E_1 | \neg E_{msk}] = 1/N$. Hence, by choosing a tuple from G_{list} uniformly at random, $\mathcal{A}_{cbdh}$ can correctly output $e(g,g)^{abc}$ with probability of at least $1/q_G \cdot 1/N \cdot \Pr[E_2 | E_1, \neg E_{msk}] \Pr[\neg E_{msk}]$.

Finally, we calculate $p_0 := \Pr[\mathcal{A}_{cbdh}(g, g^a, g^b, g^c) = e(g,g)^{abc} | \mathcal{COIN} = 0]$ from the above claims. Letting $\gamma := \Pr[\beta' = \beta | E_{msk}] - 1/2$, from Claims 1, 2 and 3, we have

$$\Pr[\beta' = \beta] - \frac{1}{2} = \Pr[\beta' = \beta | \neg E_{msk}] \Pr[\neg E_{msk}] + \Pr[\beta' = \beta | E_{msk}] \Pr[E_{msk}] - \frac{1}{2}$$

$$= \Pr[\beta' = \beta | \neg E_{msk}](1 - \Pr[E_{msk}]) + (\frac{1}{2} + \gamma) \Pr[E_{msk}] - \frac{1}{2}$$

$$\leq \Pr[\beta' = \beta | E_1, \neg E_{msk}](1 - \Pr[E_{msk}]) + (\frac{1}{2} + \gamma) \Pr[E_{msk}] - \frac{1}{2}$$

$$= (\Pr[\beta' = \beta | E_1, E_2, \neg E_{msk}] \Pr[E_2 | E_1, \neg E_{msk}]$$
$$+ \Pr[\beta' = \beta | E_1, \neg E_2, \neg E_{msk}] \Pr[\neg E_2 | E_1, \neg E_{msk}])$$
$$\cdot (1 - \Pr[E_{msk}]) + (\frac{1}{2} + \gamma) \Pr[E_{msk}] - \frac{1}{2}$$

$$\leq (\Pr[E_2 | E_1, \neg E_{msk}] + \frac{1}{2}(1 - \Pr[E_2 | E_1, \neg E_{msk}])) \cdot (1 - \Pr[E_{msk}])$$
$$+ (\frac{1}{2} + \gamma) \Pr[E_{msk}] - \frac{1}{2}$$

$$= \frac{1}{2} \Pr[E_2 | E_1, \neg E_{msk}] \Pr[\neg E_{msk}] + \gamma \Pr[E_{msk}].$$

From Claim 4, we have

$$p_0 \geq \frac{2}{q_G N}(\epsilon_{kwah} - \gamma \Pr[E_{msk}]).$$

Next, we discuss for the $\mathcal{COIN} = 1$ case. If $\mathcal{COIN} = 1$, $\mathcal{A}_{cbdh}$ responses to $\mathcal{A}_{kwah}$'s queries as follows:

Setup: $\mathcal{A}_{cbdh}$ picks random $s \in \mathbb{Z}_q^*$ and $\mathbf{b} \in \{1, 2\}$. Let $\bar{\mathbf{b}}$ be 1 (resp. 2) if $\mathbf{b} = 2$ (resp. 1). Also, $\mathcal{A}_{cbdh}$ gives $\mathcal{A}_{kwah}$ the system parameter

$$pk = \langle q, \mathbb{G}_1, \mathbb{G}_2, e, n, g, h_1, h_2, G, H \rangle,$$

where $h_{\mathbf{b}} = g_1$ and $h_{\bar{\mathbf{b}}} = g^s$ (we expect that $\mathcal{A}_{kwah}$ asks ak if $\mathbf{b} = 1$, or mk otherwise), and random oracles G, H are controlled by $\mathcal{A}_{cbdh}$ as described below.

G-queries: $\mathcal{A}_{kwah}$ issues up to q_G queries to the random oracle G. To respond to these queries algorithm $\mathcal{A}_{cbdh}$ forms a list of tuples $\langle W, x \rangle$ as explained below. We call this list G_{list}. The list is initially empty. When $\mathcal{A}_{kwah}$ gives $\mathcal{A}_{cbdh}$ a query W to the oracle G, $\mathcal{A}_{cbdh}$ responds as follows:
1. If the query W already appears on the G_{list} in a tuple $\langle W, x \rangle$, then outputs $G(W) = x$.
2. $\mathcal{A}_{cbdh}$ chooses a random $x \in \{0, 1\}^n$.
3. $\mathcal{A}_{cbdh}$ adds the tuple $\langle W, x \rangle$ to the G_{list} and outputs $G(W) = x$.

H-queries: $\mathcal{A}_{cbdh}$ picks a random $\alpha \in \{1, ..., N\}$ in advance. $\mathcal{A}_{kwah}$ issues up to q_H queries to the random oracle H. To respond to these queries algorithm $\mathcal{A}_{cbdh}$ forms a list of tuples $\langle i, u_i, r_i \rangle$ as explained below. We call the list H_{list}. The list is initially empty. When $\mathcal{A}_{kwah}$ gives $\mathcal{A}_{cbdh}$ a query i to the oracle H, $\mathcal{A}_{cbdh}$ responds as follows:
1. If the query i already appears on the H_{list} in a tuple $\langle i, u_i, r_i \rangle$, then outputs $H(i) = u_i$.

2. If $i = \alpha - 1$, $\mathbf{b} = 1$, and $\alpha = 0 \bmod \ell$, $\mathcal{A}_{cbdh}$ sets $u_i = g_2$ and $r_i = 0$.
3. If $i = \alpha$, $\mathbf{b} = 1$, and $\alpha = -1 \bmod \ell$, $\mathcal{A}_{cbdh}$ sets $u_i = g_2$ and $r_i = 0$.
4. If $i = \alpha$, $\mathbf{b} = 1$, $\alpha \neq 0 \bmod \ell$, and $\alpha \neq -1 \bmod \ell$, $\mathcal{A}_{cbdh}$ sets $u_i = g_2$ and $r_i = 0$.
5. If $i = \alpha$, and $\mathbf{b} = 2$, $\mathcal{A}_{cbdh}$ sets $u_i = g_2$ and $r_i = 0$.
6. Else, $\mathcal{A}_{cbdh}$ chooses a random $r_i \in \mathbb{Z}_q^*$ and sets $u_i = g^{r_i}$.
7. $\mathcal{A}_{cbdh}$ adds the tuple $\langle i, u_i, r_i \rangle$ to the H_{list} and outputs $H(i) = u_i$.

Challenge: Once algorithm $\mathcal{A}_{kwah}$ decides that Phase 1 is over, it outputs a target stage i^* and two messages M_0, M_1 on which it wishes to be challenged. Algorithm $\mathcal{A}_{cbdh}$ responds as follows:

1. $\mathcal{A}_{cbdh}$ sets $C^* = \langle i^*, c_0^*, c_1^* \rangle$ as $c_0^* = g_3$ and $c_1^* = \mu$ where $\mu \in_R \{0, 1\}^n$.
2. $\mathcal{A}_{cbdh}$ gives $C^* = \langle i^*, c_0^*, c_1^* \rangle$ as the challenge ciphertext to $\mathcal{A}_{kwah}$.

Exposure queries: $\mathcal{A}_{kwah}$ issues up to q_E Exposure queries. When $\mathcal{A}_{kwah}$ gives a query $\langle i, \text{class} \rangle$, $\mathcal{A}_{cbdh}$ responds as follows:

1. If $\mathbf{b} = 1$ and $\text{class} = $ "main helper", $\mathcal{A}_{cbdh}$ aborts the simulation.
2. If $\mathbf{b} = 1$ and $\text{class} = $ "auxiliary helper", $\mathcal{A}_{cbdh}$ returns s to $\mathcal{A}_{kwah}$.
3. If $\mathbf{b} = 2$ and $\text{class} = $ "main helper", $\mathcal{A}_{cbdh}$ returns s to $\mathcal{A}_{kwah}$.
4. If $\mathbf{b} = 2$ and $\text{class} = $ "auxiliary helper", $\mathcal{A}_{cbdh}$ aborts the simulation.
5. If $i = \alpha$ and $\text{class} = $ "user", $\mathcal{A}_{cbdh}$ aborts the simulation.
6. If $i = \alpha - 1$, $\text{class} = $ "user", $\mathbf{b} = 1$ and $\alpha = 0 \bmod \ell$, $\mathcal{A}_{cbdh}$ aborts the simulation.
7. If $A \leq i < A + \ell$, $\text{class} = $ "user", and $\mathbf{b} = 2$, where $A \leq \alpha < A + \ell$ and $A = 0 \bmod \ell$, $\mathcal{A}_{cbdh}$ aborts the simulation.
8. Else[1], $\mathcal{A}_{cbdh}$ runs the algorithm for responding to H-queries to obtain $\langle i, u_i, r_i \rangle$ and $\langle j, u_j, r_j \rangle$, where $j = i - 1$ if $i = 0 \bmod \ell$, or $j = L$ such that $i - \ell < L < i$, $L = 0 \bmod \ell$ otherwise.
9. $\mathcal{A}_{cbdh}$ sets $usk_i = h_1^{r_{i-1}} \cdot h_2^{r_i}$ if $i = 0 \bmod \ell$, or $usk_i = h_1^{r_i} \cdot h_2^{r_L}$ otherwise.
10. $\mathcal{A}_{cbdh}$ outputs usk_i to $\mathcal{A}_{kwah}$.

Guess: When $\mathcal{A}_{kwah}$ decides that Phase 2 is over, $\mathcal{A}_{kwah}$ outputs the guess bit $\beta' \in \{0, 1\}$. At the same time, algorithm $\mathcal{A}_{cbdh}$ terminates the simulation. Then, $\mathcal{A}_{cbdh}$ picks a tuple $\langle W, x \rangle$ uniformly at random from the G_{list}, and computes

$$
\begin{aligned}
T &= W \cdot e(g, g_3)^{-s \cdot r_\alpha} &&\text{if } \mathbf{b} = 1 \text{ and } \alpha = 0 \bmod \ell, \\
&= W \cdot e(g, g_3)^{-s \cdot r_{\alpha-1}} &&\text{if } \mathbf{b} = 2 \text{ and } \alpha = 0 \bmod \ell, \\
&= W \cdot e(g, g_3)^{-s \cdot r_A} &&\text{otherwise.}
\end{aligned}
$$

Finally, $\mathcal{A}_{cbdh}$ outputs T.

Claim 5. *If $i^* = \alpha$ and $\mathcal{A}_{cbdh}$ does not abort, then $\mathcal{A}_{kwah}$'s view is identical to its view in the real attack until $\mathcal{A}_{kwah}$ submits W^* as a G-query, where*

$$
\begin{aligned}
W^* &= e(g, g_3)^{s \cdot r_\alpha} \cdot e(g, g)^{abc} &&\text{if } \mathbf{b} = 1 \text{ and } \alpha = 0 \bmod \ell, \\
&= e(g, g_3)^{s \cdot r_{\alpha-1}} \cdot e(g, g)^{abc} &&\text{if } \mathbf{b} = 2 \text{ and } \alpha = 0 \bmod \ell, \\
&= e(g, g_3)^{s \cdot r_A} \cdot e(g, g)^{abc} &&\text{otherwise.}
\end{aligned}
$$

---

[1] Notice that in this case, class is always "user".

We note that if $i^* = \alpha$,

$$e(g,g)^{abc} = W^* \cdot e(g,g_3)^{-s \cdot r_\alpha} \qquad \text{if } \mathbf{b} = 1 \text{ and } \alpha = 0 \bmod \ell,$$
$$= W^* \cdot e(g,g_3)^{-s \cdot r_\alpha - 1} \qquad \text{if } \mathbf{b} = 2 \text{ and } \alpha = 0 \bmod \ell,$$
$$= W^* \cdot e(g,g_3)^{-s \cdot r_A} \qquad \text{otherwise.}$$

Proof. It is obvious that the responses to G are perfect. The responses to H are also as in the real attack since each response is uniformly and independently distributed in $\mathbb{G}_1$. The responses to Exposure queries are perfect if $\mathcal{A}_{cbdh}$ does not abort. Finally, we show that the response to Challenge is indistinguishable from the real attack until $\mathcal{A}_{kwah}$ submits W^*. Let the response to Challenge be $C^* = \langle \alpha, c_0^*, c_1^* \rangle$. Then, c_0^* is uniformly distributed in $\mathbb{G}_1$ due to random $\log_g g_3 (= c)$, and therefore are as in the real attack. Also, since $c_1^* = M_\beta \oplus G(W^*)$, it is information-theoretically impossible to obtain any information on M_β unless $\mathcal{A}_{kwah}$ asks $G(W^*)$.

Next, let us define by E_3, an event assigned to be true if and only if $i^* = \alpha$. Similarly, let us define by E_4, an event assigned to be true if and only if a G-query coincides with W^*, by E_5, an event assigned to be true if and only if an Exposure query coincides with $\langle \cdot, \text{"main helper"} \rangle$ if $\mathbf{b} = 1$ or $\langle \cdot, \text{"auxiliary helper"} \rangle$ if $\mathbf{b} = 2$, and by E_{msk}, an event assigned to be true if and only if an Exposure query coincides with $\langle \cdot, \text{"main helper"} \rangle$ or $\langle \cdot, \text{"auxiliary helper"} \rangle$. Notice that E_{msk} is identical to that in the case of $\mathcal{COIN} = 0$.

Claim 6. *We have that* $\Pr[\beta' = \beta | E_3, \neg E_5, E_{msk}] \geq \Pr[\beta' = \beta | E_{msk}]$.

Proof. It is clear that

$$\sum_{i \in \{1,\dots,N\}} \Pr[\beta' = \beta | i^* = i, \neg E_5, E_{msk}] \Pr[i^* = i | \neg E_5, E_{msk}] = \Pr[\beta' = \beta | \neg E_5, E_{msk}].$$

Since α is uniformly chosen from $\{1,\dots,N\}$ at random, we have

$$\Pr[\beta' = \beta | i^* = \alpha, \neg E_5, E_{msk}] \Pr[i^* = \alpha | \neg E_5, E_{msk}] \geq \frac{1}{N} \Pr[\beta' = \beta | \neg E_5, E_{msk}].$$

Therefore, we have $\Pr[\beta' = \beta | E_3, \neg E_5, E_{msk}] \geq \Pr[\beta' = \beta | \neg E_5, E_{msk}]$.
Due to $\Pr[\beta' = \beta | \neg E_5, E_{msk}] = \Pr[\beta' = \beta | E_5, E_{msk}]$, we finally have
$\Pr[\beta' = \beta | E_3, \neg E_5, E_{msk}] \geq \Pr[\beta' = \beta | E_{msk}]$.

Claim 7. *We have that* $\Pr[\beta' = \beta | E_3, \neg E_4, \neg E_5, E_{msk}] = 1/2$.

Proof. Let the response to Challenge be $C^* = \langle \alpha, c_0^*, c_1^* \rangle$. Since $c_1^* = M_\beta \oplus G(W^*)$, it is information-theoretically impossible to obtain any information on M_β without submitting W^* as a G-query. This implies that $\mathcal{A}_{kwah}$'s best strategy becomes a random guess if E_4 is false. Hence, we have the claim proven.

Claim 8. *We have that*

$$\Pr[\mathcal{A}_{cbdh}(g, g^a, g^b, g^c) = e(g,g)^{abc} | \mathcal{COIN} = 1]$$
$$\geq \frac{1}{2q_G N} \cdot \Pr[E_4 | E_3, \neg E_5, E_{msk}] \Pr[E_{msk}].$$

Proof. If $i^* = \alpha$, then $e(g,g)^{abc}$ can easily be calculated from W^*, and W^* appears in G_{list} with probability $\Pr[E_4]$. Obviously, we have

$$\Pr[E_4] \geq \Pr[E_4, E_3, \neg E_5, E_{msk}]$$
$$= \Pr[E_4 | E_3, \neg E_5, E_{msk}] \Pr[E_3 | \neg E_5, E_{msk}] \Pr[\neg E_5, E_{msk}]$$

Furthermore, $\Pr[E_3 | \neg E_5, E_{msk}] = 1/N$, and $\Pr[\neg E_5, E_{msk}] = 1/2 \cdot \Pr[E_{msk}]$. Hence, by choosing a tuple from G_{list} uniformly at random, $\mathcal{A}_{cbdh}$ can correctly output $e(g,g)^{abc}$ with probability of at least

$$1/q_G \cdot 1/N \cdot 1/2 \cdot \Pr[E_4 | E_3, \neg E_5, E_{msk}] \Pr[E_{msk}]$$

Finally, we calculate $p_1 := \Pr[\mathcal{A}_{cbdh}(g, g^a, g^b, g^c) = e(g,g)^{abc} | \mathcal{COIN} = 1]$ from the above claims. Letting $\eta := \Pr[\beta' = \beta | \neg E_{msk}] - 1/2$, from Claims 5, 6 and 7, we have

$$\Pr[\beta' = \beta] - \frac{1}{2}$$
$$= \Pr[\beta' = \beta | \neg E_{msk}] \Pr[\neg E_{msk}] + \Pr[\beta' = \beta | E_{msk}] \Pr[E_{msk}] - \frac{1}{2}$$
$$= (\frac{1}{2} + \eta) \Pr[\neg E_{msk}] + \Pr[\beta' = \beta | E_{msk}](1 - \Pr[\neg E_{msk}]) - \frac{1}{2}$$
$$\leq (\frac{1}{2} + \eta) \Pr[\neg E_{msk}] + \Pr[\beta' = \beta | E_3, \neg E_5, E_{msk}](1 - \Pr[\neg E_{msk}]) - \frac{1}{2}$$
$$= (\frac{1}{2} + \eta) \Pr[\neg E_{msk}] + (\Pr[\beta' = \beta | E_3, E_4, \neg E_5, E_{msk}] \Pr[E_4 | E_3, \neg E_5, E_{msk}]$$
$$\quad + \Pr[\beta' = \beta | E_3, \neg E_4, \neg E_5, E_{msk}] \Pr[\neg E_4 | E_3, \neg E_5, E_{msk}]) \cdot (1 - \Pr[\neg E_{msk}]) - \frac{1}{2}$$
$$\leq (\frac{1}{2} + \eta) \Pr[\neg E_{msk}] +$$
$$\quad (\Pr[E_4 | E_3, \neg E_5, E_{msk}] + \frac{1}{2}(1 - \Pr[E_4 | E_3, \neg E_5, E_{msk}])) \cdot (1 - \Pr[\neg E_{msk}]) - \frac{1}{2}$$
$$= \frac{1}{2} \Pr[E_4 | E_3, \neg E_5, E_{msk}] \Pr[E_{msk}] + \eta \Pr[\neg E_{msk}].$$

From Claim 8, we have

$$p_1 \geq \frac{1}{q_G N}(\epsilon_{kwah} - \eta \Pr[\neg E_{msk}]).$$

Claim 9. *We have that* $\epsilon_{kwah} \geq \gamma \Pr[E_{msk}] + \eta \Pr[\neg E_{msk}]$.

Proof. By the definitions of γ and η, we have $\gamma + 1/2 = \Pr[\beta' = \beta | E_{msk}]$ and $\eta + 1/2 = \Pr[\beta' = \beta | \neg E_{msk}]$, and consequently,

$$\epsilon_{kwah} + \frac{1}{2} \geq \Pr[\beta' = \beta] = (\gamma + \frac{1}{2})\Pr[E_{msk}] + (\eta + \frac{1}{2})\Pr[\neg E_{msk}].$$

Hence, we have $\epsilon_{kwah} \geq \gamma \Pr[E_{msk}] + \eta \Pr[\neg E_{msk}]$, which proves the claim.

Now, we calculate $\epsilon_{cbdh} := \Pr[\mathcal{A}_{cbdh}(g, g^a, g^b, g^c) = e(g,g)^{abc}]$. From Claim 9, we have

$$\begin{aligned}
\epsilon_{cbdh} &= \delta \cdot p_0 + (1-\delta) \cdot p_1 \\
&\geq \delta(\frac{2}{q_G N}(\epsilon_{kwah} - \gamma \Pr[E_{msk}])) + (1-\delta)(\frac{1}{q_G N}(\epsilon_{kwah} - \eta \Pr[\neg E_{msk}])) \\
&\geq \delta(\frac{2}{q_G N}(\epsilon_{kwah} - \gamma \Pr[E_{msk}])) + (1-\delta)(\frac{1}{q_G N}\gamma \Pr[E_{msk}]) \\
&\geq \frac{1}{q_G N}(2\delta \epsilon_{kwah} + (1-3\delta)\gamma \Pr[E_{msk}])
\end{aligned}$$

By letting $\delta = 1/3$, we finally have $\epsilon_{cbdh} \geq \frac{2}{3q_G N}\epsilon_{kwah}$.

From the above discussions, we can see that the claimed bound of the running-time of $\mathcal{A}_{cbdh}$ holds. This completes the proof of the theorem. $\square$

Author Index

Lecture Notes in Computer Science

For information about Vols. 1–4254

please contact your bookseller or Springer